Coastal Research Library

Volume 27

Series editor

Charles W. Finkl, Coastal Education and Research Foundation
Boca Raton, FL, USA

The aim of this book series is to disseminate information to the coastal research community. The Series covers all aspects of coastal research including but not limited to relevant aspects of geological sciences, biology (incl. ecology and coastal marine ecosystems), geomorphology (physical geography), climate, littoral oceanography, coastal hydraulics, environmental (resource) management, engineering, and remote sensing. Policy, coastal law, and relevant issues such as conflict resolution and risk management would also be covered by the Series. The scope of the Series is broad and with a unique cross-disciplinary nature. The Series would tend to focus on topics that are of current interest and which carry some import as opposed to traditional titles that are esoteric and non-controversial.

Monographs as well as contributed volumes are welcomed.

Charles W. Finkl, Coastal Education & Research Foundation, USA
Email: cfinkl@cerf-jcr.com

More information about this series at http://www.springer.com/series/8795

Lynn Donelson Wright · C. Reid Nichols
Editors

Tomorrow's Coasts: Complex and Impermanent

A collaborative synthesis promoted by the Coastal and Environmental Research Committee of the Southeastern Universities Research Association (SURA)

Springer

Editors
Lynn Donelson Wright
Southeastern Universities Research Association (SURA)
Washington, DC
USA

C. Reid Nichols
Marine Information Resources Corporation (MIRC)
Ellicott City, MD
USA

ISSN 2211-0577 ISSN 2211-0585 (electronic)
Coastal Research Library
ISBN 978-3-030-09235-1 ISBN 978-3-319-75453-6 (eBook)
https://doi.org/10.1007/978-3-319-75453-6

Softcover re-print of the Hardcover 1st edition 2019

Cover: Coastal Slum, Kolkata (Calcutta) India
Photo credit: Bertrand Celce

Printed on acid-free paper

This Springer imprint is published by the registered company Springer International Publishing AG part of Springer Nature
The registered company address is: Gewerbestrasse 11, 6330 Cham, Switzerland

In no environment is the connection between people and nature stronger than in coastal systems. Climate change, sea level rise, ecosystem evolution, altered river discharge, and changes in the size, intensity, and duration of storms will accelerate the degradation of coastal realms over the decades ahead. The coastal systems and coastal communities of the future will be different from those of today.

Foreword

In no environment is the connection between people and nature stronger than in coastal systems. Climate change, sea level rise, ecosystem evolution, hydrologic modifications of river discharge to the coasts, and changes in the intensity and duration of storms and attendant coastal erosion are likely to accelerate the alteration of coastal realms over the decades ahead. In concert with these changes, the socioeconomic environment that underpins human coastal communities is also impermanent and dynamic. The interdependence of environmental and socioeconomic processes in the Anthropocene is already giving rise to suites of highly complex and nonlinear feedbacks, many of which are counterintuitive. Improved abilities to predict, communicate, mitigate, and respond to the outcomes of future coastal processes, gradual as well as abrupt, on both long-term and event timescales are essential to the welfare of coastal communities and to the sustainability of coastal ecosystems and built infrastructures. Those interested in understanding the complex nature of coastal processes must take an interdisciplinary approach to Earth system problems, and international collaborations to strengthen the relationships between the physical and social sciences must be nurtured.

The changes that are underway can be slowed, with a global commitment to do so, but most cannot be stopped. The coastal systems of the future will be different from those of today. Preparation of this general literature synthesis was undertaken as an initiative of the *Southeastern Universities Research Association's Coastal and Environmental Research Committee* to lay a foundation for future collaborations aimed at improving predictions and strategies for responding to the complex changes that are occurring, and are expected to occur in the future, within the world's diverse coastal systems. Strategies for ensuring the joint resilience of humans and nature will rely heavily on understanding how the many complex facets of the system interconnect and how they might evolve in the foreseeable future. This understanding must extend well beyond the scientific community to include decision-makers, politicians, and the general public. This synthesis is intended to contribute to that broad understanding and stimulate pursuit of forward-looking solutions. This is not a technical or specialized treatise. The intent is to provide a conceptual roadmap to show how some of the numerous pieces of complex coastal systems intersect and might interact under changing future environmental regimes. The book is addressed to a nontechnical but environmentally literate audience that includes the lay public, policy-makers, planners, engineers, environmental scientists, and academics interested in the causes and consequences of global environmental change and its effects on coastal systems. The book also outlines some strategies for anticipating and responding to the challenges that lie ahead. It is hoped that as future changes unfold, our understandings and abilities to predict will also evolve.

Starkville, MS, USA

David R. Shaw
Chair, Coastal and Environmental Research
Committee, Southeastern Universities Research
Association (SURA)

Acknowledgements

Preparation of this work was supported in part by the Southeastern Universities Research Association (SURA) and in part by the altruism and energy of the contributing authors and their respective institutions. We especially thank Jerry Draayer, SURA's President, and CEO for his encouragement and financial support and Sara Madden of SURA for assisting with editing and proofreading. SURA (www.SURA.org) is a nonprofit organization [501(c)(3)] consisting of 60 member universities. We also thank the many authors referenced throughout this work for permissions to reprint their figures and photos. This synthesis project arose at the request of SURA's Coastal and Environmental Research Committee (CERC) in fall of 2015. The current members of this committee are as follows:

CERC Members and Institutions

Steve Boss, University of Arkansas
Gordon C. Cannon, The University of Southern Mississippi
Benjamin Cuker, Hampton University
Christopher F. D'Elia, Louisiana State University
David Eggleston, North Carolina State University
Carl Friedrichs, College of William and Mary
Monty Graham, The University of Southern Mississippi
Scott Hagen, Louisiana State University
Dawit Haile, Virginia State University
Rodger Harvey, Old Dominion University
Rick Luettich, The University of North Carolina at Chapel Hill
George Maul, Florida Institute of Technology
William Moore, Hampton University
Laura Moore, The University of North Carolina at Chapel Hill
James Morris, University of South Carolina

Anton Post, Florida Atlantic University
Joe Ramus, Duke University
Don Resio, University of North Florida
David Rouse, Auburn University
David R. Shaw, Mississippi State University
Peter Sheng, University of Florida
Carolyn Thoroughgood, University of Delaware
Joel Trexler, Florida International University
Robert Weisberg, University of South Florida
Donald Young, Virginia Commonwealth University
Dick Yue, Massachusetts Institute of Technology
Gary Zarillo, Florida Institute of Technology

Contents

Contributors

M. Bortman The Nature Conservancy, Washington, DC, USA

Arthur G. Cosby Social Science Research Center, Mississippi State University, Starkville, MS, USA

Christopher F. D'Elia School of the Coast and Environment, Louisiana State University, Baton Rouge, LA, USA

Hasna Khandekar Social Science Research Center, Mississippi State University, Starkville, MS, USA

J. Livingston The Nature Conservancy, Washington, DC, USA

S. McAfee The Nature Conservancy, Washington, DC, USA

K. McLeod The Nature Conservancy, Washington, DC, USA

James Morris Baruch Institute for Marine and Coastal Sciences, University of South Carolina, Columbia, SC, USA

S. Murdock The Nature Conservancy, Washington, DC, USA

S. Newkirk The Nature Conservancy, Washington, DC, USA

C. Reid Nichols Marine Information Resources Corporation (MIRC), Ellicott City, MD, USA

Donald T. Resio School of Engineering, University of North Florida, Jacksonville, FL, USA

Gina Rico Mendez Social Science Research Center, Mississippi State University, Starkville, MS, USA

J. P. M. Syvitski Community Surface Dynamics Modeling System, University of Colorado, Boulder, CO, USA

Bruce G. Thom School of Geosciences, University of Sydney, Sydney, NSW, Australia

N. Woiwode The Nature Conservancy, Washington, DC, USA

Lynn Donelson Wright Southeastern Universities Research Association (SURA), Washington, DC, USA

Wei Wu Department of Coastal Sciences, University of Southern Mississippi, Ocean Springs, MS, USA

Donald R. Young Department of Biology, Virginia Commonwealth University, Richmond, VA, USA

Gary Zarillo Ocean Engineering and Sciences, Florida Institute of Technology, Melbourne, FL, USA

Julie Zinnert Department of Biology, Virginia Commonwealth University, Richmond, VA, USA

List of Figures

Introduction

Life is a series of natural and spontaneous changes. Don't resist them. Let reality be reality. Let things flow naturally forward.

Lao Tzu, c. 600 BCE

To greater or lesser degrees, everything that exists or happens in the universe, on earth or within the realm of humanity somehow affects everything else. And the causal links are multidirectional, which means that there are countless feedback loops that produce never-ending change. Impermanence is a natural reality and the idea of enforced stasis is not really rational. Eastern religions and philosophies have long understood this. The scientific community has more recently come to view, explain, and model the functions and outcomes of the myriad connections and consequent dynamic changes in terms of *Complex Systems Dynamics*. The idea of complexity is now widely accepted by modelers of dynamic systems involving the nonlinear interdependence of multiple processes (Bar-Yam 1997; Liu et al. 2007). The International Geosphere-Biosphere Programme (IGBP) has articulated the importance of intersecting social and natural sciences and has evolved the "Anthropocene" paradigm that considers human and natural earth processes to be interdependent and to function and change as a unified complex system (Bondre and Gaffney 2015). The IGBP's successor, *Future Earth*, continues to promote this idea.

Global change now challenges the scientific community to evolve strategies for anticipating and adapting to possible future scenarios of natural and societal impermanence. To meet these challenges, however, alternative future circumstances must be predicted with progressively increasing confidence. Of crucial importance, the predictions must also be understood and accepted by multidisciplinary scientists, the public, and policy-makers. Observations, theory, numerical models, and education must ultimately advance human and ecosystem resilience globally, regionally, and locally.

Today, half of the world's 7 billion people live within 100 kilometers (60 miles) of the shore (Dawson 2017). Coastal communities and environments are among the most severely threatened by the myriad changes that are underway. As detailed by the *U.S. Global Change Research Program* (Wuebbles et al. 2017), among numerous other recent national and international analyses, climate change, sea level rise, storm surge, land subsidence, and coastal urbanization, along with various types of flood inundation, ecosystem evolution, hydrologic and human-induced changes to river discharge and to nutrient and chemical inputs to rivers, and changes in the intensity and duration of storms and attendant coastal erosion are likely to accelerate the alteration of natural coastal realms over the decades ahead. In concert with these natural changes, the socioeconomic environment that underpins human coastal communities is also both impermanent and dynamic.

The coastal consequences of global change are far-reaching. For example, increase in the earth's average atmospheric temperature impacts climate, which in turn alters ocean properties, such as salinity and temperature, which then affects ecosystems such as coral reefs and wetlands. In fact, human habitation, food production, potable water availability, health, safety, economic welfare, national and class conflicts, and simple enjoyment of life are all impacted. As sea levels rise, low-lying vulnerable urban areas throughout the world will be more frequently flooded by storms. Low-income people in flood-prone areas will be increasingly vulnerable to widespread tragedy. Frequent street flooding of low-lying neighborhoods can paralyze traffic, sewers can be flooded, drinking water may be contaminated, and water-borne pathogens may be spread throughout neighborhoods. And, as was the case in New Orleans in the days following Hurricane Katrina, when Cyclone Nargis swept over Myanmar (Burma), when Hurricane Sandy plowed across Caribbean nations such as Cuba, Dominican Republic, Haiti, and Jamaica before making final landfall in the United States, and when Hurricanes Harvey, Irma, and Maria made devastating landfalls in Texas, the Caribbean and Florida in 2017, extensive inundation of neighborhoods can impede rescue operations following nature-caused disasters. Model projections must support local government officials in assessing resilience, planning for humanitarian assistance and identifying the most vulnerable communities, environments, and facilities. Although humanity may take reasonable actions to slow the rate of atmospheric warming, it is highly unlikely that inevitable changes in such processes as inundation can be halted. But, with greater understanding and collaboration, we can better anticipate and adapt to future change.

Thus, there is a clear need for greater understanding, awareness, and prediction. Improved abilities to predict, communicate, mitigate, and respond to the outcomes of future coastal processes, gradual as well as abrupt, on both long-term and event timescales are essential to the welfare of coastal communities and to the sustainability of coastal ecosystems and built infrastructures. This will require collaboration among interdisciplinary scientists to integrate natural and social sciences and engage scientists, modelers, engineers, educators, and stakeholders from academia, federal agencies, local and state governments, nongovernmental organizations, and the private sector in the search for enhanced awareness of what the future may hold. *A high priority vision for future coastal science should be to enhance resilience of coastal communities by anticipating and mitigating hazards to human health, safety, and welfare and reducing economic harm to coastal industries such as tourism, fisheries, and shipping.*

The purpose of this book is not to offer a technical treatise on how to build better numerical models or to provide the *cognoscenti* with new scientific details or theories. As pointed out in the preface, the purpose is to provide a general foundation for understanding the interconnections of the myriad factors that are driving coastal change and will determine what the future may hold. The writing is non-technical, nonmathematical, and non-jargonized throughout. Wherever scientific terms are required to avoid ambiguity, a clear and simple definition is presented and those definitions are repeated in the glossary. We aim to communicate with all who care about the future of coastal environments. Additional background material can be found in a reference such as the *Encyclopedia of Marine Science* (Nichols and Williams 2017). In Part I, we present some underlying general background realities and concepts that are applicable to coastal processes and coastal change worldwide. Part II reviews, in greater depth, some of the more important physical, ecological, and societal causes and outcomes of coastal change. A selection of case studies of some prominent and highly vulnerable coastal regions is presented in Part III. Some strategies for facilitating and supporting collaboration among the global scientific community, the public, and decision-makers to enhance future coastal resilience are outlined in Part IV. Recent references are provided throughout in order to provide the interested reader with a starting point for his or her own research and to provide rigorous support for the explanations offered here.

> *If we understand what is happening and accept the fact of change,*
> *we can go with the flow and still do some steering to adapt and survive.*
> Anonymous

Lynn Donelson Wright

References

Bar-Yam, Y., 1997. *Dynamics of complex systems*. Cambridge, MA: Perseus Press. 848.

Bondre, N., and O. Gaffney, eds. *IGBP Annual Report 2014/2015*. Sweden: IGBP Secretariat, The Royal Swedish Academy of Sciences, 2015.

Dawson, A., 2017. *Extreme cities: The perils and promise of urban life in the age of climate change*. New York: Verso. 378.

Liu, J., T. Dietz, S.R. Carpenter, M. Alberti, C. Folke, E. Moran, A.C. Pell, P. Deadman, T. Kratz, J. Lubchenco, E. Ostrom, Z. Ouyang, W. Provencher, C.L. Redman, S.H. Schneider, and W.W. Taylor, 2007. Complexity of coupled human and natural systems. *Science, 317*(5844): 1513–1516.

Nichols, C. R. and R.G. Williams, 2017. *Encyclopedia of marine science*, Revised Edition, New York: Infobase Publishing.

Wuebbles, D.J., D.W. Fahey, K.A. Hibbard, B. DeAgelo, S. Doherty, K. Hayhoe, R. Horton, J.P. Kossin, P.C. Taylor, A.M. Waple, and C.P. Weaver, 2017. Executive Summary. In *U.S. Global Change Research Program*, eds. D.J. Wuebbles, D.W. Fahey, K.A. Hibbard, D.J. Dokken, B.C. Stewart, and T.K. Maycock, 12–34. Washington, D.C., USA. doi:10.7930/J0DJ5CTG.

Part I
Understanding Coastal Systems and Global Change

Chapter 1
Coastal Complexity and Predictions of Change

Lynn Donelson Wright, J. P. M. Syvitski and C. Reid Nichols

All things are connected. Whatever befalls the earth befalls the children of the earth.
—Chief Seattle of the Suquamish Native American people, c. 1855

1.1 Coasts as Dynamic Processes

Most people tend to think of coasts as material "things". What you see when you look at a coast at any instant in time may be a beach composed of sand or a coastal wetland consisting of vegetation, mud and crabs and perhaps some methane or hydrogen sulfide gas. But in previous times it may have been very different and it probably will be different in the future. In reality, coasts are not "things" but processes; they are not static but are constantly *becoming* something new. This has always been the natural way with coasts. The ever-changing coastal process involves the interplay of solid material (e.g., sand and mud), chemistry (e.g., the pH of the Earth's oceans), forces, energy fluxes and transfers (e.g., physical, chemical, biological, and solar), biological activity and ecological evolution, and, now, profound human interaction. We may reasonably expect coastal change to be accelerated in response to the climate changes that are now underway in the *Anthropocene*, a new geologic epoch in which human activities are causing profound and enduring modifications to the earth's surface (e.g., Steffen et al. 2016; Fig. 1.1; more on this in Chap. 6). To adequately anticipate what a particular coastal

L. D. Wright (✉)
Southeastern Universities Research Association (SURA), Washington, DC, USA
e-mail: ldwright@bellsouth.net

J. P. M. Syvitski
Community Surface Dynamics Modeling System, University of Colorado, Boulder, CO, USA

C. R. Nichols
Marine Information Resources Corporation (MIRC), Ellicott City, MD, USA
e-mail: rnichols@mirc-us.com

L. D. Wright and C. R. Nichols (eds.), *Tomorrow's Coasts: Complex and Impermanent*, Coastal Research Library 27,
https://doi.org/10.1007/978-3-319-75453-6_1

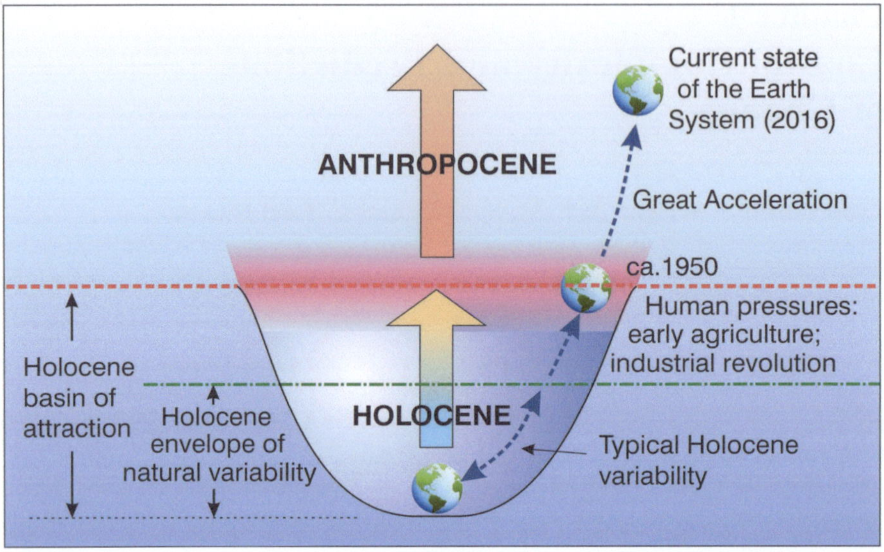

Fig. 1.1 Accelerating role of human-earth interaction in the Anthropocene (From Steffen et al. 2016)

environment may be like a few decades from now we must embrace a holistic perspective as to the complex reality of impermanent and dynamic coastal systems consisting of numerous interacting and mutually dependent agents.

Back in the middle of the 20th century, traditional science emphasized reductionist approaches in order to simplify understanding of cause/effect relationships. In addition to making assumptions that cause and effect were distinct from each other, efforts were made to limit the number of variables involved in processes of change to as few as possible. A classic example from 1960s vintage coastal science treated the height of breaking waves as "cause" and changes in beach gradient as "effect". The bidirectional interdependence whereby changing beach shape alters breaker height and shape was neglected. Further constraining understanding and prediction, disciplinary boundaries were strictly observed and physical, ecological, chemical and geographical scientists remained within their own domains and seldom shared paradigms or insights. Collaborations between natural and social scientists were almost unheard of among scientists from the "Greatest Generation."

We are now in an entirely new era and interdisciplinary research is increasingly becoming the new norm. The need for new approaches to facilitating collaborative interdisciplinary research and education was highlighted in a recent National Research Council (NRC) report on "Convergence" (National Research Council 2014). As emphasized in this NRC report, "Convergence" is intended to imply integration of knowledge, tools and ways of thinking from several disciplines. It is not simply the "patching together" of results from one single discipline as an input to another discipline. Today, high performance computing resources have eliminated the need to excessively limit the number of variables in numerical models and make it feasible to connect different types of models from multiple disciplines. “Convergence” is very much within reach. The fifth assessment report by the *Intergovernmental Panel on Climate Change* (IPCC 2013) describes the numerous changes that are presently underway and are expected to accelerate through the 21st century.

Over the next few years, advances in our ability to anticipate, plan for and mitigate the impacts of adverse changes in coastal processes and coastal communities will increasingly require not only continued refinements of natural science and social science models but also rely on development and application of complex systems models (e.g., Bondre and Gaffney 2015; Janssen 1998; Nicolis and Prigogine 1989) that account for a hierarchy of interconnections and non-linear feedbacks. Many models can now be progressively and routinely assessed by high-resolution and rapidly updated satellite data (e.g., Salisbury et al. 2017). To enable such transformational advances, the scientific community can begin by assessing:

1. Existing knowledge of human-environment complex system dynamics;
2. The ability to model socio-ecological interactions at different scales;
3. Relevance of existing models and analyses to engineering, policy making, and management practices;
4. The potential impact of legal structures on community resilience to hazards;
5. Development and assessment of “social vulnerability indices;”
6. The applicability of complexity theories to analyzing interconnectedness of socio-ecological systems and addressing coastal sustainability; and
7. New education and communication strategies to convince the general public and policy makers of the realities of what likely lies ahead.

1.2 The Reality of 21st Century Coasts: Beautiful, Ugly, Diverse

Coastal systems are not all beaches. Some are slums, some are marshes, some are agricultural lands, some are estuaries, some are mud flats, some are rocky cliffs, and some are egregiously luxurious and expensive housing developments, resorts or condominium complexes. All are complex and impermanent and most are seriously threatened by the climate and other environmental changes that are now underway and accelerating. The photos that follow in Fig. 1.2 introduce the diversity of coastal systems albeit at a very superficial level. Although these photos show what exists now, some may also be representative of what might be coming as changes occur. Another resource with powerful pictures, video, and information is Future Earth Coasts, which was formally known as the Land Ocean Interactions in the Coastal Zone (LOICZ) project. To connect with the Future Earth Coasts community

Fig. 1.2 **a** Beach, Kauai HI (Courtesy of L. D. Wright). **b** Transgressing wetlands, W. Fl (Photo: L. D. Wright). **c** Aftermath of Hurricane Rita, Mississippi Delta. This 3 October 2005 photo shows in Holly Beach, a coastal community of 300 residents in Louisiana's Cameron Parish. Credit: Win Henderson, FEMA. **d** Coastal Taro fields, Hawaii (photo by L. D. Wright). **e** Coastal Slum, Kolkata (Calcutta) India. Credit: Freelance Photographer Bertrand Celce.fotservis.typepad.com/photos/mother_India_Calcutta_var/slums.html. **f** Mangrove swamp, Florida (Photo L. D. Wright). **g** Gas lines, Plaquemines Parish, LA. (Photo: L. D. Wright). **h** Dying cypress swamp, Coastal La 2014 (Photo: L. D. Wright). **i** Fishing Village, Plaquemines Parish, LA (Photo: L. D. Wright). **j** Mouth of the Waiapu River, NZ (Photo Steve Kuehl, VIMS). **k** Recreational Fishing, Xiamen, China (Photo: L. D. Wright). **l** Nature Coast, Florida Gulf Coast 2016 (Photo: L. D. Wright). **m** Malé, capital of the Maldives 2004 (Credit: Shahee Ilyas, via Wikipedia). **n** Estuarine coast. Brisbane River, Queensland, Australia (Courtesy of C. R. Nichols). **o** Volcanic coast. Pagan Island, Commonwealth of the Northern Mariana Islands (Photo C. R. Nichols)

visit URL: http://www.futureearthcoasts.org, a network providing information on the world's coastal zones and active collaboration between nations, disciplines, programs, researchers and users of science.

Perhaps the most prominent feature of many of the photos is evidence of the intersection of human occupation with water. The World Ocean Review estimates the number of people, worldwide, currently living in low lying coastal areas subject to recurrent flooding to be roughly half a billion. However, Baker (2012) and Dawson (2017) suggest that number may be as high as two billion people. The majority of these people live in squalid, densely populated urban environments (Baker 2012; Dawson 2017). As sea levels rise and the world's population grows, these numbers are sure to rise. Increased vulnerability of coastal dwellers is one the most damaging and imminent of the expected future changes. The World Ocean Review (2015) presents data on the populations living in flood prone coastal areas today. Some highlights are summarized in Table 1.1. Notably, in a few cases, such as the Maldives (see Wadey et al. 2017; Fig. 1.2m), the entire population, though small, is subject to frequent inundation and is likely to be displaced within a few decades. A similar situation threatens the residents of the Marshall Islands (Goodell 2017).

1.3 Interconnected Agents of Coastal Change

The dynamic processes that we call *coastal systems* consist of numerous, interdependent factors that share in the roles of cause and effect. Some of the more prominent of these are listed in Table 1.2. When we consider the behavior of any coastal system we must be aware that all of these factors interact to varying degrees although some interactions are much stronger than others. In cases where the interactions are non-linear, the outcomes may be unexpectedly dramatic. The time frames of change can vary diurnally, fortnightly (e.g. two-week tides), seasonally and annually. There are also significant inter-annual fluctuations in ocean circulation, freshwater flow and weather. Coastal ecosystems are well adapted to those

(a)
04/05/2015
(b)
(c)
CAMERON 10
(d)
(e)
(f)
(g)
12/14/2014
(h)
12/14/2014

Fig. 1.2 (continued)

Fig. 1.2 (continued)

Table 1.1 Coastal populations living in low-lying, flood prone areas

Nation	Population in low-lying coastal areas (M)	Percentage of nation's total population (%)
China	127.04	10
India	63.34	6
Bangladesh	53.11	39
Indonesia	41.81	20
Vietnam	41.44	53
Japan	30.83	24
Egypt	24.41	36
USA	23.28	8
Thailand	15.69	25
Philippines	15.12	20
Maldives	0.29	100

Source World Ocean Review (2015)

cycles. However, human development, combined with global climate change, attendant sea level rise, increased storm intensity, intense and frequent droughts or floods, are now superimposing long-term trends on these more common cycles. Figure 1.3 shows the impacts of changes that are happening, or will happen, in natural ocean systems are likely to have on human economic systems.

Coastal ecosystems are particularly sensitive to changes in freshwater runoff and to the shifts in the normal patterns of salt-water inundation that sea level rise is

Table 1.2 Interacting factors in changing coastal systems

Land factors	Atmospheric factors	Ocean factors	Ecological factors	Hydrologic factors	Human factors
Elevation	Climate	Tide range	Vegetation	River discharge	Income
Sediment type and size	Wind regime	Wave climate	Benthic habitat	River-borne nutrients	Ethnicity and culture
Tectonic stability	Latitude and temperature	Coastal currents	Dissolved oxygen	River flood frequency	Health, age and mobility
Subsidence versus uplift	Precipitation	Storm surge height	Water chemistry	Ground water levels	Education and literacy
Continental shelf width	Storm frequency	Sea surface temperature	Wildlife and fish	Aquifer porosity	Employment
Mountainous versus lowland	Cloud cover	Salinity and stratification	Wetlands health	River-borne sediment	Housing
Erosion versus accretion	Tropical versus extratropical	Algal blooms	Estuarine versus marine	Urban drainage	Built infrastructure
	Albedo	Ocean circulation	Sediment geochemistry	salt water intrusion	Legal and political

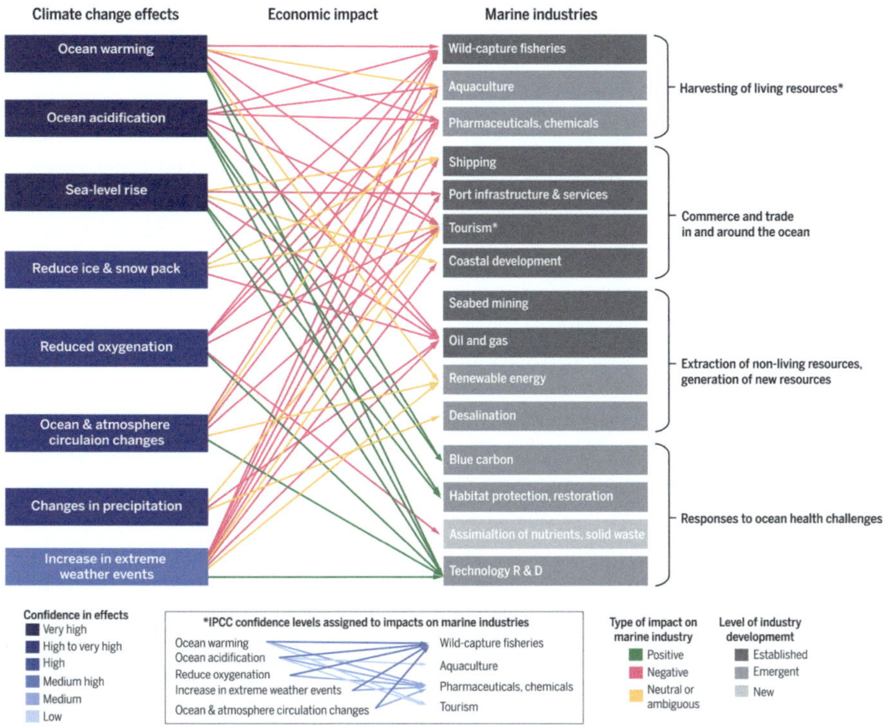

Fig. 1.3 Coastal zones are increasingly regarded as economic resources. The left column illustrates climate change effects on established or emerging marine-related industries that are listed on the right column (From Allison and Bassett 2015)

causing. Loss of coastal habitat is occurring in many cases (such as in the wetlands of Louisiana; National Research Council 2006) because of a combination of storm-induced erosion, land subsidence, sea level rise, and wetlands destruction by human activities such as canal dredging. Socio-economic consequences of coastal change include impacts on housing (especially low income housing), community cohesion, agriculture, fisheries, tourism, and human susceptibility to water–borne pathogens and toxins. But humans are not passive victims of coastal change; they are also major agents of the changes that are causing problems. Beyond the much-publicized impacts of human-induced carbon emissions on global warming, there are numerous other anthropogenic sources of coastal change. The extensive damming of rivers, the drawdown of groundwater, nutrient loading of rivers from agriculture, excessive shoreline hardening, coastal housing developments and human exacerbation of coastal subsidence by a variety of practices including water and hydrocarbon extraction (Allison et al. 2016) and sediment compaction under the weight of urban development. The extent to which these harmful practices are permitted to prevail will depend in turn on a host of political and legal factors.

1.4 Existing Discipline-Specific Short Term Predictive Models

Predictive models will necessarily underpin our ability to understand and predict future coastal circumstances and plan adaptive strategies on decadal time scales. In order to fully appreciate and trust the predictions that the scientific community offers, it is important to understand some basics of the models. It is not necessary to be a modeler to benefit from a non-technical acquaintance with models. For those who are intrigued to learn more than the basics, references offer further insight. We keep the descriptions here brief, elementary, and non-technical.

At both long-range and disturbance or event time scales, we should expect advances to be made progressively not only in modeling specific phenomena such as storm surges and demographic shifts but also in linking models and model outputs in ways that highlight feedbacks and non-linear connections. However, most operational models remain discipline specific (e.g., atmospheric physics, ocean physics, ecosystem dynamics, socioeconomic). Storm surge models in use today include two-dimensional (e.g., vertically averaged properties across a spatial domain) and three-dimensional models using both structured and unstructured (irregular) grids. One of the simplest and lowest resolution models, is NOAA's operational two-dimensional SLOSH model. SLOSH is the acronym for Sea, Lake and Overland Surges from Hurricanes. The academic community uses more accurate unstructured grid models of coupled surge-wave effects (e.g., Luettich et al. 2013). Although those models yield better results than the operational, long-standing two-dimensional SLOSH model used by NOAA for several decades, SLOSH continues to be the operational model of choice because it is well accepted, fast and does not require High Performance Computing or HPC (advanced computing) resources. The important lesson as we go forward is to remain sensitive to the ever-present tradeoff between accuracy, computational efficiency, and familiarity.

Existing models of inundation, river hydrology, water quality, coastal erosion, ecosystem dynamics, populations, and related impacts will be needed in future assessments of resilience. However, while physical and ecosystem modelers are predicting natural threats, the affected communities are also changing. The ways that their economies, health, age and cultural behaviors evolve changes the community's vulnerability. As in the case of the natural sciences, the past few years have seen significant advances in understanding and modeling societal factors and changes that can impact community resilience (e.g., Gunderson and Holling 2002). Van Zandt et al. (2012) consider neighborhood resilience in relation to social vulnerability and housing. Norris et al. (2008) offer a treatise on the psychology of community resilience as it impacts disaster readiness. Some discipline-specific models are described in the following sub-sections.

Long-term Climate Models—Climate models such as the NCAR Community Climate Model (CCM3) and its successors will probably provide the forcing inputs for models of changes in storm frequency and intensity as well as ocean responses in the form of sea level, currents, waves and storm surges. At the present time however, CCM3 has scale limitations that are now being corrected by newly-emerging "grid-within-grid" methods. Event-scale forecasts of weather events, such as hurricanes, will likely continue to depend on operational agencies such as the National Weather Service (NWS) and the National Hurricane Center (NHC), but improved tools from academia can help to make those forecasts more reliable and relevant. The *Intergovernmental Panel on Climate Change* (IPCC 2013, see https://www.ipcc.ch/) points out that "Projections of changes in the climate system are made using a hierarchy of climate models ranging from simple climate models, to models of intermediate complexity, to comprehensive climate models, and Earth System Models".

Coastal Meso-scale Atmospheric Modeling—There is much information available for incorporation into models of coastal flooding. Surface winds and atmospheric pressure drive coastal ocean circulation and generate surface gravity waves and storm surges. The operational Weather Research and Forecasting (WRF) Model at the NOAA National Centers for Environmental Prediction (NCEP) runs operational models that predict these forcing winds. Whereas CCM3 is a *climate* model, WRF is a *weather* model. The WRF predictions are available via web services. Fortunately, the WRF, model and other mesoscale atmospheric community models are available to the research community. Short-term predictions of hurricane wind fields are updated several times a day by the National Hurricane Center and these wind fields provide the inputs to ocean models that predict waves and storm surges.

Coastal Ocean Circulation Modeling—Coastal ocean forecasting is rapidly becoming a reality. Operational predictions are being performed by the Navy and NOAA's National Ocean Service (NOS) is making operational predictions for the Great Lakes and several estuaries, including Chesapeake Bay while advancing such tools for other areas. Automated (quasi-operational) predictions for the coastal ocean are also being performed by various academic research groups around the Nation. Several different community circulation models are in use. Due to their 3-dimensional structure, the baroclinic circulation modeling and prediction systems are well-positioned for utilization in ecosystem forecasting and, thus, in the analysis of eutrophication, hypoxia, harmful algal blooms, water quality, larval dispersal, marine mammal habitat characterization, etc. However, they can be critically dependent upon the chemical inputs, suspended sediments, and bed loads due to river discharges estimated from hydrological models.

Inundation and Storm Surge Modeling—Storm surge modeling is a more specialized activity. Depth-integrated 2-dimensional barotropic models that ignore density stratification have been traditionally used by operational forecaster. Operational storm surge modeling is performed by the National Weather Service (NWS) (NCEP). Storm surge modeling using more advanced 3-dimensional and unstructured grid model is performed by various federal entities as well as academia. Operational surface wave modeling is performed by NWS (NCEP). Accurate operational shallow (surface) wave and surf modeling is still an issue being addressed by a large research community. Storm surge and wave modeling, together with tidal modeling are essential for coastal erosion/deposition and coastal inundation estimates. A recent NOAA-IOOS funded Coastal Ocean Modeling Testbed (COMT) program (Luettich et al. 2013, 2017), involving over 20 academic institutions as well as several federal research centers focused on improving forecasts coastal waves, storm surge and inundation. Models were evaluated for hurricanes Rita (2005) and Ike (2008) in the northern Gulf of Mexico where both storms caused significant flooding and inland penetration of water. Models evaluated were:

1. The NWS SLOSH model was coupled to the Simulating WAves Nearshore (SWAN) spectral wave model for several available domains in the Gulf of Mexico. SWAN was developed at the Delft University of Technology in the Netherlands.
2. The Finite-Volume, primitive equation Community Ocean Model or FVCOM (circulation, surge, inundation) was coupled to the spectrum-based surface wave model SWAN and known as FVCOM-SWAVE for the study of nearshore ocean processes such as tides, circulation, storm surge, waves, sediment transport, and morphological evolution in the Gulf of Mexico.
3. SELFE (Semi-implicit Eulerian-Lagrangian Finite Element) SELFE (circulation, surge, inundation) coupled to the Wind Wave Model (waves). SELFE is an open-source community-supported modeling system, based on unstructured grids, designed for the effective simulation of 3-dimensional baroclinic circulation across river-to-ocean scales. The WWM is similar to well-known SWAN model.
4. ADvanced CIRCulation (ADCIRC) was coupled to the unstructured SWAN model (waves). ADCIRC is a finite element, time-dependent, long wave, hydrodynamic circulation numerical model for the simulation of water level and current over an unstructured gridded domain. Figure 1.4 shows ADCIRC and SWAN model predictions of storm surge for the east shore of Puerto Rico during Hurricane Georges.

Additional models not included in COMT, which may be applied to surge and wave modeling include Delft3D, WAVEWATCH III®, and GeoClaw. Delft3D developed by Delft Hydraulics (see Roelvink and Van Banning 1994), provides a 3-dimensional modeling suite that includes open source Delft3D flow (FLOW), morphology (MOR) and waves (WAVE) modules to rapidly simulate nearshore conditions. WAVEWATCH III is an operational model used by organizations such as NOAA and the U.S. Navy. This wave modeling framework uses nested storm-tracking winds to better model hurricane trajectories and intensities and predicts wind-generated waves and swell, not tsunamis, storm surge, or tides. A well-proven open-source storm surge model, which is part of the "Conservation Laws Package" or Clawpack is GeoClaw, a finite volume solver for depth-averaged flows a model with adaptive mesh refinement. GeoClaw was originally developed for tsunami modeling.

Hydrologic Models—Coastal flooding comes not only from the sea but also from rivers and torrential rain (e.g. Hurricane Harvey flooding of Houston Texas in

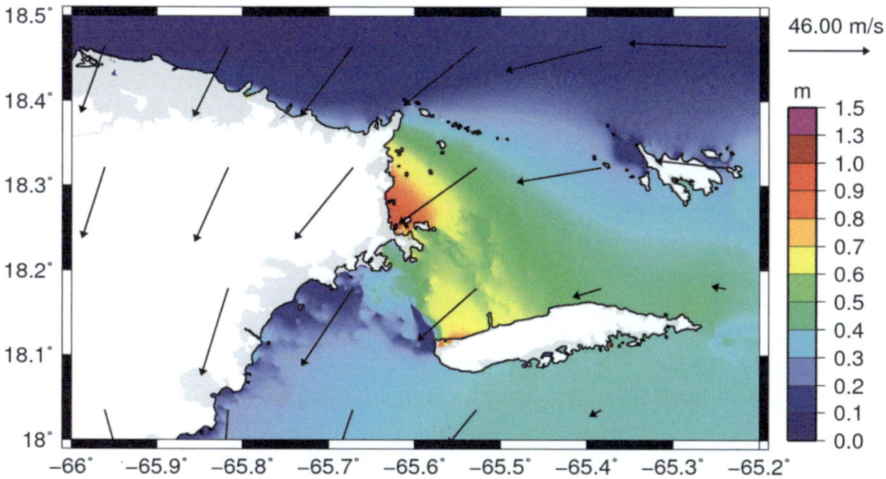

Fig. 1.4 Model results from combined ADCIRC-SWAN models for Puerto Rico with wind and wave forcing for Hurricane Georges (van der Westhuysen et al. 2015)

August 2017). The U.S. Geological Survey gauging stations provide many real-time river flows throughout the U.S., and NOAA (through the National Weather Service and the National Water Center) provides operational river forecast models, particularly the comprehensive National Water Model (NWM). The existing river forecast system is based on empirical relationships between river stage heights and flows. NCAR has recently developed a landscape hydrology model with distributed overland routing called Noah-distributed, which is data-compatible with a range of atmospheric models. The community Noah-distributed model has recently been linked directly to the National Hydrographic Dataset Plus (NHDPlus) surface water framework. NHDPlus is an integrated suite of application-ready geospatial data sets that incorporate many of the best features of the National Hydrography Dataset (NHD), the National Elevation Dataset (NED), and the Watershed Boundary Dataset (WBD). Hydrologists may use several other types of hydrological models to predict hydrologic system behavior or processes.

While these types of models can be classified in different ways they tend to be either static (steady state) or dynamic. Dynamic models include time while static models exclude time. A runoff model provides estimates of runoff as a function of the parameters that are used to describe the watershed over time steps. The static model considers the relationship between variables such as river stage and discharge along a cross-section at that same instant in time. Devi et al. (2015) provide a general review and several examples of the different types of hydrological models. Additional details of existing open-source hydrological models can be found online at the Community Surface Dynamics Modeling System (CSDMS) project, a virtual home for a vibrant and growing community of about 1000 international modeling experts and students who study the dynamic interactions of lithosphere, hydrosphere, cryosphere, and atmosphere at Earth's surface (see URL: http://csdms.colorado.edu/wiki/Hydrological_Models). CSDMS presently lists 67 hydrological models. Typically for surface runoff models they are broken down into event-models (e.g., ANUGA), daily models (e.g., HydroTrend), hydraulic models (e.g., TELEMAC), river-reach, drainage basin (e.g., TopoFlow, PIHM), and global water balance models (e.g., WBMsed, VIC). Hydrological models are an important and necessary tool for water and environment resource management. NOAA recently inaugurated the NWM to simulate observed and forecast streamflow over the entire continental United States using High Performance Computers (NOAA 2016). The NWM simulates the water cycle with mathematical representations of the different processes such as precipitation, snowmelt and infiltration and movement of water through the soil layers all which vary with changing elevations, soils, vegetation types and a host of other variables. Additionally, extreme variability in precipitation over short distances and times can cause the response on rivers and streams to change very quickly. Blodgett et al. (2016) describe how collaboration and the integration of systems such as the Great Lakes Observing System and the National Ground-Water Monitoring Network are helping to provide accurate, local, real-time flood forecasting to prevent loss of life and property during this era of changing climate and extreme weather. Future Earth system modeling is focused on the coupling of hydrological, meteorological, and oceanographic models.

Ground Water Hydrology Models—The U.S. Geological Survey (USGS) is among the world's leaders in numerical modeling of the flow of ground water through porous aquifers. The USGS HYDROTHERM program is a three-dimensional model for simulating the two-phase flow of ground water and associated heat transfer and includes predictions of the rates of ground water recharge by precipitation. The USGS Modular Ground Water Flow Model (MODFLOW) includes simulations of the effects of alternated wetting and drying within an aquifer. Linkages between MODFLOW predictions and surface water flows are predicted via the *Coupled Ground Water and Surface Water Flow Model (GSFLOW)*. Perhaps the most well-known open-source watershed flow model is PARFLOW, which results from a long, multi-institutional development history (see URL: http://csdms.colorado.edu/wiki/Model:ParFlow). Ground water behaviors are characteristically non-linear and these models accommodate these complexities.

Modeling Coastal Erosion and Shoreline Transgression—Forecasting the responses of shores and coastal lands to rising sea level will require the coupling of morphodynamic, ecological, physical, engineering and socioeconomic models. The morphology of the coast and the relative elevation (to mean sea level) of its shores, occupied land surfaces and intertidal wetlands are regulated by feedbacks among sediment supply versus removal, plant communities, tidal flooding and production of organic matter (Morris et al. 2002; Friedrichs and Perry 2001) as well as by the built infrastructure including roads and breakwaters. In order for a marsh surface to maintain its elevation relative to mean high tide, the rate of marsh growth must keep pace with rate of sea-level rise. Some of the more widely used models of beach and shoreline change include CEM, Delft3D, GEOMBEST, XBeach and

Wetland3P. Most recently, Vitousek et al. (2017) have developed a decadal-scale model of long-term coastal evolution that assimilates data from routine monitoring. A complete inventory of open-source coastal models can be accessed through the CSDMS project (see URL: http://csdms.colorado.edu/wiki/Coastal_models).

Coastal Ecosystem Modeling—Coastal ecosystems modeling efforts currently recognize that ecosystem modeling is fully multidisciplinary. It considers nutrient inputs, exposure to light, followed by primary productivity and then interactions at all higher trophic levels. With coastal ocean nutrients deriving from both deep-ocean and land drainage sources, coastal ocean ecosystem models integrate coastal circulation models and connections of the continental shelf with estuaries. Coupling circulation models with bio-chemical models helps to account for biological interactions. Recent advances have been made in developing forecast models of dissolved oxygen in terms of circulation, wind, fresh-water input and nutrient inputs for estuaries such as the Chesapeake Bay (e.g., Irby et al. 2015; Scully 2013) and the Northern Gulf of Mexico continental shelf (e.g., Fennel et al. 2016).

Social Sciences Models—Social scientists are working to predict what socio-economic changes may be coming over the next few decades. Significant advances are being made in understanding and modeling societal factors and changes that can impact community resilience (e.g., Gunderson and Holling 2002). Van Zandt et al. (2012) consider neighborhood resilience in relation to social vulnerability and housing. Cutter et al. (2010, 2014) have evolved the concept of *Baseline Resilience Indicators for Communities (BRIC)* as empirical metrics for gaging the resilience of communities to disasters. Berkes et al. (2003) offer in depth analyses of social-ecological complexity in assessing community resilience. Guillard-Gonçalves et al. (2014) have developed a "Social Vulnerability Index" (SoVI) which can be readily applied to most regionally specific communities. Typically, social science models are broken down into (1) Integrated Assessment Models such as the kind used by IPCC, (2) General Equilibrium Models (such as used by economists), (3) Agent or Individual Based Models, and (4) Land cover Land use models. Decentralized development of these models is evidenced by the Open Agent Based Modeling (OpenABM) Consortium which has grown into the Network for Computational Modeling in Social and Ecological Sciences (CoMSES). These large formal institutions provide a repository for agent based models and support researchers, educators, and professionals with a common goal of improving the way computational models in the social and life sciences are developed, shared, and utilized.

Human Health Models—The World Health Organization (WHO 2003) notes that three types of predictive models in use for assessing the likely impacts of climate change on the spread of human infectious disease are (a) statistical, (b) process-based and (c) landscape-based. Statistical models are applied to predict climate change impacts on malaria, dengue fever and encephalitis. Mathematical process-based models are used to predict how diseases are transmitted as results of given configurations of climate variables that affect vector and parasite biology. Landscape-based models combine climate-based models with spatial analytical methods to estimate, for example, how future climate-induced changes in ground cover and surface water might affect mosquitoes and hence, malaria. Similarly, for flooded coastal environments, both process-based landscape models could be applied to predicting the pathways and residence times of *Vibrio* and other water-borne pathogens as well as toxic contaminants.

Models for Supporting Emergency Management Decisions—The Hazus® Program of the Federal Emergency Management Agency (FEMA) represents the present "state of the art" in estimating potential losses from earthquakes, floods and hurricanes (FEMA 2009a, b). However, many aspects of this program are regionally specific and have not been implemented for many coastal communities. Hopefully, in the future, the academic community will be able to contribute to Hazus® in meaningful ways.

Integrative Linkages of Multiple Models—The future requires that models of the different genre as just described be connected in meaningful ways to allow the feedbacks and mutual interdependences to be considered. To begin, scientists or planners concerned with anticipating future events of coastal inundation might ask: "How will the risk of flooding during an extreme event be exacerbated in various sea level rise scenarios?". The answer will depend on where people with different vulnerabilities are living in the future. Recent advances in detailed modeling of "street-level" flooding in well-mapped neighborhoods (Blumberg et al. 2015) can contribute to answering such questions as can similar advances in modeling the timing of storm surges in relation to tides (Georgas et al. 2014). Intersecting predictions of inundation with patterns of social vulnerability are valuable contributions to disaster planning. At the present time, some fairly elementary efforts at doing this are accessible on the web.

1.5 Dynamic Change in Complex Systems

The International Geosphere Biosphere Programme (IGBP) has articulated the importance of intersecting social and natural sciences and has evolved the "Anthropocene" paradigm that considers human and natural earth processes to be interdependent and to function and change as a complex system (Bondre and Gaffney 2015; Steffen et al. 2016). The idea of complexity is now widely accepted by modelers of dynamic systems involving the non-linear interdependence of multiple processes (Bar-Yam 1997; Liu et al. 2007; Sayama 2015). Complexity science typically encompasses multiple theoretical frameworks and is highly interdisciplinary. Various patterns of coupling of societal, biogeophysical, biogeochemical, and ecological processes constitute prominent example of complexity. The complexities of such multi-faceted systems arise from the facts that causal linkages among all components are bidirectional creating positive and negative feedback loops (Fig. 1.5).

The connections in feedback loops may be either positive (i.e., an increase in A causes an increase in B and vice versa) or negative (which leads to stability). In

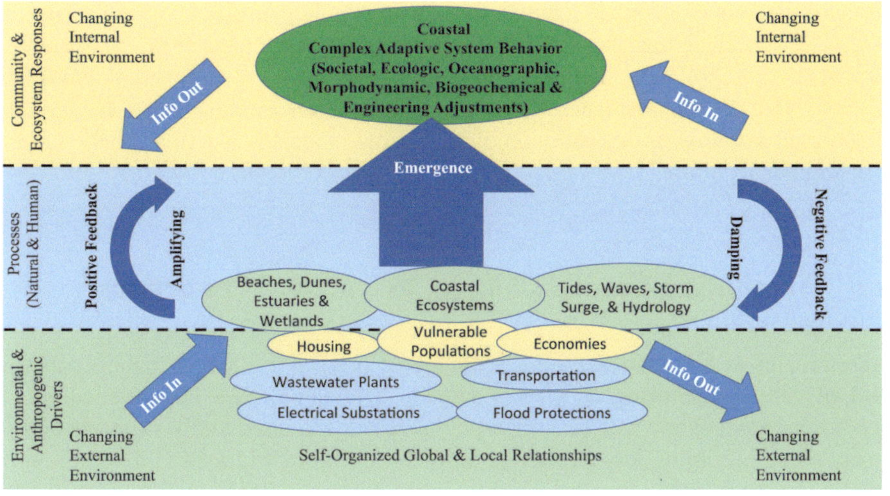

Fig. 1.5 Connections and Feedbacks in a Complex System should be considered over long time periods and across regional scales. The ability to prepare and plan for, absorb, recover from, and more successfully adapt to processes such as sea level rise will require consideration of environmental, engineering, and community factors

addition, causalities may be linear or non-linear. In the case of present-day global change processes, a very prominent example of positive feedback is found in the Arctic Ocean: decreasing sea ice cover reduces albedo (reflection of radiation) which in turn leads to more heat absorption by the seawater which in turn accelerates ice melting. Ecosystems and socioeconomic systems are *Complex Adaptive Systems* (Harris 2007; Holling 2001). Such systems are susceptible to the emergence of unexpected or abrupt changes in behavior, or state that are very different from what past trends might imply and usually have numerous interactive components. These components change and adapt as they interact (Fiksel 2006, Liu et al. 2007, Hopkins et al. 2011, 2012). In future science-based planning for change, we must seek answers to some fundamental questions about living, evolving and adapting- or vanishing-systems. The goal of a Systems Approach Framework (SAF) to future planning is to promote improved predictive modeling of the ways that complex coastal systems function and change (Fiksel 2006, Hopkins et al. 2011, 2012).

Precise quantitative forecasts of future conditions are, unfortunately, not on the horizon. However, greater confidence in estimates of the degree to which a particular combination of changed physical and ecological conditions might impact humans is already emerging from trans-disciplinary modeling activities. The challenges for the immediate future relate more to the human aspects than to the natural science aspects. To solve the problems that we face, complex system components will need to include (1) physical and natural sciences (physical models of sea level rise, river hydrology, and ecosystem models); (2) urban models;

(3) attitudes and beliefs of local regulators and policy makers; (4) the extent to which educators understand and communicate environmental science; (5) economic pressures and political ideologies; (6) pressures from environmental advocacy groups; and (7) the awareness of the general public. Finding effective ways to include all of these factors into complex systems models is a serious challenge for the immediate future.

1.6 Monitoring Future Changes to Validate Predictions

Assessments of the reliability and accuracy of predictive models require constant and sustained monitoring of what is actually happening. New instrumentation for recording atmospheric and oceanographic processes is continually evolving and in many cases becoming less expensive allowing for more widespread distribution of monitoring resources. Regional coastal observations within US waters are currently orchestrated by the U.S. Integrated Ocean Observing System (IOOS®), which is under the auspices of NOAA. Included under IOOS are arrays of in situ environmental sensors as well as the U.S. National High Frequency (HF) Radar Network. Although, improvements in resolution for coastal applications are still needed, new generations of satellite and lower altitude remote observing systems are permitting high-resolution images to be obtained around the clock (Cazenave et al. 2017). New geostationary satellite systems, such as the EPA's Proposed Geostationary Coastal and Air Pollution Events (GEO-CAPE; Salisbury et al. 2017) will soon facilitate a new level of monitoring resolution and frequency. Unfortunately, as of late 2017, funding for projects such as this is uncertain.

1.7 Resilience Is "Going with the Flow" and Doing OK

Changes in coastal systems are underway and accelerating. We may slow the processes of change if we make a global commitment to do so but we will not stop them, despite emerging plans for "climate engineering" (Kravitz et al. 2017). The future of coastal dwellers will be different from today. The role of scientists, planners, politicians and the public will not be to stop change but to understand what is likely to happen and prepare accordingly. This means constructing forward looking plans for human and ecosystem **resilience**. The extreme stimuli of advancing and retreating ice in the Pleistocene proved that humans are resilient. The time for reawakening to our roots is upon us.

"*Resilience* is the capacity of a system, be it an individual, a forest, a city or an economy to deal with change and continue to develop" (Stockholm Resilience Centre 2014; see URL: www.stockholmresilience.su.se). According to a recent National Academies report on disaster resilience (National Academies 2012), "*Resilience is the ability to prepare and plan for, absorb, recover from, and more successfully adapt to adverse events*." Resilience involves the ability to adapt to constantly changing environmental, economic, and social stressors. It does *not* imply constancy, stasis or resistance to change. It is the capacity to change and adapt continually yet remain viable.

References

Allison, E.H., and H.R. Bassett. 2015. Climate change in the oceans: Human impacts and responses. *Science* 350 (6262): 778–782. https://doi.org/10.1126/science.aac8721.

Allison, M., B. Yuill, T. Törnqvist, F. Amelung, T.H. Dixon, G. Erkens, R. Stuurman, C. Jones, G. Milne, M. Steckler, J. Syvitski, and P. Teatini. 2016. Global risks and research priorities for coastal subsidence. *EOS* 97 (19): 22–27.

Baker, J.L. 2012. *Climate change, disaster risk and the urban poor: Cities building resilience for a changing world*. Washington DC: World Bank.

Bar-Yam, Y. 1997. *Dynamics of complex systems*, 848. Cambridge, MA: Perseus Press.

Berkes, F., J. Colding, and C. Folke. 2003. *Navigating social-ecological systems; building resilience for complexity and change*. Cambridge: Cambridge University Press.

Blodgett, David, Emily Read, Jessica Lucido, Tad Slawecki, and Dwane Young. 2016. An analysis of water data systems to inform the open water data initiative. *Journal of the American Water Resources Association* 52 (4): 845–858. https://doi.org/10.1111/1752-1688.12417.

Blumberg, A.F., N. Georgas, L. Yin, T.O. Herrington, and P.M. Orton. 2015. Street-scale modeling of storm surge inundation along the New Jersey Hudson River Waterfront. *Journal of Atmospheric and Oceanic Technology* 32 (8): 1486–1497.

Bondre, N., and O. Gaffney (eds.). 2015. *IGBP annual report 2014/2015*. Sweden: IGBP Secretariat, The Royal Swedish Academy of Sciences.

Cazenave, A., G. Le Cozannet, J. Benveniste, P.L. Woodworth, and N. Champollion. 2017. Monitoring coastal zone changes from space. *Eos* 98: 2017. https://doi.org/10.1029/2017EO085581. Published on 02 November.

Cutter, S.L., C.G. Burton, and C.T. Emrich. 2010. Disaster resilience indicators for benchmarking baseline conditions. *Journal of Homeland Security and Emergency Management* 7 (1): 51. https://doi.org/10.2202/1547-7355.1732.

Cutter, S.L., K. Ash, and C. Emrich. 2014. The geographies of community disaster resilience. *Global Environmental Change* 29: 65–77.

Dawson, A. 2017. *Extreme cities: The perils and promise of urban life in the age of climate change*, 378. New York: Verso.

Devi, G., K.B.P. Ganasri, and G.S. Dwarakish. 2015. A review on hydrological models. *Aquatic Procedia* 4: 1001–1007. https://doi.org/10.1016/jj.aqpro.2015.02.126.

Federal Emergency Management Agency. 2009a. *HAZUS-MH MR4 hurricane model technical manual*, 511. Washington, D.C.: Federal Emergency Management Agency, Mitigation Division.

Federal Emergency Management Agency. 2009b. *HAZUS-MH MR4 flood model technical manual*, 432. Washington, D.C.: Federal Emergency Management Agency, Mitigation Division.

Fennel, K., A. Laurent, R. Hetland, D. Justić, D.S. Ko, J. Lehrter, M. Murrell, L. Wang, L. Yu, and W. Zhang. 2016. Effects of model physics on hypoxia simulations for the northern Gulf of Mexico: A model intercomparison. *Journal of Geophysical Research-Oceans*. https://doi.org/10.1002/2015JC011577.

Fiksel, J. 2006. Sustainability and resilience: Toward a systems approach. *Sustainability Science Practice and Policy* 15 (2): 15–21. https://doi.org/10.1109/EMR.2007.4296420.

Friedrichs, C., and J. Perry. 2001. Tidal salt marsh morphodynamics: A synthesis. *Journal of Coastal Research* 27 (27): 7–37.

Georgas, N., P. Orton, A. Blumberg, L. Cohen, D. Zarrilli, and L. Yin. 2014. The impact of tidal phase on Hurricane Sandy's flooding around New York City and Long Island sound. *Journal of Extreme Events* 1 (1): 1–32.

Goodell, J. 2017. *The water will come*, 340. New York: Little, Brown and Company.

Guillard-Gonçalves, C., S.L. Cutter, C.T. Emrich, and J.L. Zêzere. 2014. Application of social vulnerability index (SoVI) and delineation of natural risk zones in Greater Lisbon, Portugal. *Journal of Risk Research* 18 (5): 651–674.

Gunderson, Lance H., and C.S. Holling. 2002. *Panarchy: Understanding transformations in human and natural systems*. Washington: Island Press.

Harris, G. 2007. *Seeking Sustainability in an Age of Complexity*. Cambridge University Press, 374 pp.

Holling, C.S. 2001. Understanding the Complexity of Economic, Ecological and Social Systems. *Ecosystems* 4: 390–405.

Hopkins, T.S., D. Bailly, and J.G. Stttrup. 2011. A systems approach framework for coastal zones. *Ecology and Society* 16 (4): 25. https://doi.org/10.5751/ES-04553-160425.

Hopkins, T.S., D. Bailly, R. Elmgren, G. Glegg, A. Sandberg, and J.G. Støttrup. 2012. A systems approach framework for the transition to sustainable development: Potential value based on coastal experiments. *Ecology and Society* 17 (3): 39.

Intergovernmental Panel on Climate Change (IPCC). 2013. *Climate change 2013: The physical science Basis. Contribution of working group I to the fifth assessment report of the intergovernmental panel on climate change*. Ed. T.F. Stocker, D. Qin, G.-K. Plattner, M. Tignor, S.K. Allen, J. Boschung, A. Nauels, Y. Xia, V. Bex, and P.M. Midgley. Cambridge: Cambridge University Press.

Irby, I.D., M.A.M. Friedrichs, C.T. Friedrichs, A.J. Bever, R.R. Hood, L.W.J. Lanerolle, M. Li, L. Linker, M.E. Scully, K. Sellner, J. Shen, J. Testa, H. Wang, P. Wang, and M. Xia. 2015. Challenges associated with modeling low-oxygen waters in Chesapeake Bay: A multiple model comparison. *Biogeosciences* 13: 2011–2028. https://doi.org/10.5194/bg-13-2011-2016.

Janssen, M. 1998. Use of complex adaptive systems for modeling global change. *Ecosystems* 1 (5): 457–463.

Kravitz, B., A. Robock, and J.E. Kristjánsson. 2017. Understanding how climate engineering can offset climate change. *Eos*. https://doi.org/10.1029/2016EO005279.

Liu, J., T. Dietz, S.R. Carpenter, M. Alberti, C. Folke, E. Moran, A.C. Pell, P. Deadman, T. Kratz, J. Lubchenco, E. Ostrom, Z. Ouyang, W. Provencher, C.L. Redman, S.H. Schneider, and W.W. Taylor. 2007. Complexity of coupled human and natural systems. *Science* 317 (5844): 1513–1516.

Luettich Jr., R.A., L.D. Wright, R. Signell, C. Friedrichs, M. Friedrichs, J. Harding, K. Fennel, E. Howlett, S. Graves, E. Smith, G. Crane, and R. Baltes. 2013. Introduction to special section on the U.S. IOOS Coastal and Ocean Modeling Testbed. *Journal of Geophysical Research-Oceans* 118 (12): 6319–6328. https://doi.org/10.1002/2013jc008939.

Luettich Jr., R.A., L.D. Wright, C.R. Nichols, R. Baltes, M.A.M. Friedrichs, A. Kurapov, A. van der Westhuysen, K. Fennel, and E. Howlett. (2017). A test bed for coastal and ocean modeling, Eos, 98. https://doi.org/10.1029/2017EO078243.

Morris, J.T., P.V. Sundareshwar, C.T. Nietch, B. Kjerfve, and D.R. Cahoon. 2002. Responses of coastal wetlands to rising sea level. *Ecology* 83 (10): 2869–2877.

National Academies. 2012. *Disaster Resilience: A National Imperative*. Report of Committee on Increasing National Resilience to Hazards and Disasters, 260p. Washington DC: Committee on Science, Engineering, and Public Policy, National Academies Press.

National Research Council. 2014. *Convergence: Facilitating transdisciplinary integration of life sciences, physical sciences, engineering, and beyond*, 152. Washington, DC: The National Academies Press.

National Research Council, Committee on the Restoration and Protection of Coastal Louisiana. 2006. *Drawing Louisiana's New Map: Addressing Land Loss in Coastal Louisiana*, 206. Washington DC: National Academies Press.

Nicolis, G., and I. Prigogine. 1989. *Exploring complexity*, 313. New York: Freeman & Co.

NOAA. 2016. NOAA launches America's first national water forecast model. Available online. URL: http://www.noaa.gov/media-release/noaa-launches-america-s-first-national-water-forecast-model. Accessed August 30, 2017.

Norris, F., S. Stevens, B. Pfefferbaum, K. Wyche, and R. Pfefferbaum. 2008. Community resilience as a metaphor, theory, set of capacities, and strategy for disaster readiness. *American Journal of Community Psychology* 41 (1): 127–150.

Roelvink, J.A., and G.K.F.M. Van Banning. 1994. Design and development of Delft3D and application to coastal morphodynamics. In *Hydroinformatics'94*, ed. A. Verwey, A.W. Minns, V. Babovic, and C. Maksimovic. Rotterdam: Balkema.

Salisbury, Joseph, Curtiss Davis, Angela Erb, Hu Chuanmin, Charles Gatebe, Carolyn Jordan, Zhongping Lee, Antonio Mannino, Colleen B. Mouw, Crystal Schaaf, Blake A. Schaeffer, and Maria Tzortziou. 2017. Coastal observations from a new vantage point. *EOS Earth and Space Science News* 98 (1): 20–25.

Sayama, H. 2015. *Introduction to the modeling and analysis of complex systems*, 479. Binghampton: NY, SUNY Open Textbooks.

Scully, M.E. 2013. Physical controls on hypoxia in Chesapeake Bay: A numerical modeling study. *Journal of Geophysical Research-Oceans* 118 (3): 1239–1256.

Steffen, W., R. Leinfelder, J. Zalasiewicz, C. Waters, M. Williams, C. Summerhayes, A. Barnosky, A. Cearreta, P. Crutzen, M. Edgeworth, E. Ellis, I. Fairchild, A. Galuszka, J. Grinevald, A. Haywood, J. Sul, C. Jeandel, J.R. McNeill, E. Odada, N. Oreskes, A. Revkin, D. Richter, J. Syvitski, D. Vidas, M. Wagreich, S. Wing, A. Wolfe, and H. Schellnhuber. 2016. Stratigraphic and earth system approaches to defining the Anthropocene. *Earth's Future* 4 (8): 324–345. ISSN 2328–4277.

Stockholm Resilience Centre (2014) Stockholm Resilience Centre Annual Report 2014. Stockholm, 64 pp.

van der Westhuysen, André, C. Anselmi, R. Calzada, J. C. Feyen, J. Gonzalez, A. Mercado, J. Morell, E. Rodriguez, J. Smith, J. Veeramony, and J. Westerink. 2015. A wave surge and inundation modeling testbed for Puerto Rico and the U.S. Virgin Islands: Year 1 Progress. In 95th *AMS Annual Meeting, January 4–8, 2015,* 6p. Phoenix, AZ: American Meteorological Society.

Van Zandt, S., W.G. Peacock, D.W. Henry, H. Grover, W.E. Highfield, and S.D. Brody. 2012. Mapping social vulnerability to enhance housing and neighborhood resilience. *Housing Policy Debate* 22 (1): 29–55.

Vitousek, S., P.L. Barnard, and P. Limber. 2017. Can beaches survive climate change? *Journal of Geophysical Research: Earth Surface* 122: 1060–1067. https://doi.org/10.1002/2017JF004308.

Wadey, M., S. Brown, R. Nicholls, and I. Haigh. 2017. Coastal flooding in the Maldives: An assessment of historic events and their implications. *Natural Hazards*. https://doi.org/10.1007/s11069-017-2957-5.

World Health Organization (WHO). 2003. *Climate change and human health: Risks and responses*, 306. Geneva: World Health Organization.

World Ocean Review. 2015. *Hamburg*. Germany: Maribus.

Chapter 2
Global Change: More Than Climate

Lynn Donelson Wright, J. P. M. Syvitski and C. Reid Nichols

The times, they are a-changing.
—Bob Dylan, Nobel Laureate, c. 1964

2.1 The Sources of Information on Climate Change Are Impeccable

As pointed out by Forster (2017), data, models and scientific theories on the causes and realities of climate change have been subjected to tens of thousands of rigorous scientific peer reviews over the past half century. The University Corporation for Atmospheric Research (UCAR; see URL: https://www2.ucar.edu) and the National Center for Atmospheric Research (NCAR; see URL: https://ncar.ucar.edu) in Boulder Colorado, which UCAR manages with funding from the National Science Foundation (NSF) have long histories of leading the way in modeling climate change and climate linked phenomena. The Stockholm-based International Geosphere Biosphere Programme (IGBP; see URL:http://www.igbp.net) provided international leadership in these areas until the end of 2015 when the IGBP was succeeded by Future Earth, a ten-year international research platform (see URL: http://www.futureearth.org).

The United Nations Environment Programme (UNEP) and the World Meteorological Organization (WMO) are among the several governing members of Future Earth. In concert, these highly respected national and international organi-

L. D. Wright (✉)
Southeastern Universities Research Association (SURA), Washington, DC, USA
e-mail: ldwright@bellsouth.net

J. P. M. Syvitski
Community Surface Dynamics Modeling System, University of Colorado, Boulder, CO, USA

C. R. Nichols
Marine Information Resources Corporation (MIRC), Ellicott City, MD, USA
e-mail: rnichols@mirc-us.com

L. D. Wright and C. R. Nichols (eds.), *Tomorrow's Coasts: Complex and Impermanent*, Coastal Research Library 27,
https://doi.org/10.1007/978-3-319-75453-6_2

zations along with numerous universities around the world as well as agencies such as the NOAA Climate Prediction Center (see URL: http://www.cpc.ncep.noaa.gov) are the sources of data and model results on which global change projections are based. As explained by Merryfield et al. (2017), the reliability of climate predictive models is continually improving and is expected to improve significantly over the years ahead thanks in part to rigorous coordination of global climate research by the World Climate Research Programme (WCRP; https://www.wcrp-climate.org/). A key factor in assessing model accuracy involves the use of hind casting to test the skill of models in "predicting" past events for which observational data already exists. The WCRP maintains a comprehensive archive of such data to enable modelers worldwide to test and refine their models.

The series of assessment reports by the *Intergovernmental Panel on Climate Change* (IPCC; see URL: http://www.ipcc.ch) are definitive sources of current scientific consensus on the causes and consequences of global change. The IPCC draws on the work of hundreds of scientists from all over the world. Most recently, in November 2017, U.S. Global Change Research Program (USGCRP), which involves 13 federal agencies, released its Fourth National Climate Assessment (USGCRP 2017; https://science2017.globalchange.gov; Wuebbles et al. 2017). This comprehensive report points to exhaustive and conclusive evidence that air temperatures at the earth's surface as well as sea temperatures are rising leading to a host of other climate and ocean changes. The USCRP report expresses 90% confidence that warming is the result of emissions of greenhouse gases from fossil fuels.

2.2 Synopsis of Some Primary "Global Changes"

The interdependent factors and processes summarized in Chap. 1 are subject to climatic and other drivers that are changing throughout the world. Collectively, the suite of worldwide changes constitutes what is popularly referred to as "global change". While climate is an essential and prominent member of that suite of changing environmental conditions, it is not the only thing that is changing or likely to change. While it is widely understood that the human contributions to global warming involve increases in atmospheric concentrations of greenhouse gases, particularly carbon dioxide and methane, several other anthropogenic sources of change are less well-known. Among these impacts are (1) extensive damming of rivers; (2) extraction of ground water from aquifers; (3) coastal land subsidence caused by the weight of urban centers (Allison et al. 2016); (4) canal dredging (National Research Council 2006); and (5) ocean acidification among other sources of change.

The UCAR Center for Science Education, which can be accessed online at URL: https://scied.ucar.edu/longcontent/predictions-future-global-climate, offers a valuable and straightforward summary of what of the future global changes may include. A brief synopsis of these follows:

1. **Atmospheric warming**—Global temperatures have risen 1 °C as greenhouse gases associated with energy generation and use have trapped ever more heat from solar radiation. Over the next two decades, the globally averaged atmospheric temperature is expected to rise by 0.2 °C. By the end of the century, the globally averaged temperature is expected to be somewhere between 1.8 and 4.0 °C (3.2 and 7.2 °F) higher than today depending on the extent to which greenhouse gas emissions are reduced. Global average temperatures do not convey the full story since most of the warming will be concentrated in the higher latitude and polar regions rather than in the tropics. Furthermore, since most of the temperate and sub-polar lands are in the Northern Hemisphere, the consequences of rising temperatures will be most dramatic there.
2. **Ocean warming**—Of the global warming that has taken place over the past two decades or so, roughly 93% is currently estimated to be stored within the upper kilometer or so of the ocean. Ocean temperatures are expected to continue to rise significantly though by how much depends on how deeply the thermal energy is distributed within the water column. The steric effects observed since 2000 indicate that most of the excess heat resides within the upper 700 m (2300 ft) of the water surface. Although warming also affects the deeper ocean, there is a delay between surface heating and the transfer to the deep ocean.
3. **Evaporation and precipitation**—Rising temperatures are expected to increase the rates of evaporation, which will be countered by increases in precipitation. As a consequence, dry regions are likely to become drier while wet regions will experience increased rainfall. Models predict an average global increase in precipitation of 3–5%. It must be remembered, of course, that 71% of the earth's surface is ocean and most of the precipitation will be there instead of on land although the most intense floods are likely to occur over land.
4. **Ice melting**—Melting of glacial ice such as in Alaska, and the ice sheets of Greenland and Antarctica, along with the melt of permafrost and a reduction in sea ice cover across the Arctic Ocean are already taking place and are expected to accelerate. Decreasing sea ice cover reduces albedo (reflection of radiation), which in turn leads to more heat absorption by the seawater and this, in turn, accelerates ice melting- a positive feedback loop.
5. **Storm intensity**—Predictions of changes in storm frequency continue to have much uncertainty. However, tropical storm intensity is likely to increase, largely because of increased sea surface temperatures. The rapid intensification of Hurricane Matthew in 2016 and Hurricanes Harvey, Irma and Maria in 2017 has been attributed in part to unusually warm sea surface temperatures in the Caribbean. Hurricane Irma was the largest Atlantic Hurricane on record. Computational scientists are now working to improve the resolution of coupled ocean-atmosphere models from which to learn the internal dynamics of storms, before we have to experience them.
6. **Global Sea level rise**—More discussion of this will follow in the next chapter. Sea level is rising and will continue to do so throughout the 21st century. Several of the factors just described contribute to this. Melting glaciers' and thermal expansion of seawater account for 75% of sea level history. Sea level is

expected to be between 20 and 50 cm higher than today by the end of the century.

7. **Changes to the carbon cycle**—According to the UCAR Center for Science Education, warming temperatures are expected to alter the natural geographic ranges of many plant species as well as the lengths of the growing seasons. Models also suggest that, with warmer temperatures, the earth system (especially oceans) will be able to absorb less CO_2 from the atmosphere, thereby exacerbating the warming trend.
8. **Altered ocean circulation**—Major, large scale ocean current systems, such as the Gulf Stream, are *thermohaline* currents driven by gradients in water density caused by contrasts in temperature and salinity. These currents play dominant roles in effecting the transfers of heat from low to high latitudes and vice versa. Changes in ocean temperatures and salinity have the potential to alter these large-scale transport mechanisms in unexpected ways.
9. **Ocean acidification**—Derivative impacts of global warming are many and varied. Sea surface anthropogenic CO_2 is being elevated particularly in the Atlantic Ocean, but the phenomena is global. As a result, seawater acidity is being elevated in ocean surface waters, again particularly in Atlantic Ocean waters. This ocean acidification then negatively impacts marine carbonate systems, including tropical reefs.

There are other major global impacts on world coastal zones related to human activities but not directly to climate change. Some of these include the following:

10. **Reduced delivery of sediment to coastal seas**—As of February 2017, there were 58,519 large dams registered by the International Commission On Large Dams (ICOLD; see URL: http://www.icold-cigb.net/), a subset of millions of dams worldwide, although many of these are much smaller dams. According to Palmieri et al. (2003), total reservoir storage capacity is >7000 km^3 (>1679 mi^3). These dams have substantially reduced the amount of new sediment supplied to the coasts to replace the sediment lost by erosion (Syvitski and Kettner 2011; IGBP 2012). By one estimate, global reservoirs have trapped 3155 gigatons (Gt) of sediment, and in another study (Giosan et al. 2014) deltas are either no longer growing or are losing landmass, sometimes at alarming rates.
11. **Marine ecosystems impacts**—Trophic levels are being upended, partly from climate change (e.g., Best et al. 2015; Poloczanska et al. 2016; Thackeray et al. 2016), but primarily from more local to regional phenomena, including over fishing, and the transfer of marine and coastal invasive species through ship traffic (e.g., Davidson et al. 2011; Johnson et al. 2017; Ojaveer et al. 2017).
12. **Coastal dead zones**—It is relatively cheap and easy to over-fertilize an agricultural field to maximize crop production. The result is an excess of nitrogen and other fertilizing material to be washed off the farmlands and into streams where regional runoff will carry these pollutants to the coastal ocean. Upon reaching the coastal sea, the nitrogen accelerates marine production and dissolved oxygen in the water becomes rapidly consumed. With the oxygen levels

lowered, these hypoxic or dead zones can destroy entire bottom communities of sea life. Rare 40 to 50 years ago, hypoxia now occurs worldwide, particularly within American (see URL: http://www.wri.org/resources/maps/coastal-eutrophic-and-hypoxic-areas-north-america-and-caribbean) and European (see URL: http://www.wri.org/resources/maps/coastal-eutrophic-and-hypoxic-areas-europe) coastal waters.

2.3 Population and Energy

During the early part of the Holocene, human population growth rates were low, and the human environmental footprint was equally small. As agrarian communities became established (c. 5KyBP to 1250AD), population growth rates increased to about 100,000 per year. While energy use per person had changed little during these early years, an ever-growing population made the search for new energy sources an imperative. Coal was the first major energy source beyond human and animal muscle doubling the energy use per capita. Our global population was then growing at 1.4 M/y and by the start of the industrial revolution (c. 1850AD), humans numbered 1.3B. The Industrial Age was accompanied by new energy sources (petroleum and hydroelectricity) allowing our per capital energy use to double again to 25 gigajoules (GJ)/y. The amount of potential chemical energy in 160 L (approximately one US standard barrel) of oil, when combusted is 6 GJ. As humanity entered in what some call the Age of Technology and others refer to the beginning of the Anthropocene Epoch, with a start date of 1950, the world had a global population of 2.5B. Population has since been increasing at 74 M/y concomitant with a tripling of per capita energy consumption to 70 GJ/y. New energy sources included shale gas extraction, nuclear energy and new forms of renewable energy (solar, wind). Remarkably, more energy has been consumed since 1950 (to 2015) at 20.4 zettajoules (ZJ), than during the entire 11,700-year history of humans during the Holocene, by a factor of 1.6. Annual estimates of global energy use are approximately 0.5 ZJ (McGlade and Ekins 2015). With projections for future growth of the world's population, sustainability will require significant enhancements in efficiency to reduce rates of energy consumption.

2.4 Global Warming: Not a Hoax

The latest Climate Assessment Report (USGCRP 2017) is unambiguous: the earth is warming because higher concentrations of "greenhouse" gases, which absorb and emit energy within the thermal infrared range, enable the atmosphere to store more energy before the excess energy is returned to space as long wave infrared radiation. This is illustrated conceptually in Fig. 2.1. The greenhouse effect is essential to making the earth habitable. Greenhouse gases in trace quantities lead to convection, clouds, and precipitation. As illustrated in Fig. 2.1, 107 W/m^2 of the incoming 342 W/m^2 is reflected directly back to space either from the top of the atmosphere or from reflective (e.g., snow or ice covered) parts of the earth surface. The amount of radiation returned by direct reflectivity is called *albedo*. To maintain a radiation balance, the remaining 235 W/m^2 must ultimately be reradiated back to space but before this happens the energy is first converted to heat, which then results in long wave, infrared, re-radiation. The amount of heat that the atmosphere can hold before the infrared re-radiation takes place depends on the concentration of certain heat-retaining greenhouse gases in the atmosphere. The most important of these gases are water vapor, carbon dioxide and methane. Others include nitrous oxide (N_2O) and ozone (O_3). The primary constituents of air, nitrogen ($\sim$78%) and oxygen ($\sim$21%) are not greenhouse gases.

Carbon-based greenhouse gases, particularly carbon dioxide and methane, in the atmosphere have been steadily increasing at least since the middle of the 20th Century when reliable record keeping began. Since the 1950s, carbon dioxide concentration has been measured at the top of Mauna Loa, the 4,170 m (13, 680 ft above sea level) high volcanic mountain on the Big Island of Hawaii in the middle of the Pacific Ocean and thus, well removed from carbon producing cities on the various continents. The data show a steady increase in CO_2 concentration

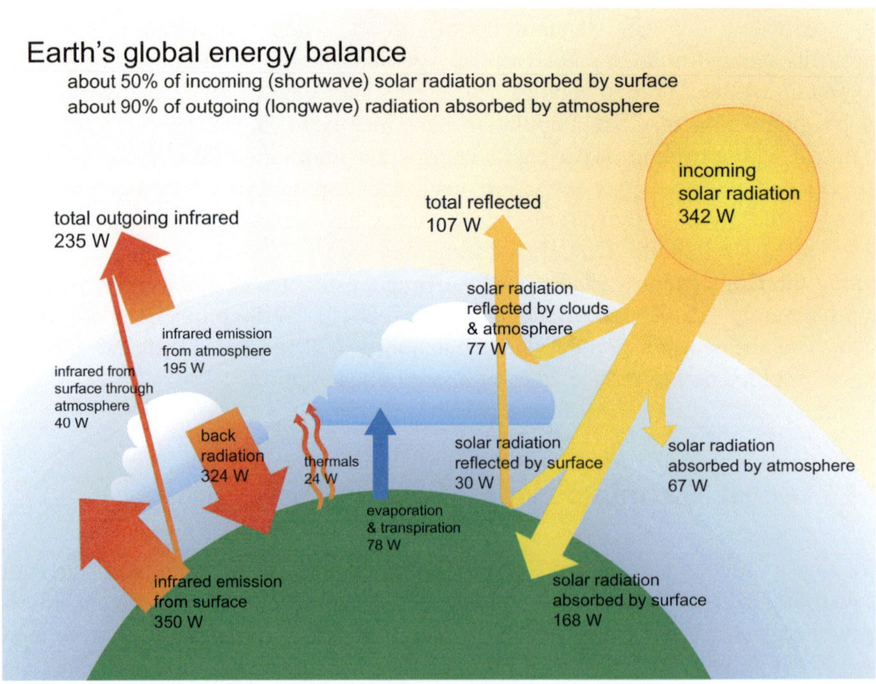

Fig. 2.1 Cartoon depicting the main elements of the global radiation balance (Figure courtesy of Christina Hulbe)

from ~320 ppm in the late 1950s to ~400 ppm in 2015. Measurements of the global temperature anomaly, averaged over both land and sea surfaces and expressed as a departure from the average temperature for the entire 20th century indicate that prior to 1940, the anomaly was negative, in other words less than the 20th century average. Since that time, however, the anomaly has been progressively increasing and reached a maximum of 0.9 °C in 2015 (NOAA-NCDC-NCEI 2016). NOAA also estimates that 93% of this excess heat is stored within the ocean waters, primarily within the upper 1500 m of the water column.

Older measurements of sea surface temperature prior to the mid 1990s relied heavily on shipboard observations made largely from engine room water intakes (but prior to passing through the engine). Water obtained this way was moderately warmed by passage through the engine room intakes. Since the 1990s, NOAA has relied more and more on data from buoy-mounted and drifter-mounted thermistors. As a result of these contrasting methods of temperature measurement, temperature anomalies based on a mix of data obtained by both old and new techniques tended to show a cooling bias as the fraction of the data derived from buoys increased with time. Because of this effect, many plots of temperature made prior to 2016 suggested a flattening or in some cases a slight reversal in the warming trend with time since 2003. This apparent trend reversal has been exploited by climate change deniers to argue that warming has ceased. However, NOAA scientists have recently reanalyzed data to base trends on uniform measurement techniques involving separate shipboard, satellite and buoy data instead of mixes of techniques (Hausfather et al. 2017; http://wwwusers.york.ac.uk/~kdc3/papers/ihsst2016/background.html). Figure 2.2 shows times series plots of temperature changes over the past 20 years. The reanalysis conclusion is that global temperatures are currently increasing at the rate of 0.12 °C per decade.

Predictions of future earth warming are not simple statistical extrapolations of recent trends but are underpinned by linked ocean-atmosphere numerical models such as such as the NCAR Community Climate Model (CCM3). Alterations of the Earth's energy budget by greenhouse gases and other processes are the drivers of climate change and models take account of these drivers in predicting future conditions. The Special Report on Emissions Scenarios (SRES) by the Intergovernmental Panel on Climate Change (IPCC Third and Fourth Assessment Reports 2001, 2007) considered different scenarios involving different degrees of greenhouse gas reductions in the future. The most common scenarios in the third and fourth assessments were those that emphasize global or regional economic growth (labeled ***SRES A1 & A2***) versus those that emphasize global or regional environmental sustainability (***SRES B1 & B2*** respectively). Models were typically run for all four scenarios. In the fifth IPCC Assessment Report (IPCC 2013), the IPCC quantified the change in energy fluxes caused by changes in climate change drivers via an index called *Radiative Forcing (RF).* Positive RF leads to surface warming and negative RF leads to surface cooling. RF estimates take account of observed conditions, properties of greenhouse gases, and numerical model characterizations of observed processes (IPCC 2013).

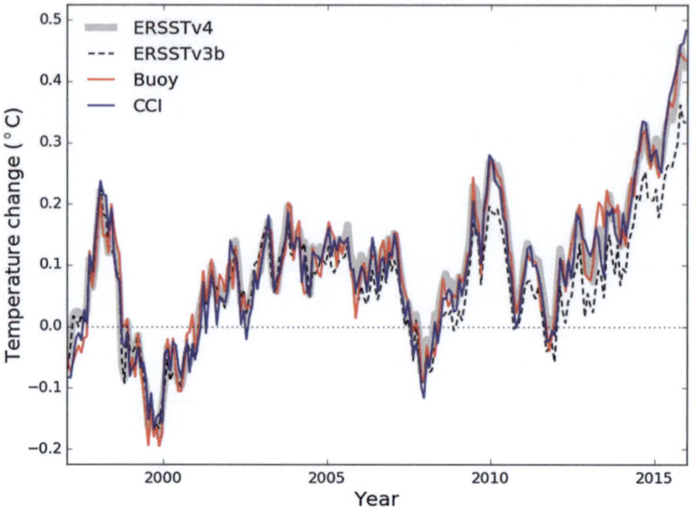

Fig. 2.2 Trends of global temperature changes over the past two decades based on uniform measurement methods (From Hausfather et al. 2017)

In the fifth assessment report the IPCC (2013) replaced the SRES scenarios used in the third and fourth assessment reports with four new scenarios in their future model predictions. They refer to these scenarios as Representative Concentration Pathways or *RCPs*. Each different RCP is distinguished by their radiative forcing (in Watts per square meter or W/m^2) in the year 2100 relative estimated 1750 values. The four scenarios are: (1) 2.6 W/m^2 for scenario ***RCP 2.6***; (2) 4.5 W/m^2 for scenario ***RCP 4.5***; (3) 6.0 W/m^2 for scenario ***RCP 6.9***; and (4) 8.5 W/m^2 for scenario ***RCP 8.5***. Model simulations carried out for each scenario assumed CO_2 concentrations by 2100 to be: 421 ppm for RCP 2.6; 538 ppm for RCP 4.5; 670 ppm for RCP 6.0; and 936 ppm for RCP 8.5. Depending on the extent to which the global emissions of carbon are reduced in the foreseeable future, global temperatures are expected to rise by somewhere between 1.8 °C (3.2 °F) and 4.0 °C (7.2 °F) by the end of the century (UCAR 2011). This temperature increase will not be evenly distributed over the earth's surface. Tropical temperatures are not expected to rise significantly; the most dramatic warming will be manifest at mid-to-high latitudes over continents (UCAR 2011). These RCPs facilitate the comparison of different modeling systems by providing a consistent set of starting conditions and historical data for use across the various branches of climate science. Since climate science is being accomplished collaboratively and internationally, RCPs provide a common, agreed upon foundation for modeling climate change, especially to provide time-dependent projections of atmospheric greenhouse gas concentrations. Additional examples of RCP use are provided in Sect. 2.9.

The most recent Climate Assessment Report (USGCRP 2017) states that "Annual average temperature over the contiguous United States has increased by

1.8 °F (1.0 °C) for the period 1901–2016 and is projected to continue to rise". Over the period 2021–20150, average annual U.S. temperatures are projected to increase by 2.5 °F (1.4 °C). However, it must be emphasized that the temperature increases will not be evenly distributed but will be greater at the higher latitudes and least at low latitudes. The actual temperature rises that ultimately result will depend on which RCP scenario prevails. For the "worst case" ***RCP 8.5*** scenario, average annual temperatures in the southeastern U.S. are predicted to increase by about 6 °F by late this century while those near the Arctic could be as much as 16 °F higher (Wuebbles et al. 2017).

2.5 Ocean Warming and Ocean Circulation

As noted in the previous subsection, 93% of excess heat is stored within the upper 1000–1500 m (0.62–0.93 mi) of the water column. This is a good thing since the ocean has a high heat capacity, is deep, and dynamic as evidenced by flows such as the Agulhas current off the south and east coast of southern Africa and the Florida Current, which runs into the Gulf Stream. The oceans play essential roles in distributing heat to higher latitudes. Averaged over the year, the incoming short-wave radiation received within the tropics exceeds the outgoing long wave radiation but at higher latitudes there is a yearly radiation deficit. Ocean currents are the primary mechanism by which thermal energy is transported toward the poles. Large-scale atmospheric Hadley Cells play major secondary roles in heat redistribution. Ocean-atmosphere coupling is a fundamental aspect of all climate change predictions and the interface between ocean and atmosphere is the sea surface. Sea surface temperature has for decades been regarded as a prime indicator of the likelihood of tropical storm occurrence in any given year.

All of the currently accepted coupled ocean-atmosphere models predict significant increases in sea surface temperatures (SST) throughout the world's oceans (i.e., at all latitudes and in all seas) by the latter 30 years or so of the 21st century (e.g., Rauscher et al. 2015). Figure 2.3 shows the late 21st century temperature anomalies (in °C), which are defined as departures from late 20th century temperatures, as predicted by eight different models. Notably, the models predict rises in SST by 2–5 °C with the largest increases occurring in the higher latitudes of the northern hemisphere. The single exception is in the case of Fig. 2.3b, which shows cooler temperatures adjacent to Greenland caused by the melting of Greenland's continental glacier. One recently reported consequence of rising ocean temperature is a 2% reduction in dissolved oxygen throughout the global ocean since 1960 (Schmidke et al. 2017). The reasons for this are two-fold. The primary reason is that oxygen solubility is less for warm water than for cold water. A secondary reason is that increased thermal stratification reduces vertical mixing of the water column (Schmidke et al. 2017).

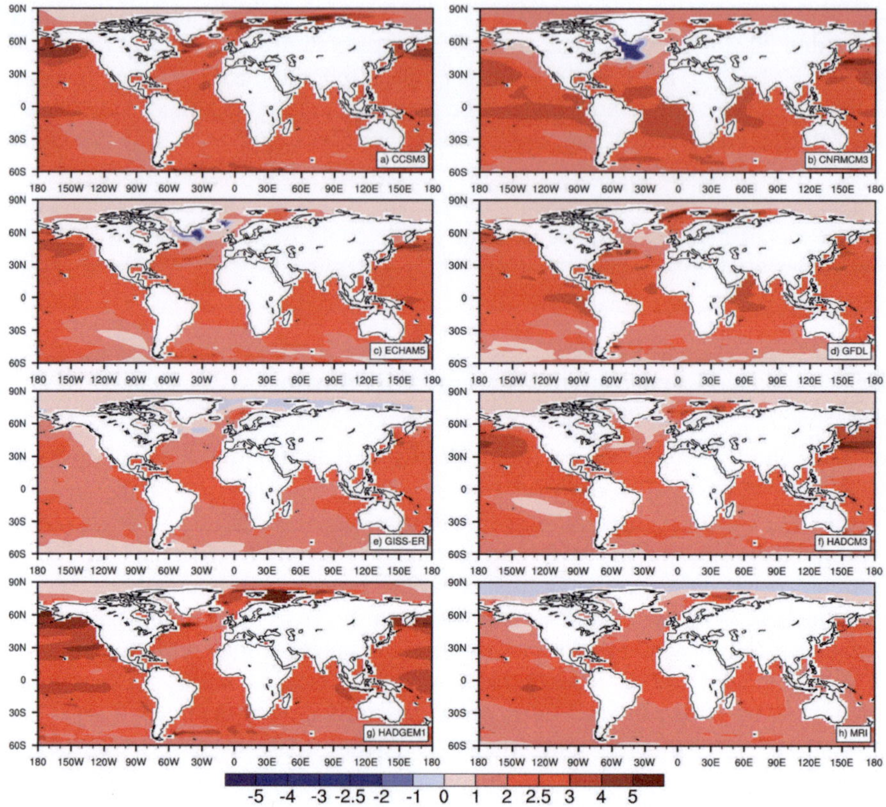

Fig. 2.3 Bias-corrected annual average SST anomalies (°C) for the end of the twenty-first century (2070–99) versus the end of the twentieth century (1970–99) from CMIP3 simulations **a** CCSM3, **b** CNRM-CM3, **c** ECHAM5, **d** GFDL, **e** GISS-ER, **f** HadCM3, **g** HadGEM1, and **h** MRI. From Rauscher et al. (2015)

Predictions of greater warming at higher latitudes have several important implications beyond the obvious thermal expansion of oceans and attendant sea level rise. One of these implications is for accelerated melting of sea ice in the Arctic Ocean. Another is the projected reduction of latitudinal *gradients* in water density that help drive thermohaline ocean circulation. Perhaps the most prominent example is the impact on the Atlantic Meridional Overturning Circulation (AMOC). This system carries lighter warm water northward within the upper layer of the North Atlantic Ocean and denser cold water southward within the deeper layers. The Gulf Stream is the best-known component of the northward flowing upper layer of the AMOC. Warming seas at higher latitudes, combined with increased runoff of fresh water from the melting Greenland ice sheet is diminishing the density gradients that drive this system. The density of the deep layer has been decreasing since 1994 and the strength of the AMOC overall has been weakening since 2004 (Frajka-Williams et al. 2016). Weakening of the Gulf Stream is one major consequence. As we explain in the next chapter, this is already causing regional rises in sea level along the US east coast. Corresponding reductions in atmospheric thermal gradients are expected to weaken the Hadley and Walker cell circulations that play essential roles in latitudinal and zonal heat exchange (Ma and Xie 2013). Other effects include a weakening of the trade winds, particularly the northeast trades, and increased precipitation in tropical regions (Ma and Xie 2013).

2.6 Changes in Storm Patterns

Coastal storms are the most damaging local and short term causes of coastal inundation and erosion because of the attendant strong onshore winds and heavy rain. In the long term, the storms are superimposed on the more gradual and less dramatic progressive rise in mean sea level. But as sea level rises in response to climate change, storm effects, particularly storm surge and destructive waves, reach farther inland. In addition, the frequency, intensity and geographic distribution of storms are also expected to change as the atmosphere and oceans warm. Because the occurrence, size and intensity of storm events all exhibit very large inter-annual variability, correlations of storms and decadal scale changes in climate (as contrasted to weather) are elusive, at least in the short term. However, as Trenberth (2012, p 283) points out, "All weather events are affected by climate change because the environment in which they occur is warmer and moister than it used to be". This is because storms are fueled by energy from a combination of warm air and the latent heat from water vapor. Coastal storms include tropical cyclones (hurricanes, typhoons, tropical storms), extra-tropical storms (e.g. northeasters), thunderstorms and monsoons. The merger of tropical and extra-tropical storms can also result in hybrid "super storms".

Tropical cyclones are characterized by central regions of very low atmospheric pressure around which strong winds circulate anti-clockwise (to the right of the pressure gradient) in the northern hemisphere and clockwise (to the left of the pressure gradient) in the southern hemisphere. Hurricanes (Atlantic) and typhoons (W. Pacific) are identical in genesis and intensity and are the most destructive type of coastal storms in terms of wind intensity, which can exceed 100 mph (45 m/s) and storm surge, which can exceed 10 ft (3 m). The wind stress felt by buildings and other structures is proportional to the wind speed squared and this fact is taken into account the in the *Saffir-Simpson* scale of 1 (74–95 mph or 119–153 km/h) to 5 (>157 mph or 252 km/h) by which the National Hurricane Center classifies hurricanes relative wind stress and the potential wind damage to structures. Hurricane Katrina (Fig. 2.4), which killed 1850 people and caused over $100 B (USD) in damage in 2005 was a category 5 storm with wind speeds of 175 mph over the Gulf

Fig. 2.4 NOAA/NASA satellite image of Hurricane Katrina in 2005 when it was a category 5 storm crossing the Gulf of Mexico on its track to an eventual landfall near the Louisiana/Mississippi state line. It had weakened to category 3 by the time of landfall

of Mexico but had weakened to a category 3 with a wind speed of 120 mph by the time it made landfall on the Louisiana coast.

Long-term records of tropical cyclones, especially Atlantic hurricanes, do not show any definitive trends with respect to the number of storms in any given year and year-to-year variability is large. There is also uncertainty in model predictions of future storm frequency. However, the connection between sea surface temperature and storm intensity is less ambiguous. Emanuel (2005) concluded that trends over the 3 decades preceding 2005 showed that tropical cyclones had become increasingly more destructive. Similar results are reported by Elsner et al. (2008). More recent analyses by Bender et al. (2010) and Holland and Bruyere (2013) indicate that global warming is causing increases in storm intensity and in the occurrence of major hurricanes of category 3, 4 or 5. Furthermore, increases in sea surface temperature contribute directly to rapid increases in storm intensification. A prominent example was Atlantic Hurricane Matthew, which in late September 2016 experienced explosive intensification from a tropical storm to a Category 5 hurricane in two days. Matthew was also unusually long lived: 12 days elapsed from the storm's formation until its dissipation. The largest and most intense Atlantic tropical cyclone ever recorded as of 2017 was Hurricane Irma, which devastated Caribbean island nations as a giant Category 5 and engulfed the entire Florida peninsula after making landfall there as an immense category 4 Hurricane. The sea surface temperatures that fueled Irma exceeded 30 °C (85 °F). An index that combines information on storm intensity, duration and total storms in a year is the Power Dissipation Index (PDI; e.g. Emanuel 2016). Figure 2.5 illustrates the close association between annual PDI values and sea surface temperatures in the North Atlantic from 1949 to 2015.

Though much less energetic, shorter lived and less damaging than tropical cyclones, severe coastal thunderstorms can cause minor erosion, localized wind

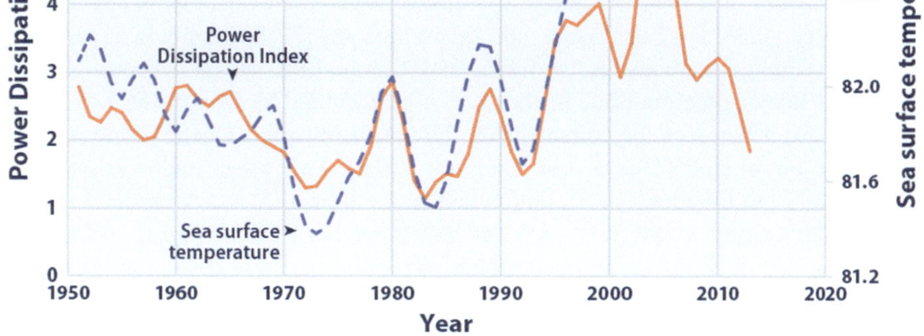

Fig. 2.5 The power dissipation index (orange curve) for North Atlantic tropical cyclone activity along with corresponding sea surface temperatures (blue dashed curve). Graphic prepared by US Environmental Protection Agency from data compiled by Emanuel (2016)

damages and brief but intense flooding. Diffenbaugh et al. (2013) offer modeling results indicating that the severity of damaging thunderstorms will likely increase progressively throughout the 21st century. Extra-tropical storms (including "northeasters") in middle and high latitudes are also expected to increase in frequency and intensity but also migrate to higher latitudes over the decades ahead (Walsh et al. 2014). Vose et al. (2014) present evidence that stronger onshore winds associated with increased storm frequency and intensity in the northern hemisphere winter have been progressing since 1950 and may be expected to continue increasing into the future. Finally, it is noted that tropical cyclones and extra-tropical storms can merge under certain circumstances to create so-called "super storms". This happened in the case of the "Halloween Storm" of 1991 (a.k.a. "Perfect Storm") when Hurricane Grace merged with a strong northeaster off the New England/Canadian coast. Hurricane Sandy, which made landfall and caused extensive storm surge damage in New Jersey and New York City in 2012 was the largest hurricane on record with a diameter of 1150 miles (1850 km). The behavior and path of Sandy involved interaction with an upper level low over the eastern US and a ridge of high pressure over Canada. At the present time, future predictions of the frequency and intensity of similar events are lacking.

2.7 The Asian Monsoon and Climate Change

The lives of roughly one half of the world's population are tightly entwined with the Asian Monsoon systems and about one fifth of all people depend on the South Asian Monsoon (Goswami et al. 2006). The Indian monsoon is perhaps the best-known part of the larger-scale Asian system as a whole, which embraces China and all of Southeast Asia. In winter, cold descending air over the Asian continent and Indian subcontinent creates a large high pressure system and dry air flows off the land and out over the warmer Indian Ocean where atmospheric pressure is lower than over land. The regional pressure gradient is reversed in summer when the land surface is hotter than over the ocean and moisture-laden air flows from sea to land. On the large scale, the East Asian Summer Monsoon (EASM) affects all of Asia, including China, Southeast Asia, India, and Bangladesh where it brings several months of heavy precipitation. In the case of the Indian Monsoon, the northward flow of moist air crosses the equator from the Southern Indian Ocean representing a large-scale circulation (Turner and Anamalai 2012). This precipitation is essential to agriculture but too much precipitation can cause damaging and sometimes deadly floods, particularly of rivers such as the Ganges, Brahmaputra, Indus and Irrawaddy. Although the annual occurrence of the life-giving wet summer monsoon is fairly reliable, there is significant inter-annual as well as intraseasonal (within a season) variability and devastating draughts can closely follow floods (Goswami et al. 2006; Loo et al. 2015).

It is generally understood that warmer air and warmer seas will increase the amount of water transfer from the sea surface to the atmosphere as well as the amount of moisture that the air can hold as it flows from ocean to land. Accordingly, model simulations for East Asia as a whole predict future increases in the amount of monsoonal precipitation accompanied by damaging floods (Turner and Anamalai 2012; IPCC 2013). This is driven by a combination of increasing contrasts in temperature between land and sea and warming of the Indian Ocean. However, the monsoon system is highly complex and is influenced by numerous other factors. These factors include variations in the Siberian High and the Arctic Oscillation (Wang et al. 2012). For the specific case of the Indian summer monsoon, strong density stratification in the upper ocean waters of the northern Indian Ocean and Bay of Bengal, caused by high freshwater input, further complicates processes of air-sea interaction and heat exchange (Goswami et al. 2006). Ashfaq et al. (2009) used a high-resolution nested climate modeling system to explore the possible effects of global climate change on the South Asian summer monsoon. Their results suggest that the summer monsoon may actually become suppressed leading to a reduction in precipitation over the Indian subcontinent as well as a delay in the onset of the wet season (Fig. 2.6). In addition, the frequency and duration of "breaks" (precipitation hiatuses during which draughts may occur) are predicted to increase as are inter-annual and intra-seasonal variability in precipitation.

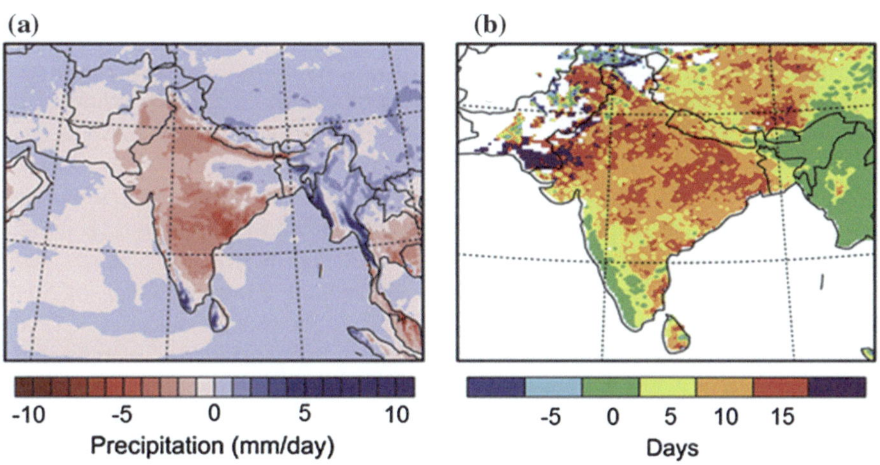

Fig. 2.6 Model simulations by Ashfaq et al. (2009) showing predicted 21st century changes in (**a**) summer monsoon precipitation and (**b**) wet season onset for India and adjacent regions influenced by the South Asian Monsoon. This graphic was reproduced from Loo et al. (2015)

2.8 Freshwater in the Anthropocene: Too Much or Too Little

It has long been accepted that warm air holds more water vapor than cooler air. Because of this, global warming is expected to cause arid regions to get drier (because of increased evaporation) and wet regions to get wetter (because of increased precipitation; e.g. IPCC 2013; Walsh et al. 2014). In North America, records for the past century show that regions to the north and northeast (including New England) have been becoming wetter while the southwest has been becoming drier (Walsh et al. 2014) and models predict those trends to continue through the 21st century. For the world, models predict the future global average precipitation to increase by 3–5% and possibly by as much as 8% if there are no reductions in greenhouse gas emissions (UCAR 2011). Models also predict increased frequency and intensity of heavy rainfall events leading to localized "flash floods" (Walsh et al. 2014). Where increased precipitation takes place over extensive portions of river catchments, river floods can be a damaging consequence. Discharge reductions may be expected where river catchments experience reduced precipitation. In 2016 and 2017, extreme pluvial flooding caused by unpredicted torrential occurred in many parts of the world including the U.S. and Asia. According to the latest Climate Assessment Report (USGCRP 2017; Wuebbles 2017), for the Eastern U.S. as a whole, the number of extreme rainfall events lasting 2 days or longer has increased by 40% since 1958.

van Vliet et al. (2013) utilized a physics-based hydrologic modeling system to predict how global river discharge and river temperature in the latter third of the 21st Century (2071–2100) might differ from the situation that prevailed a century earlier (1971–2000). They ran their models for two contrasting emissions reduction scenarios: SRES B1 which assumes future emphasis on global environmental sustainability and SRES A2 which assumes future emphasis on regional economic growth. The results are illustrated in Fig. 2.7. Notably, increased discharges under high flow and mean flow conditions are predicted for Australia, much of South America and the high latitudes of Asia as well as for India, Southeast Asia and much of Africa. Very high discharges are predicted for India and Australia under scenario SRES A2 but for most of Europe and North America dramatic flow reductions are predicted. Under the high economic growth SRES A2 scenario severe draughts are predicted for Europe and North America while India may be expected to experience alternating severe floods and severe draughts.

The most dramatic predicted discharge increase is for the Indus River: a mean discharge increase of 65% and a high discharge increase of 78%. This implies devastating floods for residents of the Indus flood plain and delta. At the other extreme, all the major European rivers including the Danube, Rhine, Loire and Rhone as well as the Mississippi and Rio Grande in North America, the Murray-Darling in Australia and Yangtze (Chiang Jiang) in China are expected to experience flow reductions. The most serious reductions will be in Europe where water shortages could have serious socio-economic consequences.

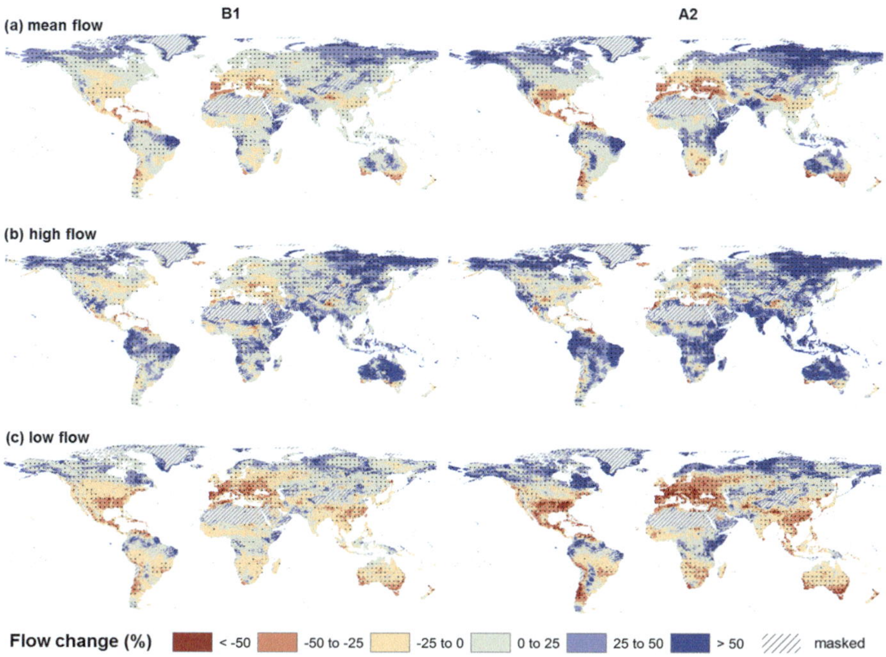

Fig. 2.7 Global projected changes in mean flow (**a**), high flow (Q95) (**b**) and low flow (Q10) (**c**) for 2071–2100 relative to 1971–2000 averaged for three selected Global Circulation Models (GCMs) for both the SRES A2 and B1 emissions scenario. Black dots indicate regions with consistent signal of change between the three GCMs. From van Vliet et al. (2013)

The future abundance or scarcity of freshwater will depend on more than climate. Anthropogenic factors are equally important. According to the now-retired International Geosphere-Biosphere Programme (IGBP 2012), there are 48,000 large dams on major rivers and many smaller ones on secondary streams. Ground water is withdrawn for industrial purposes at rates that often exceed the rate of recharge and about 70% of available water is used for agriculture including animal agriculture, which is exceptionally water intensive. Future conflicts over access to water are increasingly likely. The Three Gorges Dam on the Yangtze River is the largest such project in the world and has severed the supply of land-building sediment to the coast. Because of dams and the extraction of water along the course of the Yellow River (Huanghe) in China, that river no longer reaches the sea.

2.9 Melting Ice on Land and Sea

As would be expected, warming of the atmosphere and upper layers of the ocean have been causing both land-based glaciers and sea ice to melt for the past few decades and melting is projected to continue throughout the 21st Century. The land-based ice loss primarily involves the Greenland and Antarctic ice sheets. In its fifth assessment report, the Intergovernmental Panel on Climate Change (IPCC 2013) shows that rates of ice loss from Greenland and Antarctica have been increasing since the early 1990s. The IPCC estimates that by the end of the 21st Century, global glacial volume will be decreased by somewhere between 15 and 55% if the RCP 2.6 climate scenario prevails or by 35 to 85% for the RCP 8.5 scenario. The melt water from the glaciers will, of course, return to the sea and add to sea level rise. Snow cover in the Northern Hemisphere has decreased by an average of 1.6% per decade since the mid-20th Century. By the end of the 21st Century, Northern Hemisphere spring snow cover is projected to decrease by a further 7% assuming the RCP 2.6 scenario to 25% for RCP 8.5 (IPCC 2013).

Recent and projected reductions in the extent of sea ice cover are more dramatic. A useful indicator of the frequency and extent of ice melting is *ice age*, which indicates how long it has been since the ice in question was last fully melted and subsequently refrozen. Figure 2.8 shows a comparison of Arctic Ocean ice age in March 1987 with that in March 2011 from data compiled by NOAA. Summer ice cover in the Arctic Ocean has decreased significantly since the 1970s. According to the fifth assessment report of the IPCC (IPCC 2013) the extent of the Arctic Ocean summer sea ice minimum (corresponding to perennial ice cover) decreased at rates of 9.4% (0.73 million km^2) to 13.6% (1.07 million km^2) per decade over the period 1979–2012. Model projections are for substantial reductions in Arctic sea ice cover over the course of the 21st Century. Arctic ice cover in September is projected to decline by somewhere between 43% for the RCP 2.6 scenario and 93% for the RCP 8.5 scenario by 2100 and, for RCP 8.5 even the winter ice cover could be reduced by 34% (IPCC 2013). The IPCC expresses medium confidence that the Arctic Ocean in September could be almost ice free by midcentury. Figure 2.9 shows observed and predicted declines in Arctic sea ice over the period 1900–2100. Note that simple linear extrapolation of observed rates of sea ice decrease exceed the modeled rates and suggest the possibility of an ice free Arctic Ocean by the summer of 2050.

It must be emphasized that while the melting of glaciers on land contributes to sea level rise, the melting of sea ice does not since this ice is already part of the ocean. However, reductions in the extent of sea ice have several far-reaching consequences, not only for the Arctic coast but also for global climate. As mentioned in Chap. 1, decreasing sea ice cover in the Arctic reduces albedo (reflection of radiation), which in turn leads to more heat absorption by the seawater which in turn accelerates ice melting. This is a prominent example of positive (self-reinforcing) feedback. The impacts on the Arctic coasts, particularly the North Slope coasts of Alaska and Canada are also significant. Increased expanses of open

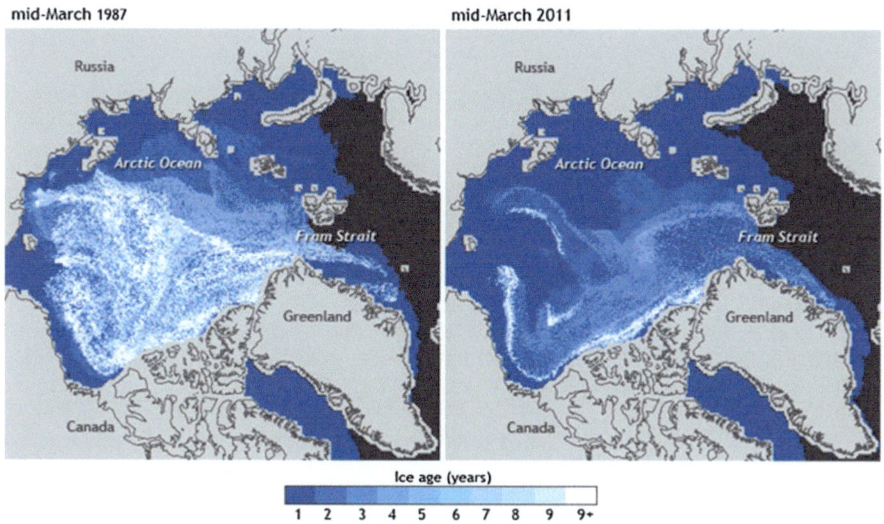

Fig. 2.8 Comparison of ice age (in years since last freeze up) for March (late winter) 1987 and March 2011. Note that ice is much younger in 2011 indicating more recent melting and refreezing. Maps based on data provided by James Maslanik, University of Colorado. *Credit* NOAA/CPO

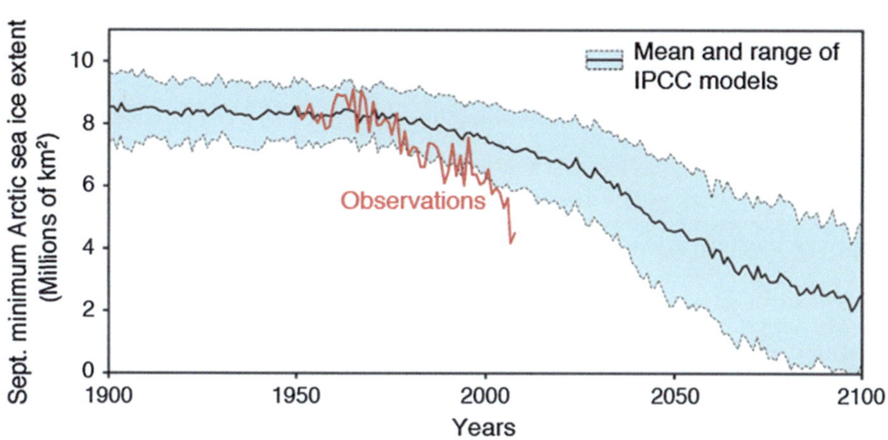

Fig. 2.9 Observed (red curve) and modeled reductions in late summer (September) sea ice cover in the Arctic Ocean over the period 1900–2100. Data and model results from the fifth assessment report of the Intergovernmental Panel on Climate Change

water in summer permits the generation of larger waves, which cause beach erosion via a combination of sediment removal and thermal erosion of shores previously secured by permafrost.

References

Allison, M., B. Yuill, T. Törnqvist, F. Amelung, T.H. Dixon, G. Erkens, R. Stuurman, C. Jones, G. Milne, M. Steckler, J. Syvitski, and P. Teatini. 2016. Global risks and research priorities for coastal subsidence. *Eos* 97 (19): 22–27.

Ashfaq, M., Y. Shi, W. Tung, R. Trapp, X. Gao, J. Pal, and N. Diffenbaugh. 2009. Suppression of south Asian summer monsoon precipitation in the 21st century. *Geophysical Research Letters* 36: L01704. https://doi.org/10.1029/2008gl036500.

Bender, M.A., T.R. Knutson, R.E. Tuleya, J.J. Sirutis, G.A. Vecchi, S.T. Garner, and I.M. Held. 2010. Modeled impact of anthropogenic warming on the frequency of intense Atlantic hurricanes. *Science* 327 (5964): 454–458.

Best, R.J., M.N. Stone, and J.J. Stachowicz. 2015. Predicting consequences of climate change for ecosystem functioning: Variation across trophic levels, species and individuals. *Diversity and Distributions* 21 (12): 1364–1374.

Davidson, A.D., A.G. Boyer, H. Kim, S. Pompa-Mansilla, M.J. Hamilton, D.P. Costa, G. Ceballos, and J.H. Brown. 2011. Drivers and hotspots of extinction risk in marine mammals. *Proceedings of the National Academy of Sciences of the United States of America* 109 (9): 3395–3400.

Diffenbaugh, N.S., M. Scherer, and R.J. Trapp. 2013. Robust increases in severe thunderstorm environments in response to greenhouse forcing. *Proceedings of the national academy of sciences USA* 110 (41): 16361–16366. 8 Oct 2013.

Elsner, James B., James P. Kossin, and Thomas H. Jagger. 2008. The increasing intensity of the strongest tropical cyclones. *Nature* 455 (7209): 92–95.

Emanuel, K. A. 2005. Increasing destructiveness of tropical cyclones over the past 30 years. *Nature* 436: 686–688.

Emanuel, K. A. 2016. Update to data originally published in: Emanuel, Kerry A. 2007. Environmental factors affecting tropical cyclone power dissipation. *Journal of Climate* 20 (22): 5497–5509.

Forster, P. 2017. Half a century of robust climate models. *Nature* 545: 296–297. https://doi.org/10.1038/545296a.

Frajka-Williams, E., J.L. Bamber, and K. Våge. 2016. Greenland melt and the Atlantic meridional overturning circulation. *Oceanography* 29 (4): 22–33.

Giosan, L., J. Syvitski, S. Constantinescu, and J. Day. 2014. Protect the world's deltas. *Nature* 516 (2014): 31–33. https://doi.org/10.1038/51603a.

Goswami, B.N., V. Venugopal, D. Sengupta, M.S. Madhusoodanan, and P.K. Xavier. 2006. Increasing trend of extreme rain events over India in a warming environment. *Science* 314: 1442–1445.

Hausfather, Z., Cowfan, K., Clarke, D.C., Jacobs, P., Richardson, M. and Rohhde 2017. Assessing recent warming using instrumentally homogeneous sea surface temperature records. *Science Advances* 3 (1): e1601207. https://doi.org/10.1126/sciadv.1601207.

Holland, G., and C.L. Bruyère. 2013. Recent intense hurricane response to global climate change. *Climate Dynamics*. https://doi.org/10.1007/s00382-013-1713-0.

International Geosphere Biosphere Programme (IGBP). 2012. Anthropocene: An epoch of our making (online newsletter). http://www.igbp.net/news/features/features/anthropoceneanepochofourmaking.5.1081640c135c7c04eb480001082.html.

International Geosphere Biosphere Programme (IGBP). 2013. Water in the Anthropocene (online multimedia video).

Intergovernmental Panel on Climate Change (IPCC). 2001. Climate change 2001: Synthesis report. In *Third assessment report*, 398pp. Cambridge University Press: Cambridge, UK.

Intergovernmental Panel on Climate Change (IPCC). 2007. Climate change 2007: The physical science basis. In *Contribution of working group I to the fourth assessment report of the Intergovernmental Panel on Climate Change*, eds. S. Solomon, D. Qin, M. Manning, Z. Chen,

M. Marquis, K.B. Averyt, M. Tignor, and H.L. Miller, 996pp. Cambridge University Press: Cambridge, United Kingdom and New York, NY, USA.

Intergovernmental Panel on Climate Change (IPCC). 2013. Climate change 2013: The physical science basis. In *Contribution of working group I to the fifth assessment report of the Intergovernmental Panel on Climate Change*, eds. T.F., Stocker, D. Qin, G.-K. Plattner, M. Tignor, S.K. Allen, J. Boschung, A. Nauels, Y. Xia, V. Bex, and P.M. Midgley. Cambridge University Press: Cambridge.

Johnson, Andrew Frederick, Marcia Moreno-Báez, Alfredo Giron-Nava, Julia Corominas, Brad Erisman, Exequiel Ezcurra, Octavio Aburto-Oropeza. 2017. A spatial method to calculate small-scale fisheries effort in data poor scenarios. *Plos One* 12 (4): e0174064, 17pp. https://doi.org/10.1371/journal.pone.0174064.

Loo, Y., L. Billa, and A. Singh. 2015. Effect of climate change on seasonal monsoon in Asia and its impact on the variability of monsoon rainfall in Southeast Asia. *Geoscience Frontiers* 6: 817–823.

Ma, J., and S.P. Xie. 2013. Regional patterns of sea surface temperature change: A source of uncertainty in future projections of precipitation and atmospheric circulation. *Journal of Climate* 26: 2482–2501.

McGlade, Christophe, and P. Ekins. 2015. The geographical distribution of fossil fuels unused when limiting global warming to 2 °C. *Nature* 517 (7533): 187–190.

Merryfield, W.J., F.J. Doblas-Reyes, L. Ferranti, J.-H. Jeong, Y.J. Orsolini, R.I. Saurral, A.A. Scaife, M.A. Tolstykh, and M. Rixen. 2017. Advancing climate forecasting, *Eos* 98. https://doi.org/10.1029/2017EO086891.

National Research Council, Committee on the Restoration and Protection of Coastal Louisiana. 2006. *Drawing Louisiana's new map: Addressing land loss in coastal Louisiana*, 206. Washington, DC: National Academies Press.

NOAA. 2016. NOAA launches America's first national water forecast model. Available online http://www.noaa.gov/media-release/noaa-launches-america-s-first-national-water-forecast-model.

Ojaveer, Henn, Sergej Olenin, Aleksas Narščius, Ann-Britt Florin, Elena Ezhova, Stephan Gollasch, Kathe R. Jensen, Maiju Lehtiniemi, Dan Minchin, Monika Normant-Saremba, and Solvita Strāke. 2017. Dynamics of biological invasions and pathways over time: A case study of a temperate coastal sea. *Biological Invasions* 19 (3): 719–813. https://doi.org/10.1007/s1053.

Palmieri, A., F. Shah, G.W. Annandale, and A. Dinar. 2003. *Reservoir conservation: The RESCON approach*, vol. I. Washington D. C.: World Bank.

Poloczanska, Elvira S., Michael T. Burrows, Christopher J. Brown, Jorge García Molinos, Benjamin S. Halpern, Ove Hoegh-Guldberg, Carrie V. Kappel, Pippa J. Moore, Anthony J. Richardson, David S. Schoeman, and William J. Sydeman. 2016. Responses of marine organisms to climate change across oceans. *Frontiers in Marine Science* 3: 62. https://doi.org/10.3389/fmars.2016.00062.

Rauscher, S., X. Jiang, A. Steiner, A. Williams, D. Cai, and N. Mc Dowell. 2015. Sea surface temperature warming patterns and future vegetation change. *Journal of Climate* 28: 7943–7961.

Schmidke, S., L. Stramma, and M. Visbeck. 2017. Decline in global oceanic oxygen content during the past five decades. *Nature* 562: 335–339. (online appendices).

Syvitski, J.P.M., and A. Kettner. 2011. Sediment flux and the Anthropocene. *Philosophical Transactions of the Royal Society of London A: Mathematical, Physical and Engineering Sciences* 369: 957–975. https://doi.org/10.1098/rsta.2010.0329.

Thackeray, S.J., et al. 2016. Phenological sensitivity to climate across taxa and trophic levels. *Nature* 535: 241–245.

Trenberth, K.E. 2012. 2012 Framing the way to relate climate extremes to climate change. *Climatic Change* 115: 283. https://doi.org/10.1007/s10584-012-0441-5.

Turner, A., and H. Annamalai. 2012. Climate change and the South Asian summer monsoon. *Nature Climate Change* 2: 587–595.

UCAR. 2011. *Predictions of future global climate.* National Center for Atmospheric Research, Online education website: https://scied.ucar.edu/longcontent/predictions-future-global-climate.

USGCRP. 2017. *Climate science special report: Fourth national climate assessment*, vol. I, eds. Wuebbles, D.J., Fahey, D.W., Hibbard, K.A., Dokken, D.J., Stewart, B.C., and Maycock, T.K. U.S. Global Change Research Program, Washington, DC, USA. https://doi.org/10.7930/j0j964j6.

van Vliet, M., W. Franssen, J. Yearsley, F. Ludwig, I. Haddeland, D. Lettenmaier, and P. Kabat. 2013. Global river discharge and water temperature under climate change. *Global Environmental Change* 23 (2013): 450–464.

Vose, R.S. et al. 2014. Monitoring and understanding changes in extremes: Extratropical storms, winds and waves. *American Meteorological Society, BAMS*, March 2014, pp. 377–386. http://dx.doi.org/10.1175/BAMS-D-12-00162.1.

Walsh, J., D. Wuebbles, K. Hayhoe, J. Kossin, K. Kunkel, G. Stephens, P. Thorne, R. Vose, M. Wehner, J. Willis, D. Anderson, S. Doney, R. Feely, P. Hennon, V. Kharin, T. Knutson, F. Landerer, T. Lenton, J. Kennedy, and R. Somerville. 2014: Ch. 2: Our changing climate. In *Climate change impacts in the United States: The third national climate assessment*, eds. J.M. Melillo, Terese (T.C.) Richmond, and G.W. Yohe. U.S. Global Change Research Program, pp. 19–67. https://doi.org/10.7930/j0kw5cxt.

Wang, P., T. Yamanaka, and G.Y. Qiu. 2012. Causes of decreased reference evapotranspiration and pan evaporation in the Jinghe River catchment, northern China. *Environmentalist* 32: 1–10.

Wuebbles, D.J., D.W. Fahey, K.A. Hibbard, B. DeAgelo, S. Doherty, K. Hayhoe, R. Horton, J. P. Kossin, P.C. Taylor, A.M. Waple, and C.P. Weaver. 2017. Executive summary. In eds. D. J. Wuebbles, D.W. Fahey, K.A. Hibbard, D.J. Dokken, B.C. Stewart, and T.K. Maycock, U.S. Global Change Research Program, Washington, D.C., USA, pp 12–34. https://doi.org/10.7930/j0dj5ctg.

Chapter 3
Sea Level Rise: Recent Trends and Future Projections

Lynn Donelson Wright, J. P. M. Syvitski and C. Reid Nichols

> *...admit that the waters around you have grown and accept it that soon you'll be drenched to the bone.*
> —Also from Bob Dylan, Nobel Laureate, c. 1964

3.1 Post-glacial Global Sea Level Rise

About 20,000 years ago, the world was in the last glacial phase of the Pleistocene ice age and vast, thick continental glaciers covered North America and Europe. A significant fraction of the earth's water was locked up in those glaciers and global sea level (referred to as glacio-Eustatic sea level) was roughly 120 m (390 ft) below the present level (Fairbanks 1989; Chappell and Shackleton 1986; Pirazzoli 1991). Beginning around 20,000 years ago, the continental ice sheets began to melt and the global (Eustatic) sea level began to rise. This rise proceeded for the ensuing 12 millennia until reaching a level a few meters below the present level about 7000 years ago (Fig. 3.1a). For the period from about 8000 years ago until about 1000 years ago, seas continued to rise much more gradually (Fig. 3.1b). Over the past millennium, changes in sea level have been dominated more by regional and global than by glacial melt.

During the glacial low stands of sea level, most of continental shelf sea floors were exposed and many were undoubtedly occupied by early peoples. Off the East Coasts of North America and China as well parts of Southeast Asia, the seashore

L. D. Wright (✉)
Southeastern Universities Research Association (SURA), Washington, DC, USA
e-mail: ldwright@bellsouth.net

J. P. M. Syvitski
Community Surface Dynamics Modeling System, University of Colorado, Boulder, CO, USA

C. R. Nichols
Marine Information Resources Corporation (MIRC), Ellicott City, MD, USA
e-mail: rnichols@mirc-us.com

L. D. Wright and C. R. Nichols (eds.), *Tomorrow's Coasts: Complex and Impermanent*, Coastal Research Library 27,
https://doi.org/10.1007/978-3-319-75453-6_3

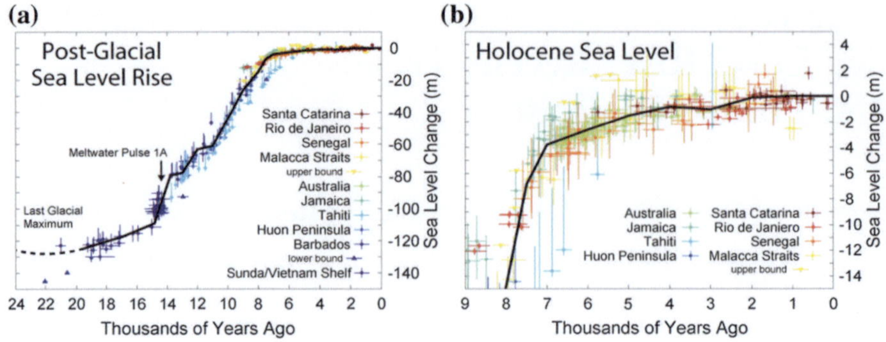

Fig. 3.1 **a** Sea level rise over past 20,000 years (courtesy of U.S. EPA). **b** Sea levels over past 8000 years (courtesy of Robert Rohde)

was up to 160 km (100 miles) seaward of its present positions. Primitive coastal residents would have been forced to relocate landward every few decades as seas rose. As the energetic surf slowly but progressively transgressed hitherto coastal plains, sediments would have been reworked into coastal dunes or deposited in migrating wetlands. Present-day continental shelves were shaped by these processes (Wright 1995) and today's coastal environments reflect this inheritance.

3.2 Observed Sea Level Changes Over the Past Century

Today, global sea level rises are caused largely by thermal expansion of seawater and by calving of Antarctic and Greenland ice sheets. Since the 1920s, coastal tide gages have provided reliable records of water levels at specific port sites. Boon and Mitchell (2015) recently carried out extensive statistical analyses of "relative mean sea level" (RMSL) for North American tide records. RMSL values are expressed relative to the average for specific locations rather than global averages. Examples from four ports on the U.S. East Coast are shown in Fig. 3.2. Over the period 1930–2014, mean sea level rose at all four sites by roughly 2 mm/year. The noticeable inter annual variability in Fig. 3.2 is attributed by Boon and Mitchell (2015) to El Niño/La Niña cycles with the highs coinciding with El Niño years. Other causes of inter-annual variability are also evident.

Figure 3.3 shows sea level records extending back to 1880 but patched together from different methods of observation; earlier data are less reliable. The latter part of the record, which covers the period since the mid 1990s is based on satellite altimetry covering the globe and is highly accurate. A world map showing spatial variations in sea level is shown in Fig. 3.4. The global averages for this period are shown in Fig. 3.5. Notably, since 1993, sea level has been rising at an average rate of 3.29 mm/year as contrasted to an average rate of 1.9 mm/year over the 75-year period prior to 1993. The IPCC (2013) concludes with high confidence that the rate

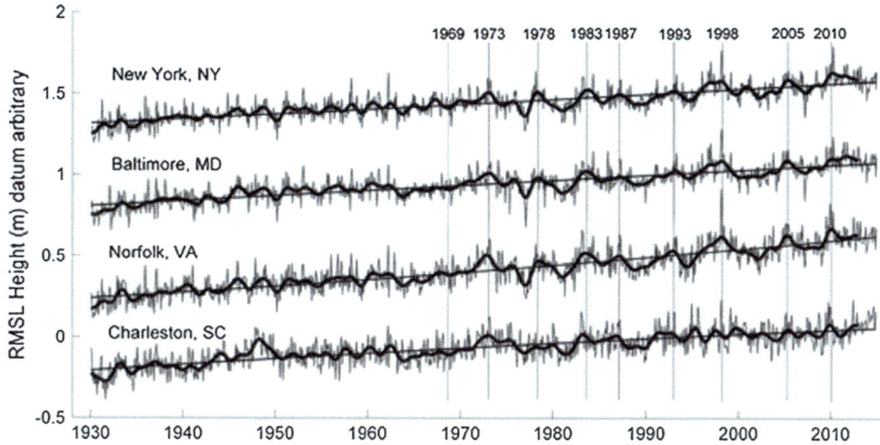

Fig. 3.2 Low-pass–filtered decadal signal (bold) superposed on monthly RMSL series at New York (the battery), New York; Baltimore, Maryland; Norfolk (Sewells Point), Virginia; and Charleston, South Carolina, from 1930 through 2014. Note decadal signal change in amplitude and frequency after 1969. Numbered years 1973–2010 correspond to El Niño events. From Boon and Mitchell (2015)

of sea level rise since the mid-19th century has been larger than the mean rate during the previous two millennia. However, an analysis by Watson (2016) using new statistical methods suggests that, at present, there is no conclusive evidence from recent trends that the rate of rise in the eastern U.S. is accelerating. Watson's conclusion is based on observed trends and takes no account of model results.

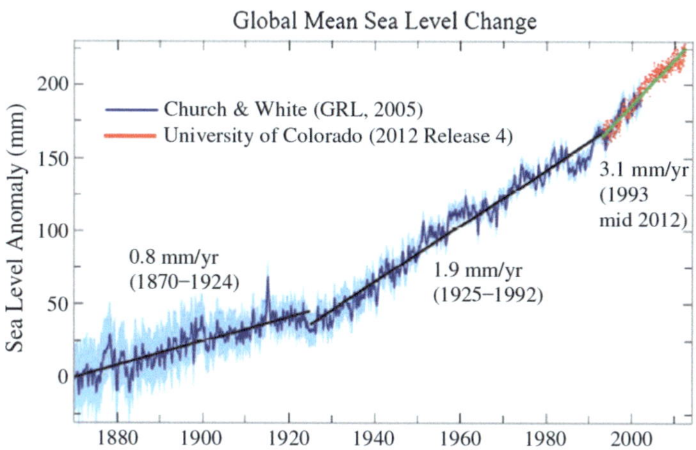

Fig. 3.3 Sea level rise 1880–2015 based on different sources of data. The most recent (since the mid-1990s) is based on satellite altimetry. From Hansen et al. (2016)

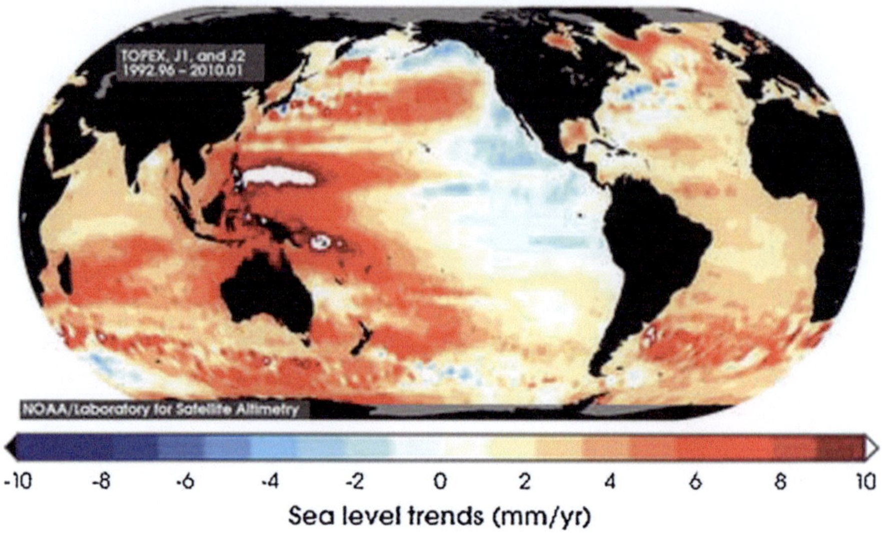

Fig. 3.4 Global mean sea levels determined from satellite altimetry. Data from NOAA Laboratory for Satellite Altimetry (Figure from Boon et al. 2010)

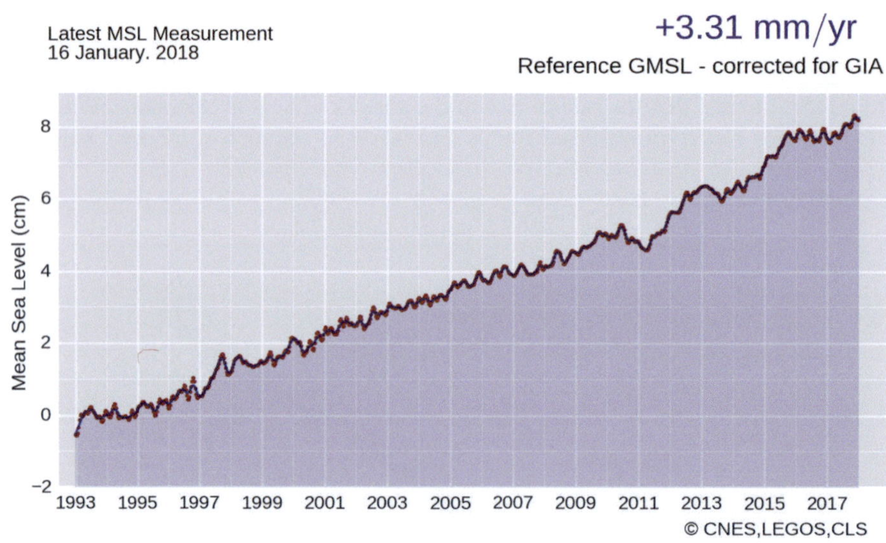

Fig. 3.5 Sea level rise 1993–2017 based on satellite altimetry data (credits CLS/CNES/Legos)

3.3 Model Predictions of 21st Century Sea Level Rise

The IPCC (2013) predicts that, for all four RCP scenarios, global mean sea levels will continue to rise throughout the 21st century at rates that exceed those of the past several decades. The estimates are based on process-based numerical models and also take account of expected glacial and ice sheet melting and calving. The expected trends are portrayed in Fig. 3.6. Global mean sea level rise for 2081–2100 relative to 1986–2005 will likely be in the ranges of 0.26–0.55 m (0.9–1.8 ft) for RCP2.6; 0.32–0.63 m (1.0–2.1 ft) for RCP4.5 and RCP6.0; and 0.45–0.82 m (1.5–2.7 ft) for RCP8.5. The IPCC expresses medium confidence in these estimates. The rate of rise for the last two decades of the 21st century are estimated by the IPCC (2013) analysis to be between 8 and 16 mm/year.

More recently, however, a joint analysis by NOAA, EPA and USGS of possible relative sea level (RSL) rises in the United States (Sweet et al. 2017) suggests a more troubling set of scenarios for American cities by the middle of the 21st century. The RSL values take account of regionally-varying factors such as land subsidence and oceanographic regime. Figure 3.7 highlights the results of that recent study for four U.S. coastal cities given a range of possible scenarios from low

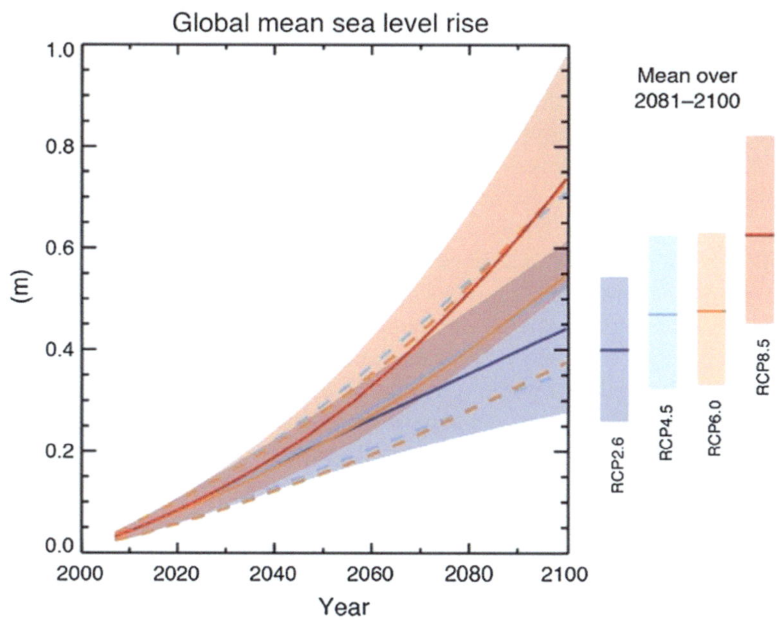

Fig. 3.6 Projections of future global mean sea level (GMSL) rises predicted by an ensemble of process-based numerical models for the four scenarios identified by the Intergovernmental Panel on Climate Change in the fifth assessment (IPCC 2013). The scenarios are explained in Chap. 2. The shaded areas indicate the range of uncertainty

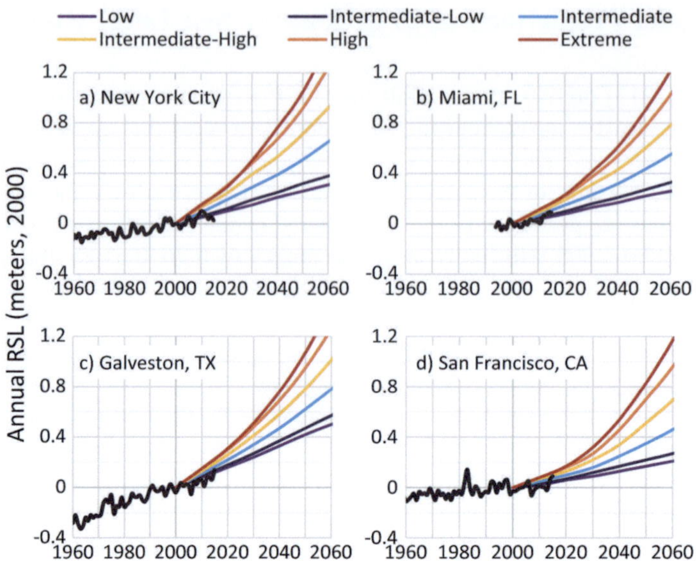

Fig. 3.7 Regional variations in projected relative sea level (RSL) rises for four American Cities. From Sweet et al. (2017)

to high severity. Note that by midcentury, relative sea levels in cities like New York and Miami may exceed those of the year 2000 by up to 1.2 m (~4 ft) in the extreme scenario cases and 40–50 cm (1.3–1.6 ft) in the least extreme cases.

On the face of it, predicted rises in mean sea level, and even relative sea level might seem fairly modest. However, it must be remembered that on very gently sloping coastal plains, wetlands, or river deltas a few inches of vertical rise can translate into hundreds of feet of horizontal transgression. In addition, saline ground water can reach landward significant distances into aquifers, contaminating the freshwater supplies of coastal communities. Also, saltwater penetration into previously forested environments can cause die back of land-stabilizing vegetation allowing for enhanced susceptibility to coastal erosion. In addition, relatively small changes in depth can alter the degrees of amplification of tides, storm surges and waves, particularly along crenulated coasts and in bays and estuaries. As Sweet et al. (2017) point out, relative sea level (RSL) provides a shifting baseline on which more extreme short-term water level fluctuations are superimposed. When high perigean spring tides ("king" tides), high storm surges and high waves coincide as they sometimes do during tropical storms or hurricanes, a couple of feet of mean sea level, plus or minus, can mean the difference between a low-lying coastal village or neighborhood being devastated or not (Goodell 2017). In such cases, it is the waves and surges that do the damage but the mean sea level that carries these forces onto the coast.

3.4 Coastal Subsidence Exacerbates Sea Level Rise

For any given coastal region, what really matters is not just the absolute mean rate of global sea level rise but the *relative* rate of rise with respect to the land surface. Many coastal lowlands are sinking and, hence the effective rate of rise of relative sea level is the rate of global rise plus the rate of local or regional sinking. A recent paper by Allison et al. (2016) highlights this with particular reference to New Orleans, in the Mississippi Delta and other deltaic coasts. Subsidence is caused by a combination of factors, including compaction of soft (mud) substrates, larger scale tectonic subsidence, withdrawal of interstitial fluids, such as ground water or hydrocarbons and loading caused by the weight of urban centers. The timescales and rates of some of these processes are illustrated in Fig. 3.8, which is from Allison et al. (2016). In many cases, the subsidence rates significantly exceed the rate of global sea level rise. This is the case for large deltas, some of which are surmounted by megacities, the weights of which are causing rapid sinking (Allison et al. 2016). For example, the Mississippi Delta is subsiding at rates that vary locally up to 35 mm (1.2 in.) per year and the Ganges-Brahmaputra Delta, which is home to 300 million people, is sinking at about 18 mm (0.6 in.) per year but some areas have sunk by over 1 m (3.3 ft) since the 1960s. In the well-known Venice Lagoon recent subsidence rates are over 40 mm/year. (Allison et al. 2016). The fastest rate of subsidence is in China's Huanghe (Yellow River) Delta, which is sinking 250 mm (10 in.) per year.

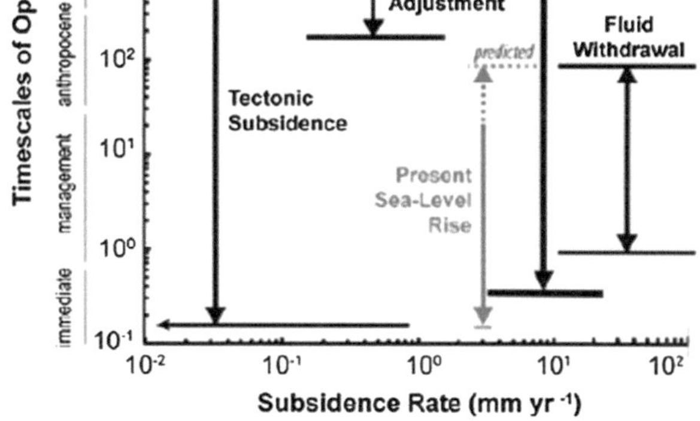

Fig. 3.8 Coastal subsidence mechanisms and their rates and timescales of operation. From Allison et al. (2016)

3.5 Regional Time-Varying Sources of High Water Levels

In addition to long-term changes in global mean sea level and land subsidence, regional contributions to water levels, include astronomical tides, atmospheric-pressure and wind induced changes, fluctuations in transport intensity of coastal currents, such as the Gulf Stream, storm surge and wave-induced set up. Tides vary on timescales ranging from hours to years due to gravitational forces exerted by the sun and moon (astronomical tides) and are typically the largest amplitude causes of time-varying water levels in most parts of the world. However, tides are highly predictable and natural coastal environments have long been in equilibrium with these fluctuations. Tide ranges vary over the course of year because of variations in the proximity of the moon and sun to the earth. The most prominent of these effects are *perigean* spring tides when the moon is closest to earth and tide ranges are at a maximum. Perigean spring tides tend to cause recurring flooding in some very low-lying areas. They provide a preview for sea level rise impacts. Incidences of flooding at the times of perigean spring tides have increased as sea level heights have risen relative to the land. In addition, if these unusually high tides coincide with tropical cyclones, which bring storm surges and high waves, excessive flooding and damage can occur (e.g., Georgas et al. 2014).

There are also important non-tidal and non-storm-induced sources of inter-annual, seasonal and intra-seasonal variability of coastal water levels. As shown in Fig. 3.2, El Niño/Southern Oscillation (ENSO) Cycles are one prominent source of inter-annual variability. Because El Niño raises temperatures, particularly in tropical and sub-tropical Pacific waters, this causes moderately higher sea levels on the Pacific coast. El Niño years also tend to bring more severe winter storms to the Atlantic East Coast of North America (e.g., Eichler and Higgins 2006), which in turn bring more frequent wind setup and storm surge. An aperiodic factor that causes inter-annual and inter-seasonal variations sea level and weather in Europe and on the East Coast of North America is the *North Atlantic Oscillation* (NAO; Stephenson et al. 2003). The NAO is related to oscillations in the pressure difference between the Azores High and the Icelandic low. Large differences, expressed as a positive index (NAO+) bring stronger prevailing westerlies, cool summers and mild but wet winters. The westerlies are weaker during negative indices (NAO−) and these are also times of increased storminess in Europe and North Africa. The reduced pressures of positive (NAO+) in the northern hemisphere cause sea level to be higher because of an inverse barometer effect.

Shankar and Shetye (1999) attribute seasonal, inter-annual and decadal variations in sea levels affecting the coast of India to the behavior of the monsoon system. On a seasonal basis, higher sea levels coincide with the summer monsoon when rainfall is heavy and winds are onshore. Similarly, on inter-annual and inter-decadal time scales, years with unusually heavy rainfall tend to also have higher sea levels. According to Sankar and Shetye (1999) the higher sea levels are more reflections of lowered salinities in coastal waters than to lowered water densities caused by warming.

Among the most prominent causes of non-tidal sea level variations on the US East coast, annual fluctuations in coastal sea level of up to 1.0 m (3.3 ft) are attributed to both long term and aperiodic variations in Gulf Stream transport with higher sea levels corresponding to times of Gulf Stream slackening (Ezer 2013; Ezer and Atkinson 2014). The reason for this is straightforward: in the northern hemisphere, the Coriolis force, which is a consequence of earth rotation, causes water to be deflected to the right with respect the direction of flow of very large scale currents such as the northerly flowing Gulf Stream and its southern component, the Florida Current. This *Ekman transport* pulls water away from the coast and causes water levels seaward of the current to rise as shown conceptually in Fig. 3.9 (from Atkinson 2016). The elevation difference between the offshore "mound" and sea level at the coast can be as much as 1.5 m ($\sim$5 ft). In the southern hemisphere, the Coriolis force deflects to the left, so the southerly flowing East Australia current off the east coast of Australia also pulls water away from the coast when it is strong and allows it to rise when it is weak. However, that current is much weaker than the Gulf Stream. Like the Gulf Stream, the Kuroshio on the western side of the North Pacific Ocean is also a western boundary current. However, it overlies the very deep Okinawa trough for much of its length, is relatively far away from the coast and no similar effects of sea level have been described although steric sea level effects caused by sea warming have been described by Lim et al. (2017) who also describe a weakening of the current as a result of warming.

As discussed in Chap. 2, weakening of the Gulf Stream is caused by reduced water density at high latitudes in the North Atlantic because of the combination of seawater warming and salinity reductions from melting of the Greenland ice sheet. These effects weaken the Atlantic Meridional Overturning and have manifestations at multiple time scales. At the longest time scale, there is evidence that Gulf Stream transport is weakening because of global warming (Ezer 2013; Ezer and Atkinson 2014). In addition to the long term progressive response, shorter term interannual, seasonal and intra-seasonal fluctuations also occur in response to ENSO cycles, the

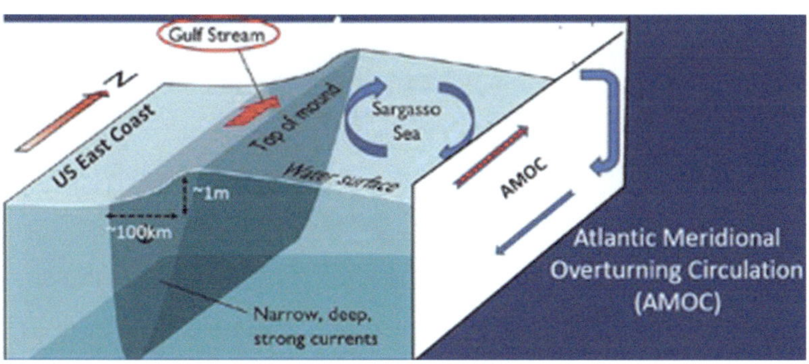

Fig. 3.9 Cartoon illustrating the relationships of sea surface elevation to Gulf Stream transport and the AMOC. From Atkinson (2016)

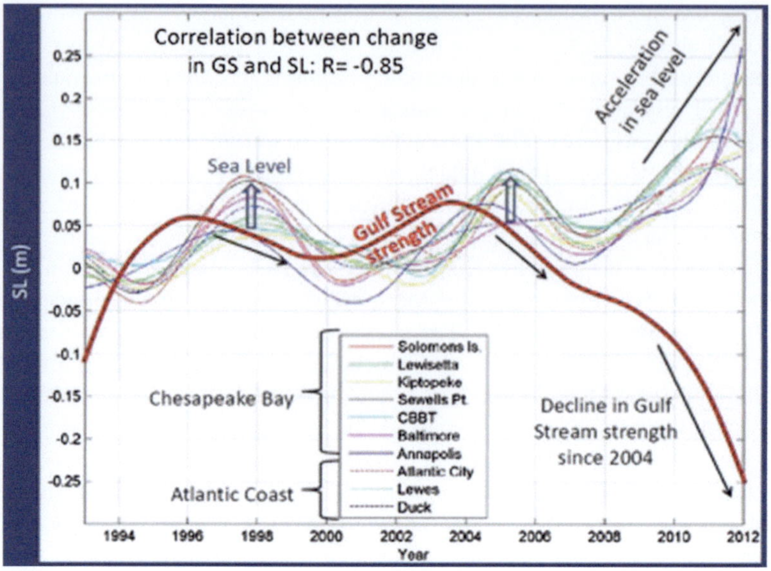

Fig. 3.10 Temporal variations in mean sea level at 10 stations in the Middle Atlantic Bight and Chesapeake Bay in relation to variations in the relative strength of the Gulf Stream over the period 1993–2012. From Ezer et al. (2013)

North Atlantic Oscillation and other weather-related events. The relationships between sea surface elevation and AMOC/Gulf Stream transport (Fig. 3.9) is also shown in Fig. 3.10 via a simplified plot of time-varying sea levels at different east coast sites along with a generalized (no scale) plot of the relative strength of the Gulf Stream (red curve).

References

Allison, M., B. Yuill, T. Törnqvist, F. Amelung, T.H. Dixon, G. Erkens, R. Stuurman, C. Jones, G. Milne, M. Steckler, J. Syvitski, and P. Teatini. 2016. Global risks and research priorities for coastal subsidence. *EOS* 97 (19): 22–27.

Atkinson, L.P. 2016. US *East Coast Sea Level and the Gulf Stream.* Weather Underground, Wunderblog Archive, 9 Mar 2016.

Boon, J. D., J. M. Brubaker, and D. R. Forrest.2010. Chesapeake Bay land subsidence and sea level change, in: *Applied Marine Science and Ocean Engineering, Report No. 425*, Virginia Institute of Marine Science, Gloucester Point, Va.

Boon, J.D., and M. Mitchell. 2015. Nonlinear change in sea level observed at North American tide stations. *Journal of Coastal Research* 31 (6): 1295–1305.

Chappell, J., and N. Shackleton. 1986. Oxygen isotopes and sea level. *Nature* 324: 137.

Eichler, T., and W. Higgins. 2006. Climatology and ENSO-related variability of North American extratropical cyclone activity. *International Journal of Climatology* 19: 2076–2093.

Ezer, T. 2013. Sea level rise, spatially uneven and temporally unsteady: Why the U.S. East Coast, the global tide gauge record, and the global altimeter data show different trends. *Geophysical Research Letters* 40 (20): 5439–5444. https://doi.org/10.1002/2013GL057952.

Ezer, Tal, and Larry P. Atkinson. 2014. Accelerated flooding along the U.S. East Coast: On the impact of sea-level rise, tides, storms, the Gulf Stream, and the North Atlantic Oscillations. *Earth's Future* 2 (8): 362–382. https://doi.org/10.1002/2014ef000252.

Ezer, T., L.P. Atkinson, W.B. Corlett, and J.L. Blanco. 2013. Gulf Stream's induced sea level rise and variability along the U.S. mid-Atlantic coast. *Journal Geophysical Research* 188 (2): 685–697.

Fairbanks, Richard G. 1989. A 17,000 year glacio-eustatic sea level record. *Nature* 342 (6250): 637–642. https://doi.org/10.1038/342637a0.

Georgas, N., P. Orton, A. Blumberg, L. Cohen, D. Zarrilli, and L. Yin. 2014. The impact of Tidal phase on Hurricane Sandy's flooding around New York City and Long Island Sound. *Journal of Extreme Events* 1 (1): 1–32.

Goodell, J. 2017. *The water will come*, 340. New York: Little, Brown and Company.

Hansen, J., et al. 2016. Ice melt, sea level rise and superstorms: Evidence from paleoclimate data, climate modeling, and modern observations that 2 °C global warming could be dangerous. *Atmospheric Chemistry and Physics* 16: 3761–3812. https://doi.org/10.5194/acp-16-3761-2016.

Intergovernmental Panel on Climate Change (IPCC). 2013. Climate change 2013: The physical science basis. In *Contribution of working group i to the fifth assessment report of the Intergovernmental Panel on Climate Change*, ed. T.F. Stocker, D. Qin, G.-K. Plattner, M. Tignor, S.K. Allen, J. Boschung, A. Nauels, Y. Xia, V. Bex, and P.M. Midgley. Cambridge: Cambridge University Press.

Lim, C., S.-H. Park, D.-Y. Kim, S.-B. Woo, and K.-Y. Jeong. 2017. Influence of steric effect on the rapid sea level rise at Jeju Island, Korea. In: *The 2nd International Water Safety Symposium. Journal of Coastal Research,* Special Issue No. 79. ed. Lee, J.L., T. Griffiths, A. Lotan, K.-S. Suh, and J. Lee, 189–193.

Pirazzoli, A. 1991. *World atlas of Holocene sea-level changes*, vol. 58, 300., Oceanography Series Amsterdam: Elsevier.

Shankar, D., and S.R. Shetye. 1999. Are interdecadal sea level changes along the Indian coast influenced by variability of monsoon rainfall? *Journal of Geophysical Research* 104 (C11): 26031–26042.

Stephenson, D.B., H. Wanner, S. Brönnimann, and J. Luterbacher. 2003. The history of scientific research on the North Atlantic Oscillation. In *The North Atlantic oscillation: Climatic significance and environmental impact*, ed. J.W. Hurrell, Y. Kushnir, G. Ottersen, and M. Visbeck, 37–50. Washington, DC: American Geophysical Union. https://doi.org/10.1029/134gm02.

Sweet, W.V., R.E. Kopp, C.P. Weaver, J. Obeysekera, R. Horton, E.R. Thieler, and C. Zervas. 2017. *Global and regional sea level rise scenarios for the United States.* NOAA Technical Report NOS CO-OPS 083 64 pp.

Watson, P.J. 2016. Acceleration in U.S. mean sea level? A new insight using improved tools. *Journal of Coastal Research* 32 (6): 1247–1261.

Wright, L.D. 1995. *Morphodynamics of inner continental shelves*, 241. Boca Raton, FL: CRC Press Inc.

Chapter 4
Complex Intersections of Seas, Lands, Rivers and People

Lynn Donelson Wright, J. P. M. Syvitski and C. Reid Nichols

Egypt is the gift of the Nile.

—Herodotus, c. 450 BCE

4.1 The Historical Significance of River Deltas

The brief introductory quote above from the Greek Historian Herodotus alludes to both the land surface upon which Egypt has rested for several millennia and the economy of the ancient Egyptian civilization. Sediments supplied to the Mediterranean Sea by the Nile River comprise the land surface of Egypt's Nile Delta while the predictable flooding of the Nile nurtured the ancient Egyptian culture by providing essential nutrients to enable agriculture. The deltas of the Nile and Tigris/Euphrates (Shatt al-Arab) provided fertile lands that "cradled" the early civilizations of ancient Egypt and Mesopotamia. The early civilizations of Babylonia and Assyria arose in Mesopotamia on the rich deltaic surface separating the lower courses of the Tigris and Euphrates Rivers as early as 5000 years ago. Deltaic sedimentation over the centuries and millennia that followed the initial establishment of Babylon (south of present-day Baghdad) and Ur caused those cities to become progressively distant from the open waters of the Persian Gulf (Aqrawi 2001; Fig. 4.1) while the newly-formed deltaic lands became the substrates for significant centers such as Basra, a major city in present-day Iraq. Deposition of river sediment continues to nourish the wetlands of Iraq's Shatt-al-Arab today

L. D. Wright (✉)
Southeastern Universities Research Association (SURA), Washington, DC, USA
e-mail: ldwright@bellsouth.net

J. P. M. Syvitski
Community Surface Dynamics Modeling System, University of Colorado, Boulder, CO, USA

C. R. Nichols
Marine Information Resources Corporation (MIRC), Ellicott City, MD, USA
e-mail: rnichols@mirc-us.com

L. D. Wright and C. R. Nichols (eds.), *Tomorrow's Coasts: Complex and Impermanent*, Coastal Research Library 27,
https://doi.org/10.1007/978-3-319-75453-6_4

(Fig. 4.2). Notably, in 2017, because of explosive population growth, uncontrolled pollution, extraction of water and extensive damming, the Nile River and its delta are facing impending demise (Schwartzstein 2017).

China's early civilization was also tightly entwined with major rivers and their alluvial deposits. The Huanghe (Yellow River) alluvial valley in particular is regarded by many historians as the "cradle of Chinese civilization" (e.g. Sinclair 1987). Sinclair (1987, p. 27) states: "The history of China is the tale of the Yellow River. … The river and the people are involved inextricably, moving together in history as certainly as the Huanghe flows toward the sea." Although Chinese agriculture began in the Neolithic Age, this took place well inland from the present day deltaic surfaces of the Huanghe and Changjiang (Yangtze) rivers which had not yet been deposited by those two major rivers (Fairbank and Goldman 1998). Progradation of the Huanghe and Changjiang Deltas over the past 6000 years has been accompanied by linked developments of Chinese society. The locale of the city of Shanghai, which dates from the 11th century A.D. and is now home to 24 million people, was submerged beneath the shallow margins of the East China Sea during the Han dynasty (c. 2000 yr B.P.; Chen 1998). Shanghai (Fig. 4.3) now rests on a substrate supplied by the Changjiang during the time of recorded Chinese history. Understanding the complex processes occurring today along the East China Sea coast is supported by satellite remote sensing, which shows the spatial extent of sediment-laden water in lakes, rivers, and bays (e.g., Yu et al. 2012; Xie et al. 2017).

4.2 Now and in the Future: Humans Versus Deltas?

Strikingly, the supplies of land-building and agriculture supporting sediments that many rivers provided to enable early civilizations to arise and flourish have now been either severed or significantly redirected. The Aswan Dam on the upper course of the Nile River now traps 98% of the sediment needed to maintain the sinking Nile Delta-the "cradle" of Egyptian civilization. Human manipulation of the course and sediment load of the Huanghe (Yellow River) dates back several millennia. Deforestation of the fine-grained *Loess* silt that covers the Mongolian plateau and began as early as 4000 years ago permitted erosion of that material into the river giving it a yellow color and high sediment load (Fig. 4.4) which nourished the Huanghe Delta and caused the lower course of the river to become elevated relative to the surrounding land. Since the 1800s, the course of the river has undergone diversions between the East China Sea and the Gulf of Bohai, which it now occupies. More recently, dams have reduced the sediment load and over the past few decades, water and sediment have been extracted along the lower course of the river to support agriculture and build land. Loads reaching the delta have been substantially reduced and some years have seen no water or sediment reaching the coast. And, as was noted in Chap. 3, the Huanghe Delta is currently sinking at a rate of 250 mm (10 in.) per year.

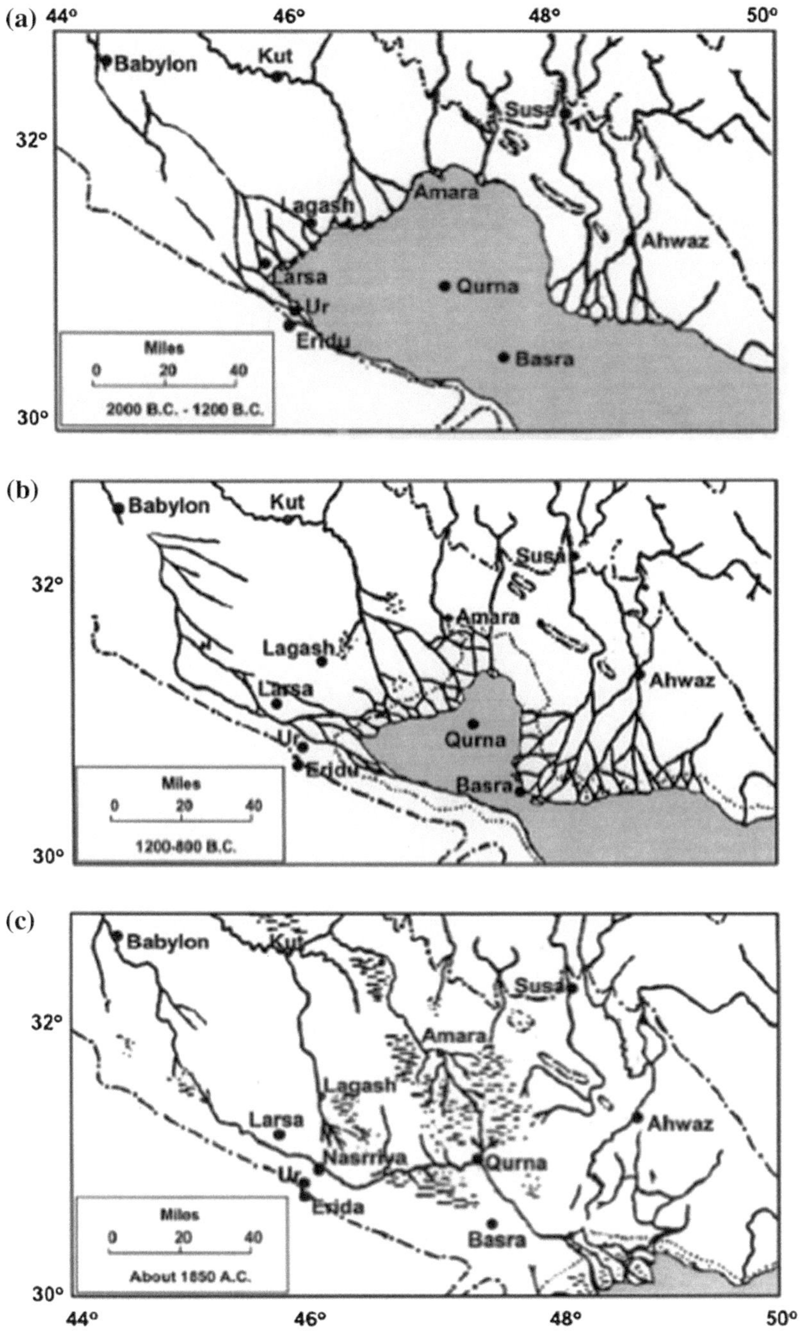

Fig. 4.1 Growth of the Shatt-al Arab (Tigris-Euphrates) delta since Babylonian time (from Aqrawi 2001) **a** 4000–3200 yr B.P.; **b** 3200–2800 yr B.P.; **c** 1850 A.D.

Fig. 4.2 Satellite image of the modern Shatt al Arab Delta, Southern Iraq. From Coleman and Huh (2004)

"Reengineering" is the future of the Changjiang (Yangtze) river and delta. The Changjiang River is the largest river in China and the third largest in the world (after Amazon and Ganges-Brahmaputra). Its drainage basin covers 1.8 million square kilometers (1/5 the area of China). The water and sediment discharges of the Changjiang pass through a large funnel shaped estuary and enter the shallow East China Sea (Donghai) a short distance to the east of Shanghai (Fig. 4.3). The nature and fate of the material debouched by the Changjiang affect Asia's coastal ocean in major ways. The sediment itself has provided the material of which the Yangtze Delta, which supports and surrounds Shanghai, is composed. The water provides buoyancy that influences coastal circulation, nutrients that affect biological productivity and pollutants and pathogens that can impact the health of marine organisms and people. The "Three Gorges Dam" (Fig. 4.5) in the middle reaches of the Changjiang (Yangtze) is reducing the supply of land-building sediment and altering the coastal transport regime over the East China Sea continental shelf. Another project involves improving maritime shipping access to Shanghai by doubling the depth of the estuarine channels and “training” or stabilizing the main shipping channel by constructing concrete jetties at the river mouth that extend 50 km into the East China Sea, thereby altering coastal circulation.

Compelling and complex coastal management issues involve profound human-induced transformations to coastal Louisiana and finding viable restoration strategies is a high priority for the state of Louisiana. Modest beginnings of such a plan have been proposed by the U.S. Army Corps of Engineers (USACE 2004). However, owing to the morphodynamic, ecological and socioeconomic complexities of Louisiana’s coastal landscape, there can be no solution that addresses the needs of all stakeholders. Indeed, the very levees that were constructed to protect low-lying coastal areas and train the river course have also deprived the coastal wetlands of much-needed sediment. A study by the National Research Council of the (US) National Academy of Sciences notes that: “It is not possible to restore the earlier extent of the delta or to maintain the present status of coastal Louisiana. However, through the selection of appropriate projects and with abandonment of

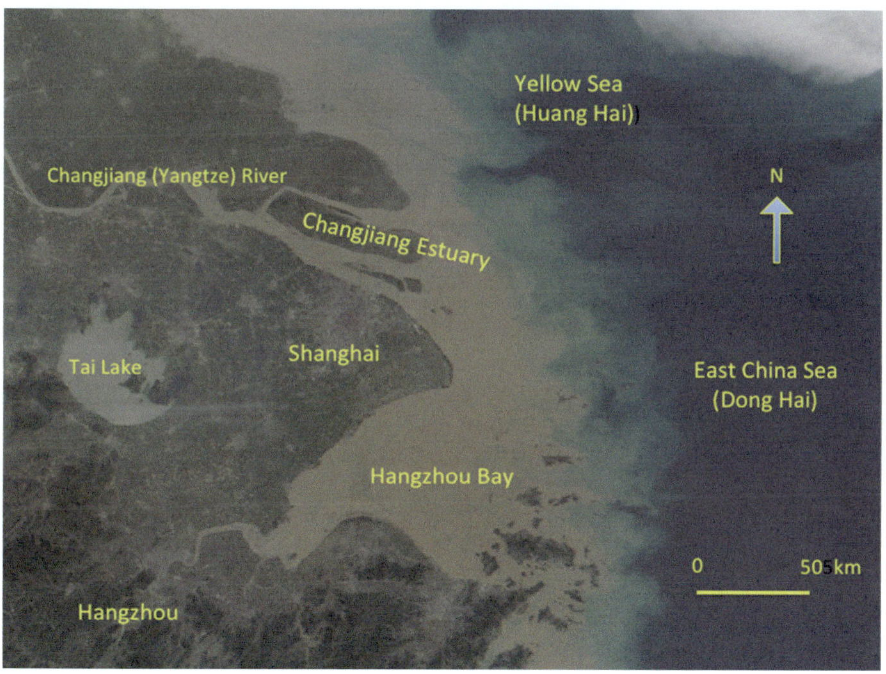

Fig. 4.3 NASA MODIS (or Moderate Resolution Imaging Spectroradiometer) image on April 13, 2002 of the Changjiang (Yangtze) River mouth and delta 2 years after completion of the Three Gorges Dam. MODIS is a sensor aboard NASA’s Terra (launched 1999) and Aqua (launched 2004) satellites. Shanghai is one of the 20 most populous cities in the world

Fig. 4.4 Silt-laden water issuing from the mouth of the Huanghe in the mid-1980s. Today, the water discharges and sediment loads reaching the sea are minimal (Photo: L. D. Wright, 1985)

areas that cannot be saved economically, a sustainable delta environment and management strategy can be achieved" (National Research Council 2006, p. 33). Coastal Louisiana will be discussed in more detail in the case study in Part 3.

The cases just described are only a few of numerous examples of river and delta systems impacted by human activities. Throughout the world, dams and reservoirs have generally reduced sediment fluxes to the sea while deforestation has increased sediment yield resulting in increased sediment discharge from rivers without dams. In addition to the sharp reduction in the supply of Nile sediment to the Mediterranean coast caused by the Aswan High Dam, the multiple European delta systems along the Mediterranean, Adriatic and Black Sea coasts (Ebro, Rhone, Po, Danube) have also experienced reductions in sediment input because of dams. Liquete et al. (2004, p. 71) conclude that: "One of the most significant factors controlling sediment discharge in the Mediterranean over the last few millennia is the development of civilization". Training structures such as levees and jetties have constrained the natural tendencies of deltaic distributaries to migrate and altered coastal and estuarine circulation. Because many deltas are sources of oil and gas, dredged canals, pipelines and fluid extraction have exacerbated subsidence. Solutions to these large-scale problems require new understandings of the interplay among physical, geological, ecological, biogeochemical and socio-economic factors.

Fig. 4.5 The "Three Gorges" dam on the Changjiang (Yangtze) River is the world's largest hydroelectric dam. Construction of this dam and others will further decrease sediment discharge and delta recession will continue to occur and impact lands surrounding Shanghai. Photo taken by Gaynor on April 6, 2013 and is licensed under the Creative Commons Attribution 2.0 Generic license. https://commons.wikimedia.org/wiki/File%3AThree_Gorges_Dam_(12280456164).jpg

4.3 Vulnerable River Deltas

The surfaces of most large river deltas rest on substrates composed of rapidly deposited soft sediments. These substrates are highly compactible and when they compact the overlying land surfaces sink. Extraction of water, oil and gas from deltaic sediments has significantly exacerbated the subsidence as was pointed out in Chap. 3. Prior to human modification, subsidence was offset by the annual supply of new sediment during river floods. But today, subsidence is the norm for large deltaic systems. Over the 20th century, deltas subsided by 80–400 cm (2.6–13.1 ft.; Syvitski et al. 2009). When sea level rise, storm surges, river floods and wave erosion are superimposed on these sinking lands, the people and ecosystems that reside there are increasingly vulnerable. Overeem and Syvitski (2009) have assessed the vulnerability of the major deltas of the world in terms of rates of sinking, exposure to storms and floods and population density. Table 4.1 lists the current population and population density following Overeem and Syvitski (2009). As Table 4.1 shows, nearly 500 million people worldwide live in low-lying deltas. The largest populations are in the Ganges-Brahmaputra (Fig. 4.6), Chao Phraya, Mekong

Table 4.1 Populations of 34 major river deltas in 2000 and 2015

Delta name	Pop. 2000	Pop. 2015	Projected growth or decline (%)	Mean pop. density (pop/sq. km.)
Amazon	318,464	522,718	64	6
Chao Phraya	14,472,900	19,541,100	35	588
Colorado	26,043	33,550	29	30
Congo	172,448	284,140	65	70
Fly	5403	7462	38	1
Ganges	147,463,000	189,175,000	28	1220
Godavari	5,339,490	5,922,290	11	849
Han	250,877	754,073	201	506
Indus	1,610,750	2,346,040	46	102
Irriwaddy	9,702,460	11,111,200	15	300
Krishna	6,115,110	6,779,550	11	580
Limpopo	2,808,180	4,436,810	58	47
Magdalena	505,615	611,870	21	139
Mahakam	20,800	29,458	42	16
Mahanadi	3,927,700	4,474,300	14	603
Mekong	28,227,700	35,209,300	25	465
Niger	21,674,400	31,468,600	45	291
Nile	39,653,300	49,227,900	24	1518
Orinoco	113,383	167,730	48	4
Parana	1,069,030	1,275,570	19	49
Pearl	13,469,200	23,848,300	77	1694
Tigris	13,479,400	19,831,000	47	114
Yangtze	44,372,400	44,803,200	1	1223
Yellow	3,842,410	8,759,240	128	335
Danube	271,407	248,162	-9	50
Mississippi	1,895,640	2,081,330	10	84
Po	61,653	56,027	-9	79
Rhone	95,059	96,618	2	62
San Francisco	67,919	69,944	3	70
Tone	3,716,990	4,028,290	8	886
Vistula	597,940	593,924	-1	256
Total	365,347,071	467,794,696	Average = 35	Average = 395

From Overeem and Syvitski (2009)

and Yangtze (Changjiang; Fig. 4.3) Deltas. The IGBP, in a 2014 "Deltasinfographic", lists the following deltas as being at "severe risk" because they have no input of new sediment to offset continued subsidence: Colorado (Mexico), São Francisco (Brazil), Rhône (France), Po (Italy), Nile (Egypt), Krishna (India), Yellow (Huanghe, China), Yangtze (Changjiang, China), Pearl (China), Chao Phraya (Thailand) and Tone (Japan).

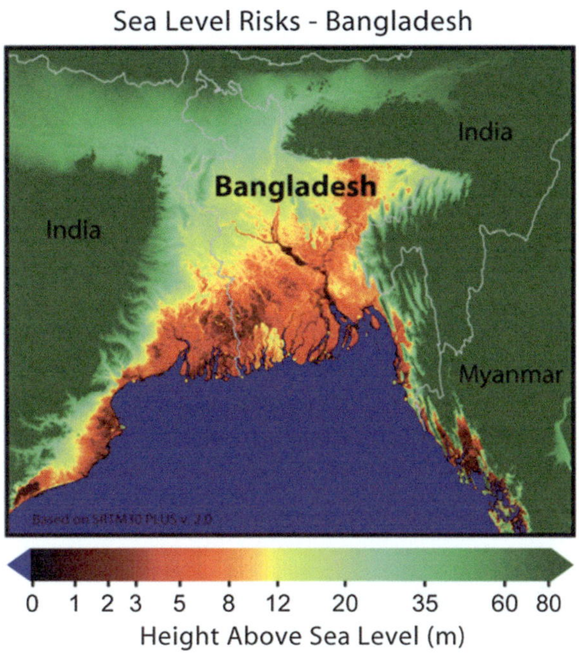

Fig. 4.6 Low-lying lands in the Ganges-Brahmaputra Delta subject to frequent flooding (red). The Ganges-Brahmaputra Delta is home to nearly 200 million people, most of them seriously impoverished. Flooding is caused by both storm surge and river flooding during the summer monsoon season (*Source* Environmental Change and Security Project and Intergovernmental Panel on Climate Change)

4.4 Invading Seas

The gradients of most deltaic surfaces are extremely gentle and this means that comparatively small vertical rises in coastal waters can result in large horizontal excursions by inundation from the sea. While the progressive rise in global mean sea level poses a long-term problem for all deltas, the most tragic consequences for delta residents are related to episodic extreme events such as storm-induced surges and destructive waves, extreme river floods and in some cases, tsunamis. Syvitski et al. (2009) report that in the decade prior to their publication, 85% of deltas worldwide experienced severe flooding and a total of 260,000 km^2 of delta lands were temporarily submerged. Inundation is certain to increase in the years ahead.

Most large river deltas are fronted by wide, low gradient continental shelves and this favors amplification of long waves such as tides and storm surges. Further amplification may occur as these waves travel upstream within funnel-shaped estuaries such as the lower course of the Changjiang (Yangtze) River (Fig. 4.3). Many Asian deltas not only possess these amplification-favorable features by also lie within zones of frequent tropical cyclone activity. When high storm surges are superimposed on high tides when the tide range is large (macro-tidal), disaster can result. For example, the spring tide ranges near the mouths of the Changjiang and the Ganges-Brahmaputra rivers are about 3.7 m ($\sim$12 ft.). Tide ranges in Hangzhou Bay just to the south of Shanghai exceed 7 m (23 ft.). In 1997, Category 5 “Super Typhoon” Winnie brought a 5.7 m (19 ft.) storm surge to the Changjiang Delta (Seavitt 2013). Breaking waves near the delta shores commonly exceed 6 m

(20 ft.) during typhoons and these waves cause significant erosion. A 6 m (20 ft.) storm surge, funneled by the Bay of Bengal, also hit the Ganges-Brahmaputra Delta during Tropical Cyclone Sidr in 2007. When high storm surges coincide with high river floods, as they have the potential to do during summer monsoons, the outcomes can be catastrophic.

References

Aqrawi, Adnan A. M. 2001. Stratigraphic signature of climate change during the Holocene evolution of the Tigris and Euphrates delta, lower Mesopotamia. *Global and Planetary Change* 28: 267–298.

Chen, X. 1998. Changjiang (Yangtze) River Delta, China. *Journal of Coastal Research* 14 (3): 839–858.

Coleman, J.M., and O.K. Huh. 2004. *Major world deltas: A perspective from space*. In eds. G.F. Hart, and J.M. Coleman, Coastal Studies Institute, and Department of Oceanography and Coastal Sciences, Louisiana State University: Baton Rouge, Louisiana. URL: http://www.geol.lsu.edu/WDD/PUBLICATIONS/C&Hnasa04/C&Hfinal04.htm and http://www.geol.lsu.edu/WDD/eS/delta_attributes.htm. The World Deltas Database Framework, URL: http://www.geol.lsu.edu/WDD.

Fairbank, J.K., and M. Goldman. 1998. *China: A new history*. Cambridge: Belknap Press (546pp).

International Geosphere Biosphere Prorgramme (IGBP). 2014. *Deltas at Risk.* Multi media infographic. http://www.igbp.net/multimedia/multimedia/deltasatrisk.5.62dc35801456272b46d351.html.

Liquete, C., M. Canals, P. Arnau, R. Urgeles, and X. Dur- rieu de Madron. 2004. The impact of humans on strata formation along the Mediterranean margins. *Oceanography* 17: 70–80.

National Research Council, Committee on the Restoration and Protection of Coastal Louisiana. 2006. *Drawing Louisiana's new map: Addressing land loss in Coastal Louisiana.* Washington, DC: National Academies Press (206pp).

Overeem, I., and J.P.M. Syvitski. 2009. *Dynamics and vulnerability of delta systems*. LOICZ Reports & Studies No. 35. Geesthacht: GKSS Research Center (54pp).

Schwartzstein, P. 2017. Death of the Nile. *BBC News Africa*. 10 Oct, 2017. https://session.bbc.co.uk/session?ptrt=http%3A%2F%2Fwww.bbc.co.uk%2Fnews%2Fresources%2Fidtsh%2Fdeath_of_the_nile&userOrigin=news&context=news.

Seavitt, C. 2013. Yangtze river delta project. *Scenario 03:Rethinking Infrastructure*. https://scenariojournal.com/article/yangtze-river-delta-project/.

Sinclair, K. 1987. *The Yellow River: A 5000 year journey through China.* Hong Kong: Intercontinental Publishing Corporation Ltd. (207pp).

Syvitski, J.P.M., A.J. Kettner, M.T. Hannon, E.W.H. Hutton, I. Overeem, G.R. Brakenridge, J. Day, C. Vörösmarty, Y. Saito, L. Giosan, and R.J. Nicholls. 2009. Sinking deltas. *Nature Geoscience* 2: 681–682.

U.S. Army Corps of Engineers. 2004. *Louisiana Coastal Area (LCA) Louisiana-Ecosystem restoration study.* New Orleans District, LA.

Xie, Dongfeng, Shu Gao, Zheng Bing Wang, Cunhong Pan, Wu Xiuguang, and Qiushun Wang. 2017. Morphodynamic modeling of a large inside sandbar and its dextral morphology in a convergent estuary: Qiantang Estuary, China. *Journal of Geophysical Research: Earth Science* 122: 1553–1572.

Yu, Qian, Yunwei Wang, Shu Gao, and Burg Flemming. 2012. Modeling the formation of a sand bar within a large funnel-shaped, tide-dominated estuary: Qiantangjiang Estuary, China. *Marine Geology* 299–302: 63–76.

Chapter 5
Coastal Morphodynamics and Ecosystem Dynamics

Lynn Donelson Wright, J. P. M. Syvitski, C. Reid Nichols and Julie Zinnert

And so castles made of sand slip into the sea, eventually….
—Jimi Hendrix, c. 1967

5.1 What Does "*Morphodynamics*" Mean?

In the introduction to this book, we promised to avoid jargon so we must begin this chapter with a simple explanation of what we are talking about. The term "morphodynamics" means simply that the shape of the land and the processes that mold the land are mutually interconnected and change together as a complex system. The concept has been used for the past few decades to imply the coupled suites of mutually-inter-dependent hydrodynamic, biologic and anthropogenic processes, seafloor and landscape morphologies, and time-dependent sequences of change. In a typical coastal morphodynamic system, changing shapes of the solid boundaries are caused by erosion in some places and re-deposition of the eroded sediments somewhere else. In many cases it is moving water or air (currents, waves, winds) that redistribute the material. The new land configurations in turn then alter the directions, intensities and gradients of the moving water or air. In other cases, it is biological rather than physical processes that build or degrade morphologic features such as coral reefs or coastal marshes. There is constant feedback among the

L. D. Wright (✉)
Southeastern Universities Research Association (SURA), Washington, DC, USA
e-mail: ldwright@bellsouth.net

J. P. M. Syvitski
Community Surface Dynamics Modeling System, University of Colorado, Boulder, CO, USA

C. R. Nichols
Marine Information Resources Corporation (MIRC), Ellicott City, MD, USA
e-mail: rnichols@mirc-us.com

J. Zinnert
Department of Biology, Virginia Commonwealth University, Richmond, VA, USA

L. D. Wright and C. R. Nichols (eds.), *Tomorrow's Coasts: Complex and Impermanent*, Coastal Research Library 27,
https://doi.org/10.1007/978-3-319-75453-6_5

multiple components of the system and when the environmental conditions change, all facets of the system also change. The morphodynamic paradigm has been applied in studies of continental shelves, surf zones, beaches, river mouths, inlets and estuaries (see Wright and Thom 1977; Cowell and Thom 1994; Wright 1995; Woodroffe 2002; Sutherland 2008; Barman, Chatterjee and Paul 2016).

Over prolonged periods of time, physical and biological process regimes and morphologic patterns co-evolve. As they do so, mutual causality is commonly manifest by way of feedback loops that may be either positive (self-enhancing) or negative (self- regulating; Wright and Thom 1977; Wright 1995). Specific examples are found throughout this book. Morphodynamic conditions as observed at any given time reflect contemporary as well as antecedent processes. Environmental factors (Table 1.2) constitute the basic setting within which the morphodynamic processes operate. Geographical and temporal variability of environmental conditions are responsible for variations in the basic morphological characteristics of coastal systems, as well as the magnitude and frequency of morphological change. It is by means of evolutionary sequences that inheritance from preceding states is propagated to succeeding states and environmental conditions are ultimately changed. The time scales of change range from hours to seasonal to decadal to millennial. Although progressive and sequential changes are the norm, abrupt shifts to different state can also take place when certain equilibrium thresholds are crossed. These abrupt "tipping points" commonly involve short-lived periods of positive feedback.

5.2 Continental Shelves: The Foundations of Coasts

The forces that change the shape of the coast are preconditioned by the continental shelves that front the coast. Most of the world's continental shelves extend from the shores to the shelf break at water depths of roughly 100–120 m (330–390 ft.). This depth corresponds to positions of low sea level during the glacial maxima of the Pleistocene ice ages (Fig. 3.1) and hence the surfaces of most shelves were exposed only 20,000 years ago. The morphologies of the shelves depend in part on large scale tectonic setting and geologic history, in part on the reworking of shelf sediment as seas transgressed across the shelf during the post-glacial rise in sea level and in part on present day processes. The widths of the shelves vary from ~100 km (60 miles) off the Atlantic and Gulf Coasts of North America (trailing edge coasts) and off many Asian coasts (marginal sea coasts) to less than 1 km off many island coasts (e.g., Canary Islands, Hawaiian Islands, and Mariana Islands) and some parts of the Pacific Coast (collision coasts) of North America. Shelves can be either sources or sinks of coastal sediments (Wright 2012). Waves frequently agitate the seafloor material over the inner shelf (depths <~30 m or 100 ft.) and, if the shelf is very wide with a low gradient, this agitation can dissipate much of the wave energy before the waves reach the shore. However, other factors such as

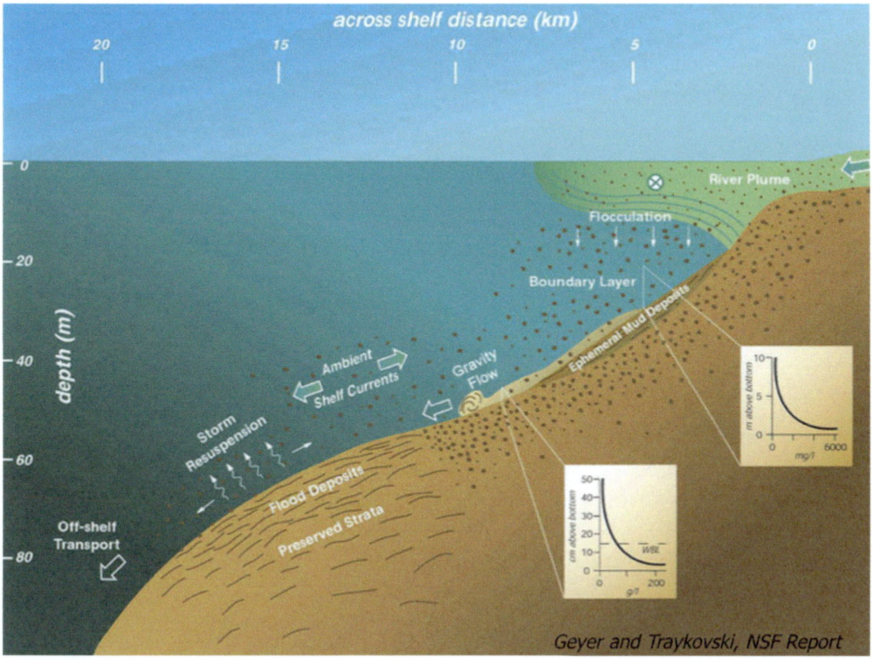

Fig. 5.1 Conceptual diagram portraying the complex interplay of multiple processes on the continental shelf. From Geyer and Traykovski (2001)

storm surges may be amplified. Strong currents may prevail at any depth depending on oceanographic regime. Figure 5.1 illustrates, conceptually, some of the processes that operate on continental shelves.

The configuration of the continental shelf is a major determinant of coastal processes. The heights of breaking waves during storms are particularly dependent on shelf gradient. For example, Fig. 5.2 shows a comparison of the inner continental shelf off the North Carolina in the Middle Atlantic Bight with that off the Louisiana coast where the Mississippi River has supplied abundant mud to build a wide low gradient, convex upward profile. The latter has the effect of greatly attenuating incoming waves by a combination of bed friction and energy absorption by the soft seabed. Much higher waves break in the surf zone of the Middle Atlantic Bight. Prior to breaking, the waves in the Middle Atlantic Bight interact with other wind generated and tidal currents during storms to transport sands into deeper water (Wright 1995). This impedes the accumulation of sediment and helps to maintain a relatively steep, concave upward shelf profile (Fig. 5.3). Notably, much steeper shelf profiles prevail along the Pacific Coasts of the Americas and Australia where breaking waves are substantially higher than in either of the cases shown in Fig. 5.2. However, the steeper shelves do not favor the amplification of storm surges the way that wider shelves do. Hence, even though the attack of shores by

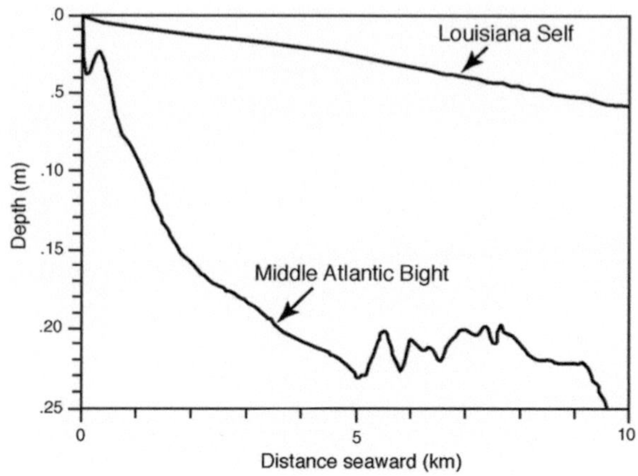

Fig. 5.2 Inner continental shelf profiles off the Middle Atlantic Bight and Louisiana. (Courtesy of L. D. Wright)

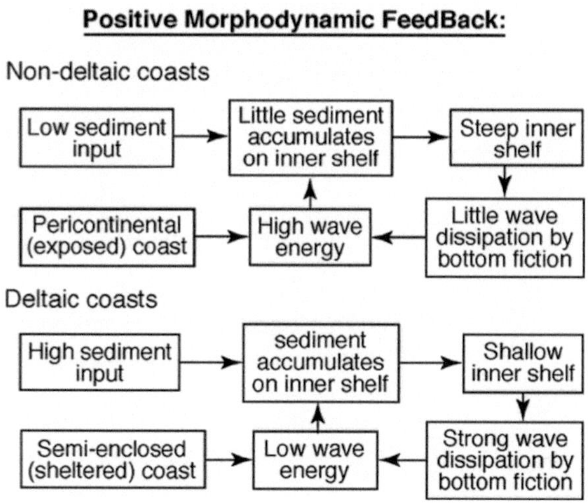

Fig. 5.3 Simplified positive feedback loops for two shelves. (Courtesy of L. D. Wright)

high breaking wind waves and swell is more severe where the inner shelf is steep and narrow, flooding by storm surges is more severe where the shelf is wide and flat.

Figures 5.1 and 5.2 suggest a simplicity that is not real. In fact, shelf topography can be very complex and may include relict ridges, such as in the case of the Middle Atlantic Bight (Fig. 5.2), relict or active coral reefs which afford partial protection from waves, or deep furrows such those found on the shelf in the Northern Indian Ocean. The Northern Indian Ocean shelf is extremely wide and low gradient and, acting in combination with the funnel-like configuration of the Bay of Bengal, causes tides to be significantly amplified as they approach the coast (e.g., Testut and Unnikrishnan 2016). A dramatic example of tidal amplification by a wide shallow shelf and a constricting seaway is found on the southern shores of the Timor Sea in northwestern Australia where tide ranges of 9.5 m (31 ft.) and very swift tidal currents prevail (e.g., Wright 1982). Other prominent examples of tidal amplification by the combination of shallow shelves, constricted gulfs or seaways and resonant interactions with shore boundaries are found in the Malacca Straits, the Bay of Fundy, the Gulf of Siam and around the British Isles.

5.3 Ephemeral Beaches and Shores

The shallow regions immediately seaward of shores exposed to open oceans with moderate to high waves and moderately steep inner continental shelves are among the earth's most energetic and dynamic realms. These are the surf zones where the incoming waves break and dissipate their energy within a relatively narrow zone (Figs. 5.4 and 5.5). The width of the surf zone increases with increasing wave height and decreasing bed slope. During high energy events, such as storms, these breaking waves can do a lot of work in a short span of time and move large volumes of beach and surf zone sand seaward over horizontal distances of hundreds of meters and to depths of 20 m or more. After the event subsides, the sands commonly return to the beach under the influence of low, long-period swell but this requires a longer period of time. Severe beach erosion can take place in a matter of hours but beach recovery can require days to weeks. Much of the wave energy dissipated within surf zones is commonly converted to surf zone currents including strong alongshore currents, which can transport large quantities of sand to other parts of the beach system as well as rip-current circulations that carry sand offshore.

Figure 5.6 shows the annual long-term averaged wind and wave regimes for the world. The easterly trade winds prevail in the tropics while stronger westerlies are dominant between north and south latitudes of 30° and 60°. These are averaged winds are not event-driven storms such as hurricanes. The prevailing westerlies generate the highest and hence most energetic waves but these waves also tend to have the longest periods. Long-period swell waves, such as those shown in Fig. 5.4 are much less destructive than waves with low period/height ratios (Fig. 5.6i). In fact, moderately (not extremely) high but long period swell are often constructive in the sense that they push sand onshore from the continental shelf. Beaches such as that shown in Fig. 5.4 are commonly in equilibrium with persistent moderate to high-energy wave regimes.

The U.S. Atlantic and Gulf of Mexico coasts are subject to the superimposition of high waves and storm surges during tropical and extra-tropical storms. This causes significant landward translations of the surf zone, often well into seaside communities. A prominent recent example of this occurred along most of the U.S. East Coast 2012 as Hurricane Sandy moved northward up the coast bringing high waves ($\sim$10 m or 33 ft.) and storm surge (>3 m or 10 ft.; Sopkin et al. 2014). The protective dune was significantly eroded and overtopped. Some of the eroded sand was carried offshore and some was pushed inland and into neighborhood streets. As shown in Fig. 5.7, Sandy shifted the surf zone landward into an amusement park in

Fig. 5.4 High, long-period swell from the Southern Ocean breaking in a wide, flat surf zone off a South Australian beach. The outer line of breakers is over an offshore bar. Waves reform within a deep intervening trough then break again closer inshore. Much energy is dissipated in this process. From Short (2012)

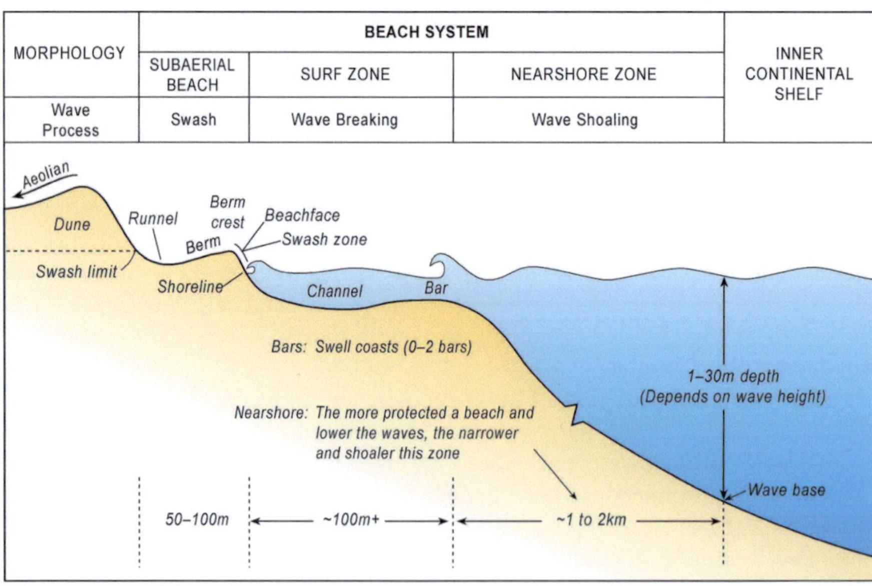

Fig. 5.5 Idealized diagram showing the components of a beach and surf zone system. From Short and Woodroffe (2009)

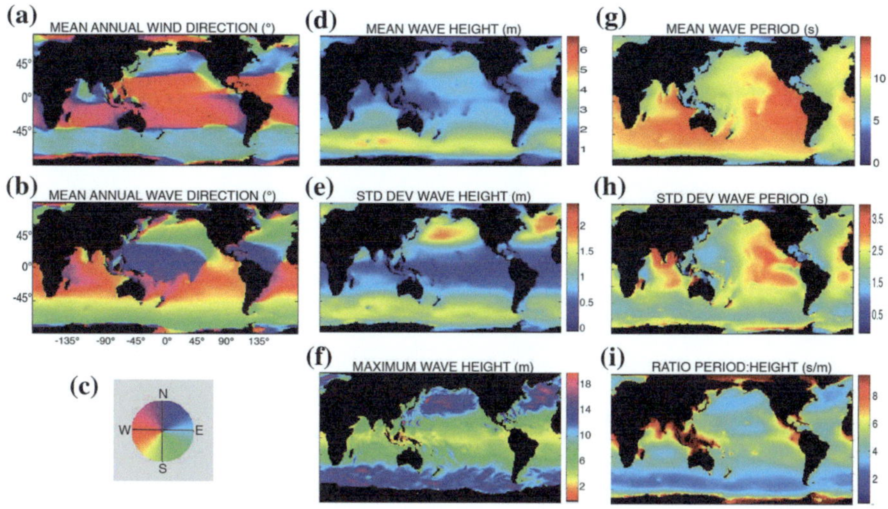

Fig. 5.6 Long-term average wave properties for the world. **a** Mean annual wind direction is the direction from which the waves are coming. **b** Mean annual wave direction is the direction of wave propagation. Wave heights are in meters and periods in seconds. Note that the highest and longest period waves are associated with the prevailing westerlies. Data are summaries from NOAA's 3-hourly WAVEWATCH III nowcasts. From Syvitski et al. (2017)

Fig. 5.7 Inundated boardwalk amusement park in Atlantic City, New Jersey in the aftermath of Hurricane Sandy (Photo of roller coaster on Casino Pier in Seaside Heights, NJ courtesy of Star Ledger)

Fig. 5.8 Large-scale beach re-nourishment project, Pompano Beach, Florida January 2016. The project took 3-months to complete (Courtesy of L. D. Wright)

Atlantic City, New Jersey. As storms become more intense and sea level rises, eroded sediments are transported offshore to depths beyond the reach of constructive fair weather waves and are thus not returned to the beach. Over time, this has the result of causing shoreline recessions over extensive lengths of coast. In many cases, especially where erosion threatens expensive beach-front property, engineering solutions such as beach re-nourishment are pursued (Fig. 5.8). However, such projects are very temporary and expensive and must be repeated often.

5.4 Migrating Barrier Islands

Morphodynamic models of the landward migration of barrier islands in response to rising sea levels are offered by Cowell and Thom (1994), Cowell, Roy and Jones (1995), Moore et al. (2010) and Lorenzo-Trueba and Ashton (2014). Those models predict that if the landward migration is unimpeded, the barrier formation may be preserved as it transgresses even though the position changes. Relatively narrow barrier islands, separated from mainland coasts by shallow lagoons fringe much of the U.S. Atlantic and Gulf coasts and the Arctic Ocean coast and are scattered around the world in places like the Netherlands where seas have recently transgressed low gradient continental shelves. The Outer Banks of North Carolina are among the most well-known barrier island chains and are described in more detail in Part 3. It is likely that barrier islands swept over the continental shelf as sea levels rose during the waning phases of the Pleistocene glaciations. Commonly, the sheltered environments in the lagoons behind the barrier islands are the homes of salt marsh or mangrove swamp ecosystems. Barrier island "roll over" involves a combination of seaward transport of eroded beach and dune sand and landward "wash over" of some dune sands by dune over-topping during storms. As the dune sands advance landward over the wetlands behind, some stabilizing vegetation is killed off creating expanded open water in the lagoons. However, as shown by Zinnert et al. (2017) some marsh vegetation is burial tolerant and resilient while other vegetation types stabilize dunes.

There is relatively limited understanding of the effects of climate change on coastal ecological communities, especially barrier islands; this is surprising considering the importance of the ecosystem services they provide (i.e. storm surge protection, creation of economically important lagoons and wetlands, etc.). Zinnert et al. (2016) documented a 27% loss of land from Virginia barrier islands over a 30 year timeframe. This dramatic land loss has been accompanied by rapidly changing ecosystem states between bare sand, interior grass and woody vegetation, and salt marsh. Vegetation type is an important functional component of coastal depositional systems and affects morphology by trapping sediment and promoting organic matter deposition. Depending on vegetation type, this can affect the resistance or resilience of coastal communities on varying timescales and disturbance types. For example, at shorter timescales and with episodic disturbance, barrier island woody vegetation such as shrub land or maritime forest reduces wave energy and can minimize impacts further inland. However, over longer timeframes, these vegetation types limit barrier island rollover, enhancing erosion and contributing to land loss (Zinnert et al. 2017).

The process of barrier island rollover is well illustrated by the case study of Cedar Island on the Eastern Shore of Virginia (Fig. 5.9; Wright and Trembanis 2003). For the past 150 years, Cedar Island has been receding landward at an average rate of over 5 m per year by "rolling over" onto the marsh, estuarine and tidal channel deposits behind the island. The low dune that caps this island is only a thin veneer of sand, which advances over the marsh during storm-driven washover events. As the thin sheet of sand migrates landward across the marsh, marsh peats emerge on the beach and in the nearshore zone. This low surface provides minimal resistance to overwash during storms. Over time, this process of coastal transgression has left behind a low gradient inner shelf profile (Fig. 5.10). This can have the effect of amplifying storm surges.

Barrier island migrations involve considerably more than the physical processes of wave-induced erosion and storm-surge wash over. Numerous ecological factors interact with physical and morphologic factors to determine barrier island stability and patterns of change. Zinnert et al. (2017) have synthesized the interacting and interdependent roles of physical, morphological and ecologic processes in

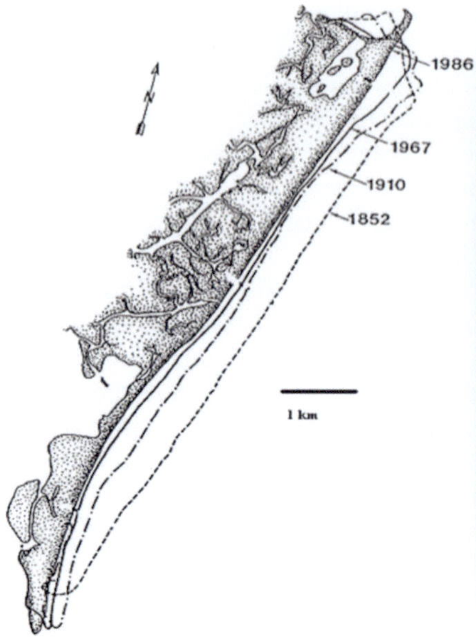

Fig. 5.9 Map view and aerial photographic view of Cedar Island, Virginia. Note the succession of shoreline positions indicated on the map. From Wright and Trembanis (2003)

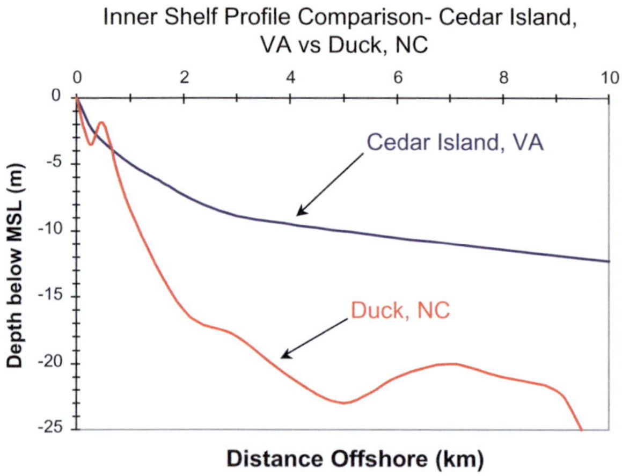

Fig. 5.10 The inner continental shelf profile fronting the transgressive Cedar Islnad is much flatter than profiles off the Outer Banks of North Carolina to the south. From Wright and Trembanis (2003)

controlling barrier island behavior and change (Fig. 5.11). They distinguish between disturbance resisting and disturbance reinforcing (positive feedback) domains. The former includes the stabilizing effects of dune grasses, which thrive by means of progressive burial by wind-blown sand, favoring the upward growth of the dunes and enhancing resistance to storm wash over. The Cedar Island example just described is an example of a disturbance-reinforcing domain. As Zinnert et al.

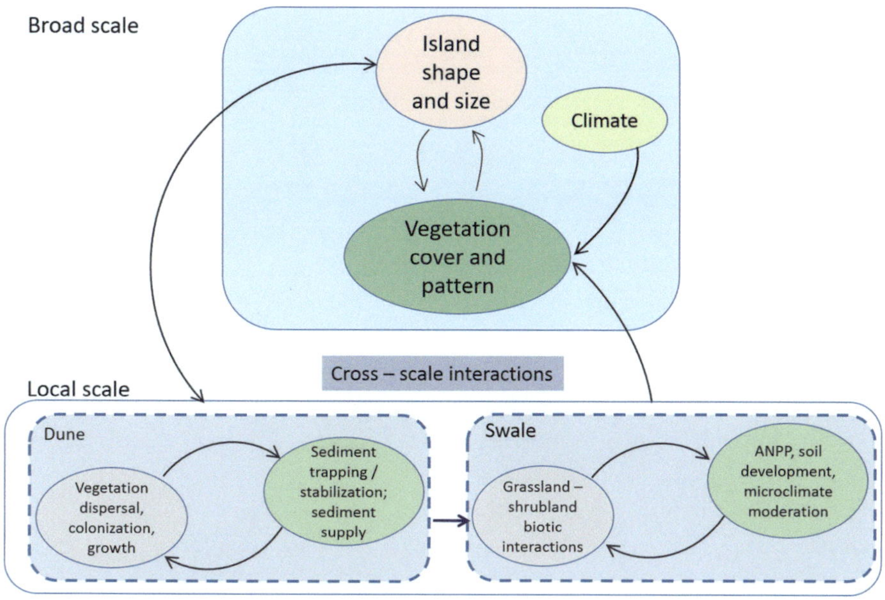

Fig. 5.11 Interdependent factors controlling barrier island shape and behavior. From Zinnert et al. (2017)

(2017) note, the *Spartina patens* marsh vegetation behind transgressing sands is tolerant to burial as the wash over process proceeds. However, *Spartina patens* is not a dune-building grass but its burial tolerance enables it to be resilient and also to bind sediments. This helps to maintain a low-lying surface that allows frequent wash over, thereby closing the positive feedback "loop".

5.5 Tidal Wetlands Dynamics

Tidal wetlands, including salt marshes, and, in tropical and sub-tropical regions, mangrove forests, constitute significant fractions of undeveloped coastal lowlands. These environments, located in brackish water realms subject to tidal rise and fall provide essential ecosystem services in the form of fisheries and fish nurseries as well as protections for coastal communities located immediately upland of the wetlands. The largest mangrove forest in the world, the Sundarbans (Fig. 5.12) of Bangladesh and India in the Ganges-Brahmaputra Delta, is home to Bengal tigers (*Panthera tigris tigris*) and to four million people. Like mangroves elsewhere in the world, this intertidal ecosystem is effective in dissipating high waves and storm surges as well as tsunamis (Ishtiaque et al. 2016). It provides valuable ecosystem services including fisheries habitat and a source of forestry products. Ishtiaque et al. (2016) report that degradation of the forest is currently underway, particularly in the

Fig. 5.12 Composite NASA Landsat 7 (Launched 1999) image of the Sundarbans mangrove ecosystem. Bangladesh. NASA Earth Observatory (Landsat 7 satellite observations from November 24, 1999, and November 17 and 26, 2000)

eastern region and that the causes are largely anthropogenic. Types of stressors include oil pollution, fires, over extraction of wood, and shrimp farming. Natural causes, many of which are also influenced by human activity include increased salinity and cyclone impacts. The combination of sea level rise and land subsidence is projected to cause significant losses of Sundarbans mangroves by the end of the 21st century. Another region of extensive mangrove forests exists along the coast of the Gulf of Papua, Papua New Guinea and covers the deltas of the Purari, Kikori and Fly Rivers. Shearman (2010) showed that, between 1973 and 2006 these forests underwent significant modification including recessions of up to 43 m (141 ft.) per year in some places, e.g. in the Kikori Delta. Future sea level rises are expected to cause continued recessions.

In middle and high latitudes, tidal wetlands consist primarily of salt marshes, tidal creek networks and shallow bays. The classic view of the stability of salt marsh ecosystems is that an equilibrium with gradual rises in relative sea level is attained when marsh macrophytes such as salt marsh cordgrass (*Spartina alterniflora*) increase productivity and increase the trapping of inorganic sediments to raise the mash surface in an attempt to keep pace with rate of sea level rise (Morris et al. 2002, 2016). The process involves interactions among sea level, land surface elevation, marsh grass production and sediment supply from the estuary. For situations where relative sea level (including subsidence) is constantly rising, a degree of disequilibrium typically prevails whereby the marsh surface lags the sea level by a small amount. Morris et al. (2002) evolved a theoretical model of this process which suggests that increases in U.S. East Coast marsh elevation may not keep pace with inter-annual anomalies in rates of sea level rise or with steady rates of rise greater than 1.2 cm/year (0.47 in./year). Declines in sediment supply by estuaries can increase disequilibrium. Schile et al. (2014) and Alizad et al. (2016) have developed more elaborate marsh equilibrium models to take into account accelerated sea level rises combined with reduced sediment supply, reduced plant productivity and increased inundation. The model results show that high marsh elevations should be maintained with relatively slow rates of sea level rise (<52 cm/century; <20.5 in./century) but with high rates of rise (>165 cm/century; >65 in./century), marsh surfaces would likely be replaced by mudflats.

5.6 Coral Reef Degradation

Fringing coral reef systems (Fig. 5.13) and offshore barrier reefs on inner continental shelves provide first lines of partial defense for many tropical coasts that are distant from muddy rivers. Reefs cause partial dissipation of storm waves, particularly the most severe waves (Sheppard et al. 2005). They are also economically important to many coastal communities for fisheries and tourism. And they are, in their own right, delicate, beautiful and valuable natural environments. Like the wetlands just described, reefs can grow upward fast enough to keep pace with gradual sea level rises; they did this successfully as post-glacial sea levels rose and as the volcanic islands of the Pacific, including the Hawaiian Islands, slowly sank. The corals composing reefs are animals and do not photosynthesize. However, the corals have a symbiotic relationship with micro algae, called *zooxanthellae,* which do photosynthesize and provide most of the nourishment to their coral hosts. The *zooxanthellae* are also the main sources of the bright colors exhibited by reefs. Corals require fairly abundant light for photosynthesis by the *zooxanthellae* and cannot thrive in turbid waters. Corals are also sensitive to water temperature; the optimal temperature for tropical corals is 26–27 °C (79–81 °F). They are rarely found at temperatures below around 18 °C (64 °F) and they are also adversely affected by temperatures greater than about 29 °C (84 °F), in part because dissolved oxygen in sea water declines as waters warm. The living portions of today's reefs commonly rest on substrates consisting of the exoskeletons of dead or relict corals.

Reefs are very complex ecosystems that involve the interdependence of numerous species, water chemistry, water temperature, water movement (currents and waves) and underlying seafloor topography. Changes in any of these factors can alter the functioning of the reef. Over the past few decades, humans have had substantial negative impacts on coral reefs and reef degradation is apparent in many parts of the world. The most conspicuous manifestations of this degradation are die back of corals and encroachment of competing macroalgae (Norström et al. 2009). There is a suite of human induced causes for this, some are global in scope such as global warming and ocean acidification and others are local such as over fishing,

Fig. 5.13 A relatively narrow fringing reef, and a surrounding artificial berm, provide the only protection from the sea for the small island community on Kurumba Island, Maldives

excessive suspended sediment, nutrient runoff from nearby urban centers and direct impact by tourists. van Hooidonk et al. (2016) stress the importance of local management strategies for protecting coral reefs in the future. However, Bruno and Valdivia (2016), in a recent study of the global changes in coral reefs, have concluded that global factors are more important than local factors in general and that degradation is not correlated with population density. Coral bleaching in particular is a consequence of global warming. Coral bleaching does not directly kill the corals but rather the symbiotic *zooxanthellae* which are the main source of food for the corals and are also responsible for the color of reefs (hence the term "bleaching"). Coral bleaching on Australia's Great Barrier Reef in 2016 was the worst on record and affected 90% of the reef and killed 20% of the corals (UNEP 2016). The event was directly correlated with corresponding record sea surface temperatures of 29.1 °C (85 °F). According to Australia's Bureau of Meteorology (BoM 2016) the sea surface temperatures over the months of February through April 2016 were the hottest on record. Continued global warming may have a devastating effect on coral reefs worldwide.

References

Alizad, K., S.C. Hagen, J.T. Morris, P. Bacopoulos, M.V. Bilskie, J.F. Weishampel, and S.C. Medeiros. 2016. A coupled, two-dimensional hydrodynamic-marsh model with biological feedback. *Ecological Modelling* 327: 29–43.

Australian Bureau of Meteorology 2016. *2016 marine heatwave on the Great Barrier Reef* 2pp.

Barman, N.K., S. Chatterjee, and A.K. Paul. 2016. Coastal morphodynamics: Integrated spatial modeling on the Deltaic Balasore Coast, India Springer Briefs in Geography ISBN: 978-3-319-33574-2 (Print), 978-3-319-33575-9 (Online).

Bruno, J. F. and Valdivia, A. 2016. Coral reef degradation is not correlated with local human population density. *Scientific Reports* 6, 29778. https://doi.org/10.1038/srep29778.

Cowell, P.J., P.S. Roy, and R.A. Jones. 1995. Simulation of large-scale coastal change using a morphological behaviour model. *Marine Geology* 126 (1–4): 45–61.

Cowell, P.J., and B.G. Thom. 1994. Morphodynamics of coastal evolution. In *Coastal evolution: Late Quaternary shoreline morphodynamics*, ed. R.W.G. Carter, and C.D. Woodroffe, 33–86. Cambridge: Cambridge University Press.

Geyer, W.R., and P. Traykovski. 2001. Wave- and density driven sediment transport. In *Ocean sciences at the new millennium*, 97. Arlington: National Science Foundation and University Corporation for Atmospheric Research.

Ishtiaque, A., S.W. Myint, and C. Wang. 2016. Examining the ecosystem health and sustainability of the world's largest mangrove forest using multi-temporal MODIS products. *Science of the Total Environment* 569–570: 1241–1254.

Lorenzo-Trueba, J., and A.D. Ashton. 2014. Rollover, drowning, and discontinuous retreat: Distinct modes of barrier response to sea-level rise arising from a simple morphodynamic model. *Journal of Geophysical Research: Earth Surface* 119: 779–801. https://doi.org/10.1002/2013jf002941.

Moore, L.J., J.H. List, S.J. Williams, and D. Stolper. 2010. Complexities in barrier island response to sea level rise: Insights from numerical model experiments, North Carolina Outer Banks. *Journal of Geophysical Research* 115: F03004. https://doi.org/10.1029/2009JF001299.

Morris, J.T., P.V. Sundareshwar, C.T. Nietch, B. Kjerfve, and D.R. Cahoon. 2002. Responses of coastal wetlands to rising sea level. *Ecology* 83 (10): 2869–2877.

Morris, J.T. et al. 2016. Contributions of organic and inorganic matter to sediment volume and accretion in tidal wetlands at steady state. *Earth's Future*. http://dx.doi.org/10.1002/2015EF000334.

Norström, A.V., M. Nyström, J. Lokrantz, and C. Folke. 2009. Alternative states on coral reefs: Beyond coral–macroalgal phase shifts. *Marine Ecology Progress Series* 376: 295–306.

Schile, L.M., J.C. Callaway, J.T. Morris, D. Stralberg, V.T. Parker, and M. Kelly. 2014. Modeling tidal wetland distribution with sea-level rise: Evaluating the role of vegetation, sediment, and upland habitat in marsh resiliency. *PLoS ONE* 9: e88760. https://doi.org/10.1371/journal.pone.0088760.

Shearman, P.L. 2010. Recent change in the extent of mangroves in the Northern Gulf of Papua. *Papua New Guinea AMBIO* 39: 181–189.

Sheppard, C., D.J. Dixon, M. Gourlay, A. Sheppard, and R. Payet. 2005. Coral mortality increases wave energy reaching shores protected by reef flats: Examples from the Seychelles. *Estuarine Coastal and Shelf Science* 64: 223–234.

Short, A.D. 2012. Coastal processes and beaches. *Nature Education Knowledge* 3 (10): 15.

Short, A.D., and C.D. Woodroffe. 2009. *The coast of Australia*. Melbourne, Australia: Cambridge University Press.

Sopkin, K.L., Stockdon, H.F., Doran, K.S., Plant, N.G., Morgan, K.L.M., Guy, K.K., and Smith, K.E.L. 2014. *Hurricane Sandy—observations and analysis of coastal change*: U.S. Geological Survey Open-File Report 2014-1088, 54pp. http://dx.doi.org/10.3133/ofr20141088.

Sutherland, J. (2008). *Coastal morphodynamics and society*. Available from http://www.coastalwiki.org/wiki/Coastal_morphodynamics_and_society.

Syvitski, J.P.M et al. 2017. Latitudinal controls on stratigraphic models and sedimentary concepts. *SEPM Special Publication No. 108*, http://dx.doi.org/10.2110/sepmsp.108.04.

Testut, L., and A.S. Unnikrishnan. 2016. Improving modeling of tides on the continental shelf off the West Coast of India. *Journal of Coastal Research* 32 (1): 105–115. https://doi.org/10.2112/JCOASTRES-D-14-00019.1.

United Nations Environment Program (UNEP). 2016. New research predicts the future of coral reefs under climate change.

van Hooidonk, R. et al. 2016. Local-scale projections of coral reef futures and implications of the Paris Agreement. Scientific Reports. 6, 39666. https://doi.org/10.1038/srep39666.

Woodroffe, C.D. 2002. *Coasts—Form and processes*. Cambridge: Cambridge University Press.

Wright, L.D. 1982. Nearshore tidal currents and sand transport in a macrotidal environment. *Geo–Marine Letters* 1: 173–179.

Wright, L.D. 1995. *Morphodynamics of Inner Continental Shelves*. Boca Raton: CRC Press, Inc. (241pp).

Wright, L.D. 2012. Recent advances in understanding continental shelf sediment transport. *International Association of Sedimentologists. Special Publication* 44: 159–172.

Wright, L.D., and B.G. Thom. 1977. Coastal depositional landforms: A morphodynamic approach. *Progress in Physical Geography* 1: 412–459.

Wright, L.D. and A. Trembanis. 2003. Complex shore face morphology of a rapidly transgressing barrier island: Cedar island Virginia. *Proceedings, Coastal Sediments 2003*, American Society Civil Engineers (10pp).

Zinnert, J.C., S.A. Shiflett, S. Via, S.N. Bissett, B.L. Dows, P.V. Manley, and D.R. Young. 2016. Spatial-temporal dynamics in barrier island upland vegetation: The overlooked coastal landscape. *Ecosystems* 19: 685–697.

Zinnert, J.C., J.A. Stallins, S.T. Brantley, and D.R. Young. 2017. Crossing scales: The complexity of barrier-island processes for predicting future change. *BioScience* 67: 39–52.

Chapter 6
Coastal Systems in the *Anthropocene*

Lynn Donelson Wright, J. P. M. Syvitski and C. Reid Nichols

All the parts incessantly work into each other's hands for the profit of man. The wind sows the seed; the sun evaporates the sea; the wind blows the vapor to the field; the ice, on the other side of the planet, condenses rain on this; the rain feeds the plant; the plant feeds the animal; and thus the endless circulations of the divine charity nourish man.

—Ralph Waldo Emerson, *Nature*—1833

But man is a part of nature, and his war against nature is inevitably a war against himself.

—Rachel Carson, *Silent Spring*

6.1 People and Coasts

People are integral parts of nature and, in many respects, are becoming dominant parts. This notion is implicit in the term "Anthropocene". In no environment is the connection between people and nature more apparent than in coastal systems. Mutual causality between humans and nature plays out there on a daily basis, sometimes in very positive ways and other times in tragic ways. The enjoyment of coastal beauty and spiritual stimulation are among the positive attractions as are access to global seaways, fisheries and recreation. Death, disease and destruction wrought by severe storms and tsunamis are paramount among the downsides. But for a multiplicity of reasons, roughly half of the world's 7 billion people live within

L. D. Wright (✉)
Southeastern Universities Research Association (SURA), Washington, DC, USA
e-mail: ldwright@bellsouth.net

J. P. M. Syvitski
Community Surface Dynamics Modeling System, University of Colorado, Boulder, CO, USA

C. R. Nichols
Marine Information Resources Corporation (MIRC), Ellicott City, MD, USA
e-mail: rnichols@mirc-us.com

L. D. Wright and C. R. Nichols (eds.), *Tomorrow's Coasts: Complex and Impermanent*, Coastal Research Library 27,
https://doi.org/10.1007/978-3-319-75453-6_6

100 km (60 miles) of the shore. And the activities of those who live much farther inland, for example within the catchments of large rivers that run to the coast, impact the coastal environment and its residents in numerous ways. Human activities that directly impact coastal systems include urbanization, agriculture, nutrient runoff, engineering works, fisheries, oil and gas production, dredging and various forms of pollution. Natural processes that impact coastal residents include sea level rise, storms and storm surges, water-borne pathogens, tsunamis, and loss of ecosystem services. Important ecosystem services include pollination, decomposition, water purification, erosion and flood control, carbon storage, and climate regulation.

6.2 Coastal Urbanization

Worldwide there are 23 *megacities* with populations of over 10 million people. Of these, 16 are in the coastal zone (Pelling and Blackburn 2013; Blackburn and Pelling 2014). In some of these cases, urban sprawl of neighboring cities has caused mergers into huge urban "agglomerations". The ten largest of these cities are: Tokyo, Japan; Mumbai, India (Fig. 6.1); São Paulo, Brazil; Dhaka, Bangladesh; New York-Newark, USA; Calcutta, India; Shanghai, China; Karachi, Pakistan; Lagos, Nigeria; and Manila, Philippines. The combined population of these 10 cities is projected to reach 216 million by the year 2025 (von Glasow et al. 2013). However, as explained in Part 3, Chap. 12, the huge agglomeration of cities surrounding Guangzhou, China in the Pearl River Delta, now has a combined population of 66 million people making it the world's largest urban center. The population of all of the 16 coastal megacities combined in 2025 will be about 323 million. Twelve of the coastal megacities occupy river deltas that are subject to subsidence (Table 6.1; Pelling and Blackburn 2013; Overeem and Syvitski 2009). In 1975, 62 million people lived in these deltaic cities. The number had grown to 140 million by 2010 and is expected to reach 242 million by 2050 (Overeem and Syvitski 2009; Pelling and Blackburn 2013). By some estimates, the number is much larger (e.g. Dawson 2017). In his recent book, Dawson (2017) describes the dire plight of the growing number of people living in flood prone coastal urban slums. Migration has been the primary cause of the population growth of these cities to date, but poverty has dictated the concentration of people in low-lying areas.

In addition to megacities, there are many other smaller cities and towns within the world's coastal zones. Among these are a growing number of tourism oriented beachfront "strip" cities. Some of the more prominent of these include many of Florida's coastal resort towns, Australia's Gold Coast, Mediterranean coastal resorts and many Caribbean islands. In many cases the wall-to-wall high rise hotels and condos of these cities have been built so close to the shore that they do not allow for sufficient buffering against storm-induced erosion and require frequent sand re-nourishment to offset the permanent offshore loss of sand.

Fig. 6.1 Coastal urban skyline, Mumbai, India. Image courtesy of Pixabay PDPics free images

Table 6.1 Megacities occupying deltas 1975–2050

City	Delta	Population (millions)		
		1975	2010	2050
Karachi	Indus	4	12	31.7
Kolkata	GBM	7.9	15.1	33
Dhaka	GBM	2.2	13	35.2
Rangoon	Irrawaddy	1.8	6.3	10.8
Bangkok	Chao Phraya	3.8	8.2	11.9
Ho Chi Minh	Mekong	2.8	10	26
Hanoi	Red	1.9	4.2	10
Guangzhou	Pearl	3.1	9.6	13
Shanghai	Changjiang	11.4	20	21.3
Tianjin	Huanghe	6.2	9.8	10.1
Cairo	Nile	6	16.9	24
Buenos Aires	Parana	10.9	14.2	15.5

Data Sources Pelling and Blackburn (2013) and Syvitski (2008)

As Blackburn and Pelling (2014) point out, "collisions" between coasts and cities are already having negative impacts on both the natural environment and the people who live in the cities. These negative consequences are expected to grow significantly over the coming decades (Duraiappah et al. 2015; UNU-IHDP 2015). The coastal environmental damages caused by extensive urban developments are manifest in numerous ways. At the lowest order, the shear weight of cities and extraction of ground water contribute to accelerated subsidence and hence to local increases in the rate of relative sea level rise (e.g., Allison et al. 2016). von Glasow et al. (2013) describe several other, less obvious impacts of cities on coasts. Urban heat islands along with the obstruction of airflow by buildings alter atmospheric circulation and may block sea-breeze/land-breeze ventilation. Runoff of nutrients and pollutants—including sewerage—directly impact estuarine, marine and wetlands ecosystems and often causes toxic algal blooms. The resulting eutrophication causes "dead zones" of low dissolved oxygen in adjacent waters. Aerosols and particles released into the local atmosphere can also affect the adjacent coastal ocean. According to von Glasow et al. (2013), the atmospheric effects can reach hundreds to thousands of kilometers beyond the cities while the estuarine and marine impacts can extend tens to hundreds of kilometers seaward or down the coast. Degradation of wetlands surrounding cities reduces natural protections against storms and increases urban vulnerability.

Natural processes, and human modifications of those processes, in turn, threaten the cities and their inhabitants. Unplanned urbanization is a major contributor to the increasing risk faced by coastal city dwellers (Blackburn and Pelling 2014). The International Human Dimensions Program on Global Environmental Change (UNU-IHDP 2015; Hallegatte et al. 2013) points out that by midcentury, the flood risk to large coastal cities will have increased by nine-fold relative to the present day. The property damage and socio-economic losses for all coastal cities combined is estimated to be in the neighborhood of US $1 trillion. It must be noted, however, that such estimates are based on damages to tangible property, such as expensive real estate, and on losses of income. In most cases the "victims" of these economic losses are wealthy or affluent corporations or individuals and are typically covered by insurance. The most tragic losses are usually less quantifiable in conventional economic terms and are often borne by low-income, and uninsured, residents. In his recent book, Ashley Dawson (2017) refers to the global convergence of climate change, urbanization and poverty as "extreme cities".

The elevations of coastal cities in general are low; but the lowest income people are commonly forced by competition with the more affluent residents to occupy the lowest and most flood-prone parts of the city. In some developing countries, the most vulnerable areas are the sites of slums (Dawson 2017). The human tragedies that unfold there during floods or extreme storms cannot be described in simple economic terms. Unfortunately, decisions about the investments of limited resources in flood mitigation structures etc. are often driven more by cost-benefit analyses than by compassion. Hanson et al. (2011) conducted a detailed analysis of predicted socio-economic and environmental changes between today and 2070. They conclude that by 2070 the monetary value of assets exposed to coastal inundation will be ten times greater than today while the exposed urban populations will increase three fold. However, the exposed assets will be primarily in more affluent western nations while the greatest increases in vulnerable populations will be in poorer Asian countries. For example, the present inundation-exposed populations of the cities of Calcutta and Mumbai, India are respectively 1.9 million and 2.8 million but are projected to be 14.0 million and 11.4 million by 2070.

As outlined by the IHDP (2015), the three main strategies for addressing future flood threats to coastal cities are: “retreat”, “resist” or “attack”. For small communities, retreat is usually the least costly and simplest option; it simply involves relocating to higher ground. But it is frequently unpopular because it means abandoning prior investments or traditional neighborhoods. For megacities, retreat is difficult to accept—but it was seriously considered after the 2011 flood of Bangkok. (For more information on the Bangkok floods of 2011 see InterNations online at URL: https://www.internations.org/bangkok-expats/guide/moving-to-bangkok-15519 and the Taylor (2011) article in *The Atlantic*.) To resist requires the construction of defensive structures such as levees, floodgates, pumps etc. Such structures directly alter surrounding environmental conditions such as estuarine circulation. Finally, the “attack” option is the most expensive because it involves the reclamation of lands surrounding the city. Hino et al. (2017) recently examined 27 cases where managed retreat from natural hazards was successful in relocating 1. 2 million people. However, these cases were relatively small scale and did not involve whole cities. In fact, those authors note that applications of the “retreat” model face serious challenges given the projected scale of climate-induced displacements and the many difficulties of resettlement. In its summary for decision makers, the UNU-IHDP (Duraiappah et al. 2015) recommends the following strategies for future adaptation: (1) seek greater understanding of what is happening; (2) consider managed retreat or relocation; (3) consider combined approaches that include engineering structures, natural enhancements such as the planting of mangroves, and improved early warning systems based on better forecasts of flood events. Dawson (2017) points out that, although retreat is typically regarded as the “last taboo”, it may be the only realistic option over the long term.

6.3 Agriculture and Its Impacts

Globally, the area of the earth’s surface covered by croplands is roughly equivalent to the size of the South American Continent and that occupied by animal agriculture is similar to the area of Africa (IGBP 2013; Nature Conservancy; Fig. 6.2). By the year 2050, the food requirements of the world will roughly double. Clearly, however, the fraction of the earth’s surface devoted to food production cannot also double without catastrophic environmental consequences: agriculture is a major contributor to climate change and animal agriculture is a major source of methane and consumer of water (Rust 1981; Crutzen et al. 1986). Most of the agriculture is not concentrated in the coastal zone but is farther inland. However, many of these inland farmlands are located within the catchments of major rivers, which flow to coasts. In addition to carrying water and sediments to the sea they also carry nutrients from farm fertilizers along with pesticides sprayed on crops. Pesticides can kill aquatic organisms and reduce biodiversity in coastal and estuarine waters and in wetlands.

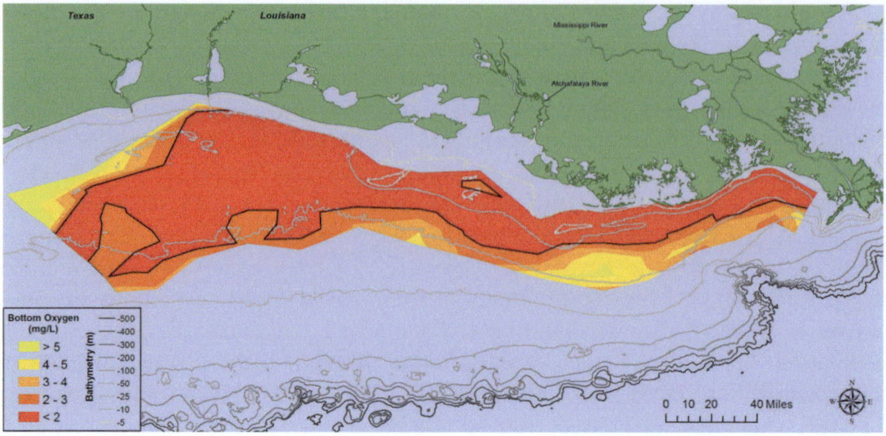

Fig. 6.2 Distribution of bottom-water dissolved oxygen, July 24–July 30, 2017. Black line denotes 2 mg l^{-1}. *Data source* N. N. Rabalais, Louisiana State University & Louisiana Universities Marine Consortium; R. E. Turner, Louisiana State University. Funding: National Oceanic and Atmospheric Administration, National Centers for Coastal Ocean Science

Nutrient over enrichment in coastal waters from non-point sources causes damaging stresses on aquatic, estuarine and marine ecosystems by fueling nuisance blooms of algae, bacteria and phytoplankton (Dale et al. 2010; Rabalais et al. 2007). Once the plant biomass sequesters the available nutrients, the plants die and fall to the seabed where they decay and, in that process, consume dissolved oxygen. This creates extensive near bottom zones of oxygen deficiency, or *hypoxia*, popularly referred to as "dead zones" because fish and other marine life cannot survive there. Diaz and Rosenberg (2008) report that there are at least 400 dead zones in coastal and estuarine waters around the world. The World Resources Institute provides an interactive map of eutrophication and hypoxia for more than 700 coastal areas worldwide (see World Resources Institute URL: https://www.wri.org/resources/data-sets/eutrophication-hypoxia-map-data-set). Perhaps the best known of these hypoxic zones is in the northern Gulf of Mexico where the Mississippi River discharges large amounts of agriculture-derived nutrients along with abundant fresh water, which stratifies the coastal ocean. Stratification suppresses vertical water column mixing and the downward flux of oxygenated water from the surface thereby exacerbating the hypoxic conditions. Figure 6.2 shows the extent of the hypoxic zone (in red) on the Louisiana continental shelf in the summer of 2014. Additional dissolved oxygen maps showing the frequency of hypoxia may be obtained online from the NOAA National Centers for Environmental Information (see URL: https://www.ncddc.noaa.gov/interactive-maps/coastal-habitats/). The Mississippi River catchment that is the source of nutrients covers 40% of the contiguous United States and extends northward into Canada. It embraces much of America's "breadbasket" including regions of corn agriculture for ethanol in Iowa. Gilbert and Burford (2017) recently presented results showing that harmful algal blooms (HABS) are increasing worldwide, in part because of increased nutrient loading to coastal waters and in part because HABS are rapidly adapting to global changes in ocean and coastal environments including altered levels of CO_2.

The conflicts between agriculture and coastal environments are highly complex and involve much more than nutrient runoff. As is the case for other coastal interactions, the cause-effect impacts are bidirectional. For example, croplands in low-lying coastal plains are subject to damage from salt-water inundation during storms or intrusion in ground water with rising sea level. Sugar cane farming in south Florida has had severe negative impacts on the highly sensitive Everglades because of nutrient runoff and also caused extreme blooms of toxic algae along Florida's Atlantic and Gulf Coasts in 2016. Sugar cane farming in Queensland Australia has also had damaging impacts on the Great Barrier Reef and solutions for this problem under active consideration include implementation of "polluter pays" laws (Queensland University of Technology 2015). In Europe, while it is well accepted that agriculture is essential, it is also understood that farming should proceed in ways that are environmentally sound and ethical. To that end, the European Union has established a *Common Agricultural Policy* (*CAP*), which requires farmers to meet certain environmental standards in order to be eligible for subsidies (e.g. Vernier 2012). In addition, in France, farmers are required to pay fees for consumption of irrigation water and for the runoff of pesticides (Vernier 2012).

The need to increase food production significantly over the next several decades cannot be denied and climate change is already negatively impacting food production (FAO 2016). To overcome these impacts and to reduce coastal damage, new, innovative and efficient agricultural practices that minimize environmental damage must be evolved. These innovations may well come at the expense of profits for large "agri-businesses" and will likely involve more emphasis on grains, vegetables and less on animal agriculture. Aquaculture will probably also play greater roles in feeding a growing population. On an optimistic note, it should be noted that an example of "*intelligent co-operation with nature*" is found in centuries old Balinese *Subak* practices of terraced rice agriculture. This system, which involves tightly interwoven cultural and environmental systems and dates back to the 9th Century, has recently been adopted by the UNESCO World Heritage Committee (2015). According to the World Heritage Committee (2015): "*The Subak system of democratic and egalitarian farming practices has enabled the Balinese to become the most prolific rice growers in the (Indonesian) archipelago despite the challenge of supporting a dense population.*"

6.4 Coastal Fisheries

Since prehistoric times, access to coastal fisheries has been a key advantage of living near the coast and seafood is still the dominant source of protein in many parts of the world. According to the UN Food and Agricultural Organization (FAO

2009, 2016), 3 billion people worldwide depend on seafood for essential parts of their nutrition and 500 million people in developing countries derive their livelihood from fisheries and aquaculture. Fish capture and aquaculture have increased steadily over the past 50 years although most of the increases since 1985 have taken place in China (FAO 2009, 2016). While the direct negative impacts of fisheries are less dramatic than some of the effects of agriculture, over fishing in particular is reducing species diversity and, of course, adversely impacts fisheries. The FAO (2016) reports that climate change threatens the sustainability of capture fisheries and aquaculture development in several ways. Ocean acidification and warming seas are slowing the ability of shellfish to produce shell through calcification, hypoxia in coastal waters limits the extent of fish habitats, more intense storms damage or destroy coastal shellfish farms and reductions in freshwater discharge by rivers have a direct effect on coastal salinity and thereby on the health of oysters.

The invasion of various tropical and alien species into higher latitude coastal waters often results in harmful competition with indigenous species. A prominent example is the ongoing migration of lionfish, which consume juveniles of other species, up the U.S. east coast. Warming seas can alter the reproductive cycles of fish, as well as growth rate and the timing of spawning (Perry et al. 2005). According to the recent synthesis prepared by the FAO (2016), models that take account of changing environmental conditions, and primary production of phytoplankton predict large-scale redistributions of marine fish catch characterized by increases in high-latitude regions and declines in the tropics. This will increase the vulnerability of small-scale tropical fisheries in poorer, less developed regions. To overcome some of these threats and remain resilient, the FAO (2016) recommends increasing the diversity of fisheries species and techniques and integrating more understanding of anticipated climate and coastal change into fisheries strategies and policies.

6.5 Coastal Oil and Gas Exploration and Production

Offshore and coastal plain oil and gas production directly impacts coastal systems around the world in numerous ways. Some of the regions affected include the Northern Gulf of Mexico, Mexican east coast, Arctic North Slope of North America; much of the Brazilian Coast; the West Coast of Africa; the North East Atlantic and North Sea Coasts of Europe, Australia's northwest coast; the Indonesian coast; and China's Bohai, Yellow Sea, East China Sea and South China Sea coasts. In addition to the obvious and highly publicized effects of oil spills, there are numerous other ways that the industry alters the coastal environment. For the case of the North Sea coast, the OSPAR Commission (2009) points out that: "*Environmental impacts may arise at all stages of oil and gas activities, including initial exploration, production and final decommissioning. There is a broad range of environmental concerns including those relating to oil discharges from routine operations, the use and discharge of chemicals, accidental spills, drill cuttings, atmospheric emissions, low level naturally occurring radioactive material, noise, and to some extent the placement of installations and pipelines on the sea bed*" (OSPAR 2009, p. 8). Direct and sometimes dramatic alterations of the coastal systems are caused by the industry's infrastructure including dredged access canals, pipelines, port facilities and, of course, the wells and platforms themselves.

Perhaps nowhere is the conflict between hydrocarbon production and nature more palpable than in the Northern Gulf of Mexico where 40% of the US wetlands coincide with activities that supply roughly one third of the nation's oil and gas (Mendelssohn et al. 2012). The Gulf of Mexico continental margin is a mature offshore oil and gas production area generating more than 1.7 million of barrels of oil per day, through more than 3500 oil platforms. Currently the northern Gulf has 45,000+ km (28,000+ mile) of underwater pipes exposed to structural damage from extreme oceanic and atmospheric events. In 2010, British Petroleum's *Deepwater Horizon* production platform in the northern Gulf of Mexico exploded, killing 11 people. This event was followed by an uncontrolled blowout that persisted for 87 days during which 4.9 million barrels of crude oil were discharged directly into the sea. The wetlands of the Mississippi delta and the Louisiana coast, which account for a third of the US fish production, were affected as toxic oil slicks spread over large areas of coastal marshland (Mendelssohn et al. 2012). Fortunately, the oil did not penetrate very far into the interior of the marshes. However, the adverse effects on the benthic habitats of the continental shelf are becoming more apparent with time. As is typical of processes in the *Anthropocene*, causal connections between humans and nature in the Mississippi Delta are bidirectional: the oil industry's $100 billion dollar investment in infrastructure is now under serious threat from land sinking and relative sea level rise (Traywick 2017).

The National Petroleum Reserve-Alaska, which includes the large Prudhoe Bay oil field discovered in the 1960s, lies immediately south of the coasts of the Beaufort and Chukchi Seas on Alaska's North Slope, an expansive coastal plain covered by permafrost. To the east of this reserve near the Canadian border is the Arctic National Wildlife Reserve. A summary report released in 2003 by the National Research Council (2003) assessed the cumulative effects of oil production from the 1960s to 2001 on the coastal environment. As of the date of that report, no significant oil spill had occurred. However, cumulative damage had resulted from the extensive infrastructure including roads, raised pipelines and helipads, which have largely remained since their original construction. Tundra damage from off-road vehicle traffic was also reported. More recently, NOAA's National Marine Fisheries Service (NOAA-NMFS 2016) carried out an extensive environmental impact study for alternative future scenarios of future oil and gas exploration utilizing seismic surveys and exploratory drilling. Among numerous minor to moderate impacts, the study concludes that acoustic impacts on beluga and bowhead whales would be significant and that, although unlikely, a large oil spill caused by an exploratory drilling accident could result in widespread environmental damage. Temporary disruption to wildlife is also caused by exploratory drilling installations such as Mars Island, which was built completely from spray ice for exploratory drilling by Amoco Exploration in January and February of 1986 in western

Fig. 6.3 Mars Ice Island, Beaufort Sea Alaska. Image shows a 60-day exploratory well built 8 km (5 mi) offshore of Cape Halkut near the National Petroleum Reserve in Alaska (NPRA), an area of land on the Alaska North Slope owned by the United States federal government and managed by the Department of the Interior, Bureau of Land Management (BLM). (Photo courtesy of Bureau of Ocean Energy Management)

Harrison Bay in Alaska. The island was built on the land-fast first-year ice in a water depth of 7.6 m (25 ft) to provide grounded platforms in shallow water. Mars Island was approximately 8 m (26 ft) thick and 213 m (700 ft) in diameter and used to support a drill rig as shown in Fig. 6.3.

6.6 Tourism Impacts

Sandy beaches and beach resorts are among the world's most popular tourist destinations. The throngs of tourists that frequent these sites each year are commonly accommodated in the hotels, condominiums and resorts that comprise beach-front "strip cities". Some of the impacts that these complexes have on coastal environments were described in Sect. 6.2. But, as described in detail by Davenport and Davenport (2006), there are myriad other impacts and some of these are long lasting. Coastal tourism began in the 19th century and has accelerated since then. Today, 63% of Europeans choose coastal sites for their vacations. The threats are from mass tourism and transport. According to Davenport and Davenport (2006) "Tourism is now the largest single economic sector in the World. Impacts of leisure transport and tourism on the coastal environment are considerable, have increased (and are currently scheduled to continue increasing) in non-linear fashion, and are extremely difficult to manage or limit". One of the most problematic factors involves mass tourism and extensive use automobiles for transport. Pollution and runoff from parking lots cause damage. Boats, cruise liners and various types of personal water-craft (e.g., jet skis) damage reefs and grass beds and wakes cause beach erosion (e.g., Parnell and Koefoed-Hansen 2001; Soomere and Rannat 2003; Bilkovic et al. 2017). Trampling of intertidal habitat and coral reef damage inadvertently inflicted by scuba divers are also harmful. Fortunately, although highly sensitive environments such as marshes and mangrove swamps are visited by small numbers of low impact eco-tourists, those environments do not suffer from mass tourism.

6.7 Ports, Engineering and Infrastructure

Roughly two thirds of the Netherlands surface area is low lying and would be inundated on each high tide if it were not for the extensive system of artificial dikes that have protected the country for nearly a millennium. An outcome of this is that the Netherlands has been world leader in coastal engineering for many decades. Today, the Netherlands has 350 km of coastline, 27% of the country is below sea level and 55% of it is flood prone (Santinelli 2016). Protection is provided by an elaborate system of dunes, dikes, floodgates and frequent land reclamation by onshore pumping of shelf sediments. The Netherlands model is commonly invoked by coastal managers all over the world as an example to be explored as a potential solution to coastal flooding and erosion. Such considerations are under serious review for protecting lower Manhattan. Elsewhere on the U.S. coast and around the world, the variety of coastal protection structures include seawalls, jetties and armored airport runway extensions.

The levee system on the lower Mississippi River and delta has long provided protection from flooding to Louisiana's lowlands and to the city of New Orleans but has also deprived the delta of much needed sediment re-nourishment. The history of this system and the roles, some of which were tragic, that it played during the Mississippi River flood of 1927 are described in eloquent detail by Barry (1997). Levee failure during the storm surge produced by Hurricane Katrina in 2005 was a major cause of the deadly flooding of the Ninth Ward and other parts of New Orleans. The impact of the levee system on the Mississippi Delta was understood as far back as the 19th century. E.L. Corthell wrote the following in 1897: "*The conditions are very different now from those existing prior to the construction of levees. There are at present no annual accretions of sedimentary matters from the periodical overflows of the river. These accretions formerly were a little more than equal to the annual subsidence of the lands...No doubt the great benefit to the present and two or three following generations accruing from a complete system of absolutely protective levees, excluding the waters entirely from the great areas of the lower delta country, far outweighs the disadvantages to future generations from the subsidence of the Gulf delta lands below the level of the sea and their gradual abandonment due to this cause.*"

Ports and harbors around the world involve significant human alterations of bays, estuaries and the lower courses of rivers by combinations of levee and breakwater construction, hardened shores, cargo loading facilities, rail and heavy truck access, warehouses and oil storage tanks, pipeline terminals, cruise liner terminals and channel deepening by dredging. Wakes from boat and ship traffic and spills of oil and gas are common disruptions. Improper port design can allow harbor resonance, which can cause excessively high oscillations and overtopping during storms. The recurring need for maintenance dredging in many ports has significant impacts on benthic habitats and can also lead to erosion of the adjacent coasts and beaches. Healy et al. (2002) report an example of a New Zealand port that was redesigned with guidance from numerical models to accommodate anticipated demands for ship traffic while minimizing ecological impacts and allowing dredged material to be used to nourish adjacent beaches. Figure 6.4 shows an example of a modern, multi-purpose port complex, Port Everglades, in southeastern Florida. Facilities such as Port Everglades are in increased jeopardy of damage or closure from storms caused by structural damage or entrance shoaling. On a much larger scale, it is the world's major port cities that are most acutely threatened by future inundation (Hanson et al. 2011). These threats will be considered in more detail in cases described in Part 3.

Fig. 6.4 Port Everglades, Broward County Florida (near Ft. Lauderdale). Photo courtesy of, Broward County, FL Environmental Planning and Community Resilience Division

References

Allison, M., B. Yuill, T. Törnqvist, F. Amelung, T.H. Dixon, G. Erkens, R. Stuurman, C. Jones, G. Milne, M. Steckler, J. Syvitski, and P. Teatini. 2016. Global risks and research priorities for coastal subsidence. *EOS* 97 (19): 22–27.

Barry, J.M. 1997. *Rising tide: The great mississippi flood of 1927 and how it changed America*, 524. New York: Touchstone.

Bilkovic, D., M. Mitchell, J. Davis, E. Andrews, A. King, P. Mason, J. Herman, N. Tahvildari, and J. Davis. 2017. *Review of boat wake wave impacts on shoreline erosion and potential solutions for the Chesapeake Bay*. The Scientific and Technical Advisory Committee (STAC) Publication Number 17-002, 68. MD: Edgewater.

Blackburn, S., and M. Pelling. 2014. Coastal megacities: Risks and opportunities. *IGBP Global Change Magazine* 82: 8–11.

Burnett, John. 2015, Aug 28. Billions Spent on flood barriers, but New Orleans still a 'fishbowl'. *NPR*. Available online. http://www.npr.org/2015/08/28/432059261/billions-spent-on-flood-barriers-but-new-orleans-still-a-fishbowl. Accessed on 25 Sept 2017.

Corthell, E.L. 1897. The delta of the Mississippi River. *National Geographic Magazine*, December 1897.

Crutzen, P.J., I. Aselmann, and W. Seiler. 1986. Methane production by domestic animals, wild ruminants, other herbiverous fauna, and humans. *Tellus* 38B: 271–284.

Dale, V.H., and 23 members of the U.S. EPA Hypoxia Advisory Panel. 2010. *Hypoxia in the northern Gulf of Mexico*, 284. New York: Springer.

Davenport, J., and J.L. Davenport. 2006. The impact of tourism and personal leisure transport on coastal environments: A review *Estuarine*. *Coastal and Shelf Science* 67: 280–292.

Dawson, A. 2017. *Extreme cities: The perils and promise of urban life in the age of climate change*, 378. New York: Verso.

Diaz, R.J., and R. Rosenberg. 2008. Spreading dead zones and consequences for marine ecosystems. *Science* 321 (5891): 926–929. https://doi.org/10.1126/science.1156401. PMID 18703733.

Duraiappah, A.K., C. Scherkenbach., B. Solich, et al. 2015. *Coastal zones and urbanization: Summary for decision makers*, Bonn: UN University—International Human Dimensions Programme (UNU-IHDP). ISBN 978-3-44-944535-32-6.

Food and Agricultural Organization of the United Nations (FAO). 2009. *The state of world fisheries and aquaculture*, 84, Rome: FAO Fisheries and Aquaculture Department. ISSN 1020-5489.

Food and Agricultural Organization of the United Nations (FAO). 2016. Climate change and food security: Risks and responses, 98. Rome: Fisheries and Aquaculture Department. ISBN 978-92-5-108998-9.

Fisheries and Aquaculture and Climate Change. 2016. *FAO fisheries and aquaculture department*. Rome. Available online: http://www.fao.org/fishery/climatechange/en. Updated 2 Nov 2016.

Gilbert, P., and M. Burford. 2017. Globally changing nutrient loads and harmful algal blooms: Recent advances, new paradigms and continuing challenges. *Oceanography* 30: 58–69.

Hallegatte, S., C. Green., R.J. Nicholls., and J. Corfee-Morlot. (2013). Future Flood losses in major coastal cities. *Nature Climate Change* 3: 802–806. Available at: http://www.nature.com/nclimate/journal/v3/n9/full/nclimate1979.html.

Hanson, Susan, Robert Nicholls, N. Ranger, S. Hallegatte, J. Corfee-Morlot, C. Herweijer, and J. Chateau. 2011. A global ranking of port cities with high exposure to climate extremes. *Climatic Change* 104: 89–111.

Healy, T., S. Stephens, K. Black, R. Gorman, R. Cole, and B. Beamsley. 2002. Port redesign and planned beach renourishment in a high wave energy sandy-muddy coastal environment, Port Gisborne, New Zealand. *Geomorphology* 48: 163–177.

Hino, M., C.B. Field, and K.J. Mach. 2017. Managed retreat as a response to natural hazard risk. *Nature Climate Change*. https://doi.org/10.1038/NCLIMATE3252.
International Geosphere Biosphere Programme (IGBP). 2012. *Anthropocene: An epoch of our making (online newsletter)* http://www.igbp.net/news/features/features/anthropoceneanepochofourmaking.5.1081640c135c7c04eb480001082.html.
International Geosphere Biosphere Programme (IGBP). 2013. *Water in the Anthropocene (online multimedia video).*
Mendelssohn, I., et al. 2012. Oil impacts on coastal wetlands: Implications for the mississippi river delta ecosystem after the deepwater horizon oil spill. *BioScience* 62: 562–574. ISSN 0006-3568.
National Research Council. 2003. Cumulative effects of oil and gas activities on Alaska's North Slope, 4. Report in Brief National Academies.
NOAA National Marine Fisheries Service. 2016. *Environmental impact statement (EIS) on the effects of oil and gas activities in the arctic ocean, Final EIS, Oct. 2016.* http://www.nmfs.noaa.gov/pr/permits/eis/arctic.htm.
OSPAR Commission. 2009. Assessment of impacts of offshore oil and gas activities in the North-East Atlantic London, OSPAR, 39. Publication Number: 453/2009. ISBN 978-1-906840-93-8.
Overeem, I., and J.P.M. Syvitski. 2009. Dynamics and vulnerability of delta systems. In: *LOICZ reports & studies no. 35*, 54. Geesthacht: GKSS Research Center.
Parnell, K.E., and H. Koefoed-Hansen. 2001. Wakes from large high-speed ferries in confined coastal waters: Management approaches with examples from New Zealand and Denmark. *Coastal Management* 29 (3): 217–237. https://doi.org/10.1080/08920750152102044.
Pelling, M., and S. Blackburn (eds). 2013. *Megacities and the coast: Risk, resilience and transformation*, 248. London, New York: Routledge. www.routledge.com/books/details/9780415815123/.
Perry, A.L., P.J. Low, J.R. Ellis, and J.D. Reynolds. 2005. Climate change and distribution shifts in marine fishes. *Science* 308 (5730): 1912–1915.
Queensland University of Technology. 2015. *Managing the impacts of sugarcane farming on the great barrier reef: An evaluation of the implementation of the polluter pays principle.* Final Report, 42.
Rabalais, N.N., R.E. Turner, B.K. Sen Gupta, D.F. Boesch, P. Chapman, and M.C. Murrel. 2007. Hypoxia in the northern Gulf of Mexico: Does the science support the plan to reduce, mitigate, and control hypoxia? *Estuaries and Coasts* 30 (5): 753–772. https://doi.org/10.1007/BF02841332.
Rust, F. 1981. Ruminant methane delta ($^{13}C/^{12}C$) values: Relation to atmospheric methane. *Science* 211 (4486): 1044–1046. https://doi.org/10.1126/science.7466376.
Santinelli, G. 2016. *Coastal protection in The Netherlands Tech Report*, 14, Deltares.
Soomere, Tarmo, and Kalev Rannat. 2003. An experimental study of wind waves and ship wakes in Tallinn Bay. *Proceedings of the Estonian Academy of Science and Engineering* 9 (3): 157–184.
Syvitski, J.P.M. 2008. Deltas at risk. *Sustainability Science* 3: 23–32.
Taylor, Alan. 2011, Oct 12. Worst flooding in decades swamps Thailand. *The Atlantic*. Available online: http://www.theatlantic.com/infocus/2011/10/worst-flooding-in-decadesswamps-thailand/100168/.
Traywick, C. 2017. Louisiana's sinking coast is a $100 billion nightmare for big oil. *Bloomberg News*.
UNESCO. 2015. *Convention concerning the protection of the world cultural and natural heritage*, 254. Thirty-ninth session report WHC-15/39.COM/19. Bonn, Germany: United Nations Educational, Scientific and Cultural Organization, World Heritage Committee.
UNU-IHDP. 2015. *Coastal zones and urbanization: Summary for decision-makers*, 19. Bonn, Germany: Secretariat of the International Human Dimensions Programme on Global Environmental Change (UNU-IHDP).

Vernier, F. 2012. *Agriculture in coastal areas: Environmental issues, impacts and regulation tools*, 19. France: Cemagref.

von Glasow, R., T.D. Jickells, A. Baklanov, G. Carmichael, Church, T.L. Gallardo, C. Hughes, M. Kanakidou, P. Liss, L. Mee, R. Raine, P. Ramachandran, R. Ramesh, K. Sundseth, U. Tsunogai, M. Uematsu, and T. Zhu. 2013. Megacities and large urban agglomerations in the coastal zone: Interactions between atmosphere, land, and marine ecosystems. *Ambio* 42: 13–28. https://doi.org/10.1007/s13280-012-034.

Part II
Causal Processes, Their Consequences and Their Mitigation

Chapter 7
Causes and Impacts of Coastal Inundation

Lynn Donelson Wright, Donald T. Resio and C. Reid Nichols

> *You never really know what's coming. A small wave, or maybe a big one. All you can really do is hope that when it comes, you can surf over it, instead of drown in its monstrosity.*
>
> — Alysha Speer

7.1 Rising Damp: Many Causes

According to NOAA's Office of Coastal Management, inundation events are the dominant causes of natural-hazard-related deaths in the U.S. and are also the most frequent and costly of the natural hazards affecting the nation. The effects of inundation in other nations such as Bangladesh, Indonesia, Thailand and India are often devastating. While the long-term rises in mean sea level as discussed in Part 1, Chap. 3 are instrumental in allowing inundation to reach farther inland and to ever higher elevations, it is the short-lived episodic, non-tidal, events that cause the most damage. Included among these inundation causes are tsunamis, storm surges, coastal flooding caused by onshore winds and wave-induced set up, river and inland flooding and extreme rainfall events. The deepest flooding occurs when two or more of these phenomena reinforce each other and coincide with perigean spring high tides (aka. "king tides"). For example, it is common for tropical cyclones to

L. D. Wright (✉)
Southeastern Universities Research Association (SURA), Washington, DC, USA
e-mail: ldwright@bellsouth.net

D. T. Resio
School of Engineering, University of North Florida, Jacksonville, FL, USA
e-mail: don.resio@unf.edu

C. R. Nichols
Marine Information Resources Corporation (MIRC), Ellicott City, MD, USA
e-mail: rnichols@mirc-us.com

L. D. Wright and C. R. Nichols (eds.), *Tomorrow's Coasts: Complex and Impermanent*, Coastal Research Library 27,
https://doi.org/10.1007/978-3-319-75453-6_7

bring storm surges along with heavy rainfall and wave induced set up of mean water level. Structures such as levees that are designed to protect can also impede the return of floodwaters once overtopped. Navigation channels can also provide funneling pathways for surges.

In addition to extensive loss of life, there are many other negative consequences of coastal inundation. Some of the more obvious impacts include destruction of houses and buildings and damage to built infrastructure such as roads and bridges. Less visible are mold infestations inside homes that are flooded but not destroyed. Floodwaters can be contaminated with sewage and lingering inundation can spread diseases throughout neighborhoods and contaminate drinking water. Salt water can penetrate into aquifers and ground water and high salinities can kill vegetation and degrade wetlands. In some cases, severely inundated communities may be forced to relocate. Details of the human health impacts of coastal inundation are presented in Chap. 10.

7.2 Tsunamis

Sea floor seismic events such as earthquakes, volcanic activity and underwater landslides that cause sudden and large displacements of the overlying water column are the cause of tsunamis (previously called tidal waves). The vertically displaced water creates a long gravity wave that then travels away from the area of disturbance at speeds that are proportional to the square root of the water depth. In the deep ocean these speeds may exceed 500 mi/h (800 km/h). The wave is not a single wave but a train of multiple succeeding waves separated by distances (wave lengths) of 300–400 km (190–250 mi) and time intervals (periods) of 30–40 min. In the deep ocean, the heights of the tsunamis are typically on the order of 1 m or less (<3.3 ft) and so this rise over the large horizontal distance (wave length) and time interval (period) makes these waves imperceptible to ships at sea. However, as the tsunami propagates into the shallow waters over the continental shelf, their speeds and wavelengths decrease dramatically while their periods remain constant. The energy that was previously distributed over a very large horizontal distance now becomes more concentrated and the tsunami undergoes what is generally referred to as a shoaling amplification that causes the tsunami height to increase many times relative to its deep ocean height.

The Asian tsunami of December 26, 2004 (NOAA/NGDC 2014; Pomonis et al. 2006; Titov et al. 2005), that killed 230,000 people and caused $10 billion (USD) in damages originated from a seafloor seismic disturbance at a depth of ~4000 m (13,000 ft), had a deep sea amplitude of 70 cm (2.3 ft) and initially travelled away from the epicenter at a speed of 720 km/h (450 mi/h). It had a period of 34 min and a deep-water wavelength of 410 km (256 mi). By the time it reached nearshore depths of 5 m (17 ft) its length would have decreased to about 14 km (8.75 mi).

Because of the complex configuration of the shelf and coast around the area of greatest devastation in Indonesia and Thailand the height of the "wall of water" that made landfall varied considerably but exceeded 10 m (33 ft) in places of greatest impact. The magnitude 9 Tohoku earthquake (Dunbar et al. 2011) that occurred east of Honshu Island Japan on March 11, 2011 generated a 15 m (50 ft) tsunami that devastated Fukushima Japan, killed 19,000 people and inundated a nuclear power plant. Since the epicenter was adjacent to the coast, the arrival of the tsunami was immediate with minimal warning. Following the impact on Japan, the wave crossed the Pacific Ocean at speeds proportional to the square root of the abyssal water depths as shown in Fig. 7.1. Note from Fig. 7.1 that the tsunami reached southern California about 10 h after the earthquake, roughly about the same as the time required for a jet liner to make the same trip.

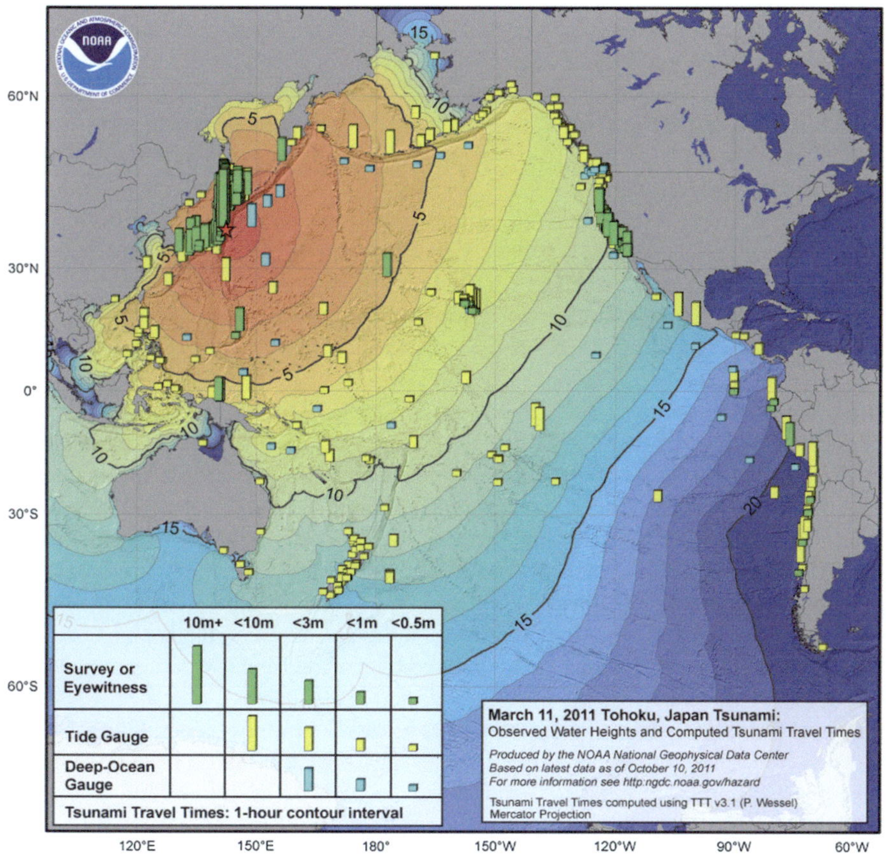

Fig. 7.1 Observed heights and travel times of the Tohoku Tsunami that devastated Fukushima Japan in March, 2011. Graphic from the NOAA National Geophysical Data Center

7.3 Storm Surges and Storm Tides

Storm surges are elevated water levels near the coast primarily caused by severe atmospheric disturbances such as tropical storms and hurricanes as well as less severe storms such as mid latitude "northeasters", all of which involve strong winds. Dean and Dalrymple (2002) describe four major components that jointly contribute to storm surges. The most important of these factors is wind stress (or frictional force) of the wind blowing over the sea surface and pushing the water shoreward. Added to this is the secondary "inverse barometer" effect whereby low barometric pressure causes a rise in the water level below (or "barometric tide"; Dean and Dalrymple 2002). Typically, this effect accounts for only about 5% of the total surge height. A third component is the Coriolis force related to the earth's rotation, which, causes water to be deflected to the right in the northern hemisphere and to the left in the southern hemisphere. In the northern hemisphere, this results in south-flowing currents on east coasts and north-flowing currents on west coasts (such the Florida Gulf Coast) causing rises in water levels at the shore and near shore. The fourth component is a water level "setup" caused by the dissipation of breaking waves (surf) and is often equal to 10% or more of the breaker heights themselves. The storm waves are then superimposed on this surge and can reach much farther inland than would be the case without the surge. *Storm tides*, as distinct from storm surges, are the water levels that result from the addition of normal astronomical tide levels and the superimposed storm surges. The total impact of a storm surge coinciding with a high tide, particularly a perigean spring tide ("King tide") will be much greater than if the predicted tide is low at the time the storm surge reaches its maximum.

The magnitudes of the four main components of storm surges as introduced above provide an approximate basis for quantifying the heights of storm surges at the coast. Unike tsunamis- and tides- storm surges are not freely propagating waves emanating from a source off the coast. Instead, these waves are locally forced (Resio and Westerink 2008). Consequently, surges are not strongly affected by processes such as refraction and shoaling. As shown by Resio and Westerink (2008), primary factors affecting surges depend strongly on the wind speed, the width and slope of the continental shelf and on the configuration and irregularity of the shores, bays and crenulations that can amplify or diminish the surge. Given the same forcing, the largest surges are generated over wide, shallow and gently sloping shelves and the smallest would be generated over narrow, steep shelves. For example, the storm surge that Category-5 Hurricane Andrew generated in Miami, which is fronted by a steep and very narrow shelf, was less than 2 m (6.6 ft) high in most places although it reached 5 m (16.4 ft) in parts of Biscayne Bay (Rappaport 1993). In contrast, an 8.5 m (28 ft) surge was generated by the weaker, but much larger Hurricane Katrina over the shallow shelf of the northern Gulf of Mexico in 2005. Some of the greatest amplification of surges can be affected significantly by alongshore processes when the surge "wave" propagates along the coast at roughly the same speed as the surge-generating storm advances. It is primarily because of the wide, shallow shelves that front the U.S. Atlantic and Gulf coasts that storm surges are much more consequential there than they are on the Pacific Coast. In addition to surge amplification by propagation across wide continental shelves, other amplification processes may operate. Among these, are amplification of surges travelling up funnel-shaped bays or estuaries and resonant seiching within some harbors or bays.

The operational storm surge prediction model used by NOAA's National Weather Service (NWS) is the 2-dimensioal *Sea, Lake and Overland Surges from Hurricanes (SLOSH)* model (Jelesnianski et al. 1992). Even though there are numerous more sophisticated and accurate models available and used by academia and other federal agencies (described in Chap. 1), SLOSH continues to be preferred by the NWS because of its speed of execution and low computational requirements and because forecasters are familiar with it. For purposes of long-range emergency planning, assessing evacuation routes and assessing relative risk of flooding of specific localities, NOAA's National Hurricane Center runs SLOSH thousands of times for different hypothetical hurricanes to generate vulnerability maps of worse case scenarios of flooding. These maps portray the quantities *Maximum Envelopes of Water (MEOW)* and the *Maximum of MEOWs (MOM).* Figure 7.2 shows a map of MOMs for hypothetical Category 3 hurricanes hitting the southern part of the Florida peninsula. Note that predicted inundation maxima along the open coast, neglecting wave setup, are fairly minimal for the Miami area and southeast Florida where the shelf is very narrow and steep but exceeds 2.7 m (9 ft) on the western side where a wide, low gradient shelf prevails (see Fig. 14.2 in Chap. 14). It should be noted that a Category 3 hurricane is not really extreme over decadal time frames. Consequently, the surge heights predicted by FEMA to have a statistical 100-year return interval are larger than those shown in Fig. 7.2. Although the predicted surge heights on the Atlantic side are low, flooding by wave-induced setup and runup may be greater but surf effects are not included in these analyses.

An interesting comparison between Hurricanes Camille in 1969 and Katrina in 2005 shows the inherent importance of hurricane size on storm surges at the coast. Camille was a Saffir-Simpson category 5 hurricane when it struck the Mississippi coast, while Katrina was only a category 3 hurricane when it reached this coast. Because of this, many people underestimated the potential danger of Katrina and chose not to evacuate. As noted by Irish et al. (2008), an article on the rising death toll in Hurricane Katrina found in Biloxi, Mississippi's Sun Herald (Norman 2006), "an oft heard refrain … is Hurricane Camille killed more people in 2005 than it did in 1969. Many officials and locals believed those … who had survived what was then the strongest recorded hurricane were lulled into a false sense of security that kept them in harm's way."

Even today, many people still echo the sentiment that it would have been much worse if a Saffir–Simpson category 5 storm had struck this area rather than Katrina. Irish et al. (2008) developed evidence showing that the Saffir–Simpson scale is not a particularly good indicator of storm surge along the coast and that storm size, along with bottom slope, is also a critical factor in the generation of large coastal surges. Figure 7.3 shows a plot of the surge contours for an idealized 1:10,000

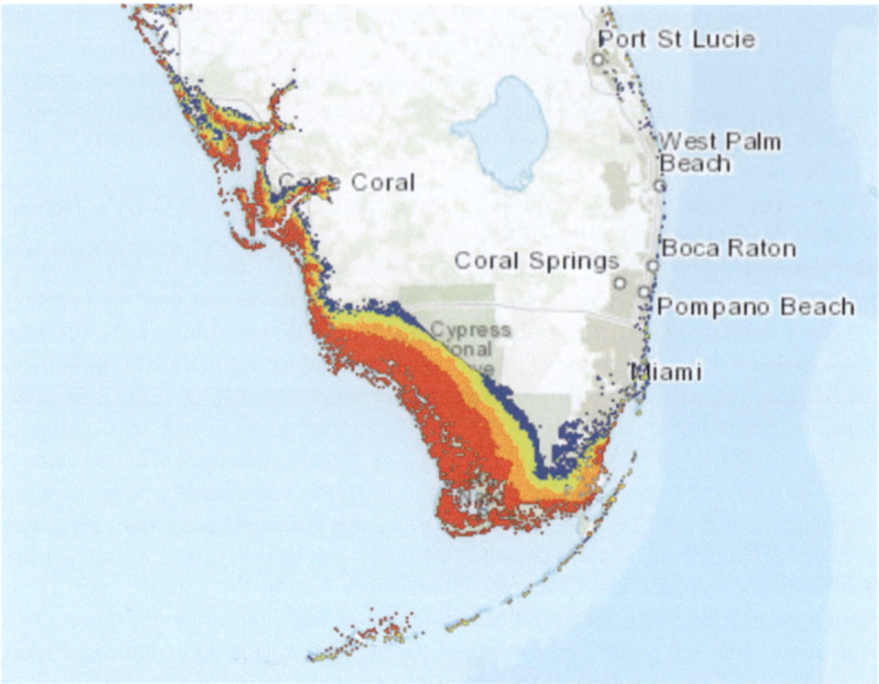

Fig. 7.2 Predicted maxima of maximum envelope of water (MOMs) for a category 3 Hurricane affecting south Florida. Red=greater than 9 ft (2.7 m) above ground; Orange=greater than 6 ft (1.8 m) above ground; yellow=greater than 3 ft (0.9 m) above ground; blue= less than 3 ft (0.9 m) above ground *Source* NOAA/NWS/NHC/Storm Surge Unit, NOAA/NOS/Office for Coastal Management. http://noaa.maps.arcgis.com/apps/MapSeries/index.html?appid=d9ed7904dbec441a9c4dd7b277935fad&entry=1

slope, generated by computer simulations with the ADICRC code. Different symbols superimposed on the contours represent estimates of potential surge heights for historical hurricanes in the Gulf of Mexico. As can be seen in this figure, and consistent with actual surge heights, the surge generated by Katrina was significantly larger than that of Camille. In simple terms, larger storms generate higher surges.

7.4 Wave-Induced Setup, Runup and Overtopping

In the foregoing discussion on storm surge, we noted that, according to Dean and Dalrymple (2002) one of the four components of storm surge, as manifest at the shore is *wave setup* (Stockton et al. 2005; FEMA 2015; Fig. 7.4). In fact, this setup represents the transfer of momentum from the wave field into the mean current flow, which occurs when waves break and dissipate. The setup elevation can be

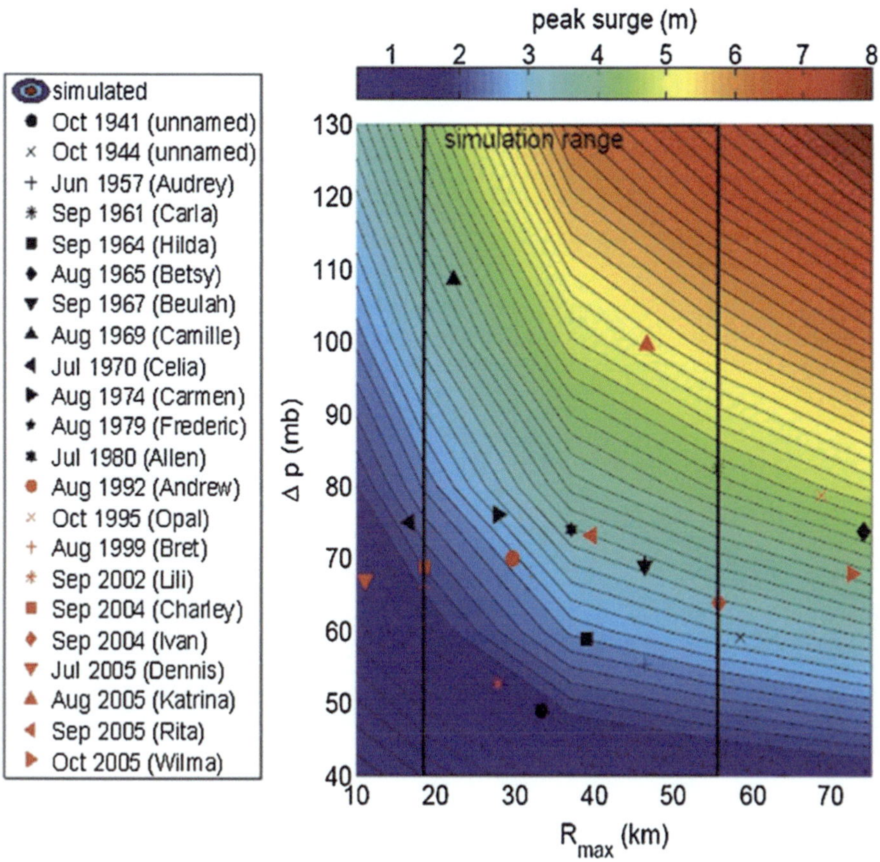

Fig. 7.3 Simulated peak surge as a function of hurricane size (denoted here by the radius to maximum winds, Rmax) and intensity (taken as the pressure differential from the center of the storm to its periphery, Δp) for a 1:10,000 bottom slope. Historical size and central pressure observations are superimposed on the numerical results to indicate the potential peak surge potential of historical storms made landfall in a region characterized by a bottom slope of 1:10,000, which is similar to the Mississippi Gulf Coast

10%- 20% of the height of the waves breaking in nearshore areas (FEMA 2015). Because breaking waves commonly arrive in varying "packets" of high and low waves, the setup height is not constant but usually forces "infragravity" periods of 50–100 s, traditionally called "surf beat" along many swell-dominated coasts (Guza and Thornton 1985). After transiting the surf zone, waves that have not been fully dissipated then surge up and down a beach or coastal engineering structure as *wave runup* the height of which increases with wave height and length and with beach slope (Fig. 7.5; Mignone and Maine 2016; Jones et al. 2005; Senechal et al. 2011). *Overtopping* occurs when the height of runup exceeds the height of the beach berm,

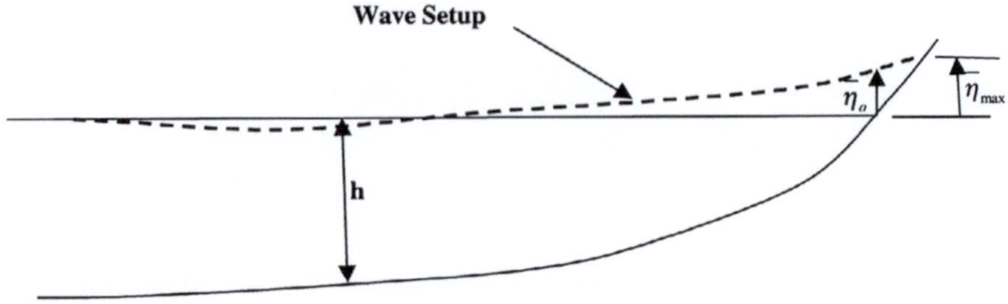

Fig. 7.4 Wave setup is a rise in the mean water level within the surf zone and on the beach caused by the dissipation that accompanies wave breaking. Diagram is from FEMA (2015), Guidance document 44 for the simple case of monochromatic waves

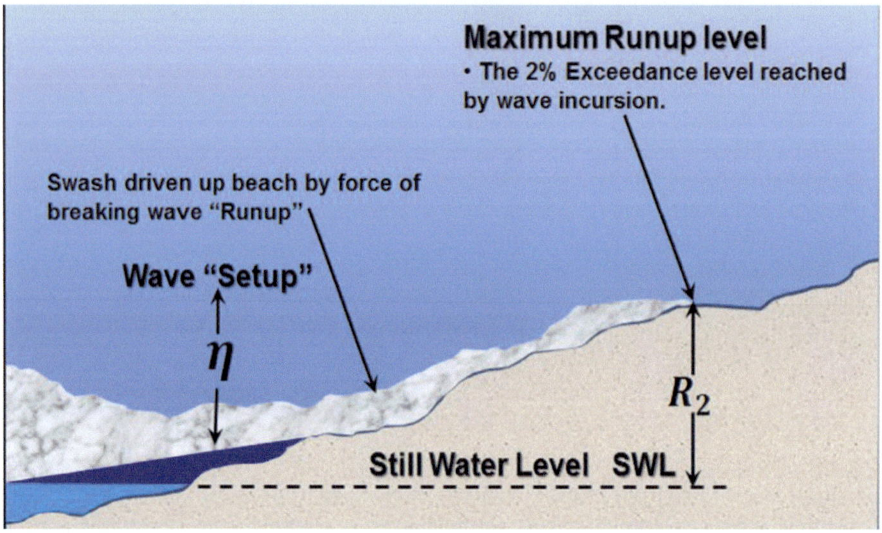

Fig. 7.5 Wave runup is superimposed on wave setup and reaches higher up the beach than setup. Figure from Mignone (2016), NOAA, National Weather Service, based on the model of Stockton et al. (2005)

protective dune or seawall and can result in flooded roadways or land surfaces (Fig. 7.6).

While the largest storm surges occur on coasts fronted by wide, gently sloping shelves, the heights and impacts of wave setup, runup and overtopping are greater on the more exposed shores fronted by steep and narrow shelves which favor high breakers because of less attenuation. It is in fact fortunate that the regimes that favor the highest surf do not also favor the highest storm surges. As is indicated by Fig. 7.2, the Atlantic coast of south Florida is much less vulnerable to storm surge than is Florida's Gulf Coast but the surf-related inundation is much greater for Miami-Dade and Broward Counties. And this was the case for the entire Atlantic

Fig. 7.6 Splash-over and wave overtopping at Crescent Beach in the Town of Hull, MA during the Patriot's Day Storm of 2007. High winds, heavy rainfall, and high tides during this nor'easter caused flooding, storm damages, power outages, evacuations, and disrupted traffic and commerce (photo from Applied Coastal Research and Engineering, Inc. 2015)

seaboard as Hurricane/Superstorm Sandy moved northward in 2012. With specific reference to Sandy, Mignone and Maine (2016) reports that: "NWS (National Weather Service) lacks sufficient forecast guidance on inundation associated with wave runup and coastal rivers making it difficult to forecast impacts from coastal storms". This points to the urgent need to couple storm surge, nearshore (surf zone) and hydrology models for future forecasting capabilities.

7.5 Minor, but Frequent, Coastal Flooding (a.k.a. "Nuisance Flooding")

Severe events such as tsunamis and tropical cyclones are the causes of the most dangerous and damaging inundation. However, important problems in many areas are associated with much smaller, much more frequent events, which can produce recurrent and problematic flooding even under fair weather conditions. Rising sea levels and subsiding land levels make their presence felt in troublesome ways during times of high, but normal, astronomical tides. On coasts where land surfaces rise abruptly with distance from the sea, these effects may be minimal. But in other cases where land elevations rise slowly or where urban and suburban areas are near or even below mean sea level, high tides are increasingly accompanied by shallow, but sometimes paralyzing, flooding of streets, neighborhoods and occasionally the ground floors of homes and buildings. NOAA (Sweet et al. 2014) refers to these situations as "nuisance flooding". The threshold for an excess water level rise to be a nuisance varies with location depending primarily on surface elevation and secondarily on several other factors including socioeconomic vulnerability, and public infrastructure including roads and storm water drainage.

As we might expect, given local topography, in the U.S. the threshold is fairly high on the generally higher and steeper west coast and lower (more troublesome) on the Atlantic and Gulf coasts (Sweet et al. 2014). The Atlantic and Gulf coasts are low lying and, for the most part, experiencing significant subsidence. On the U.S. Atlantic coast, the aperiodic variations in sea level are related to occurrences of strong onshore winds as well as fluctuations in the flow of the Gulf Stream (Ezer 2013; Ezer and Atkinson 2014; see Chap. 3, Sect. 3.5; Chap. 14). Tide ranges also vary somewhat over the course of a year and are greatest during perigean tides when the earth and moon are closest (e.g. at times of "harvest moons" in October). Perigean spring tides are locally referred to as "King Tides" because the range is above normal. King tides alone do not constitute nuisance flooding. But it is when higher sea levels combine with onshore winds and effects related to the slackening of the Gulf Stream that these nuisances happen. Since sea levels are progressively rising, the frequency of occurrence of nuisance floods is increasing and when torrential rainfalls and clogged storm drains coincide with nuisance flood events, the effects sometimes become somewhat more than a nuisance (Sweet et al. 2014). As shown in Fig. 7.7, the number of days with nuisance flooding in Atlantic City, New Jersey have steadily increased from 0 to 5 days per year in the 1950s to over 30 days in 2012. Similar trends are evident at other U.S. Atlantic and Gulf Coast locations.

7.6 Compound Ocean, Fluvial and Pluvial Flooding

Considerations of coastal flooding tend to place heavy emphasis on storm surge and model predictions of inland flooding during tropical cyclones produced by storm surge models, such as SLOSH, as described in subsection 7.3. Storm surges are, of course, a major concern but their effects are increasingly magnified by those of less

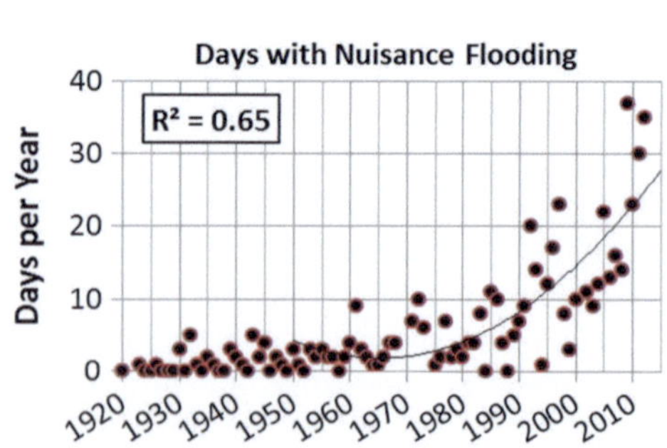

Fig. 7.7 Graph and scatter plot showing the number of days per year with nuisance flooding at Atlantic City, New Jersey from 1920 to the present based on NOAA tide gage records. From Sweet et al. (2014, NOAA Technical Report NOS CO-OPS 073)

Fig. 7.8 Aerial image of a pluvial flood in Louisiana in 2016. Image created using data collected by National Oceanic and Atmospheric Administration for the National Geodetic Survey. *Credit* Jason Burton/USGS. From Witman (2017)

well predicted phenomena such river, or fluvial, flooding and pluvial flooding (Fig. 7.8) by extreme and often unexpected torrential and prolonged rainfall. If the ground is already saturated from previous rains, the effects can be further exacerbated. When all three, or even two, of these inundation phenomena coincide, the compound effects can be catastrophic as was recently demonstrated when Hurricane Harvey made landfall on the coast of Texas and caused widespread flooding of Houston in late August, 2017. The coasts of south Asia are subject to compound flooding on a near annual basis. As explained in Chap. 2, flooding of the Ganges-Brahmaputra delta, home to 200 million people, is caused by the combination of storm surge, river flooding and torrential rain during the summer monsoon season. In Guangzhou, China and the Pearl River delta, severe rainstorms, which may or may not accompany typhoons are responsible for 63% of the flood-related economic losses while storm surges caused by typhoons directly account for 33% (He and Yang 2011; Chap. 12). Lian et al. (2013) describe the increase of similar threats from compound flooding in Fuzhou, China, with the greatest threat attributable to heavy rainfall.

Wahl et al. (2015) have examined the joint probability of severe storm surge and pluvial flooding coinciding in U.S. coastal cities and concluded that the occurrence of such compound events has increased significantly over the past century. Ikeuchi et al. (2017) have developed a large-scale global coupled river-coast flood model. Their results indicate that the greatest risks from compound flooding are in mega-deltas, particularly the large Asian deltas, and estuaries where storm surges can propagate over 200 km (124 mile) upstream. A crucial aspect of compound floods is that the storm surge not only contributes its own component of inundation,

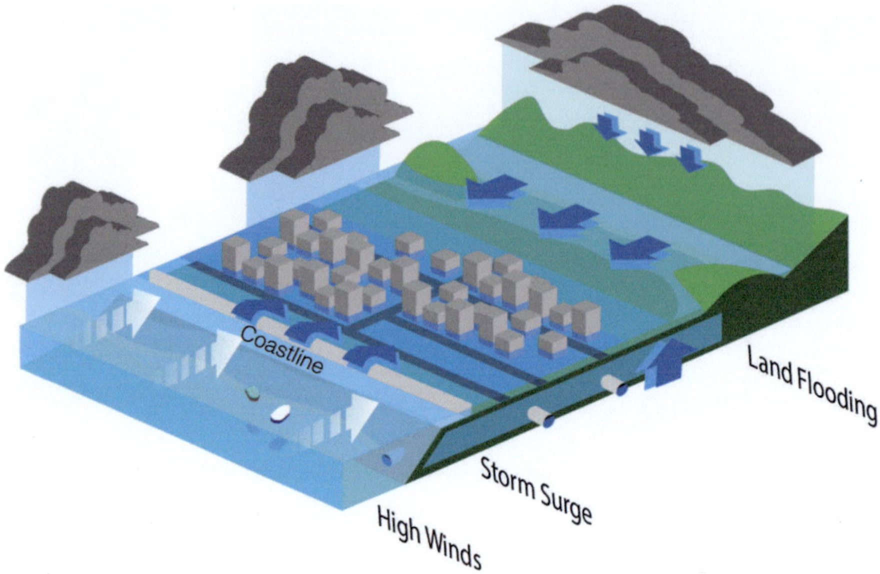

Fig. 7.9 The joint occurrence of torrential rainfall and storm surge causes compound flooding in coastal regions because storm surge not only causes inundation but also slows or blocks freshwater drainage. From Wahl and Jains (2015). How storm surges and heavy rainfall drive coastal flood risk in the U.S. *Carbon Brief* 27 July 2015. *Diagram Credit* Theodore Scontras, University of Maine

it also opposes the drainage of the pluvial floodwaters from the land. This is illustrated conceptually in Fig. 7.9. Reed et al. (2015) conclude that the risk of such compound flooding in New York City is increasing.

High sea surface temperatures fuel tropical cyclones in several ways. Most importantly they favor rapid storm intensification and increased amounts of rainfall and in combination, these factors can lead to compound flooding. As it traversed the northern Gulf of Mexico *en route* to eventual landfall on the east Texas coast, Hurricane Harvey was nourished by sea surface temperatures in excess of 30 °C (85 °F). Then, prior to landfall it encountered an eddy consisting of even warmer water and, overnight, it intensified from a depression to a hurricane (Fischetti 2017) and incorporated excessive amounts of water vapor into its core before making landfall as a Category 4 Hurricane and then stalling a short distance inland. Although the storm surge generated by Harvey was fairly modest in most places, the rainfall was extreme: peak accumulations reached 64.58 in (164.0 cm) making Harvey the wettest tropical cyclone to ever hit the U.S. Flooding of Houston was uncommonly widespread and prolonged. As discussed in Part 1, Chap. 2, rising ocean temperatures are likely to increase the severity, but not necessarily the frequency, of future such events.

7.7 Flood Mitigation, Storm Water Drainage and Runoff

Extensive engineering works to protect cities and communities from floods have been constructed in many developed countries, particularly in Europe. As pointed out in Chap. 6, the low-lying Netherlands has been a world leader in coastal engineering for many decades. A complex system of dunes, dikes, floodgates and frequent land reclamation maintains lands that would be frequently or continually submerged. Subsequent to the devastation caused by Hurricane Katrina, the U.S. Army Corps of Engineers has completed a \$14.5B flood protection system intended to withstand a 100-year flood event. This system consists of levees, a 26ft (8 m) high storm surge barrier and high-volume pumps (Fig. 7.10; Burnett 2015). According to the Southeast Louisiana Flood Protection Authority, the pumping system is the largest in the world and can fill an Olympic size swimming pool in 3 s. But despite its cost and size, the system still does not meet the design criteria specified by Congress in 1965 and New Orleans continues to be threatened. And chronic failures of the system occurred in summer, 2017.

Following the lessons that New York learned from Hurricane Sandy, plans are underway to surround lower Manhattan with a ten-foot (3 m) high wall extending from the end of E. 42nd St. to W. 57th St. at a cost of more than US \$3 billion (Goodell 2017).

It is often easier to predict the level of coastal inundation by storm surge than it is to predict how long it will take for floodwaters to subside. This is because there

Fig. 7.10 Storm surge barrier and pumping pipes at the 17th Street Canal in New Orleans (photo by David Gilkey/NPR)

are numerous natural and anthropogenic factors that control storm water drainage and runoff. Natural factors include the degree to which the ground is already saturated, whether or not the wind blows from a direction that opposes runoff, local topography and the stage levels and proximity of local river and streams. Human factors include the extent of paved surfaces, the effectiveness of drainage systems, engineered flood control systems such as pumps and—perhaps most difficult of all-how managers triage decisions regarding the opening or closing of flood control dams and floodgates. For example, decisions by the U.S. Army Corps of Engineers concerning where and when to relieve the rising waters of the Mississippi River flood of 1927 were beneficial for some and devastating-and fatal- for others (Barry 1997).

7.8 The Road Ahead for Flood Mitigation

Goodell (2017) in his popular book *The Water Will Come*, discusses changing climate, increasing coastal populations, and significant flood damage that is occurring worldwide. He provides illustrations of flooding in Miami-Dade County, where inundation and sewer overflows are attributed to sea level rise and the conversion of pervious surfaces into impervious (i.e. paved) surfaces, result in increased volumes of runoff which exceed the capacity of sewer systems. Major coastal construction efforts are planned to combat sea level rise and sinking cities such as Venice, the capital of northern Italy's Veneto region, is a prime example. Venice, which is reportedly subsiding approximately 1–2 mm per year (0.04–0.08 in. per year), was built on small islands in the Adriatic Sea (e.g., Bock et al. 2012; Tosi et al. 2015). Complex structures and solutions to protect Venice include installation of mobile barriers in the lagoon that will lie on the sea floor, but inflate during high tides. Perhaps the most ambitious and futuristic proposal would involve injection of billions of gallons of seawater into porous sediments under the canal-crossed city, to raise the city (e.g., Comerlati et al. 2004; Teatini et al. 2010). Other options include retreat, which involves no effort to protect the land from rising sea levels and inundation. This implies acceptance of the reality that people will not prevent the land from flooding, but must adapt by erecting emergency flood shelters, elevating buildings, or moving structures inland.

References

Applied Coastal Research and Engineering, Inc. 2015, June 30. Environmental notification form: Crescent beach seawall and revetment repair hull. 38pp, plus appendices. Prepared for Town of Hull Conservation Department.

Barry, J.M. 1997. *Rising tide: The great Mississippi flood of 1927 and how it changed America*, 524. New York: Touchstone.

Bock, Yehuda, Shimon Wdowinski, Alessandro Ferretti, Fabrizio Novali, and Alfio Fumagalli. 2012. Recent subsidence of the venice lagoon from continuous GPS and interferometric synthetic aperture radar. *Geochemistry, Geophysics, Geosystems* 13 (3): 1–13. https://doi.org/10.1029/2011GC003976.

Burnett, John. 2015, Aug 28. Billions spent on flood barriers, but New Orleans still a 'fishbowl'. *NPR*. Available online http://www.npr.org/2015/08/28/432059261/billions-spent-on-flood-barriers-but-new-orleans-still-a-fishbowl. Accessed on 25 Sept 2017.

Comerlati, Andrea, Massimiliano Ferronato, Giuseppe Gambolati, Mario Putti, and Pietro Teatini. 2004. Saving venice by seawater. *Journal of Geophysical Research: Earth Surface* 109 (F3): F03006. https://doi.org/10.1029/2004JF000119.

Dean, R.G., and Dalrymple R.A. 2002. *Coastal processes with engineering applications.* Cambridge University Press. 475pp.

Dunbar, P., H. McCullough, G. Mungov, J. Varner, and K. Stroker. 2011. 2011 Tohoku earthquake and tsunami data available from the national oceanic and atmospheric administration/national geophysical data center. *Geomatics, Natural Hazards and Risk* 2 (4). https://doi.org/10.1080/19475705.2011.632443.

Ezer, T. 2013. Sea level rise, spatially uneven and temporally unsteady: Why the U.S. East Coast, the global tide gauge record, and the global altimeter data show different trends. *Geophysical Research Letters* 40 (20): 5439–5444. https://doi.org/10.1002/2013GL057952.

Ezer, T., and L.P. Atkinson. 2014. Accelerated flooding along the U.S. East Coast: On the impact of sea-level rise, tides, storms, the Gulf Stream, and the North Atlantic Oscillations. *Earth's Future* 2 (8): 362–382. https://doi.org/10.1002/2014EF000252.

FEMA. 2015. *Guidance for flood risk analysis and mapping. Coastal wave setup.* Guidance document 44.

Fischetti, M. 2017. Hurricane harvey: Why is it so extreme? *Scientific American.* 28 Aug 2017.

Goodell, J. 2017. *The water will come*, 340. New York, NY: Little Brown and Company.

Guza, R.T., and E.B. Thornton. 1985. Observations of surf beat. *Journal Geophysical Research* 90: 361.

He, C., and L. Yang. 2011. Urban development and climate change in China's pearl river delta. *Land Lines.* Lincoln Institute of Land Policy. 7pp.

Ikeuchi, H., Y. Hirabayashi, D. Yamazaki, S. Muis, P.J. Ward, H.C. Winsemius, M. Verlaan, and S. Kanae. 2017. Compound simulation of fluvial floods and storm surges in a global coupled river-coast flood model: Model development and its application to 2007 Cyclone Sidr in Bangladesh. *Journal of Advances in Modeling Earth Systems* 9: 1847–1862. https://doi.org/10.1002/2017MS000943.

Irish, J.L., D.T. Resio, and J.J. Ratcliff. 2008. The influence of storm size on hurricane surge. *Journal of Physical Oceanography* 38 (11): 2003–2013.

Jelesnianski, C.P., J. Chen, and W.A. Shaffer. 1992. *SLOSH: Sea, lake, and overland surges from hurricanes.* NOAA Technical Report NWS 48. Silver Spring, MD: National Weather Service, NOAA. 77pp.

Jones, C., et al. 2005. Wave runup and overtopping. In *FEMA coastal flood hazard analysis and mapping guidelines: Focused study report.*

Lian, J.J., K. Xu, and C. Ma. 2013. Joint impact of rainfall and tidal level on flood risk in a coastal city with a complex river network: A case study of Fuzhou City, China. *Hydrology and Earth System Sciences* 17: 679–689. https://doi.org/10.5194/hess-17-679-2013.

Mignone, A., and Caribou Maine. 2016. *Forecasting wave runup and overtopping*. National Weather Service VLabForum.

NOAA National Geophysical Data Center (NGDC). 2014. Tsunami Event Database. http://www.ngdc.noaa.gov/nndc/struts/form?t=101650&s=70&d=7.

Norman, J. 2006. Katrina's dead. *Sun Herald* A1, A8–9. 17 Feb 2006.

Pomonis, A., T. Rossetto, N. Peiris, S. Wilkinson, D. Del Re, R. Koo, R. Manlapig, and S. Gallocher. 2006. *The Indian ocean tsunami of 26 December 2004: Mission findings in Sri Lanka and Thailand*, 180. London, UK: Technical Report of Earthquake Engineering Field Investigation Team.

Rappaport, E. 1993. *Preliminary report Hurricane Andrew 16–28 August 1992.* National Center for Environmental Prediction (NCEP) Technical Report.

Reed, A.J., et al. 2015. Increased threat of tropical cyclones and coastal flooding to New York city during the anthropogenic era. *Proceedings of the National Academy of Sciences.* https://doi.org/10.1073/pnas.1513127112.

Resio, D.T., and J.J. Westerink. 2008. Modeling the physics of storm surges. *Physics Today* 61: 33.

Senechal, N., G. Coco, K.R. Bryan, and R.A. Holman. 2011. Wave runup during extreme storm conditions. *Journal Geophysical Research* 116: C07032. https://doi.org/10.1029/2010JC006819.

Stockton, H., R.A. Holman, P. Howd, and A.H. Sallenger Jr. 2005. Empirical parameterization of setup, swash, and runup. *Coastal Engineering* 53 (7): 573–588. https://doi.org/10.1016/j.coastaleng.2005.12.005.

Sweet, W., J. Park, J. Marra, C. Zervas, and S. Gill. 2014. Sea level rise and nuisance flood frequency changes around the United States. NOAA Technical Report NOS CO-OPS 073. 58pp.

Teatini, Pietro, Massimiliano Ferronato, Giuseppe Gambolati, D. Baù, and Mario Putti. 2010. Anthropogenic Venice uplift by seawater pumping into a heterogeneous aquifer system. *Water Resources Research* 46 (11): W11547. https://doi.org/10.1029/2010wr009161.

Titov, V.V., A.B. Rabinovich, H.O. Mofjeld, R.E. Thomson, and F.I. González. 2005. The global reach of the 26 December 2004 Sumatra Tsunami. *Science* 309 (5743): 2045–2048.

Tosi, Luigi, Tazio Strozzi, Cristina Da Lio, and Pietro Teatini. 2015. Regional and local land subsidence at the Venice coastland by TerraSAR-X PSI. *Proceedings of the International Association of Hydrological Sciences* 372: 199–205. https://doi.org/10.5194/piahs-372-199-2015.

Wahl, T., and S. Jain. 2015. How storm surges and heavy rainfall drive flood risk in the U.S. *Carbon Brief* 27 July 2015. https://www.carbonbrief.org/how-storm-surges-and-heavy-rainfall-drive-coastal-flood-risk-in-the-us.

Wahl, T., S. Jain, et al. 2015. Increasing risk of compound flooding from storm surge and rainfall for major US cities. *Nature Climate Change.* https://doi.org/10.1038/nclimate2736.

Witman, S. 2017. Deciphering deluges. *EOS* 98 (11): 42.

Chapter 8
Degradation of Coastal Ecosystems: Causes, Impacts and Mitigation Efforts

C. Reid Nichols, Julie Zinnert and Donald R. Young

> *The real conflict at the beach is not between the sea and the shore [...] but between man and nature. On the beach nature has achieved a dynamic equilibrium that is alien to man and his static sense of equilibrium. Once a line has been established, whether it is a shoreline or a property line, man unreasonably expects it to stay put.*
>
> —Soucie (1973)

8.1 Synopsis of Coastal Ecosystems at Risk

The coastal zone is shaped by abiotic and biotic forces as evidenced by climate, tides, waves, longshore currents, geology, aerosols, light penetration, sediment supply, the seasonal growth and decay of plants, and man-made features that are observed worldwide. Primary forces such as wind and waves and secondary forces such as friction combine to form coastal features such as moraines and drumlins, rivers, bays, cuspate forelands, wave-cut cliffs, barrier islands, coastal inlets, lagoons, mangrove swamps, coral reefs, deltas, seaports, and even coastal cities. The processes associated with these coastal features impacts the manner in which associated coastal ecosystems develop. Examples of the delicate ecosystems that form in the areas where land and water meet are listed in Table 8.1. These dynamic locations have a distinct structure, diversity, and flow of energy, which provide homes for many different types of plants and animals. As explained in Chap. 1, coasts are complex systems formed by the interaction of a diverse community of organisms with their environment. They tend to provide easy access to the deep ocean, a high concentration of nutrients, and various types of protection from predators (e.g. mangrove swamps, submerged aquatic vegetation, and oyster reefs).

C. R. Nichols (✉)
Marine Information Resources Corporation (MIRC), Ellicott City, MD, USA
e-mail: rnichols@mirc-us.com

J. Zinnert · D. R. Young
Department of Biology, Virginia Commonwealth University, Richmond, VA, USA

L. D. Wright and C. R. Nichols (eds.), *Tomorrow's Coasts: Complex and Impermanent*, Coastal Research Library 27,
https://doi.org/10.1007/978-3-319-75453-6_8

Table 8.1 There are many types of ecosystems and the following are influenced by oceanic processes. These coastal ecosystems are listed from the hinterlands across the nearshore and out into the coastal ocean

Maritime forests	Wooded habitat found on ground usually higher than dune areas and within range of salt spray. Trees, bushes, and other plants in maritime forests stabilize the shoreline and are able to withstand strong winds and periodic flooding. They provide sheltering effects for the marsh behind the maritime forest
Dunes	Coastal mounds or ridges of sand or other loose sediment formed by the wind. Plants and wrack are barriers to wind flow causing sand to accumulate. Most coastal dunes are vegetated with grasses and/or small shrubs. Specific species such as sea oats (*Uniola paniculata*) and marram grass (*Ammophila arenaria*) influence dune formation and growth. "Wrack" is the term for seaweed, driftwood, and other organic materials produced by coastal ecosystems that wash ashore and make a high water mark along the shore
Beaches	Unstable, constantly changing environments owing to erosion and accretion by winds, waves, tides, and sea level rise. Organisms living in and on the beach usually have well-adapted appendages and streamlined bodies
Salt marshes	Herbaceous vegetative community occurring in temperate intertidal zones. Soils tend to be composed of deep mud and peat. These intertidal habitats provide essential food, refuge, or nursery habitat for many commercial important fisheries species
Mangroves	Halophytic shrub and tree community found where land and ocean meet along subtropical and tropical coastlines, worldwide. Mangrove tree species with their roots submerged in water, thrive in hot, muddy, salty conditions. Mangroves absorb the force of waves as a natural buffer between the land and sea. They provide essential food, refuge, or nursery habitat for many commercial important fisheries species
Mud flats	Un-vegetated intertidal zones with substrates ranging from sand to mud found in low energy and shallow-sloped sheltered environments, e.g., between subtidal channels and vegetated saltmarshes. Mudflats help to dissipate wave energy and protect salt marshes from eroding. They are characterized by high biological productivity and an abundance of organisms. They provide a feeding ground for many species of wildlife
Rocky shores	Intertidal habitats with hard substrate, complex topography, and diverse biota. Owing to tides and waves, the rocky shore is characterized by erosional features such as steep cliffs and boulder rubble. It is often a biologically rich environment inhabited by organisms that cope with extremes of both temperature and salinity and flooding and drying
Wetlands	Land areas that are saturated with water, either permanently or seasonally. They are characterized by hydric soils and hydrophytic vegetation. Wetlands, besides being biologically diverse and highly productive ecosystems, are nursery areas for many commercially important fisheries species and critical habitat for migratory birds and waterfowl
Seagrass beds	Community of flowering plants that are fully-submerged in salty and brackish waters with a sediment substrate in many parts of the world, from the tropics to the poles. Seagrasses provide shelter and food to a diverse community of animals, from tiny invertebrates to marine mammals
Estuaries	Semi-enclosed areas where saltwater is diluted by freshwater from land drainage. Water filtered through estuaries brings in nutrients from the

(continued)

Table 8.1 (continued)

	surrounding watershed. Estuarine habitats support unique plant and animal communities adapted to brackish water. Estuaries also buffer streams, river channels and coastal shores from excessive erosion caused by wind, waves, and ice
Coral reefs	Layers of calcium carbonate in clear, saline, and tropical waters from corals that have formed into fringing and barrier reefs as well as atolls. Corals are marine invertebrates that live in compact colonies of many identical individual polyps. The largest coral reef is the Great Barrier Reef, which spans 2600 km (1600 mi) off the east coast of Australia. The largest atoll is the Great Chagos Bank with an area of 12,642 km^2 (4881 mi^2) in the Indian ocean. Coral reef ecosystems feed, shelter and provide habitat for fish and protect the shoreline from wave and beach erosion

An ecosystem consists of living organisms of a particular habitat together with the physical, non-living environment in which they live. Coastal ecosystems differ from the freshwater ecosystems in places such as the Great Lakes since they contain saltwater, which usually supports different types of species than does freshwater

8.2 Damage Assessments of Coastal Ecosystems

Coastal bays, river mouths, and other wetlands such as marshes are all physically contiguous to the ocean, but are affected by land drainage. These simple facts led Cameron and Pritchard (1963) to define estuaries as (a) a semi-enclosed and coastal body of water, (b) with free communication to the ocean, and (c) within which ocean water is diluted by freshwater derived from land. Freshwater flow from the rivers and saltwater flow from the ocean are the main drivers for circulation within an estuary. Other important factors impacting the salinity of estuaries are evaporation, precipitation, and withdrawals of water for agriculture. Estuaries provide a variety of important ecosystem services and functions such as serving as natural filters for runoff and providing nursery grounds for many species of birds, fish, and other animals. They provide migration stopovers for fish such as American Shad (*Alosa sapidissima*), Barramundi (*Lates calcarifer*), Chinook salmon (*Oncorhynchus tshawytscha*), European eel (*Anguilla anguilla*), Striped Bass (*Morone saxatilis*), and White sturgeon (*Acipenser transmontanus*). Large assessment programs such as the National Estuary Program in the United States have shown that these regions where land meets the sea are vital for many commercially important fish species and deserving of protection. Many countries have established estuarine protected areas to safeguard essential habitat for wildlife, offer educational opportunities for students, and serve as living laboratories for scientists.

Ecosystems by definition are networks or communities of living and nonliving things, where these biotic and abiotic components are linked together through nutrient cycles and energy flows. In the coastal zone, ecosystems usually encompass specific, limited spaces and provide services such as provisioning (the production of food and water), regulating (control of climate and disease), supporting (nutrient cycles and crop pollination), and cultural (spiritual and recreational).

Numerous researchers (e.g., Ackerman et al. 2017; Barr et al. 2009; Bramanti et al. 2015; Krumhansla et al. 2016; Muller-Karger et al. 2014; Wieski and Pennings 2014; Zinnert et al. 2011) studying ecosystems through the U.S. Long-Term Ecological Research (LTER) program have shown that even the smallest change can impact the entire ecosystem. The LTER network consists of 28 sites with a rich history of ecological inquiry, collaboration across a wide range of research topics, and engagement with students, educators, and community members. Several of these sites are focused primarily on coastal areas including Beaufort Lagoon Ecosystem, California Current Ecosystem, Florida Coastal Everglades, Georgia Coastal Ecosystems, Moorea Coral Reef, Northern Gulf of Alaska, Palmer Antarctica, Plum Island Ecosystems, and the Virginia Coast Reserve.

Through sustained data collection, ecosystem manipulation experiments, and modeling, researchers have found numerous causes for habitat losses and species decline. Ecosystem degradation is often blamed on the consumption of assets, for example, air, water and soil; the destruction of environments and the eradication of wildlife. However, primary causes can be attributed to a combination of natural phenomena and man. Natural causes include complex weather patterns such as El Niño and La Niña and extreme events such as tropical cyclones, droughts, wild fires, tsunami, and climate change. Anthropogenic causes include coastal population explosion, alteration of river flows and sediment loads, mechanical habitat destruction, overfishing, and pollution. Coastal erosion, as a case in point, may be caused by waves generated by storms and fast moving motor craft. These forces contribute to either long-term losses of sediment and rocks, or the temporary redistribution of coastal sediments from one location (erosion) to another nearby location (accretion).

8.3 Natural Causes of Degradation

Ecosystem degradation manifested by natural events is much more complicated to define than changes resulting from man-made causes. New understandings of the world's changing climate and the likely implications for coastal ecosystems continues to challenge scientists, worldwide. Episodic events such as natural disasters are much easier to define than the slow changes, which are occurring in synchronization with processes such as global warming. Short term, episodic disturbances, or pulses, in coastal environments are often dramatic and are typified by shoreline erosion, flattening of dunes, and large sand deposits (overwash) on island interiors. These changes are most always accompanied by large changes in vegetation, both community structure and even extirpations of species and entire ecological communities. It should be noted, however, that these disturbances also create opportunities for species to colonize newly formed sites. Due to the extreme physical nature of the short term processes, especially wave-related erosion and deposition, vegetation may reduce storm effects by reducing wave energy and water velocity.

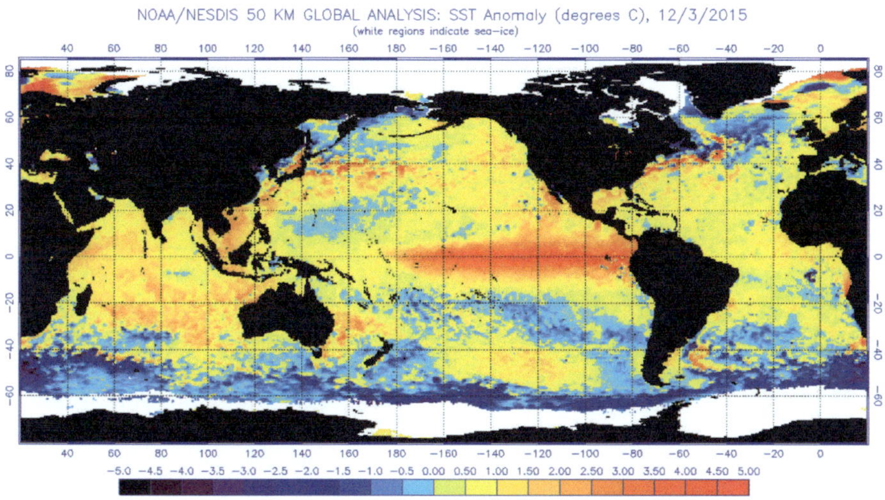

Fig. 8.1 El Niño captured in sea surface temperature imagery. During a strong El Niño, as occurred during 2015, there are resultant weather conditions such as floods that can significantly affect agriculture and the economy. NOAA/NESDIS image

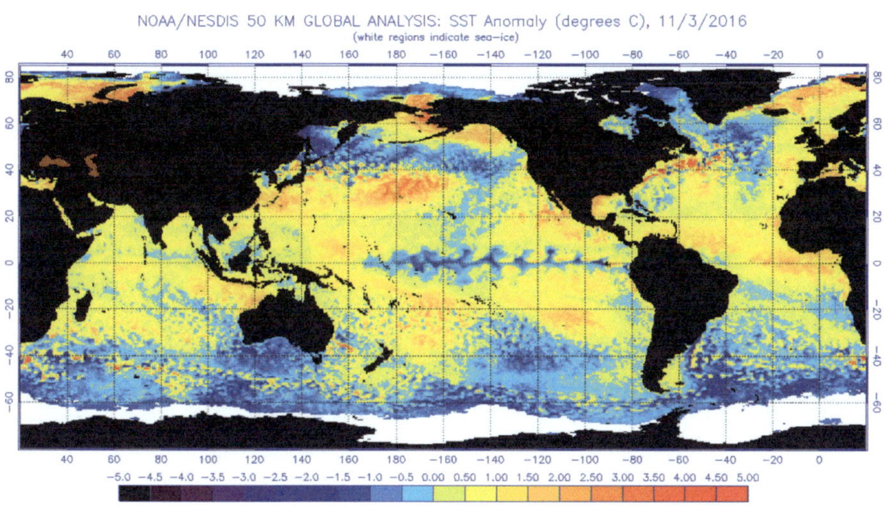

Fig. 8.2 La Niña captured in sea surface temperature imagery. During this weak La Niña weather pattern air temperatures and precipitation are lower than normal. NOAA/NESDIS image

Climate fluctuations such as the El Niño-Southern Oscillation (ENSO) are naturally occurring phenomena that involve fluctuating ocean temperatures in the equatorial Pacific. The cycle is illustrated by the Sea Surface Temperature anomalies in Fig. 8.1 (El Niño, the warm phase) and Fig. 8.2 (La Niña, the cold phase). During El Niño, water temperatures in the Pacific Ocean are unusually high, as much as 7 °F above average. La Niña causes drier than normal atmospheric conditions and is less frequent. Southern Oscillation is the accompanying atmospheric component, where El Niño is associated with high, and La Niña with low air surface pressure in the tropical western Pacific. El Niño and La Niña have a significant impact on coastal processes, especially in a country such as Chile (Gale 2016; Quintana 2000). On average, these conditions occur every two to seven years. Social and economic impacts of extreme La Niña events in Chile include stress to the vineyards, timber, mining, and energy sectors. ENSO is a phenomenon that drives the weather and contributes to flooding and droughts.

Natural hazards such as tropical cyclones, earthquakes, landslides, volcanic eruptions, tsunamis and wild fires all affect coastal ecosystems. These episodic events negatively affect the biodiversity of coastal systems through the spread of invasive species, mass species mortality and loss of habitat. In the short term, ecosystem degradation reduces ecosystem services such as the ability of kelp forests to sequester carbon, which exacerbates climate change (Wilmers et al. 2012). Natural factors influencing kelp forest stability include grazing by fishes, sea urchins, and crustaceans; plant competition; storms; El Niño events; and sedimentation. Because of their spectacular growth rates, kelp tend to recover quickly from physical disturbances such as storms that might uproot the fragile plants. Extreme natural events such as earthquakes can cause the loss of forests due to landslides, shifting river basins, and even the creation of lakes when landslides dam rivers and streams. Environmental damage resulting from a disaster may have consequences on the ecosystem services that are provided.

The destructive effects of a tropical cyclone will depend on the storm's intensity, size, and location. Environmental effects include strong wind, heavy precipitation, large storm surges at landfall, and tornadoes. Impacts extend far beyond the storms track. Heavy rainfall and floods erode soil, divert rivers, and cause severe damage to habitats, infrastructure, and entire populations. Imagery following the passage of a tropical cyclone depicts the losses of forest canopy as well as changes in coastal landscape such as beach erosion. Heavy inland rainfall can lead to mudslides and landslides. Destruction of critical infrastructure such as roads, bridges, and powerlines may hamper clean-up and rescue efforts. Standing water can cause the spread of disease such as West Nile virus, which is most commonly spread to people by mosquito bites. This was a threat to those people who are spending more time outside living in inadequate shelters in the aftermath of Hurricane Katrina. Despite these devastating effects, tropical cyclones are also beneficial, by moving heat from the tropics toward the poles, bringing rain to dry areas, and re-oxygenating hypoxic regions. Coastal ecosystems such as mangroves and coral reefs may act as buffers to lessen the impacts of typical storm-generated waves (Kerr and Baird 2007; Sheng et al. 2012; Sakib et al. 2015).

Underwater landslides, earthquakes and volcanic eruptions can all cause tsunamis, by displacing water that creates long period waves that are sustained by gravity as they travel towards land at great speeds. In deep water, tsunamis have a small wave height which rises as the wave reaches shallow water and land. The resulting damage to coastal ecosystems such as coral reefs, mangrove forests and wetlands, which are all dependent on each for nutrient supplies, can be severe. As an example, the Indian Ocean earthquake and tsunami occurring on December 26, 2004, triggered a series of devastating tsunamis along the coasts of most landmasses bordering the Indian Ocean, killing hundreds of thousands of people in 14 countries, and inundating coastal communities with reported wave heights of 30 m (100 ft). The tsunami caused accumulated debris in lagoons and other coastal ecosystems, plus salt intrusion inland, which negatively affected local fisheries and agricultural productivity and increased vulnerability to erosion. The Indian Ocean earthquake and tsunami of 2004 was one of the deadliest natural disasters in recorded history (Chap. 7).

Active volcanoes such as on Pagan Island in the Marianas Island spew gas and ash causing damage to the plant life, the soil, polluting water bodies (e.g., Laguna Sanhalom and Sanhiyon), and possibly severe damage to people visiting the island. Active volcanoes can erupt explosively or ooze magma for weeks or even years. Regions near an active vent are incompatible with life. North of the Marianas is a more dangerous volcanic island, Iwo To (aka Iwo Jima), which has a growing magma chamber underneath. An eruption could cause a tsunami that would impact southern Japan and coastal China including megacities Shanghai and Hong Kong.

Landslides and debris flows cause damage to coastal ecosystems by overturning soil horizons, burying vegetation, polluting water bodies and damaging entire habitats and infrastructure. Heavy waves and precipitation following other natural phenomena, such as an earthquake, wildfire, or flood, might trigger landslides and debris flows. Vulnerable areas include cliffs and coastal bluffs worldwide. Leshchinsky et al. (2017) describe landslides occurring along the U.S. Pacific coastline in Washington, Oregon, and California. The loss of life and property is attributed to structures that are located in areas that are geologically prone to slope instability. Mitigation might involve restricting development in landslide-prone areas and developing and installing monitoring and warning systems.

Strong winds, ice, snow and tornadoes are natural occurrences that can cause extensive damage to maritime forests by uprooting, wounding, bending and breaking trees. Hailstorms accompanied by heavy rain, as is the case with some thunderstorms, may cause large amounts of soil to be eroded. Excessive rainfall events can produce a large amount of water in a short period of time across local areas. This excess of water overwhelms the local watershed and leads to flooding. Sediment laden waters following these storms may smoother oyster and coral reefs and block sunlight from submerged aquatic vegetation.

By comparison, long-term variability due to natural processes (i.e. longshore current erosion or deposition of sediments, Aeolian transport of sediments) or changing climate (i.e. temperature induced range shifts, sea level rise, ocean acidification) is subtler, but is essential for the development and stability of coastal systems. Climate change impacts all coastal ecosystems from estuaries and coral reefs to rocky shores and mangrove forests. It is leading to spatial variations in species composition partly due to range expansion, which creates new ecosystemsthat are outside the range of historical dominance. Researchers are studying the warmer bottom water temperatures found along the continental shelf off North Carolina and how these temperatures are affecting the area's species composition. Likewise, researchers are also studying temperature induced range shifts in both plants and animals of coastal areas (Perry et al. 2005; Harley et al. 2006; Zinnert et al. 2011; Harris et al. 2017; Huang et al., in review).

Sea level rise in coastal areas and small islands causes seawater to reach further inland. The impacts which are being documented more and more are destructive erosion, wetland flooding, aquifer and agricultural soil contamination, and lost habitat for many animal and plant species. Saltwater from rising sea levels and storm surges threatens freshwater supplies. Several studies in recent years have indicated accelerating sea level rise along the coastal mid-Atlantic region (Sallenger et al. 2012; Ezer et al. 2013). A northeast hotspot of sea level rise has been identified ranging from north of Cape Hatteras NC to Cape Cod, MA. Here sea level rise is approximately 3–4 times higher than the global average (Sallenger et al. 2012). As a result of this accelerated regional sea level rise, many mid-Atlantic coastal communities have experienced increased frequency in flooding (Ezer et al. 2013). A challenge to analyzing sea level data arises from the spatial and temporal variability in ocean dynamics that affect coastlines (Ezer 2013; Ezer et al. 2013), and ultimately confounds predictions in coastal communities. Sea level rise may reduce the spatial extent of available habitat or cause changes between ecosystem states by outpacing the accretion rates of coastal ecosystems (i.e., barrier islands, marshes, coral reefs) (Harley et al. 2006; Zinnert et al. 2016).

Ocean acidification is likely to negatively impact calcifying animals and plants. Decreases in carbonate ions can make building and maintaining shells and other calcium carbonate structures difficult for organisms such as oysters, clams, sea urchins, shallow water corals, deep sea corals, and calcareous plankton. Researches such as Azevedo et al. (2015) have studied the impact of increased ocean acidity on the growth, reproduction, and survival of calcifying organisms.

8.4 Anthropogenic Causes

Man has contributed to the disintegration of coastal zones since areas such as marshes, mangroves, and locations near seagrass beds were first inhabited. Deterioration of the environment occurs through unsustainable consumption of natural resources, pollution, the mechanical destruction of habitat, and the introduction on nonindigenous species, which all contributes to ecosystem alterations. Overhunting of the Caribbean monk seal (*Neomonachus tropicalis*), which has not been seen since 1952, contributed to the declaration of this marine mammal's extinction in 2008 by the International Union for Conservation of Nature, the International Convention on Trade in Endangered Species, and the National Oceanic and Atmospheric Administration.

Many anthropogenic factors tend to be interconnected and are difficult to isolate. For example, over population causes all types of human and coastal ecosystem issues and has been linked to global warming and habitat loss. This is illustrated by the heat island effect, where built up areas have been shown to be hotter than nearby rural areas. According to the EPA (2008), the annual mean air temperature of a city with 1 million people or more can be 1–3 °C (1.8–5.4 °F) warmer than its surroundings and in the evening, the difference can be as high as 12 °C (22 °F). Heat islands can affect communities by increasing summertime peak energy demand, air conditioning costs, air pollution and greenhouse gas emissions, heat-related illness and mortality, and water quality. Social problems from overpopulation include urban sprawl, overproduction, and the over consumption of finite natural resources, such as fresh water, arable land and fossil fuels. Urban sprawl includes the expansion of cities and roads that contribute to drainage issues that cause stress factors to ecosystems, especially those which are highly fragile as a result of their weather conditions and the nature of their soil and water. Overpopulation is a major factor in ecosystem degradation through excessive exploitation of wildlife species either to feed the population or for other commercial objectives such as in the trade of animal furs or skins. In some cases, the use of inappropriate harvesting techniques such as dynamite fishing endangers the existence of many species, altering the food chain, and consequently deteriorating coastal ecosystems. The introduction of nonindigenous species for agriculture and exotic species as pets has been shown to endanger the existence of indigenous species. Non-native species if left to spread are likely to cause economic or environmental harm or harm to human and animal health. They are a threat to biodiversity.

Fishing constitutes an important source of livelihood and food for millions of people, but if done in unsustainable ways is a major contributor to degradation in coastal ecosystems. Overfishing is a form of overexploitation where marine species are reduced to below acceptable levels, i.e., fish are taken from the sea at rates too high for the fished species to replace themselves. Jackson et al. (2001) clearly illustrate the significance of overfishing to coastal ecosystem degradation by highlighting changes to food webs. It is well known that overfishing occurred to whale species such as the Blue (*Balaenoptera musculus*), Humpback (*Megaptera novaeangliae*), North Atlantic Right (*Eubalaena glacialis*) whales during the 19th century when whale blubber was sought for lamp oil. In the 20th century, especially with development of more effective high-technology fishing gear, it became readily apparent that once productive stocks from fishing grounds such as the Georges Bank, located between Cape Cod, Massachusetts, and Cape Sable Island, Nova Scotia, were not inexhaustible as evidenced by drastic declines in Atlantic halibut (*Hippoglossus hippoglossus*) and Cod (*Gadus morhua*) catches. The decline in fish stocks poses a disturbing, and potentially dangerous, threat to life in the ocean. Depletion of fish stocks results in a decrease in food supply from the sea, economic loss, hardship to fishers and disruption of traditional ways of life. Overfishing thus threatens the ecosystem, the sustainable use of fishing grounds and the livelihood of fishing communities. Some types of fishing techniques or gear may damage coral reef ecosystems. Spraying or dumping sodium cyanide on reefs to catch aquarium fish can damage and kill corals. Anchors, gill nets, and bottom trawl nets can break shallow- and deep-water coral, causing irreparable damage. Unsustainable fishing practices illustrate how entire ecosystems may be degraded, and even destroyed, by human actions.

Deforestation, usually through cutting or burning, to advance agriculture contributes to the degradation of ecosystems. In Central and South America, more tropical rainforest lands are deforested for commercial agriculture, live-stock rearing, and for wood then in any other location. For example, estimates for lost rainforest coverage from 1990 to 2005 were Honduras 37.1%, Nicaragua 20.6%, Costa Rica lost 6.8%, Panama 1.9%, and Colombia 1.2%. Assessments were made using satellite imagery and are published by organizations such as the Food and Agriculture Organization of the United Nations. Deforestation processes make the land more susceptible to erosion by both winds and water which increases sediment load to estuaries and the coastal ocean. Another major cause of the deterioration of coastal ecosystems is the draining of wide areas of wetlands to give access to agricultural zones or expanding housing.

Land reclamation, dike reinforcement, and deepening of shipping channels is common practice in the Netherlands and has resulted in substantial tidal marsh loss. For example, in the Scheldt estuary, marshes have been removed and eroded over the last several hundred years, with 2500 ha (6178 acres) lost since 1900 (Eertman et al. 2002). Deepening of shipping channels is occurring worldwide in estuaries and has significant consequences to the ecosystem as well as increased flood potential inland. Channel deepening enhances turbidity of sediments in the water column, increases salinity upstream, and increases tidal amplitude and flow. More recently, the recognition of natural marsh systems in reducing tidal energy have led to the protection and restoration of tidal salt marshes in many areas.

Ecosystem degradation may be attributed to land and water pollution. Land pollution is the contamination of land with hazardous waste like garbage and other waste materials. Heavy metals in the soil may be consumed by plants and animals and then when the next consumer feeds on either the plant or the animal, it accumulates and contaminates the body. Land pollution contributes to excess sedimentation, which causes changes in water quality and negatively impacts many marine species. Water Pollution is the contamination of the aquifer, streams, rivers, estuaries, and oceans by substances harmful to living things. Impure water kills phytoplankton, zooplankton, and many larger plants and animals. It contributes to known diseases, child-birth defects, and cancer. Pollution, in whatever form, whether it is air, water, land or noise is harmful to coastal ecosystems.

The mining of natural resources may cause drastic changes in the natural landscape while degrading its valuable ecosystems. Strip mines that remove top soil, subsoil, and rock cause extensive damage such as erosion. Further, sulfur deposits which may be associated with coal mining pollute nearby waterbodies. The sulfur reacts with water in the presence of air to create sulfuric acid. The resultant sediment laden and acidic runoff enters streams and rivers below. Strip mines should be regraded and then covered with topsoil and replanted. Oil exploration and production has also polluted the environment owing to eroded and broken pipelines, blowouts, spills, and oil fires. Reed and Fitzgerald (2011) describe the horrific blowout of the Macondo well in the Gulf of Mexico on April 20, 2010. By the time the well was capped on July 15, 2010 (87 days later), an estimated 3.19 million barrels of oil had leaked into the Gulf of Mexico making this the worst oil spill in U.S. history. Once the oil reached the coast it coated plants such as cordgrass (*Spartina alterniflora*) that were along the marsh edge, which damaged the plants and wildlife that depend on those plants. Oil sinking to bottom has a similar effect on submerged aquatic plants, coral and oyster reefs, and other members of the food chain. Rangoonwala et al. (2016), described how the combined impacts of oil spills and storm surge increase shoreline recession.

8.5 Vulnerable Coral Reefs

Coral reefs are among the most diverse and productive communities on Earth and we rely on reef ecosystems for food, shelter, tourism and recreation. Degradation of these tropical ecosystems is described in Chap. 5. As explained in that chapter, rising sea surface temperatures are causing coral bleaching on the Great Barrier Reef of Australia and elsewhere. International research programs such as CoML, European Land-Ocean Interaction Studies, International Coral Reef Initiative, and Land-Ocean Interactions in the Coastal Zone are working to understand the extent of and reasons for the decline of coral reefs, worldwide. Research results will help coastal zone managers to implement more effective ways to protect these resources. In shallow water, healthy coral reefs absorb wave energy and protect our coasts from waves, storms, and floods. There are also important coralline habitats occurring at depths below 50 m (164 ft) that provide vertical structure above the seafloor for other species. These cold water corals are found in many different places around the world and include Antipatharia (black corals), Gerardia (gold corals), Stylasteridae (lace corals), and Gorgonians (sea fans and sea whips). Black coral, the official state gem of Hawaii, have been harvested for charms and medicine since antiquity. Cold water and slow-growing corals are increasingly threatened by a variety of activities ranging from bottom fishing to energy exploration. Lumsden et al. (2007), Wilkinson (2008), and resources provided by the NOAA Coral Reef Conservation Program also document coral threats, declines, and their societal impact.

8.6 Preserving Threatened Ecosystems

The coastal ocean and its many ecosystems are a valuable public good. One's enjoyment of a saltmarsh or mangrove forest does not diminish the ability of other people to enjoy these open spaces. Similarly, people should not be prevented from enjoying, learning, being protected, and benefiting from these coastal habitats. Louv (2008), describes the societal importance of nature (physical, emotional, and spiritual wellbeing) and the costs of alienation from nature. Owing to overexploitation of coastal ecosystems and growing knowledge about their importance, communities are starting to balance the future land-use needs of wildlife and people. Through careful planning, healthy ecosystems are being maintained for people as well as wildlife. Restoration activities include returning abiotic factors (e.g., soil chemistry, water quality, beach extent) and biotic factors (e.g., species composition, food webs, submerged aquatic vegetation beds, oyster reefs) to historical levels. Organizations such as the Chesapeake Bay Foundation have evaluated different habitat restoration techniques for seagrass beds and oyster beds in order to implement estuarine restoration and stewardship programs. Progress is evidenced by imagery of expanding seagrass beds and recent improvements in the spawning of Atlantic sturgeon (*Acipenser oxyrinchus oxyrinchus*), an endangered species that was historically overfished and is susceptible to poor water quality.

From an economic standpoint, estuaries are especially important to provide natural buffers to sea level rise, for supporting commercial fisheries, and are therefore vital to the economy. This makes research into these ecosystems especially important. In the United States, NOAA has established the National Estuarine Research Reserve System as a network of protected estuarine areas or natural field laboratories to develop and implement coordinated programs of research, monitoring, education and volunteer activities. These reserves are actually partnerships among NOAA, non-governmental organizations such as The Nature Conservancy, coastal states, and territories. Each reserve is managed on a day-to-day basis by a lead state agency or university, with input from local partners. These reserves, which are listed in Table 8.2, promote coastal and estuarine stewardship. From an international perspective, the Man and the Biosphere Programme of the United Nations Educational, Scientific and Cultural Organization has established a World Network of Island and Coastal Biosphere Reserves. Research is focused on studying and implementing coastal strategies to preserve biodiversity and heritage, promote sustainable development, and adapt to and mitigate the effects of climate change. The World Network of Biosphere Reserves currently counts 669 sites in 120 countries.

Table 8.2 The U.S. National Estuarine Research Reserve System is comprised of a network of coastal sites designated to protect and study estuarine systems. These reserves serve as sentinel sites to better understand the effects of climate change

Reserve	Location	URL
Kachemak Bay	Southern end of the Kenai Peninsula in Southcentral Alaska	http://accs.uaa.alaska.edu/kbnerr/
Padilla Bay	Bay View, Washington	http://www.ecy.wa.gov/programs/sea/padillabay/index.html
Lake Superior	St. Louis river Estuary between Superior, Wisconsin, and Duluth, Minnesota	http://lakesuperiorreserve.org/
Wells Reserve at Laudholm	Wells, Maine	http://www.wellsreserve.org/
Great Bay	Greenland, New Hampshire	https://www.greatbay.org/
South Slough	Coos estuary on the south coast of Oregon	http://www.oregon.gov/DSL/SS/Pages/About.aspx
Waquoit Bay	Cape Cod, Massachusetts	http://www.waquoitbayreserve.org/
Narragansett Bay	Prudence, Patience, Hope and Dyer Islands in Rhode Island	http://nbnerr.org/
Hudson River	Along Hudson River in New York	https://www.hrnerr.org/
Old Woman Creek	Along Lake Erie near Huron, OH	http://naturepreserves.ohiodnr.gov/oldwomancreek
Jacques Cousteau	Atlantic coastal plain in New Jersey	https://jcnerr.org/
San Francisco Bay	North of Golden Gate along the western shore of San Francisco Bay	http://www.sfbaynerr.org/
Delaware	Upper Blackbird Creek and Lower St. Jones River, which drain into Delaware Bay	http://www.dnrec.delaware.gov/coastal/DNERR/Pages/DelawareNationalEstuarineResearchReserve.aspx

(continued)

Table 8.2 (continued)

Reserve	Location	URL
Chesapeake Bay	Sites located in upper, middle, and lower Chesapeake Bay	http://dnr.maryland.gov/waters/cbnerr/Pages/default.aspx
Elkhorn Slough	Monterey Bay, California	http://www.elkhornslough.org/
N.C. Coastal Reserve & National Estuarine Research Reserve	Barrier islands and lagoons along the Carolina Capes	http://www.nccoastalreserve.net/web/crp
Tijuana River	California coast bounded by Tijuana, Imperial Beach, and San Diego	http://trnerr.org/
North Inlet-Winyah Bay	Along the coast near Georgetown, South Carolina	http://www.northinlet.sc.edu/
Ashepoo-Combahee-Edisto Basin	Beaufort, Colleton and Charleston Counties in Southeastern South Carolina	http://www.dnr.sc.gov/marine/NERR/index.html
Sapelo Island	One of approximately 15 of Georgia's major barrier islands	http://sapelonerr.org/
Guana Tolomato Matanzas	North of St. Augustine and south Jacksonville, Florida	http://www.gtmnerr.org/
Weeks Bay	Along the eastern shore of Mobile Bay in Baldwin County, Alabama	http://www.outdooralabama.com/weeks-bay-reserve
Grand Bay	Along the Mississippi River Delta on the southeastern coast of Mississippi	http://grandbaynerr.org/

(continued)

Table 8.2 (continued)

Reserve	Location	URL
Apalachicola	Lower Apalachicola River and floodplain, as well as most of Apalachicola Bay in Northwest Florida	https://floridadep.gov/FCO/NERR-Apalachicola
He'eia	Along He'eia estuary, Moku o Lo'e (Coconut Island), and a portion of Kāne'ohe Bay	http://www.nerra.org/nerra-news/aloha-he%CA%BBeia-reserve/
Mission-Aransas	Western Gulf of Mexico estuarine system which includes the Aransas and Mission rivers	http://missionaransas.org/
Rookery Bay	Northern end of Ten Thousand Islands on the gulf coast of Florida and considered the westernmost extent of the Florida Everglades ecosystem	https://rookerybay.org/
Jobos Bay	Southern coast of Puerto Rico, between the municipalities of Guayama and Salinas	http://www.nerra.org/reserves/jobos-bay-national-estuarine-research-reserve/

8.7 Conclusions

Coastal ecosystem degradation is a worldwide threat and can occur through numerous natural or anthropogenic processes or, more often, a combination of both. It leads to depletion of resources such as clean air, water and soil and the reductions ecosystem diversity and in the goods and services that ecosystems offer (e.g., negatively affecting indigenous people and/or migratory species). The impacts of degradation are evidenced by habitat loss, the extinction of wildlife such as the Caribbean monk seal (*Neomonachus tropicalis*), and pollution. Ecosystem changes must be measured and understood in order to devise techniques that can be implemented to slow or prevent degradation, including science-based coastal ecosystem protection efforts. A healthy coastal ecosystem represents a source of wealth for society, hence the importance to restore and sustain coastal ecosystems.

Research, monitoring, and assessments are required to fully-understand and manage factors that stress coastal ecosystems (Busch and Trexler 2003). International programs such as the Census of Marine Life (CoML) and regional projects such as Long-term Coastal Ecosystem Monitoring Program at the Cape Cod National Seashore are critical to understanding ecosystem status and trends. Progress is made through integrated assessments that describe the ecosystem, assess its current condition or health, forecast future ecological health based on current management, and evaluate alternative management options and their consequences. Historical information, imagery, in situ sensors and model output are all useful to understand loss of biodiversity by elimination of indigenous species, alteration of trophic dynamics, degradation of habitats, and diminution of fisheries productivity.

References

Ackerman, Daniel, Daniel Griffin, Sarah E. Hobbie, and Jacques C. Finlay. 2017. Arctic shrub growth trajectories differ across soil moisture levels. *Global Change Biology* 23 (10): 4294–4302.

Azevedo, Ligia B., A.M. De Schryver, A. Jan Hendriks, and Mark A.J. Huijbregts. 2015. Calcifying species sensitivity distributions for ocean acidification. *Environmental Science & Technology* 49 (3): 1495–1500. https://doi.org/10.1021/es505485m.

Barr, Jordan G., Jose D. Fuentes, Vic Engel, and Joseph C. Zieman. 2009. Physiological responses of red mangroves to the climate in the Florida Everglades. *Journal of Geophysical Research* 114: G02008. https://doi.org/10.1029/2008JG000843.

Bramanti, Lorenzo, Mimmo Iannelli, Tung-Yun Fan, and Peter J. Edmunds. 2015. Using demographic models to project the effects of climate change on scleractinian corals: *Pocillopora damicornis* as a case study. *Coral Reefs* 34 (2): 505–515.

Busch, David E., Joel C. Trexler, and Monitoring Ecosystems. 2003. *Interdisciplinary approaches for evaluating ecoregional initiatives*. Washington, D.C.: Island Press.

Cameron, W.M., and D.W. Pritchard. 1963. Estuaries. In *The Sea*, vol. 2, ed. M.N. Hill, 306–324. New York: John Wiley & Sons.

Eertman, Richard H.M., Bart A. Kornman, E. Stikvoort, and Verbeek Harm. 2002. Restoration of the Sieperda tidal marsh in the Scheldt estuary, The Netherlands. *Restoration Ecology* 10 (3): 438–449.

Ezer, Tal. 2013. Sea level rise, spatially uneven and temporally unsteady: Why the U.S. East Coast, the global tide gauge record, and the global altimeter data show different trends. *Geophysical Research Letters* 40 (20): 5439–5444.

Ezer, Tal, Larry P. Atkinson, William B. Corlett, and Jose L. Blanco. 2013. Gulf Stream's induced sea level rise and variability along the U.S. mid-Atlantic coast. *Journal of Geophysical Research Oceans* 118 (2): 685–697.

Gale, Alana. 2016, May 19. El Niño in Chile, Southern Pacific Review. Available online. http://southernpacificreview.com/2016/05/20/el-nino-chile/. Accessed 24 Jan 24, 2017.

Harley, Christopher D.G., A. Randall Hughes, Kristin M. Hultgren, Benjamin G. Miner, Cascade J.B. Sorte, Carol S. Thornber, Laura F. Rodriguez, Tomanek Lars, and Susan L. Williams. 2006. The impacts of climate change in coastal marine systems. *Ecology Letters* 9 (2): 228–241.

Harris, April L., Julie C. Zinnert, and Donald R. Young. 2017. Differential response of barrier island dune grasses to species interactions and burial. *Plant Ecology* 218: 609–619. https://doi.org/10.1007/s11258-017-0715-0.

Huang H, J.C. Zinnert, L.K. Wood, D.R. Young, and P D'Odorico (in review). A non-linear shift from grassland to shrubland in temperate barrier islands. *Ecosystems.*

Jackson, Jeremy B.C., Michael X. Kirby, Wolfgang H. Berger, Karen A. Bjorndal, Louis W. Botsford, Bruce J. Bourque, Roger H. Bradbury, Richard Cooke, Jon Erlandson, James A. Estes, Terence P. Hughes, Susan Kidwell, Carina B. Lange, Hunter S. Lenihan, John M. Pandolfi, Charles H. Peterson, Robert S. Steneck, Mia J. Tegner, and Robert R. Warner. 2001. Historical overfishing and the recent collapse of coastal ecosystems. *Science* 293 (5530): 629–638.

Kerr, Alexander, and Andrew H. Baird. 2007. Natural barriers to natural disasters. *BioScience* 57 (2): 102–103.

Krumhansla, Kira A., Daniel K. Okamoto, Andrew Rassweiler, Mark Novak, John J. Bolton, Kyle C. Cavanaugh, Sean D. Connell, Craig R. Johnson, Brenda Konar, Scott D. Ling, Fiorenza Micheli, Kjell M. Norderhaug, Alejandro Pérez-Matus, Isabel Sousa-Pinto, Daniel C. Reed, Anne K. Salomon, Nick T. Shears, Thomas Wernberg, Robert J. Andersond, Nevell S. Barrett, Alejandro H. Buschmann, Mark H. Carr, Jennifer E. Caselle, Sandrine Derrien-Courtel, Graham J. Edgar, Matt Edwards, James A. Estes, Claire Goodwin, Michael C. Kenner, David J. Kushner, Frithjof E. Moy, Julia Nunn, Robert S. Steneck, Julio Vásquez, Jane Watson, Jon D. Witman, and Jarrett E.K. Byrnes. 2016. Global patterns of kelp forest change over the past half-century. *Proceedings of the National Academy of Sciences of the United States* 113 (48): 13785–13790.

Leshchinsky, Ben, Michael J. Olsen, Curran Mohney, Kira Glover-Cutter, Geoff Crook, Jonathan Allan, Michael Bunn, Matthew O'Banion, and Nicolas Mathews. 2017. Mitigating coastal landslide damage. *Science* 357 (6355): 981–982.

Louv, Richard. 2008. *Last child in the woods: Saving our children from nature-deficit disorder.* Chapel Hill, NC: Algonquin Books.

Lumsden, S. Elizabeth, Thomas F. Hourigan, Andrew W. Bruckner, and Gabrielle Dorr (eds.). 2007. *The state of deep coral ecosystems of the United States.* NOAA Technical Memorandum CRCP-3. Silver Spring, MD: National Marine Fisheries Service.

Muller-Karger, Frank E., Maria T. Kavanaugh, Enrique Montes, Willaim M. Balch, Mya Breitbart, Francisco P. Chavez, Scott C. Doney, Elizabeth M. Johns, Ricardo M. Letelier, Michael W. Lomas, Heidi M. Sosik, and Angelicque E. White. 2014. A framework for a marine biodiversity observing network within changing continental shelf seascapes. *Oceanography.* 27 (2): 18–23.

Perry, Allison L., Paula J. Low, Jim R. Ellis, and John D. Reynolds. 2005. Climate change and distribution shifts in marine fishes. *Science* 308 (5730): 1912–1915.

Quintana, Juan. 2000, July 1. The drought in Chile and La Niña. Drought network news (1994–2001). Paper 71. Available online. http://digitalcommons.unl.edu/droughtnetnews/71. Accessed 24 Jan 2017.

Rangoonwala, Amina, Cathleen E. Jones, and I.I.I. Elijah Ramsey. 2016. Wetland shoreline recession in the Mississippi river from petroleum oiling and cyclonic storms. *Geophysical Research Letters* 43 (22): 11652–11660.

Reed, Stanley, and Alison Fitzgerald. 2011. *In too deep: BP and the drilling race that took it down*. Hoboken, NJ: Bloomberg Press.

Sallenger Jr., H. Asbury, Kara S. Doran, and Peter A. Howd. 2012. Hotspot of accelerated sea-level rise on the Atlantic coast of North America. *Nature Climate Change* 2: 884–888. https://doi.org/10.1038/nclimate1597.

Sakib, Mohiuddin, Fatin Nihal, Anisul Haque, Mansur Rahman, and Mansur Ali. 2015. Sundarban as a buffer against storm surge flooding. *World Journal of Engineering and Technology* 3 (3): 59–64.

Sheng, Y. Peter, Andrew Lapetina, and Gangfeng Ma. 2012. The reduction of storm surge by vegetaion canopies: Three-dimensional simulations. *Geophysical Research Letters* 39 (L20601). https://doi.org/10.1029/2012gl053577.

U.S. Environmental Protection Agency. 2008. Reducing urban heat islands: Compendium of strategies. Available online. https://www.epa.gov/heat-islands/heat-island-compendium. Accessed on 6 Nov 2017.

Soucie, Gary. 1973. Where beaches have been going: Into the Oceans. *Smithsonian* 4 (3): 55–61.

Wieski, K., and S. Pennings. 2014. Climate drivers of *Spartina alterniflora* Saltmarsh production in Georgia, USA. *Ecosystems.* 17 (3): 473–484. https://doi.org/10.1007/s10021-013-9732-6.

Wilkinson, Clive (ed.). 2008. *Status of the coral reefs of the world: 2008*. Townsville, Australia: Global Coral Reef Monitoring Network and Reef and Rainforest Research Centre. 296pp.

Wilmers, C.C., J. Estes, M. Edwards, K. Laidre, and B. Konar. 2012. Do trophic cascades affect the storage and flux of atmospheric carbon? An analysis of sea otters and kelp forests. *Frontiers in Ecology and the Environment* 10 (8): 409–415.

Zinnert, J.C., Sheri A. Shiflett, Jaclyn K. Vick, and Donald R. Young. 2011. Woody vegetative cover dynamics in response to recent climate change on an Atlantic coast barrier island: A remote sensing approach. *Geocarto International* 26 (8): 595–612.

Zinnert, J.C., S.A. Shiflett, S. Via, S. Bissett, B. Dows, P. Manley, and D.R. Young. 2016. Spatial-temporal dynamics in barrier island upland vegetation: The overlooked coastal landscape. *Ecosystems* 19 (4): 685–697.

Chapter 9
Coastal Erosion and Land Loss: Causes and Impacts

Lynn Donelson Wright, Wei Wu and James Morris

We speak of course of that narrow strip of land over which ocean waves and moon-powered tides are masters-that margin of territory that remains wild despite the proximity of cities....
—W. J. Dakin—Australian Seashores

9.1 Shore "Volatility" Versus Net Erosion and Retreat

The margins of the sea are encroaching landward throughout most of the world. This is happening not simply because of sea level rise (SLR) and the inundation phenomena discussed in Chap. 7, but also because the solid material-sand, mud, gravel-composing the shore and the subaerial and subaqueous lands immediately adjacent to it is being displaced. It has long been accepted that shores—particularly sandy beaches—are cut back during high wave energy events but recover following periods of moderate to low energy during which long-period swell waves return sediments shoreward. Shepard (1950) offered the original, and classic, description of these beach "cycles" with reference to the beach at Scripps Institute of Oceanography in Southern California. More recent discussions of beach mobility can be found in most texts on coastal processes (e.g. Komar 1998; Dean and Dalrymple 2002). Such quasi-cyclic changes may be regarded as a type of *volatility* but not necessarily as indicative of chronic or irreversible coastal retreat.

The quasi-cyclic seasonal erosion and recovery of beaches are commonly superimposed on longer-term trends. When the sediments composing the shores,

L. D. Wright (✉)
Southeastern Universities Research Association (SURA), Washington, DC, USA
e-mail: ldwright@bellsouth.net

W. Wu
Department of Coastal Sciences, University of Southern Mississippi,
Ocean Springs, Hattiesburg, Mississippi, MS, USA

J. Morris
Baruch Institute for Marine and Coastal Sciences, University of South Carolina,
Columbia, SC, USA

L. D. Wright and C. R. Nichols (eds.), *Tomorrow's Coasts: Complex and Impermanent*, Coastal Research Library 27,
https://doi.org/10.1007/978-3-319-75453-6_9

dunes and shallow sea floor are transported seaward by extreme events to depths beyond the maximum depth from which constructive or fair weather processes can return them, then erosion may be permanent, the shores will retreat landward and seas will encroach on the coastal hinterlands. Similarly, sediments cast landward by high waves and storm surge or moved along the coast by shore parallel currents will lead to retreat. In the simplest terms, when the rate of removal of coastal material exceeds the rate of input, recession will be the result and this recession will add to and exacerbate the landward translation of the land-sea interface that accompanies SLR alone. Based on a recent literature review and numerical modeling analysis, Vitousek et al. (2017) argue that because of the combination of rising sea level and anthropogenic severance of natural sediment to the coast, many beaches, particularly those on the Southern California coast, are likely to vanish by the end of the century.

9.2 Erosion of Coastal Barrier Islands and Contiguous Wetlands

According to Stutz and Pilkey (2001), coastal barrier islands and barrier spits fringe roughly 7% of the world's coasts. Such barrier systems dominate the shores of U.S. Atlantic and Gulf of Mexico coasts. Although coastal barrier islands are not usually the sites of dense habitation, except in cases of high tourist-oriented communities such as the Outer Banks of North Carolina, Florida's resort coasts and Australia's Gold Coast, they provide essential protection for communities and ecosystems situated immediately landward of the shallow lagoons or wetlands that back the islands. Erosion of the barrier systems removes or reduces the shelter provided by the islands and recent studies indicate that barrier systems are presently receding at alarming rates by way of different modes. Without the protection of the barriers, the wetlands that often exist behind the barrier islands or spits are subject to rapid, and often permanent loss. According to NOAA (2016. *U.S. Climate Resilience Toolkit*), between 1998 and 2009, the U.S. lost a total wetlands area greater than the area of the state of Rhode Island.

Cowell et al. (1995) and subsequently Lorenzo-Trueba and Ashton (2014) developed morphodynamic behavioral models to explain how barrier islands may retreat as sea levels rise. These models typically assume that some kind of equilibrium profile is maintained as the barriers migrate shoreward. Lorenzo-Trueba and Ashton (2014) describe four general modes of barrier retreat as predicted by their model. In the first, and simplest, mode, dynamic equilibrium is maintained and the shape and volume of the barrier are preserved as the barrier advances landward by overwash of dune sands onto the land surface or wetlands behind. The second possibility, "height drowning" prevails when the rate of overwash is insufficient to keep pace with landward migration and the barrier eventually becomes submerged. "Width drowning" occurs when the landward delivery of sand from the shoreface (or surf zone) is inadequate to maintain the barrier volume or when the net movement of sediment is offshore rather than shoreward. The fourth mode of retreat is characterized by episodic landward migration with periods of rapid overwash alternating with periods of relative stability. Crucial to these fairly simple models, however, is an adequate availability of sediment from the inner continental shelf. When the shelf is unable to nourish the beach and dunes, the barrier will ultimately vanish.

In the simple "roll-over" model, it is assumed that the barrier is composed of a deeply-rooted body of sand. This is the case in some instances but in many other cases the "barrier" is little more than a thin veneer of sand migrating over an eroding surface of salt marsh or relict marsh peat as in the Cedar Island case described in Chap. 5 (Fig. 5.9). In a recent analysis of the retreat of Virginia's barrier islands, Deaton et al. (2017) describe the how the processes of barrier island retreat are reducing the area of wetlands behind the islands and are also causing the tidal prism of the back-barrier estuaries to diminish. The cases they describe are experiencing both wash over and shore erosion and are thus narrowing. They attribute the loss of salt marsh to burial by sands from the migrating barrier islands. Figure 9.1a shows an example of washover fans burying the marsh behind and Fig. 9.1b shows the dead marsh peat being exhumed on the "beach" as the sands migrate landward. As discussed in Chap. 5, Zinnert et al. (2017) point out that the marsh grass, *Spartina patens* is tolerant of burial and can survive periods of burial beneath transgressing sands. However, the marsh cannot thrive in the high-energy surf zone environment once the protective barrier has migrated landward.

The worst losses of coastal lands in the United States are taking place on the Gulf of Mexico coast, particularly in coastal Louisiana where the rate of land loss is popularly described as equivalent to "a football field every hour". "*The barrier islands of Louisiana are eroding at an extreme rate. In places up to 100 feet of shoreline are disappearing every year. Though it has long been assumed that this*

Fig. 9.1 The outcomes of barrier island "roll over" on Cedar Island, Virginia's Eastern Shore. A. Washover fans of the narrow and thin barrier burying the marsh and tidal channels to the west. B. Relict marsh peats compose the foreshore left behind by the migrating barrier. (Photos: L. D. Wright)

erosion was due to the area's rapid rate of relative sea level rise, recent studies by the U.S. Geological Survey show that other coastal processes, such as the long-shore redistribution of sediments, are responsible for this erosion."—Dr. Jeffrey H. List, U.S. Geological Survey (USGS Fact Sheet *Louisiana Barrier Islands: A Vanishing Resource*). We will examine the unique situation of Louisiana in more detail in Part 3. Rosati and Stone (2009) describe the recent evolution of the barrier islands along the northern Gulf of Mexico (NGOM) coast, which includes the Florida Panhandle, Alabama, Mississippi and Louisiana. In contrast to most of the U.S. Atlantic coast barriers, the NGOM barriers are backed by extensive bays, sounds or open water areas such as Santa Rosa Bay, Mobile Bay, Mississippi Sound and Barataria Bay. Storm surges can significantly raise water levels within these bays and cause erosion to shores behind the barriers. The width and height of the dunes comprising most barrier systems are not uniform along the coast and Houser et al. (2007) show, with reference to Santa Rosa Island in west Florida, that barrier island responses to the high waves and storm surges associated with severe storms are critically dependent on dune height and width. In the case of Louisiana, subsidence is progressively enlarging the open water areas (Morton et al. 2002) while the width and height of transgressive barriers like the Chandeleur Islands are diminishing.

9.3 Retreating Marshes and Wetlands

Coastal wetlands provide critical ecosystem services, including carbon sequestration, protection from storms, flood control, habitat for a variety of species, fisheries, water quality improvement, recreational and aesthetic opportunities, and cultural values etc. (Costanza et al. 1997; Engle 2011). However, wetlands have been disappearing at an unprecedented rate over the past several decades. Coleman et al. (2008) analyzed geologic and geomorphic data on 42 world deltas and found a total of 15,845 km^2 (6118 mi^2) of wetlands were irreversibly lost at an average rate of 95 km^2/year (37 mi^2) in 14 of these deltas from the middle of the 1980s to the early 2000s. With a similar trend in the other deltas, a total wetland loss would be on the order of 364,000 km^2 in the 42 deltas. Dixon et al. (2016) found the world's marine/coastal wetlands declined at an average rate of 38% from 1970 to 2008, with the highest decline rate of 50% in Europe, followed by 41% of decline in Asia, 28% in North America, 19% in Africa, and 17% in Oceania. This dramatic loss of coastal wetlands is due to natural causes and anthropogenic conversion of wetlands for agricultural and industrial uses (Coleman et al. 2008). In the United States, the vast majority of wetland loss occurred in the northern Gulf of Mexico coasts of Louisiana and Texas (Dahl 2006; Dahl and Stedman 2013). About 99% of the losses of estuarine emergent wetlands between 2004 and 2009 were caused by coastal storms, land subsidence, SLR, salt-water intrusion, wave erosion etc. The conversion of coastal wetlands to other land uses are rare as they are protected by various State and Federal coastal regulatory measures and are not allowed to be converted to other land uses (Dahl and Stedman 2013). A retreating marsh is shown in Fig. 9.2.

Stability of the coastal wetland platform under SLR reflects the balance between inputs due to allochthonous (externally derived) matter deposition and in situ vegetation production, versus losses through subsidence, erosion, and organic matter decomposition (Neubauer 2008). Feedback among inundation, sediment trapping, and vegetation productivity help maintain coastal wetlands facing SLR (Morris et al. 2002). However, when the relative rate exceeds a threshold, coastal wetlands can collapse (Wu et al. in press; Loucks 2006; Kirwan et al. 2010; Ratliff et al. 2015).

A suite of models have been developed for predicting the response of coastal wetlands to relative SLR (e.g., Morris and Bowden 1986; Park et al. 1989; Costanza et al. 1990; Martin et al. 2000, 2002; Reyes et al. 2000; Morris et al. 2002, 2016; D'Alpaos et al. 2007; Kirwan and Murray 2007; Mudd et al. 2009; Ross et al. 2009; Stralberg et al. 2011; Fagherazzi et al. 2012; Rogers et al. 2012; Hagen et al. 2013; Schile et al. 2014; Ratliff et al. 2015; Wu et al. 2015; Enwright et al. 2016; Clough et al. 2016; Wu et al. in press). These models vary in structure, complexity, and ease of application. Simpler models empirically capture the key characteristics of wetland dynamics, require less data, and are easily applied, but interactions and feedbacks between geo-morphological and ecological processes are missing or overly simplified (e.g. Park et al. 1989; Wu et al. 2015; Kirwan and Guntenspergen 2009). For example, the Sea Level Affecting Marshes Model (SLAMM) model uses digital elevation data and other information to simulate potential impacts of long-term SLR on wetlands and shorelines. More sophisticated models mechanistically account for the important interactions and feedback mechanisms among

Fig. 9.2 Salt marsh retreat on the Louisiana Coast due to wave erosion and sea-level rise Photo by Wei Wu

vegetation, sediment, hydrology, and sea level factors, but generally require more input data, and are difficult to implement, especially at broader spatial scales (e.g., Costanza et al. 1990; Martin et al. 2000; Reyes et al. 2000; Morris et al. 2002). The various models highlight the importance of accounting for feedback among inundation, sediment trapping, and vegetation productivity in predicting wetland change. They also emphasize the need to integrate the environmental drivers other than sea-level rise in predictions, such as increased concentration of CO_2 and rising temperature, as they interact with sea-level rise to affect coastal wetland dynamics.

9.4 Disappearing River Deltas

Deltas owe their existence to the fact that rivers have been able to supply sediment to the coast at rates that have exceeded the rate of removal by oceanographic processes or the rates of relative submergence by subsidence and rises in relative sea level. Unfortunately, today, sediment supply has been sharply reduced by multiple anthropogenic factors as described in Chap. 4 while rates of subsidence have been increased, also by humans. Already, subaerial deltaic lands are rapidly disappearing and the disappearance is sure to accelerate as the rate of SLR accelerates over the coming decades. According to Overeem and Syvitski (2009), in the year 2009 roughly 26,000 km^2 (10,039 mi^2) of the world's deltaic land surfaces were below sea level and about 96,000 km^2 (37,066 mi^2) were highly vulnerable at elevations less than 2 m (6.6 ft.). As pointed out in Chaps. 4 and 6, deltas currently support twelve megacities and 500 million people so relocation is not a realistic option in many or most cases.

When Hurricane Katrina made landfall on the Louisiana coast in 2005, it caused a storm surge of up to 8.5 m (28 ft) that devastated New Orleans, particularly in the Ninth Ward, and killed over 1800 people (NHC). A comparable storm surge was generated by Hurricane Camille in 1969 but the impact on New Orleans was minimal. The difference between the 1969 and 2005 impacts was attributable largely to loss of protective wetlands separating the city from the Gulf of Mexico over the intervening 36 years. The Barras 2006 reported that just prior to Katrina, the Mississippi Delta, was losing land at the rate of 65–90 $km^{2/}yr$ (25–35 mi^2) (Fig. 9.3). Numerous human activities, including the dredging of canals as well as the isolation of the delta plain from renewed sediment supply contributed to the land loss. Then, in 2005, Hurricanes Katrina and Rita together converted an additional 15% of Louisiana's wetlands to open water (Barras 2006; Xing 2015) thereby exposing even more wetlands to wave action and erosion. Hurricane Katrina alone caused over 250 km^2 (97 mi^2) of coastal lands to disappear in two days (Barras 2006). Today, extensive engineering works (see Chap. 6) protect New Orleans even though it is sinking. But the surrounding lands and wetlands are more vulnerable than ever. According to Barnes and Virgets (2017), over 5180 km^2

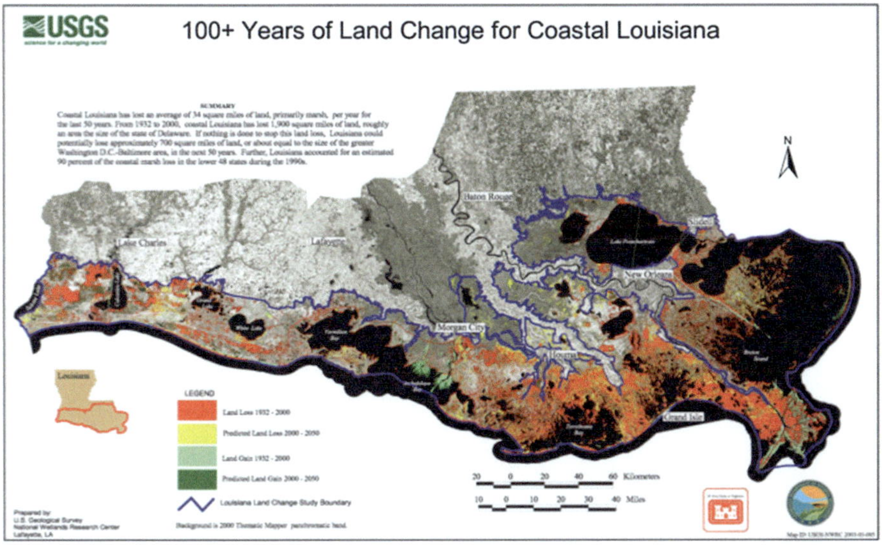

Fig. 9.3 Coastal land loss in Louisiana in the century prior to Hurricane Katrina (from the U.S. Geological Survey 2004)

(2000 mi^2) of coastal land have been lost over the past century and that much again could be lost within the next 50 years. The Gulf of Mexico is progressively invading the Mississippi Delta.

9.5 Thermal Erosion of Arctic Coasts

The dramatic reduction of sea ice in the Arctic Ocean, described in Chap. 2 (see Fig. 2.8and Chap. 16) is being accompanied by equally dramatic erosion of the Beaufort Sea and Chukchi Sea coasts. This coastal realm is backed by the Arctic North Slope coastal plain of Alaska and Canada. The flat, relatively low-elevation, coastal plain, north of Alaska's Brooks Range, is a tundra surface comprised of a thin active (alternate freezing and thawing) layer of mosses and lichens overlying a much thicker substrate of permafrost, or permanently frozen soil. The interstitial ice normally holds the soil together but thawing causes the soil to lose its "cement" which makes it easily erodible. As the Arctic Ocean has increasing opened up (become ice free) during the summer, higher waves and warmer water have washed the shores. The exposed permafrost at the shore has gradually melted, releasing the sediments previously secured by the ice to the more energetic waves, which carry this sediment away causing coastal recession. The thermal erosion at the shore typically allows the permafrost to collapse in large blocks (Fig. 9.4) to be further eroded by waves and waters that are above freezing temperatures.

Fig. 9.4 Thermal erosion of the permafrost shoreline on Alaska's Beaufort Sea Coast. (Photo From USGS; Photo Credit: U.S. Department of the Interior|U.S. Geological Survey URL: https://walrus.wr.usgs.gov/climate-change/hiLat.html. Contact: Laura Zink Torresan

The U.S. Geological Survey has recorded the erosion rates of the Arctic coast over several decades. Jones et al. (2009) reported that, between the years 2002 and 2007, a 64 km (40 mi.) section of the Beaufort Sea coast retreated at an average rate of 14 m (45 ft) per year. Of course, erosion rates vary appreciably along the coast. In a more recent USGS Open File Report, Gibbs and Richmond (2015) conclude that for the entire 1600 km coastal reach from the Canadian border to Icy Cape on the Chukchi Sea coast, the average rate of retreat is 1.4 m/year (4.6 ft/year) and the local maximum rate is 18.5 m/year. (60 ft/year). As pointed out in Chap. 2, the Arctic Ocean could be completely ice free in summer by mid century. The coasts would be subject to attacks from larger waves and warmer water bathing the frozen permafrost shores. According to Gibbs and Richmond (2015) in the years ahead "—Arctic coasts will be more vulnerable to storm surge and wave energy, potentially resulting in accelerated shoreline erosion and terrestrial habitat loss in the future." This represents a classic case of positive (self reinforcing) morphodynamic feedback: accelerated ice melting and coastal retreat will increase exposure to wave attack by increasing the open water fetch distance over which wave-generating winds can blow. And as pointed out previously, the decreased albedo resulting from the increased area of ice-free open water will be followed by continued warming of the earth-and coastal recession, which in turn will further increase the open water area. Coastal retreat in the Arctic is likely to accelerate in the years ahead. The socioeconomic and ecological impacts Arctic coastal land loss are discussed in more detail in Chap. 16.

9.6 "Tipping Points" for Coastal Submersion

So long as the rates of SLR have remained relatively slow, or at least below certain critical levels, it has been possible for coastal lands to keep pace and remain emergent, or at least remain within the intertidal zone, by means of a combination of sediment accumulation and organic production. However, it has long been understood that there must be limits to the rate of drowning beyond which accretion cannot cope and permanent submersion will prevail. Several recent studies (e.g., Kirwan et al. 2010; Morris et al. 2002, 2016; Ratliff et al. 2015; Schile et al. 2014; Turner et al. 2017; Watson et al. 2017a, b; Wu et al. in press) have focused on determining what these critical inundation rate "tipping points" might be under different circumstances. The studies have involved various combinations of methodologies including statistical analyses of historical trends, numerical modeling and stratigraphic analyses of sedimentary cores. Morris et al. (2016) determined that intertidal wetlands on the U.S. East Coast and Gulf Coast can accrete at rates up to 3 mm/year (.12 in./year) via organic production and up to 2 mm/year (.08 in./year) as a result of deposition of inorganic sediment for a total maximum accretion rate of 5 mm/year (.20 in./year). Watson et al. (2017a, b) arrived at a similar conclusion with respect to the maximum accretion rate of New England salt marshes and reported that, in that particular region, the rates of relative SLR may have already reached the tipping point for wetlands loss. Relative SLR at rates in excess of $\sim$5 mm/year (>.2 in./year) would thus likely result in wetlands loss and submersion of the intertidal realm.

Turner et al. (2017 *in press*) recently reanalyzed stratigraphic data and radiocarbon ages for deposits associated with 36 of the world's deltas spanning the Holocene SLR period of the past 22,000 years. Their results indicate that for 90% of the delta deposits examined, the onset of significant delta formation took place roughly 8000 years ago at a time when the rate of SLR had slowed to somewhere between 10 mm/year (.39 in./year) and 5 mm/year (.2 in./year), typically <6 mm/year (.24 in./year). They argue that if this rate of SLR is the tipping point for the onset of delta growth as the rate of rise decelerates, it may also be the tipping point for delta destruction as the rate of rise accelerates.

As shown by Morris et al. (2016), the accretion rates vary depending on plant productivity, root:shoot ratio, suspended sediment concentration, sediment-capture efficiency and episodic events. This leads to high spatial variability in wetland accretion rates and SLR thresholds on sustainability of coastal wetlands. At a river dominated estuary on the Mississippi Coast—lower Pascagoula Bay, the measured accretion rates is up to 8.6 mm/year (.34 in./year) with a standard deviation of 3.7 mm/year (.15 in./year) using fallout radionuclides (137Cs and 210Pb) (Wu et al. 2015). Some other studies showed higher sea-level rise threshold than 5 mm/year (.2 in./year) based on different settings of sediment concentration in water columns and tidal range. Kirwan et al. (2010) predicted the threshold of SLR rate as 10 mm/year (.39 in./year) in a typical estuary in the southeastern US and western Europe with suspended sediment concentrations greater than 20 mg/L (20,000 ppb) and tidal ranges great than 1 m (3.28 ft). Wu et al. (in press) predicted a SLR threshold of 8.4 mm/year (.33 in./year) for the Grand Bay National Estuarine Research Reserve which receives limited freshwater input, has an average of total suspended sediment concentration as 18 mg/L (18,000 ppb) and microtidal range of 0.6 m (1.96 ft) (Wu et al. in press). Wu et al. (in press) proposed a new threshold of accelerated sea-level rise to study the tipping point for coastal wetlands under the scenario of accelerating SLR. They also raised the point that other landscape metrics that quantify spatial patterns of coastal wetland distribution in addition to total area, for example, indices to represent fragmentation, should be considered when SLR tipping point is derived.

Notably, the present rate of global SLR of 3.1 mm/year (.12 in./year) (Fig. 3.3) is below the thresholds described in the above studies if regional subsidence is neglected. However, as discussed in Chap. 3, the IPCC (2013) analyses indicate that the rate is likely to be between 8 and 16 mm/year (.31 and .63 in./year) by the mid to latter part of the 21st century, neglecting any potential collapse of major Antarctic ice sheets. And, of course, in the case of deltas, subsidence is added to the effects of SLR. It would thus seem wise to prepare for the possibility that, world wide, vast areas of wetlands and deltaic lands may be replaced by open water before the end of this century.

9.7 Socioeconomic Consequences of Coastal Land Loss

For some low-lying small island nations, such as the Marshall Islands and Maldives, entire populations are at risk of becoming "climate change refugees" because relocation to high ground within their country is not possible. Notably, northern European coasts fronting the North Sea, particularly the Netherlands, have for centuries relied on complex coastal engineering programs to protect lands that are increasingly below sea level-in some cases by more than 2 m (6.6 ft; Dronkers and Stojanovic 2016). These programs are essential to the Netherlands survival but are extremely costly. In the United States, Hauer et al. (2016) predicted that by 2100 SLR of .9 and 1.8 m (2.95–5.9 ft) would place land areas projected to house 4.2 million and 13.1 million people respectively at risk of inundation. This amounts to a three-fold increase in the vulnerable population under the scenario of 1.8 m (5.9 ft) SLR by 2100 relative to today. In addition to monetary costs, population growth and economic development are severely constrained. Fortunately, however, the affected Northern European countries are relatively affluent and their prospects of resilience are reasonably good. This is not the case for the populations of many coastal megacities such as Mumbai, Dhaka, Calcutta and Lagos where the loss of habitable land enhances misery and disease and forces some homeless people to live in the bottoms of boats and canoes (e.g. in Lagos, Nigeria; McDonnell 2017). Perhaps most dramatically of all coastal land loss worldwide, the Marshall Islands are experiencing serious inundation on a regular basis and are likely to become uninhabitable in the foreseeable future causing the displacement of an entire nation (Goodell 2017).

The most serious losses of coastal lands in the U.S. are taking place in Louisiana (see Sect. 9.3 and Chap. 13) and are bringing the Gulf of Mexico closer and closer to cities like New Orleans, Houma and Lake Charles. From a recent economic impact study supported by the Environmental Defense Fund, Barnes and Virgets (2017) conclude that $3.6 billion in business, residential and infrastructure assets are presently at risk in Louisiana and the loss of these assets could cause an additional loss of $7.6 billion per year in nationwide economic activity. With the continued disappearance of protective wetlands, increased exposure of homes, industries and infrastructure could allow a single hurricane to destroy as much as $138 billion in assets (Barnes and Virgets 2017). On the positive side, however, Barnes and Virgets (2017) note that if plans for coastal restoration can be implemented, up to 10,500 jobs could be created. There are numerous other consequences of wetlands loss that cannot be readily expressed in monetary terms. The traditional culture and ways of life of Cajun and Creole residents of coastal Louisiana are among these. Two threatened "cultural islands", Tangier Island, VA and Smith Island, MD, also exist in the Virginia and Maryland portions of the Chesapeake Bay and are experiencing severe erosion and frequent inundation. High rates of SLR in the Chesapeake Bay are evidenced by erosion of marsh and salt water intrusion that is slowing killing woodland trees and shrubs that are not adapted to the high marsh. In these locations, many homeowners who do not want to leave are hoping for their States to build seawalls and breakwaters.

References

Barnes, S.R., and S. Virgets. 2017. *Regional Impacts of Coastal Land Loss and Louisiana's Opportunity for Growth*, 39. Baton Rouge: LSU E. J. Ourso College of Business Economics & Policy Research Group.

Barras, J.A. 2006. Land area change in coastal Louisiana after the 2005 hurricanes—A series of three maps: *U.S. Geological Survey Open-File Report 06–1274*, Available online. URL: http://pubs.usgs.gov/of/2006/1274/.

Clough, J., A. Polaczyk, and M. Propato. 2016. Modeling the potential effects of sea-level rise on the coast of New York: Integrating mechanistic accretion and stochastic uncertainty. *Environmental Modelling and Software* 84: 349–362.

Coleman, J.M., O.K. Huh, and D. Brud Jr. 2008. Wetland loss in world deltas. *Journal of Coastal Research* 24 (sp1): 1–14.

Costanza, R., F.H. Sklar, and M.L. White. 1990. Modeling coastal landscape dynamics. *BioScience* 40: 91–107.

Costanza, R., R. dArge, R. de Groot, S. Farber, M. Grasso, B. Hannon, K. Limburg, S. Naeem, R. V. Oneill, J. Paruelo, R.G. Raskin, P. Sutton, and M. van den Belt. 1997. The value of the world's ecosystem services and natural capital. *Nature* 387: 253–260.

Cowell, P.J., P.S. Roy, and R.A. Jones. 1995. Simulation of large-scale coastal change using a morphological behaviour model. *Marine Geology* 126 (1–4): 45–61.

Dahl, T.E. 2006. *Status and Trends of Wetlands in the Conterminous United States, 1998 to 2004*. Fish and Wildlife Service, Washington, DC: US Department of the Interior.

Dahl, T.E., and S.M Stedman. 2013. Status and trends of wetlands in the coastal watersheds of the Conterminous United States 2004 to 2009. U.S. Department of the Interior, Fish and Wildlife Service and National Oceanic and Atmospheric Administration, National Marine Fisheries Service.

D'Alpaos, A., S. Lanzoni, M. Marani, and A. Rinaldo. 2007. Landscape evolution in tidal embayments: Modeling the interplay of erosion, sedimentation, and vegetation dynamics. *Journal of Geophysical Research, Part F, Earth Surfaces* 112: 17.

Dean, R.G. and R.A Dalrymple. 2002. *Coastal Processes with Engineering Applications*, 475 pp Cambridge: Cambridge University Press.

Deaton, C.D., C.J. Hein, and M.L. Kirwan. 2017. Barrier island migration dominates ecogeomorphic feedbacks and drives salt marsh loss along the Virginia Atlantic Coast, USA. *Geology* 45 (2): 123. https://doi.org/10.1130/G38459.1.

Dixon, M.J.R., J. Loh, N.C. Davidson, C. Beltrame, R. Freeman, and M. Walpole. 2016. Tracking global change in ecosystem area: The wetland extent trends index. *Biological Conservation* 193: 27–35.

Dronkers, J., and T. Stojanovic. 2016. Socio-economic impacts—coastal management and governance. *North Sea Region Climate Change Assessment.* pp. 475–488.

Engle, V.D. 2011. Estimating the provision of ecosystem services by gulf of Mexico coastal wetlands. *Wetlands* 31 (1): 179–193. https://doi.org/10.1007/s13157-010-0132-9.

Enwright, N.M., K.T. Griffith, and M.J. Osland. 2016. Barriers to and opportunities for landward migration of coastal wetlands with sea-level rise. *Frontiers in Ecology and the Environment* 14 (6): 307–316. https://doi.org/10.1002/fee.1282.

Fagherazzi, S., M.L. Kirwan, S.M. Mudd, G.G. Guntenspergen, S. Temmerman, et al. 2012. Numerical models of salt marsh evolution: Ecological, geomorphic, and climatic factors. *Reviews of Geophysics* 50: RG1002.

Gibbs, A.E., and B.M. Richmond. 2015. National assessment of shoreline change—Historical shoreline change along the north coast of Alaska, U.S.–Canadian border to Icy Cape. *U.S. Geological Survey Open-File Report 2015–1048*, 96 p., https://dx.doi.org/10.3133/ofr20151048.

Goodell, J. 2017. *The Water Will Come: Rising Seas, Sinking Cities and the Remaking of the Civilized World*, 340. New York: Little, Brown and Company.

Hagen, S.C., J.T. Morris, P. Bacopoulos, and Jf Weishampel. 2013. Sea-level rise impact on a salt marsh system of the lower St. Johns River. *J Waterway, Port and Ocean Engineering*. 139: 118–125.

Hauer, M.E., J.M. Evans, and D.R. Mishra. 2016. Millions projected to be at risk from sea-level rise in the continental United States. *Nature Climate Change* 6 (7): 691–695.

Houser, C., C. Hapke, and S. Hamilton. 2007. Controls on coastal dune morphology, shoreline erosion and barrier island response to extreme storms. *Geomorphology* 100 (3–4): 223–240. https://doi.org/10.1016/j.geomorph.2007.12.007.

Intergovernmental Panel on Climate Change (IPCC), 2013 Climate Change 2013: *The Physical Science Basis. Contribution of Working Group I to the Fifth Assessment Report of the Intergovernmental Panel on Climate Change* Stocker, T.F., D. Qin, G.-K. Plattner, M. Tignor, S.K. Allen, J. Boschung, A. Nauels, Y. Xia, V. Bex and P.M. Midgley (eds.). Cambridge University Press, Cambridge.

Jones, et al. 2009. Increase in the rate and uniformity of coastline erosion in Arctic Alaska. *Geophysical Research Letters* 36 (3): L03503. https://doi.org/10.1029/2008GL036205.

Kirwan, M.L., G.R. Guntenspergen, A. D'Alpaos, J.T. Morris, S.M. Mudd, and S. Temmerman. 2010. Limits on the adaptability of coastal marshes to rising sea level. *Geophysical Reseach Letters* 37: L23401.

Kirwan, M.L., and A.B. Murray. 2007. A coupled geomorphic and ecological model of tidal marsh evolution. *PNAS* 104 (15): 6118–6122.

Kirwan, M.L., and G.R. Guntenspergen. 2009. Accelerated sea-level rise—A response to Craft al. *Frontiers in Ecology and the Environment* 7 (3): 126–127.

Komar, P.D. 1998. *Beach Processes and Sedimentation*. Prentice Hall 544 pp.

Lorenzo-Trueba, J., and A.D. Ashton. 2014. Rollover, drowning, and discontinuous retreat: Distinct modes of barrier response to sea-level rise arising from a simple morphodynamic model. *Geophysical Research Earth Surface* 119: 779–801. https://doi.org/10.1002/2013JF002941.

Loucks, D.P. 2006. Modeling and managing the interactions between hydrology, ecology, and economics. *Journal of Hydrology* 328 (3–4): 408–416.

Martin, J.F., M.L. White, E. Reyes, G.P. Kemp, H. Mashriqui, and J.W. Day. 2000. Evaluation of coastal management plans with a spatial model: Mississippi Delta, Louisiana, USA. *Environmental Management* 26: 117–129.

Martin, J.F., E. Reyes, G.P. Kemp, H. Mashriqui, and J.W. Day. 2002. Landscape modeling of the Mississippi delta: Using a series of landscape models, we examined the survival and creation of Mississippi Delta marshes and the impact of altered riverine inputs, accelerated sea-level rise, and management proposals on these marshes. *BioScience* 52 (4): 357–365.

McDonnell, T. 2017. Slum Dwellers In Africa's biggest megacity are now living in Canoes. *NPR,* 15 May 2017.

Morris, J.T., and W.B. Bowden. 1986. A mechanistic, numerical model of sedimentation, mineralization, and decomposition for marsh sediments. *Soil Science Society of America Journal* 50: 96–105.

Morris, J.T., P.V. Sundareshwar, C.T. Nietch, B. Kjerfve, and D.R. Cahoon. 2002. Responses of coastal wetlands to rising sea level. *Ecology* 83 (10): 2869–2877.

Morris, J.T., et al. 2016. Contributions of organic and inorganic matter to sediment volume and accretion in tidal wetlands at steady state. *Earth's Future*. https://doi.org/10.1002/2015EF000334.

Morton, R.A., N.A. Buster, and M.D. Krohn. 2002. Subsurface controls on historical subsidence rates and associated wetland loss in Southcentral Louisiana. *Transactions. Gulf Coast Association of Geological Societies* 52: 767–778.

Mudd, S.M., S.M. Howell, and J.T. Morris. 2009. Impact of dynamic feedbacks between sedimentation, sea-level rise, and biomass production on near-surface marsh stratigraphy and carbon accumulation. *Estuarine, Coastal and Shelf Science* 82 (3): 377–389.

Neubauer, S.C. 2008. Contributions of mineral and organic components to tidal freshwater marsh accretion. *Estuarine, Coastal and Shelf Science* 78: 78–88.

Overeem, I., and J.P.M. Syvitski. 2009. Dynamics and vulnerability of delta systems. *LOICZ Reports & Studies No. 35*. Geesthacht: GKSS Research Center, 54 pp.

Park, R.A., M.S. Trehan, P.W. Mausel, and R.C. Howe. 1989. The effects of sea level rise on US coastal wetlands. In *The Potential Effects of Global Climate Change on the United States*. Appendix B. Sea level rise. U.S. EPA Office of Policy, Planning, and Evaluation, Washington, D.C., USA.

Ratliff, K.M., A.E. Braswell, and M. Marani. 2015. Spatial response of coastal marshes to increased atmospheric CO2. *Proceedings of the National Academy of Sciences of the United States of America* 112: 15580–15584. https://doi.org/10.1073/pnas.1516286112.

Reyes, E., M.L. White, J.F. Martin, G.P. Kemp, J.W. Day, and A. Aravamuthan. 2000. Landscape modeling of coastal habitat change in the Mississippi Delta. *Ecology* 81: 2331–2349.

Rogers, K., N. Saintilan, and C. Copeland. 2012. Modelling wetland surface elevation dynamics and its application to forecasting the effect of sea-level rise on estuarine wetlands. *Ecological Modelling* 244: 148–157.

Rosati, J.D., and G.W. Stone. 2009. Geomorphologic evolution of Barrier Islands along the Northern U.S. Gulf of Mexico and implications for engineering design in barrier restoration. *Journal of Coastal Research* 25 (1): 8–22.
Ross, M.S., S. Mitchell-Brucker, J.P. Sah, S. Stothoff, P.L. Ruiz, D.L. Reed, K. Jayachandran, and C.L. Coultas. 2006. Interaction of hydrology and nutrient limitation in ridge and slough landscape of southern Florida. *Hydrobiologia* 569: 37–59. https://doi.org/10.1007/s10750-006-0121-4.
Ross, M.S., J.J. O'Brien, R.G. Ford, K. Zhang, and A. Morkill. 2009. Disturbance and the rising tide: The challenge of biodiversity management for low island ecosystems. *Frontiers in Ecology and the Environment* 9: 471–478.
Schile, L.M., J.C. Callaway, J.T. Morris, D. Stralberg, V.T. Parker, and M. Kelly. 2014. Modeling tidal marsh distribution with sea-level rise: Evaluating the role of vegetation, sediment, and upland habitat in Marsh Resiliency. *PLoS ONE* 9 (2): e88760. https://doi.org/10.1371/journal.pone.008876.
Shepard, F.P. 1950. *Beach Cycles in Southern California.* U.S. Army Corps of Engineers, Beach Erosion Board. Technical Memorandum No. 20.
Stralberg, D., M. Brennan, J.C. Callaway, J.K. Wood, L.M. Schile, et al. 2011. Evaluating tidal marsh sustainability in the face of sea-level rise: A hybrid modeling approach applied to San Francisco Bay. *PLoS ONE* 6: e27388.
Stutz, M.L., and O.H. Pilkey. 2001. A review of global barrier island distribution: *Journal of Coastal Research*, Special Issue 34, ICS 2000 Proceedings, 15–22.
Turner, R.E., M.S. Kearney, and R.W. Parkinson. 2017. Sea-level rise tipping point of delta survival. *Journal of Coastal Research*. https://doi.org/10.2112/JCOASTRES-D-17-00068.1. (in press; published online October 2017).
Vitousek, S., P.L. Barnard, and P. Limber. 2017. Can beaches survive climate change? *Journal of Geophysical Research Earth Surface* 122: 1060–1067. https://doi.org/10.1002/2017JF004308.
Watson, E.B., K.B. Raposa, J.C. Carey, C. Wigand, and R.S. Warren. 2017a. Anthropocene survival of Southern New England's salt marshes. *Estuaries and Coasts* 40: 617–625.
Watson, E.B., C. Wigand, E.W. Davey, H.M. Andrews, J. Bishop, and K.B. Raposa. 2017b. Wetland loss patterns and inundation-productivity relationships prognosticate widespread salt marsh loss for southern New England. *Estuaries and Coasts* 40: 662–681.
Wu, W., K. Yeager, M. Peterson, and R. Fulford. 2015. Neutral models as a way to evaluate the Sea Level Affecting Marshes Model (SLAMM). *Ecological Modelling* 303: 55–69.
Wu, W., P. Biber, and M. Bethel. in press. Thresholds of sea-level rise rate and sea-level rise acceleration rate in a vulnerable coastal wetland. *Ecology and Evolution*.
Xing, F. 2015. *Deltas' Responses to Fluvial and Marine Forces.* Ph.D Dissertation, Submitted to the Graduate School of the University of Colorado, 165 pp.
Zinnert, J.C., J.A. Stallins, S.T. Brantley, and D.R. young. 2017. Crossing scales: the complexity of barrier-island processes for predicting future change. *BioScience* 67: 39–52.

Chapter 10
Impacts of Coastal Waters and Flooding on Human Health

Lynn Donelson Wright, Christopher F. D'Elia and C. Reid Nichols

> *"Go to the banks of the great grey-green, greasy Limpopo River, all set about with fever-trees, and find out"* (*what the crocodile has for dinner*).
>
> —Rudyard Kipling-The Elephant's Child

10.1 Rising Water, Rising Health Hazards

As detailed in Chap. 7, floods are among the most common natural hazards with complex and far-reaching impacts. Coastal floods are most often caused by storm surge (coastal), rivers that exceed their flood stage capacity (fluvial), and torrential rainfall (pluvial). Increasingly, compound flooding by all three causes is the most severe. The adverse consequences of flood events, especially coastal flooding, to human health have been evaluated by the World Health Organization (WHO 2003), the European Centre for Disease Prevention and Control (ECDC) (see http://www.ecdc.europa.eu/) and the U.S. Global Change Research Program (Trtani et al. 2016). Drowning is the major cause of death, followed by heart attacks, hypothermia, blunt trauma caused by wind-borne objects and vehicle-related accidents. A few fatal injuries also occur during evacuations and cleanups. Snakebites, electrocution and wound infections are also causes of death. Less obvious health impacts involve diseases and toxins spread by water and water-nurtured vectors (e.g. mosquitos). Some of the most prominent of these include:

L. D. Wright (✉)
Southeastern Universities Research Association (SURA), Washington, DC, USA

C. F. D'Elia
School of the Coast and Environment, Louisiana State University,
Baton Rouge, LA, USA

C. R. Nichols
Marine Information Resources Corporation (MIRC), Ellicott City, MD, USA

L. D. Wright and C. R. Nichols (eds.), *Tomorrow's Coasts: Complex and Impermanent*, Coastal Research Library 27,
https://doi.org/10.1007/978-3-319-75453-6_10

- Enteric infections from drinking water contamination and sewerage disruption;
- Vector-borne diseases, such as malaria, dengue and dengue hemorrhagic fever, yellow fever, West Nile fever and leptospirosis;
- Infections from numerous diseases contracted through direct contact or wound exposure with filthy, polluted flood waters (Fig. 10.1);
- Water-borne pathogens particularly *Vibrio* bacteria which cause intestinal, skin, ear, eye infections and septicemia (blood stream infection);
- Contamination by toxic chemicals during and after floods;
- Mold in homes following inundation.

Impacts from Hurricane Harvey on the Texas coast, from August 17, 2017 to September 3, 2017, included displacement of wild animals such as raccoons, skunks, foxes, and coyotes that might carry rabies, domesticated animals from farms and households, and some dangerous animals such as fire ants, poisonous snakes and American alligators. In the aftermath of Hurricane Harvey, animal shelters were overflowing with rescued pets. Press reports following the landfall of Hurricane Harvey indicated high concentrations of *E. coli* and even some flesh-eating bacteria in the urban floodwaters flowing through the streets and neighborhoods of Houston.

According to Doocy et al. (2013), throughout the world, 2.8 billion people were adversely affected by floods of all sorts between 1980 and 2009. Of those affected,

Fig. 10.1 Contaminated floodwaters surrounding houses in a Memphis TN neighborhood following Mississippi River flooding in 2011. Photo Credit: Sophia Ronan. Similar situations prevailed in Houston following Hurricane Harvey in 2017. The lack of potable water early in the recovery following large-scale natural disasters is a serious issue as was the case throughout Puerto Rico after Hurricane Maria in 2017

Table 10.1 Public health consequences of cyclones and tsunamis

Consequence	Cyclones	Tsunamis
Death rates	Developed nations- low Developing nations- high	High
Severe injuries (among survivors)	Few	Few to moderate
Loss of clean water	Widespread	Local to widespread
Loss of shelter	Widespread	Local to widespread
Loss of personal property	Widespread	Local to widespread
Loss of routine hygiene	Widespread	Local to widespread
Loss of sanitation	Widespread	Local to widespread
Disruption of solid waste removal	Widespread	Local to widespread
Increased pests and disease vectors	Widespread	Local to widespread
Worsening of chronic illnesses	Widespread	Local to widespread
Damage to health care system	Widespread	Local to widespread
Power outages	Widespread	Local to widespread
Exposure to toxic substances	Possible	Possible
Food scarcity	Common in low lying islands	Common in early stages

Source Keim (2006)

540,000 died and 4.6 million were rendered homeless; roughly 50% of flood related fatalities in recent years have been in Asia. Lane et al. (2013) list several secondary health hazards that are particular problems in flooded urban areas. Included are impeded or delayed evacuations of health care facilities, particularly nursing homes, and complications related to the transportation of frail, elderly or incapacitated patients. Prolonged power outages can shutdown life-support medical equipment and incapacitate elevators in high-rise buildings. Prison populations are sometimes at high risk of being isolated from food, water and adequate ventilation for prolonged periods. Table 10.1, from Keim (2006) summarizes some of adverse consequences of inundation events involving cyclones (hurricanes/typhoons) and tsunamis. A similar list is offered by the European Centre for Disease Prevention and Control as problematic in Europe (e.g. Vasconcelos 2006).

10.2 Drowning Deaths by Coastal Inundation

In 1970, a category 4 tropical cyclone, "The Bhola Cyclone", generated a storm surge that was funneled up the Bay of Bengal flooding the coastal Ganges-Brahmaputra Delta and killing somewhere between 300,000 and 500,000 people in what is now Bangladesh (Frank and Husain 1971). Storm surge associated with Cyclone Nargis killed over 138,000 people in Myanmar in 2008 (Fritz et al. 2009) and "Super Typhoon" Haiyan took 6000 lives in the Philippines in 2013 (Mas et al. 2015). The day after Christmas, 2004, the Asian Tsunami of 2004 killed 230,000 people, more than half of them (170,000) in Indonesia, 35,000 in Sri Lanka, 18,000 in India and 8,000 in Thailand. Poor people who lived in flimsy houses were easily swept away during the 2004 tsunami (Hays 2008). The eruption of Krakatoa and the ensuing tsunami in 1883 killed 36,000 people and 19,000 people were killed in Japan by the March 2011 Fukushima earthquake and tsunami. Most of the victims drowned; others were killed by debris.

Less than one year after the Asian Tsunami of 2004, hurricane Katrina hit the Louisiana/Mississippi Coast and killed over 1800 people. The majority living in New Orleans 9th Ward died by drowning, many within their own homes. The hurricane of 1900 that totally submerged the city of Galveston Texas, killed somewhere between 6000 and 10,000 residents and many of those people drowned (Larson 1999). Of the 540,000 flood-related deaths reported by Doocy et al. (2013) for the period 1980–2009, the majority drowned. Fortunately, the death toll of 43 in New York City from Hurricane Sandy in 2012 was modest thanks to accurate forecasting and the fact that warnings were heeded. In the future, improved predictions combined with more effective communication, assisted by social media, should substantially reduce mortality from cyclones. Unfortunately warning times for tsunamis are often much shorter but community education to enhance awareness of the general public will help to reduce the loss of lives. It is crucial, however, that emergency managers and the public at large understand- and trust-forecasts and warnings.

10.3 Snakes and Alligators

During and in the aftermath of the hurricane of 1900 that devastated Galveston Texas, numerous deaths were attributed to bites from poisonous snakes trying to escape the floodwaters. "All over Galveston freakish things occurred….Venomous snakes spiraled upward into trees occupied by people" (Larson 1999, p. 202). And in 1957 when Hurricane Audrey made landfall in Louisiana, snakes were abundant along the roads, in trees, in the remains of buildings and on other high features (Henry et al. 1982). In Florida, Louisiana and other states in the southeastern U.S., alligators along with poisonous snakes represent hazards in low-lying neighborhoods during and following storm surges and heavy rainfall events. Cottonmouth water moccasins (*Agkistrodon piscivorus*) are particularly dangerous and may be present in wet areas during post-storm cleanup. In northwestern Australia and Papua New Guinea, saltwater crocodiles may be found in places where they are not welcome following the passage of tropical cyclones. And they bite!

10.4 Water-Borne Pathogens

According to the U.S. Global Change Research Program (Trtani et al. 2016), water-borne pathogens are responsible for somewhere between 8.5 and 12% of acute gastrointestinal illnesses in the United States; 12 million to 19 million people are affected annually. The main sources of disease are human and animal waste and agricultural runoff. Transmission pathways for illnesses are most commonly through contaminated drinking water, direct exposure of skin surfaces and wounds to water and through consumption of shellfish. Trtani et al. (2016) report that eight climate-dependent pathogens account for approximately 97% of waterborne illnesses in the United States. These are: the enteric viruses norovirus, rotavirus, and adenovirus; the bacteria *Campylobacter jejuni*, *E. coli* O157:H7, and *Salmonella enterica*; and the protozoa *Cryptosporidium* and *Giardia.* Rising water temperatures expand the seasonal windows of growth and the geographic ranges of these pathogens as well as those of toxic algae. Trtani et al. (2016) report medium

Table 10.2 Climate impacts on water related illnesses

Pathogen or toxin	Exposure pathway	Health outcomes	Major climate driver
Toxic marine algae	Shellfish; aerosolized toxins and water vapor	Shellfish poisoning; respiratory irritation	Warming waters; ocean acidification
Cyanobacteria (blue-green algae)	Drinking water; recreational waters	Liver & kidney damage; gastroenteritis; respiratory arrest; neurological ailments	Increasing temperature; changing precipitation patterns
Enteric bacteria and protozoan parasites (e.g. *Salmonella, Giardia, Cryptosporidium*)	Drinking water; recreational waters; shellfish	Gastroenteritis, possibly severe	Heavy precipitation, flooding, and water temperature (+ or −)
Enteric viruses: enteroviruses; rotaviruses; noroviruses; hepatitis A and E	Drinking water; recreational waters; shellfish	Gastroenteritis, possibly severe, possible paralysis; infection of organs	Heavy precipitation, flooding, and temperature changes
Leptospira and *Leptonema* bacteria	Recreational waters	Flu-like illness, meningitis, kidney or liver damage	Flooding, increased water temperature, heavy rain
Vibrio bacteria species (*V. parahaemolyticus, V. cholera, V. vulnificus, V. alginolyticus)*	Recreational waters; shellfish	Gastroenteritis, septicemia (bloodstream infection); eye and ear infections	Increased water temperature, sea level rise, changing salinity of coastal waters

Source Trtani et al. (2016)

confidence that future warming will increase risk of exposure to these pathogens and toxins. Table 10.2 summarizes the health outcomes and climate change drivers for key pathogens and toxins.

There are growing concerns about increases in the spread of and exposure to *Vibrio* bacteria, which cause intestinal, skin, ear, eye infections and septicemia (blood stream infection). The specific *Vibrio* species of concern are *V. vulnificus, V. parahaemolyticus, and V. alginolyticus. Vibrio cholerae,* which is carried by planktonic hosts and requires marine salt water for growth has shown strong correlation with climatic events such as El Niño (Colwell 1996). Figure 10.2, from the U.S. Global Change Research Program (Trtani et al. 2016), diagrams the pathways by which humans are exposed to *Vibrio* and the climate drivers that are likely to increase future exposure. Like the other pathogens just described, longer seasons for growth and expanding geographic range of occurrence increase the risk of exposure to *Vibrio*. It must be noted that, according to Trtani et al. (2016) illnesses from water-borne pathogens may be underestimated by as much as 43-fold on average, and by up to 143 times for *Vibrio*. Following Hurricane Katrina, *Vibrio vulnificus,* which had infected patients through wounds, caused significant mortality (Lemonick 2011).

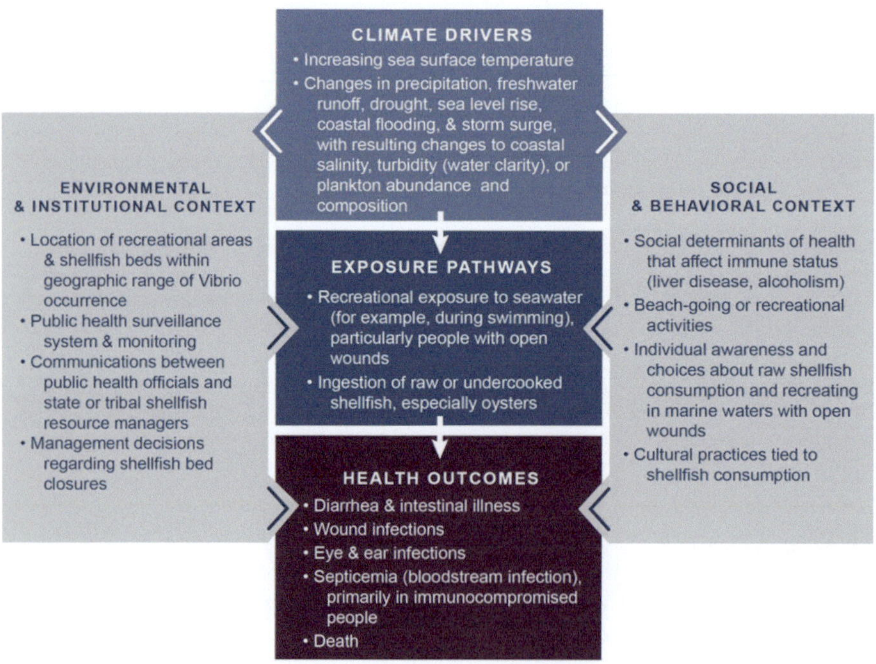

Fig. 10.2 Climate change and exposure to *Vibrio*—conceptual diagram of pathways and possible impacts (from Trtani et al. 2016)

10.5 Vector-Borne Pathogens

Water does not have to be the direct pathway of pathogen delivery for floods to cause spread of disease. Standing water following floods is a breeding ground for mosquitos and other vectors of disease transmission, including rats, flies, fleas and ticks. Vector-borne diseases in flooded coastal areas include malaria, dengue fever, dengue hemorrhagic fever, yellow fever, West Nile fever and *leptospirosis,* which is transmitted by rodents (Vasconcelos 2006). Most recently, *Zika* virus has joined the list of vector-borne diseases. It is spread specifically by *Aedes* mosquito species, which are the same mosquitos that transmit dengue fever and yellow fever. As is the case for the water-borne diseases just discussed, increasing temperatures are expanding the seasonal and geographic ranges of many of these diseases. For example, West Nile fever has recently emerged in Europe and *Zika* has spread throughout much of the tropics, particularly in low-lying regions of developing countries (Smith 2017). In the case of West Nile Virus, birds are typically the hosts and transmission to humans is attributed to mosquitoes that have first bitten a hosting bird (Beard and Eisen 2016).

Beard and Eisen (2016), in a report of the U.S. Global Change Research Program, point out that rising temperatures are likely to accelerate all aspects of the biologic processes that lead to human infections by vector borne diseases. Included are increases in the seasonal and geographic ranges of mosquito breeding, accelerated growth of the viruses themselves and enhanced contacts between vectors and hosts (e.g. birds or other humans). According to the World Health Organization (2003), in addition to warming temperatures, there are many other anthropogenic global changes that are increasing the spread of vector-borne diseases. Among these, dams, canals and irrigation increase exposure to malaria, *Schistosomiasis* and *Helminthiasies.* Urbanization and crowding causes water contamination and reduces hygiene and can enhance the likelihood of cholera. Malaria and hemorrhagic fever outbreaks can also be exacerbated by increases in rodents related to agricultural intensification (WHO 2003). Although these problems are not specifically coastal in nature, they can be intensified in coastal environments, particularly low-lying coastal plains where standing waters following floods can linger for prolonged periods. In addition, the high poverty levels of coastal "megacities", especially those in river deltas, significantly raises the level of vulnerability to vector-borne diseases.

10.6 Harmful Algal Blooms

"It is now well recognized that there are more harmful algal blooms (HABs), more often, in new and different places, often lasting longer, and with a range of toxicities, and that many of these blooms are related to nutrient pollution." (Glibert and Burford 2017, p. 58). The most well known of the HABs are the so-called "red tides" because of the discoloration (red or orange) of coastal waters, but HABs can also cause other colors or no color at all depending on the nature of the bloom. In all cases, they involve photosynthetic algae that require nutrients. These include single-celled phytoplankton, cyanobacteria (also called blue-green algae) and macroalgae. Algae serve as the base of the food web in healthy aquatic systems and are not harmful in low concentrations (e.g. Kudela et al. 2017). They become harmful when "explosive" blooms take place in response to an over abundance of nutrients along with complex interactions of a number of other factors including changes in freshwater runoff from rivers and climate change (Glibert and Burford 2017). Some, but not all, of these algae are toxic. Whether toxic or not, however, eutrophication can lead to hypoxic (low dissolved oxygen) or anoxic (oxygen depletion) conditions when the blooms decay. The adverse human health impacts of HABs include paralytic shellfish poisoning, respiratory ailments from aerosolized HAB toxins and, in the case of cyanobacteria, liver and kidney damage, gastroenteritis, respiratory arrest and neurological ailments. In 2015, an extensive HAB affected the entire west coast of the U.S. and Canada and resulted in extensive closure of coastal fisheries (Fig. 10.3).

Harmful algal blooms can occur along open ocean coasts (Fig. 10.3) or in bays and semi-enclosed waterways (Berdalet et al. 2017). A detailed review of the causes and toxicities of harmful algal blooms is offered by Anderson (2017). They also occur at sharp vertical density gradients (pycnoclines) in coastal waters that are

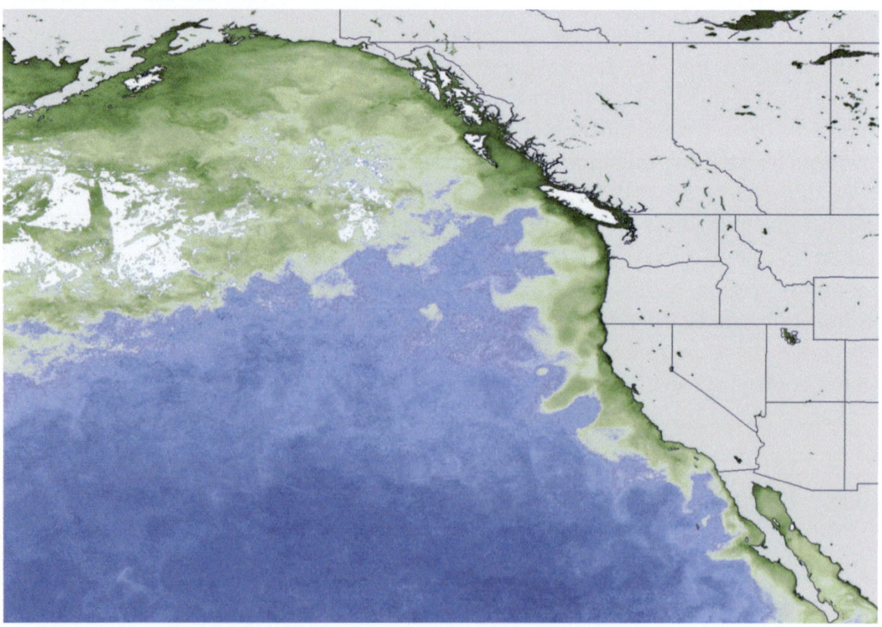

Fig. 10.3 Map compiled from satellite images showing the algal bloom in the western U.S. and Canada in 2015. NOAA public domain image

stratified by temperature and salinity (Berdalet et al. 2017). Perhaps the most dramatic and widespread increases in HABs over the past decade have taken place in the Arabian Sea and have resulted from several complex factors including warming climate and the effects of urbanization. Goes et al. (2005) report that warming of the Eurasian land mass and reduction of Himalayan snow cover have caused intensification of the southwest summer monsoons, which in turn cause increased upwelling of nutrient-rich deep water off the coasts of Somalia, Oman and Yemen on the western side of the Arabian Sea. This caused a 350% increase in phytoplankton productivity between 1997 and 2005. On the eastern side of the Arabian Sea, nutrient runoff from India has increased and contributed to HABs on that coast (Swaney et al. 2014). In addition to agricultural runoff, the population of Mumbai has doubled over the past decade without corresponding improvements to sewerage treatment facilities and increases in nitrogen inputs to the sea are a consequence (Glibert and Burford 2017). The east and west coasts of Florida have in recent years experienced severe coastal blooms of the Florida "red tide" species *Karenia brevis* (Maze et al. 2015), which cause fish kills and can cause respiratory irritation in humans when the brevitoxins become aerosolized in the surf zone. Maze et al. (2015) report a strong correlation between blooms of *K. brevis* on the West Florida continental shelf and the position of the regional scale Loop Current in the Gulf of Mexico. In 2016, the release of nutrient-rich agricultural runoff from Lake Okeechobee into rivers flowing to the coast was implicated in severe *K. brevis* blooms on Florida's east and west coasts.

10.7 Airborne Illnesses

Though not directly dependent on coastal inundation for their spread, there are several airborne ailments that affect coastal residents and appear to have emerged or spread as a result of climate change. Some of these are, of course, well known and include flu and the common cold. However, climate change is already affecting air in several ways including increased production of allergens, fine particles, dust, fungal spores and various pollutants, which can cause distress in children and the elderly (Mansour 2013). Morfin et al. (2014) have described the airborne dissemination of fungal spores on the California coast and conclude that the densities of such spores probably depend on temperature and humidity and are likely to increase as climates change. Fungal spores can trigger various respiratory ailments in humans. *Cryptococcus gattii* is an airborne fungus that originally prevailed in tropical environments, particularly in Northern Australia and Papua New Guinea, but has most recently been affecting residents of British Columbia and the U.S. Pacific Northwest coast and is particularly dangerous for persons with HIV or other immune deficiencies (Springer et al. 2014). It can also lead to *Meningitis* in patients with healthy immune systems (Cookman and Huqi 2013).

Griffin et al. (2001) analyzed samples of atmospheric aerosols obtained from the U.S. Virgin Islands in the Caribbean Sea that apparently originated in the Saharan

Desert of Africa and were carried across the Atlantic in dust storms. The transit times for crossing the Atlantic were estimated to be 5–7 days. The microorganisms associated with the dust include bacteria, viruses and fungal spores. Griffin et al. (2001) suggest that these airborne microorganisms from Africa may be implicated in sharp increases in asthma and other respiratory ailments in the Caribbean

10.8 Flood-Disseminated Chemical Pollution in Coastal Waters

Coastal cyclones, floods and tsunamis can release garbage, sewerage, toxic chemicals and, in rare cases, radioactive material into coastal seas, rivers and estuaries. When Hurricane Katrina hit the Louisiana coast in 2005, it burst open oil storage tanks at numerous coastal locations spilling eight million gallons of oil. The spill, in aggregate, was comparable in magnitude to the 1989 Exxon Valdez spill in Alaska. The worst of these spills involved the spread of one million gallons of crude oil among 10,000 homes in Plaquemines Parish, southeast of New Orleans (Sturgis 2015). Onshore oil and petrochemical storage facilities worldwide are similarly vulnerable to future storm surges superimposed on sea level rises. Many of these, on the U.S. Atlantic coast, represent potential sites of environmental disaster as seas rise and storms potentially intensify. In addition to oil, there are numerous other chemicals and toxins that are spread by floods. Included are sediments that may contain heavy metals or pesticides, various types of hydrocarbons from spills at gas stations and fuel depots, heavy metals, road salts, fertilizers, pesticides and herbicides (Euripidou and Murray 2004). Volatile organic compounds (VOCs), lead, and arsenic were detected in the air, soil, and sediment samples of Louisiana and Mississippi in the aftermath of Hurricane Katrina (Lane et al. 2013). Overflows of sewers and leaching of landfills can spread a variety of toxic substances including asbestos and mercury. Common household items that pollute flood waters are listed in Table 10.3. As noted by Euripidou and Murray (2004), the most harmful impacts typically occur where residential neighborhoods are in close proximity to industrial or agricultural sites. With specific reference to the United Kingdom, Euripidou and Murray (2004) point out that public health professionals often have limited experience in managing chemical contamination incidents. Various toxic chemicals were reportedly carried throughout some areas of Houston during the unprecedented floods caused by Hurricane Harvey in 2017.

On March 11, 2011, a magnitude 9.0 earthquake occurred offshore to the east of Honshu Island, Japan. In addition to causing direct onshore damage, the quake generated a 15-m (50 ft) tsunami which devastated the city of Fukushima and killed 19,000 people. The following description of what ensued with regard to the nuclear power plant is summarized from a recent update by the World Nuclear Association (2017). The tsunami inundated the Fukushima nuclear power plant disabling the power and cooling systems of three nuclear reactors, which melted down within

Table 10.3 Hazardous household items contributing to pollution

Household Item	Pollutant
Drain cleaner	Sodium hydroxide, also known as lye or caustic soda
Paint, lacquer, and varnish removers	Benzyl alcohol; ethanol; formic acid; methyl alcohol; methylene hydrochloride; naphtha; and xylene
Petroleum, oil, and lubricants	Polycyclic aromatic hydrocarbons such as naphthalene and metals such as lead, zinc, chromium, barium, and arsenic
Toilet-bowl cleaner	Chlorine, hydrochloric acid, sodium hydroxide, sodium hypochlorite, or phosphoric acid
Antifreeze	Ethylene glycol or propylene glycol
Pesticides	Aldrin, atrazine, chlordane, dichlorodiphenyltrichloroethane, dieldrin, endrin, endosulfan, heptachlor, lindane, mirex, toxaphene, trifluralin
Fertilizers	Nitrogen and phosphorus
Vehicle batteries	Cadmium, lead, mercury, nickel, lithium and electrolytes
Sewage	Cloth, detergents, dirt, food residues, human excrement, micro-organisms, paper, soap, etc.

Many of the products are of toxicological concern

three days. A series of explosions demolished much of the building and cast radioactive debris on the grounds surrounding the plant. People living within a 20 km (12.43 mile) radius of the site were evacuated. Inundating seawater was radioactively contaminated and much of the contaminated water was ultimately released into the coastal environment. Radioactivity was also released into the air. The main radionuclides involved were iodine-131, which has a half-life of 8 days and caesium 137 with a half-life of 30 years; caesium 134 with a half-life of 2 years was also dispersed. The highest levels of radiation were from debris rather than water or air. In April 2016, five years after the event, radiation had diminished significantly but was still present at non-lethal levels. However, the iodine-131 with its short 8-day half-life was gone by April 2011. Notably, no deaths were directly attributed to radiation, either immediately following the event or several years thereafter. Ironically, however, roughly 1000 deaths were indirectly attributed to the dispersal of radioactivity. These deaths primarily involved elderly people and were apparently caused by the enforced and prolonged evacuation during which those displaced from their homes were forced to live in shelters (World Nuclear Association 2017). It is worth noting that some debris from the disaster eventually washed ashore on the west coast of the U.S. Buesseler (2014) discusses this triple disaster - earthquake, tsunami, and radiation release.

10.9 Health-Related Impacts of Infrastructure Damage

Modern humans are heavily dependent on the built infrastructure for their health and well-being and this is especially the case in urban environments. In 2012, “Superstorm” Sandy shut down power to several hundred thousand residents of New York City and its environs, caused the closure of five hospitals and numerous nursing homes and long-term care facilities, disrupted water supplies, closed roads and public transport and seriously hampered evacuations of elderly and ailing patients from hospitals and nursing homes (Lane et al. 2013). New Orleans experienced comparable, or in many cases more severe, disruptions following Hurricane Katrina. Power outage was prolonged in both cases. Virtually all hospitals and health care facilities in New Orleans were seriously impacted and some were closed permanently or for extended periods. Closure of stores and pharmacies in New York City and New Orleans restricted access to food and medication. Limited access of emergency responders to vaccines, oxygen supplies and medical equipment is listed by Lane et al. (2013) as another of the problems experienced after Sandy and Katrina. Prolonged and extensive power outages in Puerto Rico following Hurricane Maria in 2017 were continuing more than two months after the storm. Numerous post-storm deaths in Puerto Rico were reportedly related to lack of power for life-supporting medical equipment for chronically ill patients as well as the lack of potable water.

Extended power outages can have several adverse health-related consequences. Lack of refrigeration can lead to food poisoning. Elevator shutdowns in high-rise buildings during power outages have caused cardiac arrest in older or infirm people attempting to climb stairs. Hypothermia may result from lack of heat. There have been numerous cases of carbon monoxide poisoning during power outages because of attempts at using unsafe cooking methods such as charcoal grills or camp stoves or from operating gasoline generators in enclosed or inadequately ventilated areas. Shutdown of electrical medical equipment like ventilators can cause fatalities. Lane et al. (2013) also report numerous physical accidents, such as tripping or falling because of darkness. For situations of prolonged post-event disruptions, there are some chronic outcomes. For example, homes and shelters subject to moisture and mold contamination can cause respiratory ailments including asthma attacks.

10.10 Mental Health Impacts of Coastal Inundation

While physical security is a priority in dealing with natural hazards, Fernandez et al. (2015) point out that detrimental short-, medium- and long-term effects on well-being, relationships and physical and mental health are common following floods. These types of disasters may precipitate mental illness such as post-traumatic stress disorder, depression, anxiety, and substance abuse in some people. Survivors of traumatic events such as floods may begin to experience symptoms such as:

- interpersonal changes like withdrawal
- irritability and mood swings
- a sense of isolation
- trouble sleeping
- confusion.

Hurricane Katrina caused several mental health problems (e.g. Picou and Hudson 2010). Among other mental illnesses, displacement from homes and extended stays in shelters following Katrina caused acute stress disorder (Mills et al. 2007). Shelters were also responsible for the spread of several communicable diseases and for the onsets of substance abuses. For many, the mental anguish of losing family members, neighborhoods, homes and employment may persist for a lifetime.

Resilient communities will create safe spaces where well-trained workers offer medical and health care supplies, food and diapers, and a warm and dry place for flood victims to rest their heads. This support helps victims to take the time to mourn their losses. Resources such as transportation, clothes, shoes, and toothbrushes can restore the victim's sense of safety. Emergency response personnel and neighbors all need to be aware of the psychological effects caused by floods, and know how to respond appropriately.

10.11 Vulnerable Populations

By far, the people who are most likely to be drowned, devastated, sickened or severely distressed by coastal flooding are poor. Worldwide, a majority of these people live in coastal slums, particularly in megacities such as Mumbai, Dhaka (Rashid 2000), and other southeastern Asian cities. World Bank data analyzed by Doocy et al. (2013) showed that 45% of the flood-related fatalities in the three decades after 1980 involved low-income people and 41% involved those with low-to-mid incomes. Only 4% of the deaths involved those with high incomes. World Health Organization statistics indicate that over that same three-decade period, residents of Southeast Asia were the most vulnerable accounting for 33% of flood-related deaths; the West Pacific region was close behind at 32% (Doocy et al. 2013). Only 3% of the deaths were in Europe and 15% were in the Americas (North and South).

The most vulnerable people during and after Sandy and Katrina were those over 65 years of age and low-income African Americans (Lane et al. 2013). And, of course, in the Katrina case, New Orleans Ninth Ward was the most vulnerable neighborhood and was the center of the most casualties. Future decisions as to how to triage the distribution of limited mitigation resources, post-storm recovery assets and disaster relief should be prioritized on the basis of community vulnerability and predicted storm *impact*, not simply storm intensity. The Saffir-Simpson scale of 1 to 5, by which the National Hurricane Center classifies hurricanes, indexes relative wind stress and the potential wind damage to structures. However, these levels apply only to wind intensity and likely impacts on structures and do not distinguish among the contrasting impacts that these storms may have on human communities. For example, a Category 3 hurricane can be devastating to a low-lying coastal city subject to widespread inundation by storm surge, with a poor community with low societal and economic resilience. The same storm may be but a short-lived nuisance to a more affluent community on high ground, surrounded by robust infrastructure.

Fortunately, advances are being made in methodologies for gaging vulnerability of human communities. Van Zandt et al. (2012) consider neighborhood resilience in relation to social vulnerability and housing. Norris et al. (2008) offer a treatise on the psychology of community resilience as it impacts disaster readiness. More recently, Cutter et al. (2014, 2010) have evolved the concept of *Baseline Resilience Indicators for Communities (BRIC)* as empirical metrics for gaging the resilience of communities to disasters. Guillard-Gonçalves et al. (2014) have developed a "Social Vulnerability Index" (SoVI) which can be readily applied to most regionally specific communities. Application of these and other approaches should become routine in planning for future disasters.

References

Anderson, D.M. 2017. Harmful algal blooms. In *Harmful algal blooms (HABs) and desalination: A guide to impacts, monitoring and management*, ed. D.M. Anderson, S.F.E. Boerlage, M.B. Dixon, 17–52. Paris, Intergovernmental Oceanographic Commission of UNESCO, 2017. 539pp. (IOC Manuals and Guides No.78.) (English.) (IOC/2017/MG/78).

Beard, C.B., and R.J. Eisen. 2016. Chapter 5. Vector borne diseases. In *The Impacts of climate change on human health in the United States: A scientific assessment.* U.S. Global Change Research Program, Washington, DC.

Berdalet, E., M. Montresor, B. Reguera, S. Roy, H. Yamazaki, A. Cembella, and R. Raine. 2017. Harmful algal blooms in fjords, coastal embayments, and stratified systems: Recent progress and future research. *Oceanography* 30 (1): 46–57. https://doi.org/10.5670/oceanog.2017.109.

Buesseler, Ken O. 2014. Fukushima and ocean radioactivity. *Oceanography* 27 (1): 92–105. http://dx.doi.org/10.5670/oceanog.2014.02.

Colwell, R.R. 1996. Global climate and infectious disease: The cholera paradigm. *Science* 274 (5295): 2025–2031. https://doi.org/10.1126/science.274.5295.2025.

Cookman, L., and M. Huqi. 2013. Meningitis secondary to *Cryptococcus gattii*, an emerging pathogen affecting immunocompetent hosts. *World Journal of Emergency Medicine* 4 (2): 151–153. https://doi.org/10.5847/wjem.j.issn.1920-8642.2013.02.012.

Cutter, S.L., K. Ash, and C. Emrich. 2014. The geographies of community disaster resilience. *Global Environmental Change* 29: 65–77.

Cutter, S.L., C.G. Burton, and C.T. Emrich. 2010. Disaster resilience indicators for benchmarking baseline conditions. *Journal of Homeland Security and Emergency Management* 7(1), Article 51. https://doi.org/10.2202/1547-7355.1732.

Doocy, S., A. Daniels, S. Murray, T.D. Kirsch. 2013. The human impact of floods: A historical review of events 1980–2009 and systematic literature review. *PLOS Currents Disasters.* 2013 Apr 16. 1st ed. https://doi.org/10.1371/currents.dis.f4deb457904936b07c09daa98ee8171a.

Euripidou, E., and V. Murray. 2004. Public health impacts of floods and chemical contamination. *Journal of Public Health* 26 (4): 376–383. https://doi.org/10.1093/pubmed/fdh163.

Fernandez, Ana, John Black, Mairwen Jones, Leigh Wilson, Luis Salvador-Carulla, Thomas Astell-Burt, and Deborah Black. 2015. Flooding and mental health: A systematic mapping review. *PLoS One* 10 (4): e0119929. https://doi.org/10.1371/journal.pone.0119929.

Frank, N.L., and S.A. Husain. 1971. The deadliest tropical cyclone in history? *Bulletin of the American Meteorological Society* 52 (6): 438–44.

Fritz, H.M., C.D. Blount, S. Thwin, M.K. Thu, and N. Chan. 2009. Cyclone Nargis torm surge in Myanmar. *Nature Geoscience* 2: 448–449. https://doi.org/10.1038/ngeo558.

Glibert, P.M., and M.A. Burford. 2017. Globally changing nutrient loads and harmful algal blooms: Recent advances, new paradigms and continuing challenges. *Oceanography* 30 (1): 58–69.

Griffin, D.W., V.H. Garrison, J.R. Herman, and E.A. Shinn. 2001. African desert dust in the Caribbean atmosphere: Microbiology and public health. *Aerobiologia* 17: 203–213.

Goes, J.I., P.G. Thoppil, H. Gomes, and J.T. Fasullo. 2005. Warming of the Eurasian landmass is making the Arabian Sea more productive. *Science* 308 (5721): 545–547. https://doi.org/10.1126/science.1106610.

Guillard-Gonçalves, C., S.L. Cutter, C.T. Emrich, and J.L. Zêzere. 2014. Application of social vulnerability index (SoVI) and delineation of natural risk zones in Greater Lisbon. *Journal of Risk Research* 18 (5): 651–674.

Hays, J. 2008. Great Tsunami of 2004 in Thailand and Malaysia *Facts and Details*. http://factsanddetails.com/asian/cat63/sub411/item2544.html#chapter-0.

Henry, W.K., D.M. Driscoll, and J.P. McCormack. 1982. *Hurricanes on the Texas Coast center for applied geosciences*. Texas: Texas A&M University.

Keim, M.E. 2006. Cyclones, tsunamis, and human health: The key role of preparedness. *Oceanography* 19 (2): 40–49.

Kudela, R.M., E. Berdalet, H. Enevoldsen, G. Pitcher, R. Raine, and E. Urban. 2017. GEOHAB–The global ecology and oceanography of harmful algal blooms program: Motivation, goals, and legacy. *Oceanography* 30 (1): 12–21. https://doi.org/10.5670/oceanog.2017.106.

Lane, K., K. Charles-Guzman, K. Wheeler, Z. Abid, N. Graber, and T. Matte. 2013. Health effects of coastal storms and flooding in urban areas: A review and vulnerability assessment. *Journal of Environmental and Public Health* 2013: 13, Article ID 913064, https://doi.org/10.1155/2013/913064.

Larson, E. 1999. *Isaac's storm*, 323. New York: Crown Publishers.

Lemonick, D.M. 2011. Epidemics after natural disasters. *American Journal of Clinical Medicine* 8 (3): 144–153.

Mansour, S.A. 2013. Impact of climate change on air and water borne diseases. *Air and Waterborne Diseases* 3, https://doi.org/10.4172/2167-7719.1000e126.

Mas, E., J. Bricker, S. Kure, B. Adriano, C. Yi, A. Suppasri, and S. Koshimura. 2015. Field survey report and satellite image interpretation of the 2013 Super Typhoon Haiyan in the Philippines. *Natural Hazards and Earth System Sciences Discussions* 15: 805–816.

Maze, G., M.J. Olascoaga, and L. Brand. 2015. Historical analysis of environmental conditions during Florida Red Tide. *Harmful Algae* 50: 1–7. https://doi.org/10.1016/j.hal.2015.10.003.

Mills, M.A., D. Edmondson, and C.L. Park. 2007. Trauma and stress response among Hurricane Katrina evacuees. *American Journal of Public Health* 97 (Supplement 1): S116–S123.

Morfin, J., S.G. Crandall, G.S. Gilbert. 2014. Seasonal trends in airborne fungal spores in coastal california ecosystems. *American Geophysical Union, Fall Meeting 2014, abstract #A23B-3235.*

Norris, F., S. Stevens, B. Pfefferbaum, K. Wyche, and R. Pfefferbaum. 2008. Community resilience as a metaphor, theory, set of capacities, and strategy for disaster readiness. *American Journal of Community Psychology* 41 (1): 127–50.

Picou, J.S., and K. Hudson. 2010. Hurricane Katrina and mental health: A research note on Mississippi Gulf Coast residents. *Sociological Inquiry* 80 (3): 513–524.

Rashid, S.F. 2000. The urban poor in Dhaka city: Their struggles and coping strategies during the floods of 1998. *Disasters* 24 (3): 240–253.

Smith, L. 2017. Zika virus: Symptoms, facts and affected countries. *Medical News Today*, April 24, 2017.

Springer, D.J., R.B. Billmyre, E.E. Filler, K. Voelz, R. Pursall, P.A. Mieczkowski, et al. 2014. *Cryptococcus gattii* VGIII isolates causing infections in HIV/AIDS patients in Southern California: Identification of the local environmental source as Arboreal. *PLoS Pathogens* 10 (8): e1004285. https://doi.org/10.1371/journal.ppat.1004285.

Sturgis, S. 2015. *The Katrina oil spill disaster: A harbinger for the Atlantic Coast?*. Durham: NC Institute for Southern Studies.

Swaney, D.P., B. Hong, A. Paneer Selvam, R.W. Howarth, R. Ramesh, and R. Purvaja. 2014. Net anthropogenic nitrogen inputs and nitrogen fluxes from Indian watersheds: An initial assessment. *Journal of Marine Systems* 141: 45–58. https://doi.org/10.1016/j.jmarsys.2014.09.004.

Trtani, J., L. Jantarasmi, J. Brunkard, T. Collier, J. Jacobs, E. Lipp, S. McLellan, S. Moore, H. Paerl, J. Ravenscroft, M. Sengco, and T. Thurston. 2016. *The impacts of climate change on human health in the United States: A scientific assessment Capt. 6 in climate impacts on water related illness*. Washington, DC: U.S. Global Change Research Program.

Van Zandt, S., W.G. Peacock, D.W. Henry, H. Grover, W.E. Highfield, and S.D. Brody. 2012. Mapping social vulnerability to enhance housing and neighborhood resilience. *Housing Policy Debate* 22 (1): 29–55.

Vasconcelos, P. 2006. Flooding in Europe: A brief review of the health risks. *Euro Surveill* 11(16): 2947. Available online: http://www.eurosurveillance.org/ViewArticle.aspx?ArticleId=2947Eurosurveillance, 20 April 2006.

World Health Organization. 2003. *Climate change and human health—Risk and responses summary*, 36.

World Nuclear Association. 2017. *Fukushima accident (April 2017 update).*

Chapter 11
Natural Infrastructure to Mitigate Inundation and Coastal Degradation

J. Livingston, N. Woiwode, M. Bortman, S. McAfee, K. McLeod, S. Newkirk and S. Murdock

...the sense of alienation between man and nature leads to the use of technology in a hostile spirit—to the conquest of nature instead of intelligent co-operation with nature.
—Alan W. Watts, philosopher, 1968

11.1 In Search of Harmony Among Water, Coasts and People

People have long been drawn to the water's edge. Coastal waters provided early humans with abundant resources. Once humans began to explore beyond their own shores, journeying by water became the most efficient means of travel and trade—connecting cultures and people across the globe for the very first time and pulling even more to the coast. Eventually, megacities such as Seoul, Mumbai, New York, São Paulo, Lagos, and London arose around seaports where there were increased opportunities for trade, jobs, and transportation.

However, in attempts to satisfy our appetite for sea breezes and ocean views, we've forgotten the ocean is a thing to be feared and respected. Consequently, we've placed ourselves in locations highly susceptible to the impacts of coastal storms.

The ocean sustains life, but it also poses a great danger to us. And, continued development of houses, roads, and commercial infrastructure is slowly degrading and destroying the coastal features—such as wetlands, beaches, and reefs—which protect us from the ocean's force. Healthy wetlands and marshes shield communities by reducing wave energy. Beaches and sand dunes limit coastal flooding, and, in near shore and estuarine areas, oyster and coral reefs buffer wave energy.

J. Livingston (✉) · N. Woiwode · M. Bortman · S. McAfee
K. McLeod · S. Newkirk · S. Murdock
The Nature Conservancy, Washington, DC, USA
e-mail: jacquelyn.livingston@TNC.ORG

L. D. Wright and C. R. Nichols (eds.), *Tomorrow's Coasts: Complex and Impermanent*, Coastal Research Library 27,
https://doi.org/10.1007/978-3-319-75453-6_11

The ocean's enticement will never fade; our communities must invest in a future where people and water live in harmony, no matter how high it may rise. Smart investments in natural infrastructure can provide a pathway to a safer, healthier, more prosperous and more resilient future.

11.2 Grey Infrastructure

For as long as we've lived near water, we've been trying to control it—constructing barriers such as dams, levees, seawalls, groins, and jetties to keep it at bay. However, "grey infrastructure" tends to block natural processes and degrade natural features, such as wetlands and beaches, which provide important economic, environmental, and recreational benefits. The *Coastal No Adverse Impacts (CNAI) Handbook* (NOAA and the Association of State Floodplain Managers 2007) describes at least nine adverse impacts of flood "protection" structures. Furthermore, these structures are often very expensive to build and maintain. While grey infrastructure can be the right- or only- choice in some locations, it has well-documented adverse consequences, many of which are incompatible with the natural systems that support healthy coastal communities. As explained in Chap. 18, poorly planned "grey infrastructure" can increase, rather than decrease, the vulnerability of coastal communities. For example, shoreline structures built to alter the effects of ocean waves, currents, and sand movement, tend to cause down-coast erosion leading to associated costs that exceed the original cost of construction.

A 2017 study conducted by The Nature Conservancy, compared economic costs and benefits of nine coastal climate change adaptation strategies at four locations in Southern Monterey Bay, California. The analysis also spanned four time horizons (2010, 2030, 2060, and 2100). The study concluded that in *all* cases the least economically beneficial alternative, especially over the long-term, involved shoreline armoring—that is, grey infrastructure (Leo et al. 2017). To determine the costs and benefits of alternative strategies for different coastal reaches, Leo et al. 2017 examined the physical impacts by modeling expected shoreline changes for each proposed adaptation strategy under a range of sea level rise projections (using the High and Medium projections recommended by the Intergovernmental Panel on Climate Change Fifth Assessment Report (IPCC 2013)) and time horizons (2010, 2030, 2060 and 2100). Results for the Del Monte reach are shown in Fig. 11.1 as an example. The economic costs of each strategy were estimated by gathering information on the engineering costs of implementation of various adaptation measures (e.g. sand placement, construction of groins, etc.), as well as an economic analysis of the recreational and ecological value of coastal and upland resources that could be affected by coastal hazards. The relative ecological condition of the beach within the study area was ranked using several metrics to score the physical, biotic, and human impacts conditions of km^2 blocks of southern Monterey beaches.

The resulting Beach Ecological Index Score was then combined with estimates of beach restoration (replacement) costs to provide a monetized ecological value. The estimates for all these costs and benefits were combined and expressed in terms of net present value using a 1% discount rate, which is appropriate for long-term climate change modeling. Results were expressed as net present value of the shoreline. "Net Present Value" refers to the sum of all the benefits (e.g., the recreational and ecological value of beaches) minus the costs (e.g., engineering costs of armoring and nourishment). Loss of land, buildings, roads, and other infrastructure, as well as the cost of adaptation (e.g., elevating roads), were incorporated as costs in the analysis.

A similar, previous economic assessment of shoreline management strategies in Monterey Bay (ESA PWA 2012) examined various erosion-control alternatives at three of the same locations. This study also found armoring strategies were generally not cost-effective. And yet, in the face of such research, U.S. shorelines continue to be armored at a rate of 200 km (125 mi) per year. If this rate continues, nearly 1/3 of our nation's shoreline will be hardened by 2100 (Gittman et al. 2015).

As we continue to further harden our shores, we struggle to maintain existing grey structures. In 2017 the American Society of Engineers gave U.S. infrastructure a D+ rating (www.infrastructurereportcard.org). The price tag to repair and fortify just 225 km (140-mi) of coastal structures in Massachusetts is over one billion dollars (The Nature Conservancy 2017). As these unmaintained structures continue to age, tens of millions of people living along U.S. coastlines are put at risk and the potential for significant economic loss increases.

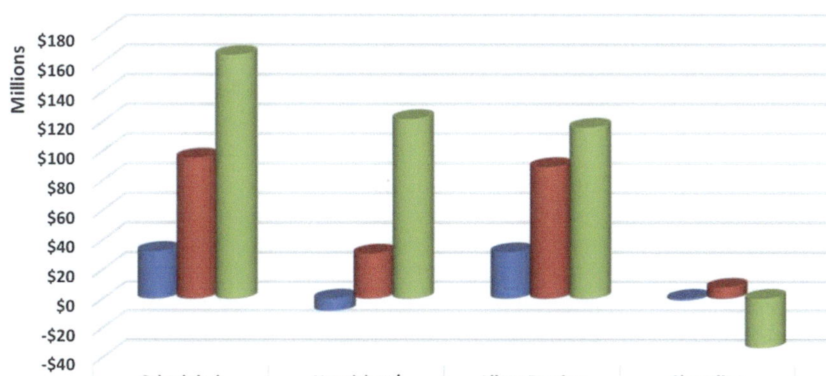

	Scheduled Nourish	Nourish w/ Groins	Allow Erosion	Shoreline Hardening
2030 (22 cm)	$32,400,000	-$8,500,000	$31,600,000	-$1,200,000
2060 (72 cm)	$95,400,000	$30,700,000	$89,100,000	$7,800,000
2100 (159 cm)	$164,900,000	$121,700,000	$115,900,000	-$33,500,000

Fig. 11.1 Economic benefits of adaptation approaches for Del Monte reach (Leo et al. 2017)

Not only is grey infrastructure expensive to build and maintain, it encourages development in areas which are particularly susceptible to flooding and storm damage. According to a 2013 study by Arkema et al., 16% of the U.S. coastline comprises high-hazard areas, harboring 1.3 million people and $300 billion in residential property value.

History has shown the worst damages occur when these structures fail—and fail they have. In 2005, infrastructure failures during Hurricane Katrina, and the resulting loss of nearly 2000 lives plus over $100 billion in damages, captured headlines. Then, in 2016, Hurricane Matthew overtopped levees and defeated dams leading to catastrophic flooding which again stole lives and cost billions of dollars in damage. It's becoming increasingly clear that grey infrastructure alone cannot sufficiently minimize the risk and damage of natural catastrophes. Storm-driven waters surging over barriers and floods overtopping levees have repeatedly shown it's time to accept the limits of our engineered-solutions. Fortunately, the very source of devastation—nature—offers a promising solution.

11.3 Natural Infrastructure

Figure 11.2 shows the relative exposure of U.S. coasts to sea level rise. As will be described in the case studies in Part III, greatest exposure exists on the East and Gulf coasts. As sea levels continue to rise and storm impacts intensify, there is growing pressure on coastal communities to identify cost-effective strategies for mitigating societal, economic, and environmental impacts. There is evidence that shorelines with intact natural features, such as wetlands, dunes, mangroves or reefs, experience less damage from severe storms and are more resilient than hardened shorelines (Arkema et al. 2013; Gittman et al. 2015). Natural infrastructure aims to protect, restore, and mimic these natural features to reduce our vulnerability to climate change related hazards and increase resilience of coastal areas by capitalizing on their protective qualities.

Natural infrastructure (Fig. 11.3) protects coastlines by employing nature-based solutions to reduce wave energy, trap sediment, reduce erosion, store excess water, and weaken storm surge (Ferrario et al. 2014; Gedan et al. 2011; Möller et al. 2014; Narayan et al. 2016; Pinsky et al. 2013; Shepard et al. 2011; Wamsley et al. 2010). A growing body of evidence shows that natural infrastructure solutions can reduce risk and meet the multiple needs of communities better than grey infrastructure alone.

In addition to protecting coastal communities, these natural features also provide wildlife habitat, opportunities for recreation, and improved water quality. Furthermore, many of these natural infrastructure solutions are more cost-effective because they are largely self-sustaining and grow stronger—not weaker—over time (The Nature Conservancy, *Natural Solutions for Reducing Flood Risk*). Depending on location and needs, there are many types of natural infrastructure which can be used alone or in combination to achieve desired outcomes. These include:

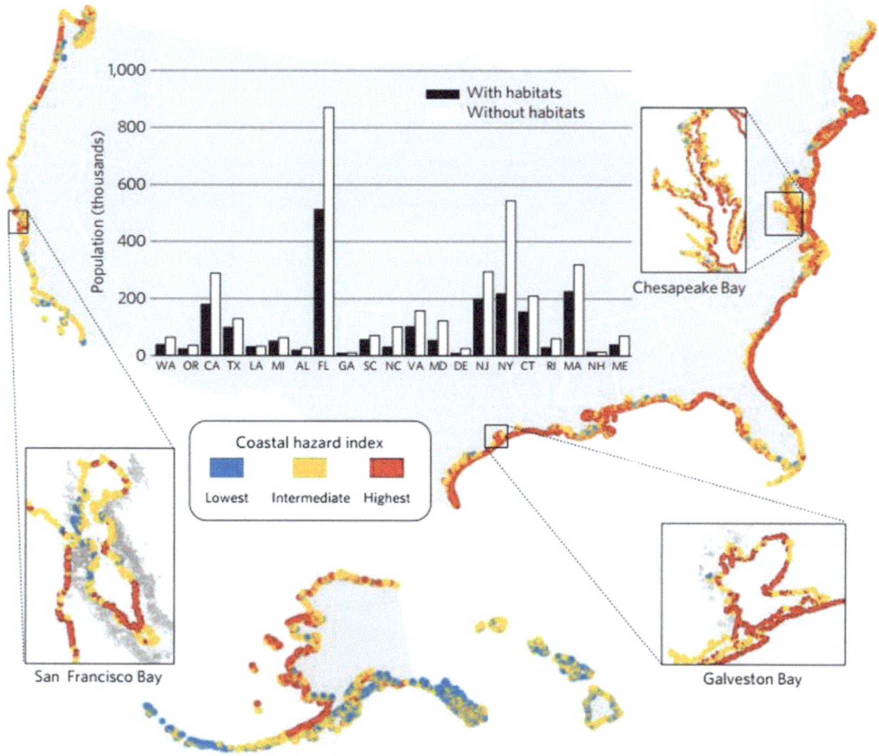

Fig. 11.2 Exposure of the US coastline and coastal population to sea-level rise in 2100 (modeled future scenario) and storms. Warmer colors indicate regions with more exposure to coastal hazards (index > 3.36). The bar graph shows the population living in areas most exposed to hazards (red 1 km^2 coastal segments in the map) with protection provided by habitats (black bars) and the increase in population exposed to hazards if habitats were lost owing to climate change or human impacts (white bars). States are indicated on the x axis. Data depicted in the inset maps are magnified views of the nationwide analysis (Arkema et al. 2013)

- Wetlands and marshes;
- Mangroves;
- Beaches and Dunes;
- Coral Reefs;
- Oyster Reefs; and
- Living Shorelines.

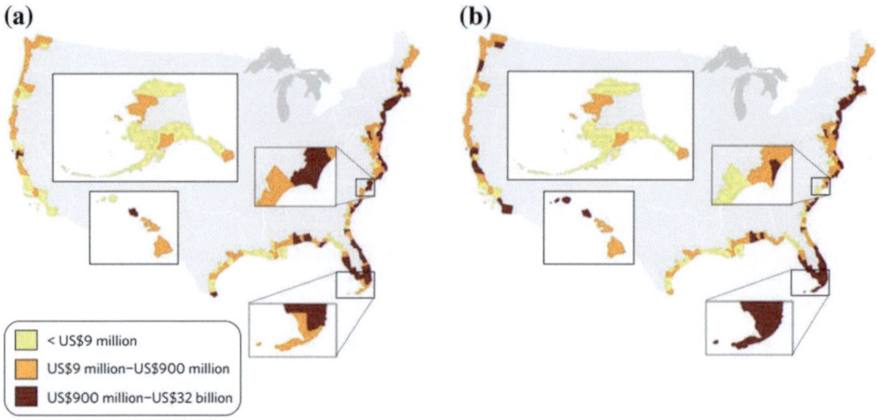

Fig. 11.3 Nature's shield for total residential property value. **a**, **b** Total property value for which habitats reduce exposure to storms and sea-level rise in each coastal county of the United States for the current (**a**) and future (**b**) sea-level-rise scenarios. Insets show Monroe County in Florida, Georgetown and Horry counties in South Carolina and Brunswick and Pender counties in North Carolina. Reduction in the total value of property exposed to coastal hazards is the difference in the total value of property exposed to coastal hazards with and without habitats included in the hazard index. Estimates for each 1 km^2 segment in the highest hazard category (index > 3.36) are summed by county (Arkema et al. 2013)

11.4 Wetlands and Marshes

Coastal wetlands and marshes are among the most productive habitats on earth. They play critical roles in the life cycles of an enormous number of species and offer tremendous recreational opportunities. By some estimates, over 50% of commercial fish and shellfish species in the United States rely on coastal wetlands for survival (Naturally Resilient Communities). Coastal wetlands and marshes are also vital for the safety of our coastal communities. Often referred to as "horizontal levees" or "nature's sponges", coastal wetlands and marshes provide flood protection, erosion control, and water quality protection by reducing and absorbing wave energy and flood waters. U.S. wetlands are valued at $23.2 billion per year in storm protection (Costanza et al. 2008).

Through partnership with Risk Management Solutions, Guy Carpenter & Company, and others, The Nature Conservancy analyzed benefits provided by coastal marshes during Superstorm Sandy. The analysis found coastal wetlands reduced property damages by more than $625 million across the impacted area, with an average of 10% damage-reduction per state (Narayan et al.).

Furthermore, these natural defense systems are capable of keeping pace with sea level rise; thus, maintaining their protective functions into the future without the additional investments that would be necessary to maintain grey structures. And yet, as storm impacts—and related social, economic, and environmental costs—continue to increase, these natural defenses continue to be lost at alarming rates.

The EPA has estimated losses as high as 80,000-acres per year between 2004 and 2009. New York has lost 50% of its wetlands on Long Island Sound (Office for Coastal Management 2017).

In Louisiana, the loss of protective wetlands was seen in the aftermath of Hurricane Katrina—the deadliest and costliest Atlantic hurricane. Although less intense than Hurricane Camille of 1969, Katrina caused far more damage, loss of life, and general devastation. The increased impact of Katrina is primarily attributed to wetland loss sustained over the 36-year period between storms (Chap. 14).

Case Study: Broward County, Florida

An assessment of the world's metropolitan areas with the most to lose from hurricanes and sea level rise, placed southeast Florida at the top of the list. Accelerated sea-level rise and predictions of stronger hurricanes have led to an increased demand for action and willingness to participate from both the public and private sectors. As a result, numerous leaders from around the region have invested in nature-based solutions to protect coastal communities.

One project, undertaken at West Lake Park in Broward County, garnered attention from environmental groups and regulatory agencies nationwide. In 2011, Broward County completed a $10 million mitigation and habitat restoration project to replace 15-acres of wetland habitat lost when Fort Lauderdale-Hollywood International airport was expanded. In addition to 15-acres of restored wetland habitat, the project also created 13-acres of new mangrove wetlands and preserved nearly 4.5-acres of additional park land. The project was so successful that it is now used by the University of Florida in its Master Naturalist Program training classes.

The restoration, which is the largest of its kind in the area, provides valuable recreational opportunities to residents and visitors.

11.5 Mangroves

Mangroves, a type of coastal or estuarine wetland characterized by the presence of salt adapted trees and shrubs, are among the most effective natural forms of coastal protection found in the world. Their complex root system and physical structure significantly absorbs wave energy and reduces wave heights. Studies suggest wave heights can be reduced by 13–66% over a 100-m-wide mangrove belt or as much as 50–100% over a 500-m-wide mangrove belt. Furthermore, the vegetation found within mangroves reduces wind speeds thereby lessening the likelihood of waves increasing in height within mangrove areas.

The same physical structure that reduces wave energy also serves to settle sediment out of the water column. The combination of lessened wave energy and increased sedimentation reduces coastal erosion and, in some cases, allows mangroves to keep pace with sea level rise through accumulation of sediment and vertical growth, providing ongoing protection as sea levels increase.

In addition to reducing the impacts of storm events and limiting erosion, mangroves provide habitat for an enormous number of fish and bird species which creates opportunities for recreational fishing and bird watching and outdoor tourism activities such as kayaking and wildlife viewing. Furthermore, mangroves provide habitat for many economically important species. In Florida, bone fish, tarpon and red snapper recreational fisheries—all mangrove reliant species—are estimated to produce as much as $1 billion in annual value (Naturally Resilient Communities).

11.6 Beaches and Dunes

For many, the thought of vacation and relaxation brings to mind a beach getaway. Not surprisingly, studies have shown beaches to be the largest driver of coastal tourism in the United States. It is estimated nearly 85% of annual, tourist-related dollars received by coastal states are due specifically to trips to beaches and associated coastal areas. In 2002, James Houston of the United States Army Corps of Engineers estimated beaches contributed over $250 billion to the U.S. economy in 1999 alone (Naturally Resilient Communities).

Beaches and dunes also provide important habitat for a number of species, including migratory birds which contribute to the birdwatching industry—estimated to contribute more than $40 billion annually to the national economy. In addition, beaches and dunes protect coastal communities from storm surges and other forces of the ocean. These natural features dissipate the energy of breaking waves by offering 'sacrificial sediment'. When sand and sediment are transported offshore, the surf zone and sandbar systems are widened which causes waves to break further from shore; thus, weakening them before landfall (see examples in Chap. 5). By absorbing and dissipating waves, beaches act as buffers between the sea and inland areas. However, hard structures, such as breakwaters and seawalls, deny storm waves access to sacrificial sediment needed to create wide, dissipative surf zones.

A study of barrier islands in Louisiana showed barrier islands could reduce wave heights by 30% or more and delay the propagation of surges by up to two hours during hurricanes (Grzegorzewski et al. 2011). Healthy barrier islands often consist of multiple dunes fronted by wide beaches. A beach or dune's size, width, slope, shape, and sand volume determine how well the beach can protect a developed area during a storm. The wider a beach or dune system is, and the more space between the sea and any developed or populated areas, the more effective and efficient the system will be at reducing the impacts of coastal hazards.

A beach's popularity is often its downfall. As indicated in Table 2, many of the world's most popular beaches are located near megacities. Unfortunately, the sprawl of these cities causes beaches to lose their natural characteristics. Grey infrastructure stops the movement of sand and sediment, effectively starving a beach or dune system of sand and exacerbating impacts of coastal erosion.

Fig. 11.4 South Cape may Meadows during the dune storm breach in October 1991. *Credit* U.S. Army Corps of Engineers

Fig. 11.5 Beach replenishment on the state park beach and around Cape May Point, before (left) and after (right) the restoration. *Source* U.S. Army Corps of Engineers 2007

Fortunately, efforts to restore these systems have been shown to be successful in renewing and enhancing the protective services they offer communities. A prominent example is Cape May, New Jersey (Figs. 11.4 and 11.5).

Case Study: South Cape May Meadows, Cape May Point, New Jersey
The communities in the Cape May area have a history of falling victim to flooding, coastal storms and coastal erosion. The once vibrant, Victorian resort town of South Cape May was abandoned following the Great Atlantic Hurricane of 1944 and has since eroded away into the ocean. Other areas of Cape May were being lost to erosion at a rate of 4.6 m/year (15 ft per year). But, it was the "Perfect Storm" of

Halloween of 1991 that truly fueled the urgency for a coastal risk reduction project. The storm caused an estimated $75 million dollars in damages along the New Jersey coastline—$10 million of damages to the beach in Cape May alone. Then, early in 1992, the Cape May area was hit again. These two storms ignited a growing sense of urgency for more effective flood protection strategies.

Following the October storm of 1991 (aka "Perfect Storm"), a diverse group of stakeholders, including the U.S. Army Corps of Engineers (USACE), New Jersey Department of Environmental Protection, The Nature Conservancy, and the local governments, met to discuss a potential comprehensive ecological restoration project. By 1998, the USACE completed the feasibility study, funding was secured, and implementation was underway.

During the first phase of the project nearly 1,070,377 m^3 (1,400,000 yd^3) of sand were used to construct a 1.6 km (1 mile) long, 5.5 m (18 ft) tall sand dune and widen 3.2 km (2 mile) of beach. During the second phase, freshwater wetlands were restored and drainage culverts were added to improve water flow, quality and draining—thereby increasing their ability to prevent flooding. The project was completed in 2007 and restored nearly 460 acres of coastal habitats.

The community of Cape May Point has experienced significant economic benefits since the implementation of these nature-based solutions. In 2012, the Cape May peninsula was hit by Superstorm Sandy, but the newly created dune was not breached; while damage to the surrounding area was estimated at $640 million dollars, the newly restored area of Cape May suffered virtually no damage. Prior to the restoration, flood claims in Cape May Point following major storm events averaged $143,713. That number has now dropped to $3713, leading to a projected flood claims savings of $9.6 million dollars over the next 50 years.

Furthermore, birds have been flocking to the newly restored habitats—and bird-watchers, and the dollars they spend, are following. An analysis conducted in 2014 by Elizabeth Schuster of The Nature Conservancy predicted that the ecotourism from these birders will add more than $310 million dollars per year to the county's revenue.

The restoration of South Cape May Meadows offered a unique opportunity to conduct a before-and-after analysis of nature-based solutions, and emerged as a prime example of the significant economic, ecological and social benefits such solutions can provide. The community of Cape May Point is now better protected from the dangers and economic costs of coastal storms, boasts newly restored wildlife habitat and a thriving ecotourism industry, and has even more recreation opportunities for residents and visitors alike.

11.7 Offshore Coral and Oyster Reefs

Offshore coral and oyster reefs serve as natural breakwaters. Their physical structure and rough surfaces absorb and dissipate the force of waves; creating calmer waters on the shoreline side and reducing impacts of erosion and flooding.

Corals are marine invertebrates that permanently attach themselves to the ocean floor while living in compact colonies of many identical, individual polyps. According to Spalding et al. (2001), most coral reefs are found in tropical waters less than 50 m (164 ft) deep where temperatures range from 23 to 29 °C (73–84 °F).

The reef systems built by coral polyps can reduce as much as 97% of wave energy, greatly decreasing the impacts to communities and creating calm waters on the backside of the reef which protects beaches and dunes from erosion. Worldwide, nearly $6 billion of built capital protected from flooding by coral reefs annually (The Nature Conservancy). Studies show onshore damages triple with as little as a one-meter loss of coral reef (Tercek 2017).

Coral reefs are also popular with tourists, especially snorkelers and scuba divers. A study conducted by Nature Conservancy scientists for the Journal of Marine Policy, found coral reef tourism generates $36 billion globally each year (Spalding et al. 2017). Coral reef based tourism in the Florida Keys alone has been estimated to provide more than $1 billion of economic activity annually.

Additionally, coral reefs provide shelter, habitat, and food for numerous species. Some estimates have found nearly 25% of all marine species rely on coral reefs during some phase of their life (Naturally Resilient Communities). Many of the species that rely on coral reefs are recreationally or commercially important. According to the National Oceanic and Atmospheric Administration (NOAA) the commercial value of coral reef-based fisheries is more than $100 million annually in the United States.

Oysters, bivalve mollusks which inhabit coastal waters with variable salinity and are generally found between 64° North and 44° South latitudes (Bayne 2017), also live in compact colonies which form reefs. Oyster reefs are capable of meaningfully reducing wave energy. Studies from the Gulf of Mexico have found oyster reefs can decrease wave energy by as much as 76–93%.

In addition to providing protection, oysters are very economically important. The oyster fishery is still one of the most valuable in the U.S. In the Chesapeake Bay, dockside value for oysters was approximately $44 million from 2013 to 2014. Dockside value, however, only accounts for a small portion of the overall value of the fishery—one analysis found each $1 million of dockside value in the Chesapeake Bay equated to $36 million in total sales and almost 1000 jobs (Naturally Resilient Communities). Rich oyster fisheries also flourish in the bays of coastal Louisiana and Apalachicola Bay, Florida.

Like wetlands and mangroves, offshore reefs are generally expected to be able to keep pace with modest amounts of sea level rise if their environmental conditions are good. However, there is growing concern about the health of our globe's reef systems. Studies suggest more than 85% of historic oyster reefs have been lost, and it is estimated that 70% of the world's coral reefs could be lost by 2050. These vital ecosystems are in desperate need of protection and restoration. Fortunately, there are a number of recent success stories which give inspiration and hope.

For instance, a review of oyster reef restoration projects in Alabama showed every mile of restored reef would provide revenue of $2.3 million and create 25 jobs. Additionally, the construction of these self-sustaining oyster reefs could save Alabama property owners as much as $150 million over rip-raps and bulkheads (Kroeger 2012; NOAA 2012).

Furthermore, a 2012 study by economists from The Nature Conservancy, titled "Economics of Oyster Reef Restoration in the Gulf of Mexico", found an investment of $150 million towards oyster reef restoration will:

- Build 100 miles of oyster reefs
- Create 380 jobs per year for 10 years, or rather, 3800 jobs during the decade-long construction phase
- Boost regional household income by $9.7 million a year during the 10-year construction period
- Increase revenues and sales of crab, fish, and oyster harvests by $7.87 million yearly
- Increase annual sales by $7.3 million in the commercial seafood supply chain.

To date, more than 200 jobs have been supported by NOAA-funded oyster reef restoration projects implemented by the Conservancy across the Gulf of Mexico. Coastal habitat restoration produces jobs at higher rates than sectors like road infrastructure or oil and gas (NOAA 2012).

Case Study: Half Moon Reef, Texas

The Half Moon Reef oyster colony is one of the most productive fisheries for blue crabs, oysters and shrimp in Texas. Once massive, Half Moon Reef had to be reconstructed from the ground up. Completed by The Nature Conservancy in 2014, the 54-acre restoration project is one of the nation's largest.

Surveys of the reef show the Conservancy's complex design is working. Oysters have attached to roughly 70% of the reef's total surface and between January 2014 and May 2016 the size of those oysters increased an astonishing 551%. Biodiversity is 40% higher at Half Moon Reef when compared to the adjacent bay floor; biomass, which helps measure the level of sea life in and around the reef, is 1014% greater than at nearby areas.

Better still, this innovative restoration project has spurred some exciting social and economic benefits. A 2016 survey by the Conservancy and Texas Sea Grant found:

- Increased recreational fishing at Half Moon Reef added $691,000 to Texas' gross domestic product year over year and generated an additional $1.27 million in annual economic activity
- Half Moon Reef has created a dozen new jobs and $465,000 in annual labor income

- 94% of fishermen reported that the restored habitat at Half Moon Reef offered a more satisfying experience than other fishing locations.

Using the success of Half Moon Reef as a blueprint, The Nature Conservancy is spearheading three new large-scale oyster reef restoration projects.

Case Study: Monroe County, Florida
In Monroe County more than 33,000 jobs, accounting for more than half of the local economy, are supported by ocean recreation and tourism. But, the reefs of the Florida Keys have been declining since the 1970s. So, Monroe County, NOAA, and The Nature Conservancy are collaborating to protect the remaining healthy coral and restore populations. Tens of thousands of corals are being grown in underwater nurseries and outplanted to the degraded reefs off Monroe County.
The project's primary restoration and recovery approach is to take small fragments of live tissue from healthy coral colonies of known genetic stock, grow them in nurseries over time to create multiple colonies of each genetic type, and then outplant genetically distinct individuals in proximity to one another so they spawn and help reseed surrounding reefs.
Each outplanting site directly enhances live coral cover, wave-breaking structure, fisheries habitat and tourism value.
Nearly 10,000 colonies have been re-established since 2004 with survival rates averaging above 80%.

11.8 Living Shorelines

Living shorelines are an important coastal management tool designed to reduce the impacts of erosion while maintaining other vital coastal processes. Living shorelines consist of a broad suite of shoreline erosion control techniques combined with natural or engineered means of dissipating wave energy. Applications of living shorelines range from purely vegetation, to pairing vegetation with offshore sills of some sort, to a truly hybridized approach which utilizes both green and grey infrastructure features.

Unlike traditional erosion control structures, such as bulkheads or seawalls, which focus on deflecting wave energy away from a site (and may actually increase erosion), living shorelines reduce energy onsite by breaking-up and absorbing wave energy as they roll into the shoreline. Although primarily implemented as an erosion control feature, when properly designed living shorelines can provide protection during low level storm surge events.

In addition to erosion control, living shorelines allow for the continuation of important natural processes that maintain the health of the broader coastal system. Furthermore, living shorelines help maintain coastal habitat which would otherwise be lost if traditional coastal management approaches, such as grey infrastructure, were utilized.

(a) (b) (c)

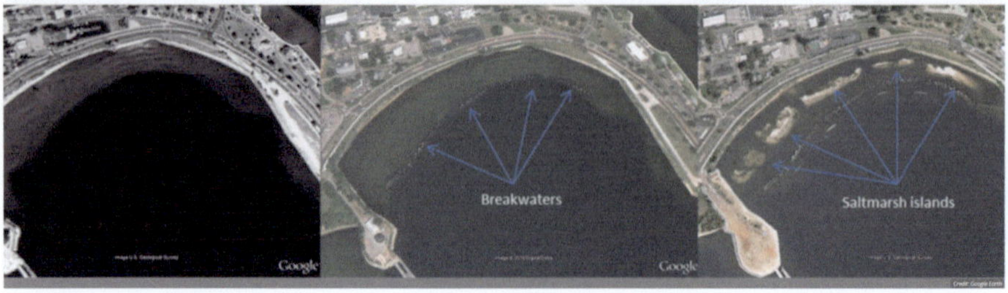

Fig. 11.6 Project greenshores, Pensacola Florida. *Credit Google earth*

At sites where shoreline armoring has been previously implemented, hybridized living shorelines can be a strategic means of recapturing functionality of aged grey structures while maintaining or improving the ecosystem services—such as improved water quality, enhanced habitat, and increased aesthetics—provided by natural coastal features. Pensacola Florida provides an example of a successful hybrid shoreline (Figs. 11.6, 11.7 and 11.8). In 2007, Project Greenshores began construction at Site 2 that featured a similar, but smaller-scale design. Here, citizens requested the breakwaters be completely submerged underwater to improve the site's aesthetics. The breakwaters were constructed from 25,000 cubic yards of recycled concrete donated after a runway was decommissioned at a nearby Naval air station. Three islands were then created from 16,000 cubic yards of dredge spoil that was removed from the Escombia River during a separate project. Site 2 was

Fig. 11.7 View of project greenshores showing breakwaters. *Credit* Darryl Boudreau, The Nature Conservancy

Fig. 11.8 View of project greenshores showing salt marsh islands and breakwaters. *Credit* Darryl Boudreau, The Nature Conservancy

completed in 2008 with the planting of 30,000 plants. Plans are being developed to enhance the breakwaters, to provide better wave reduction, at Site 2 and increase marsh areas.

Case Study: Hybrid Living Shoreline, Pensacola, FL

Pensacola's location in the western Panhandle of Florida, combined with its low, flat elevation, means that intense wave action, rising water levels and storm surges pose significant threats to city. In addition to the threat of loss of life, storm surges can damage property and public infrastructure, force businesses to close, delay emergency responses and erode coastlines.

In 2000, the City of Pensacola began a project to re-establish lost marsh habitat along its downtown shoreline through "Project Greenshores".

The first phase of the project was construction of a breakwater. The structure sits on the bay's floor and extends above the surface, creating a wall that reduces the height and force of waves. The breakwaters were constructed using 14,000 tons of limestone, 6000 tons of recycled concrete, and 40 preformed concrete blocks. Next, workers formed 35,000 cubic yards of sand into a series of five islands just offshore. Some 41,000 *Spartina Alterniflora* plants, a native species of common saltmarsh grass, were then planted to produce eight acres of marsh habitat. With the combination of sand, natural vegetation, and offshore breakwaters the site increased protection from storm surge by attenuating the waves, stabilizing sediment and reducing erosion during storms.

Project Greenshores has been a resounding habitat restoration success and has exceeded expectations in terms of its potential benefits. Bird species are now so abundant at Project Greenshores that the Audubon Society added it as a site along the Great Florida Birding Trail, which has boosted tourism. Oysters have grown on the breakwaters, which enhances habitat for fish. Crab fishermen have benefitted from increased harvests of blue crabs, which were scarce in the bay before the project. The breakwaters have reduced wave action and turbidity, while the marsh habitat better retains nutrients before they wash into bay, which has improved water quality.

Shortly after completion, the effectiveness of Project Greenshores was tested by two consecutive, category 3 hurricanes (Ivan 2004; Dennis 2005). During both events, the breakwaters and marsh islands helped protect the shore from storm surge, proving its utilization in coastal resiliency.

"This project has clearly demonstrated that living shorelines have tremendous benefits not only in terms of our ecosystem, our economy, and our recreational activities, but also in mitigating the impact of tropical storms and hurricanes. Experts have found that areas with living shorelines experience less coastal erosion as a result of storm events."—*Pensacola Mayor Ashton Hayward.*

11.9 Implementing and Protecting Natural Infrastructure

Protect and Restore Natural Systems—A fundamental philosophy of municipal governance is investing in and enhancing what is already present in the community —building on the community's existing strengths as a way to promote long-term growth and success. This same philosophy should be applied to the natural infrastructure communities possess.

As with any type of infrastructure, the most effective and efficient means for a community to continue receiving the benefits provided by natural infrastructure is to invest in their protection and maintenance. Intact, healthy, unprotected stretches of coastal or estuarine habitat can be thought of as investment opportunities to ensure ongoing delivery of the services they provide. Degraded or destroyed habitats can be considered as infrastructure in need of upkeep or repair. To invest in the restoration or enhancement of these features is much the same as one would invest in road repairs to keep transportation ways safe and effective.

To understand the scale of opportunity and need, communities should identify their existing, unprotected natural infrastructure as well as the areas that are already protected but in need of restoration. Mapping and inventorying their coasts should be followed by an effort to identify and prioritize the most important systems to guide investment decisions. Many municipalities already have maps of potential open space and identified targets based on criteria such as biodiversity or proximity to existing protected lands, but assessing these inventories for their potential protective benefits can provide additional data for consideration when deciphering where and how to invest.

One means of protecting natural infrastructure is acquisition of property. This usually takes one of two forms—open space acquisition of an undeveloped parcel or voluntary buy-out of owned, developed property. Functionally these processes are quite similar. The community identifies a property of interest, finds a willing seller, and negotiates an agreement of just compensation for the parcel. Open space acquisition is a long-standing practice and it is quite likely that a community already engages in this effort, even if historically the reason for acquisitions was not related to flood and storm reduction (Naturally Resilient Communities).

Voluntary buy-outs of vulnerable properties offer communities a two-fold opportunity—moving people out of a vulnerable situation and preserving or restoring the underlying natural infrastructure. A common practice in river communities to address flooding, voluntary buy-outs have begun to gain more traction in coastal communities as frequency and severity of coastal storms increases. In the aftermath of Superstorm Sandy, hundreds of homes were acquired by New York State, creating opportunities to restore wetlands and coastal habitats that had been degraded by development (Naturally Resilient Communities). However, communities do not always have to rely on property acquisition as the only means of protecting existing natural systems. Establishing regulations on allowable land uses, limiting the amount of clearing on individual parcels, designating no-build zones, and other regulatory strategies can promote healthier, more effective natural infrastructure by guiding land-use decisions (Naturally Resilient Communities).

Each coastal habitat is unique and its protection or restoration should be accompanied by a comprehensive plan to ensure the greatest likelihood of success. Focusing additional efforts on restoring or enhancing the natural processes these systems rely upon—clean water, sufficient sediment supply, and enough space to persist now and in the future—further increases the likelihood these features will continue to benefit the community for years to come.

These systems, and the benefits they provide, cannot continue to be taken for granted. Communities that take a strategic approach to identifying their most important natural infrastructure and investing in its current and future health will be well-rewarded.

Influence Public Policy—Not surprisingly, existing financial, legislative and cultural norms favor grey infrastructure over natural defenses. But, studies have shown natural infrastructure often inspires greater public/local support because of the attractive and highly valued community amenities, such as restored river channels, river parkways, and beaches, they often provide. In 1995 a $115 million "grey" Napa River flood protection proposal from the Army Corps of Engineers was rejected amidst strong local opposition. Two years later, Napa County voters approved a local sales tax increase to fund a "Living River" design, despite its higher projected cost of $163 million (The Nature Conservancy, *Reducing Climate Risks with Natural Infrastructure; The Case for Green Infrastructure* 2014).

To most effectively promote natural solutions, a suite of policies, programs and funding is needed to facilitate investments in natural infrastructure. Fortunately, there are examples of innovative strategies emerging. NOAA has led in this arena by providing technical assistance and grant funding for coastal resilience projects and creating the Digital Coast Partnership, a collaboration of organizations providing science, tools and assistance to advance natural infrastructure work.

In Congress, some progress has been made through legislation, such as Section 1184 of the 2016 Water Infrastructure Improvements for the Nation (WIIN) Act, which calls on the U.S. Army Corps of Engineers to consider natural and nature-based techniques when scoping flood management projects. Additionally, the Federal Emergency Management Agency (FEMA) has adopted policies to help account for the value of natural infrastructure when conducting flood risk mitigation work. While these examples are encouraging and progressive, there is much more that needs to be done to bring natural infrastructure into mainstream public policy.

Assess the Value of, and Insure, Nature—By assigning value to nature in risk assessments, insurance companies are beginning to see potential opportunities to stabilize, or even reduce, losses and increase the resilience of communities in the face of climate change and heightened disaster risks.

Swiss Re, a global reinsurance industry provider and leader in understanding climate risk, and Risk Management Solutions (RMS), one of the largest catastrophe risk modelers in the market, are some of the first agencies to account for nature's protective services in their risk modeling. A recent study by Swiss Re found the island of Barbados loses the equivalent of 4% of its GDP each year in hurricane disaster costs. But, the study also found for every $1 spent on protecting and restoring mangroves and coral reefs, $20 can be saved in future hurricane losses. When asked what he thought the potential for natural infrastructure was in places like Barbados, Swiss Re's Vice President of Global Partnerships, Alex Kaplan, said simply, "It's about allowing economic progress to continue."

Insurance companies are not the only ones working to insure nature. The Nature Conservancy has launched new approach to protecting and restoring coastal ecosystems and coral reefs. The goal is to create a shared governance trust fund with a reliable source of funding which will be used to manage the health of a 60-km stretch of the Mesoamerican Reef, as well as the beaches, of greater Cancún, México.

In addition to funding management of the reef and beaches, a key purpose of the fund is the creation of an insurance policy for the beaches and reef. The policy will pay out for reef and beach restoration while the shared trust fund supports perpetual beach and reef maintenance and "First Response" teams for storm-ravaged ecosystems. Key stakeholders to this approach are the local Cancún hotel association, state government, and science community (represented by The Nature Conservancy). The Nature Conservancy began this project in 2015 and expects it will culminate in early 2018. The hope is to create a profitable insurance mechanism that can be replicated across reef-dependent communities, a sustained source of public and private funding, and new science protocols which help calculate the value of coastal resources.

The Nature Conservancy is also looking at a range of other insurance mechanisms, including reduced premiums for property owners protected by healthy natural infrastructure—which would hopefully incentivize those parties to help keep the infrastructure healthy—and nature-based captives, where companies create their own insurance companies to cover the protective services of natural infrastructure. As the world prepares for more expensive climate-related disasters, novel insurance strategies provide opportunity for reducing our climate costs and increasing the resilience of people and economies.

Move People out of Harm's Way: Planned Retreat—In some cases, relocating away from areas vulnerable to flooding will increasingly become a necessary choice for health, safety, economic, and environmental reasons. Sea level rise in particular has the potential to drive net migration away from vulnerable coastal areas (Hauer 2017). A key component of any relocation plan is that it be a resident and stakeholder-based participatory process with a shared vision based on natural and social science and urban planning principles. The CNAI Handbook (NOAA and the Association of State Floodplain Managers 2007) recognizes relocation as the "preferred" strategy for reducing vulnerability of residents living in flood prone areas.

Analyses of sea level rise projections, storm surge scenarios, physical, structural, economic, and social susceptibility, areas of repetitive loss, and conservation or restoration potential are useful for prioritizing areas of retreat or redevelopment. Behavioral economics are also an important consideration in planning so as to fully understand issues of loss aversion, community culture, fairness, and uncertainty about retreat and relocation (Buchanan et al. in review; Masterson et al. 2017).

Community planning can also address the potential change in local tax revenue from property acquisitions and the benefit or gain from cost reduction for community services and creation of open space and public access in the retreat areas. These participatory community plans can disseminate critical information and foster trust, relationships, and collaboration among stakeholders (Nelson et al. 2007); all of whom are needed to ensure long-term resilience of their community.

The mechanics of planned retreat require acquisition of willing sellers' private property and removal of homes, roads, telephone poles, electrical and sewage connections, and other vulnerable infrastructure. Acquisitions take time and are done on an individual basis. Ideal planned retreat areas are assemblages of numerous willing sellers to create a coherent, contiguous area of future open space. Over the short-term, however, there may be a patchwork of homes and infrastructure amid undeveloped properties which can temporarily limit creation of parks and public access. For this reason, it is important to implement short, medium, and long-term planning for the properties which considers maintenance, access, and use. Property acquisition can happen many ways, including purchase of fee title, easements, rolling easements, or life tenancy agreements. There may also be opportunities to provide tax incentives such as bargain sales.

A key component for successful planned retreat is providing fair market value (FMV) for acquired properties, which will vary depending on the different programs' rules for determining FMV. Factors such as pre- or post-storm value of similar properties, presence of hazardous materials, underground storage, or utilities are some considerations when determining FMV for a property. Depending on funding programs, different criteria are followed to complete acquisitions. A major source of funds for acquisition is post-disaster funding from federal programs. For example, after Superstorm Sandy in 2012, New York received $4 billion in disaster recovery aid primarily through FEMA and the US Department of Housing and

Urban Development (HUD) Community Development Block Grant Disaster Recovery program (CDBG-DR). A portion of these funds is being used to acquire approximately 500 homes for retreat where the land will be returned to open space and deed-restricted, while another approximately 500 homes are being acquired for "resilient re-development" to be re-sold, with the potential for some relocation support. Resilient redevelopment includes elevating homes and other flood-prevention-related structural changes.

There are also pre-disaster, flood mitigation, and other federal, state, and local funds that can be used for acquisition of developed and undeveloped land or for easements to limit development in floodplains and protect wildlife, habitat, water quality, and current and future wetlands that will result from flooding and inland wetland migration.

The benefits associated with property buyouts accrue in two areas. First, moving people out of harm's way and removing vulnerable infrastructure reduces the overall risk and vulnerability in a community, thereby reducing the costs associated with ensuring the health and well-being of those communities, particularly during storm events, and the costs associated with rebuilding in vulnerable areas. By moving people out of vulnerable areas and replacing the built infrastructure with natural infrastructure, communities are provided with ecosystem service benefits that can include storm protection, access to open space, recreation, economic development, and improved water quality and habitats.

Sidebar note: A large percentage of the federal payouts for flood damages are to repetitive loss properties (King 2013). For example, in 2013, 9000 homes equal to only 1% of properties insured by the National Flood Insurance Program (NFIP) made up approximately one-third of the National Flood Insurance Program's claim costs since 1978 (Daley and Wang 2015).

11.10 Conclusion

Nature provides numerous, quantifiable services and benefits; however, communities are not often educated or incentivized to protect their natural systems. Bridging this divide is paramount for communities across the globe to survive in a climate changed future. Nature must be prioritized in our regulatory, financial, and social systems in ways which empower, encourage, and incentivize smart and necessary investments in natural infrastructure.

We cannot stop the seas from rising or the storms from coming. If we are to live safely in these areas, at harmony with nature and water, tomorrow's coasts will need to look very different than those of today. We must make choices today that enable succeeding generations to stand at the waters' edge and feel the same, deep affinity that has drawn so many.

References

Arkema, et al. 2013. Coastal habitats shield people and property from sea-level rise and storms. *Natural Climate Change*. https://doi.org/10.1038/nclimate1944.

American Society of Engineers. www.infrastructurereportcard.org.

Bayne, Brian L. (ed.). 2017. *Biology of oysters,* vol. 41. Developments in Aquaculture and Fisheries Science. Academic Press.

Buchanan, M.K., M. Oppenheimer, and A. Parris (in review). 2017. Drivers of household adaptive behavior in response to coastal flood risk and external stressors. Climatic Change.

Bruckner, Andrew W. 2002. Life-saving products from coral reefs. *Issues in Science and Technology* 18 (3): 35.

Cooper, Edwin L., Kyle Hirabayashi, Kevin B. Strychar, and Paul W. Sammarco. 2014. Corals and their potential applications to integrative medicine. *Evidence-Bases Complementary and Alternative Medicine* 2014, Article ID 184959 (9pp). http://dx.doi.org/10.1155/2014/184959.

Costanza, R., O. Pérez-Maqueo, M.L. Martinez, P. Sutton, S.J. Anderson, and K. Mulder. 2008. The value of coastal wetlands for hurricane protection. *Ambio* 37: 241–248. https://doi.org/10.1579/0044-7447(2008)37[241:tvocwf]2.0.co;2.

Daley, B., and S. Wang. 2015. A call to cull homes threatened by the sea. *Boston Globe*, Feb 8.

Demers, Caroline, C. Reggie Hamdy, Karin Corsi, Fatiha Chellat, Maaryam Tabrizian, and L'Hocine Yahia. 2002. Natural coral exoskeleton as a bone graft substitute: a review. *Bio-Medical Materials and Engineering* 12 (1): 15–35.

ESA-PWA. 2012. Evaluation of erosion mitigation alternatives for Southern Monterey Bay. Report prepared for the Monterey Bay Sanctuary Foundation and the Southern Monterey Bay Coastal Erosion Working Group. 30 May 2012. http://montereybay.noaa.gov/research/techreports/tresapwa2012.html.

Ferrario, F., M.W. Beck, C.D. Storlazzi, F. Micheli, C.C. Shepard, and L. Airoldi. 2014. The effectiveness of coral reefs for coastal hazard risk reduction and adaptation. *Nature Communications* 5: 3794. https://doi.org/10.1038/ncomms4794.

Gedan, K.B., M.L. Kirwan, E. Wolanski, E.B. Barbier, and B.R. Silliman. 2011. The present and future role of coastal wetland vegetation in protecting shorelines: Answering recent challenges to the paradigm. *Climate Change* 106: 7–29. https://doi.org/10.1007/s10584-010-0003-7.

Gittman, R.K., et al. 2015. Engineering away our natural defenses: an analysis of shoreline hardening in the US. *Frontiers in Ecology and the Environment*. https://doi.org/10.1890/150065.

Grzegorzewski, A.S., M.A. Cialone, T.V. Wamsley. 2011. Interaction of barrier islands and storms: Implications for flood risk reduction in Louisiana and Mississippi. *Journal of Coastal Research* 156–164.

Hauer, M.E. 2017. Migration induced by sea-level rise could reshape the US population landscape. *Nature Climate Change* 7: 321–327. Published online www.nature.com/natureclimatechange 17 Apr 2017. https://doi.org/10.1038/nclimate3271.

King, R.O. 2013. The national flood insurance program: Status and remaining issues for Congress. Congressional Research Service. 6 Feb 2013 (39pp).

Kroeger, T. 2012. Dollars and sense: Economic benefits and impacts from two oyster reef restoration projects in the Northern Gulf of Mexico. Arlington, MD.

Leo K, R. Battalio, W.N. Heady, P. King, A. McGregor, B. Cohen, J. Calil, E. Vandebroek, J. Jackson, F. DePaolis, D. Revell, R. Vaughn, J. Giliam, S. Newkirk. 2017. *Economic impacts of climate adaptation strategies for Southern Monterey Bay.* Technical Report prepared for the California State Coastal Conservancy by The Nature Conservancy. 2017/07. ClimateReadyGrant#13–107.

Masterson, V.A., R.C. Stedman, J. Enqvist, M. Tengö, M. Giusti, D. Wahl, U. Svedin. 2017. The contribution of sense of place to social-ecological systems research: A review and research agenda. *Ecology & Society* 22 (1): 49. https://www.ecologyandsociety.org/vol22/iss1/art49/#responding.

Möller, I., M. Kudella, F. Rupprecht, T. Spencer, M. Paul, B.K. van Wesenbeeck, G. Wolters, K. Jensen, T.J. Bouma, M. Miranda-Lange, and S. Schimmels. 2014. Wave attenuation over coastal salt marshes under storm surge conditions. *Nature Geoscience* 7: 727–732. https://doi.org/10.1038/ngeo2251.

Narayan, S., M.W. Beck, B.G. Reguero, and I.J. Losada, B. van Wesenbeeck, N. Pontee, J.N. Sanchirico, J.C. Ingram, G-M. Lange, L.A. Burks-Copes. 2016. The effectiveness, costs and coastal protection benefits of natural and nature-based defences. *PloS One* 11 (5), Available online http://journals.plos.org/plosone/article?id=10.1371/journal.pone.0154735.

Nelson, M., R. Ehrenfeucht, and S. Laska. 2007. Planning, plans, and people: professional expertise, local knowledge, and governmental action in post-hurricane Katrina New Orleans. *Cityscape: A Journal of Policy Development and Research* 9 (3): 23–52.

NOAA. 2012. Oyster restoration creates jobs, protects coasts in the gulf [WWW Document]. *Habitat* Conservation. *National Marine Fisheries* Service. http://www.habitat.noaa.gov/highlights/oysterrestorationcreatesjobs.html. Accessed 24 March 2017.

NOAA and the Association of State Floodplain Managers. 2007. *Coastal no adverse impact (CNAI) handbook* (165pp).

Pinsky, M.L., G. Guannel, K.K. Arkema. 2013. Quantifying wave attenuation to inform coastal habitat conservation. Ecosphere 4, art 95. https://doi.org/10.1890/es13-00080.1.

Reef Resilience Network. http://www.reefresilience.org/.

Shepard, C.C., C.M. Crain, and M.W. Beck. 2011. The protective role of coastal marshes: A systematic review and meta-analysis. *PLoS ONE* 6: e27374. https://doi.org/10.1371/journal.pone.0027374.

Spalding, Mark D., Edmund P. Green, and Corinna Ravilious. 2001. *World atlas of coral reefs*. Berkeley: University of California Press.

Spalding, M., L. Burke, S.A. Wood, J. Ashpole, J. Hutchison, P. Zu Ermgassen. 2017. Mapping the global value and distribution of coral reef tourism. https://doi.org/10.1016/j.marpol.2017.05.014.

Tapsell, S.M., E.C. Penning-Rowsell, S.M. Tunstall, and T.L. Wilson. 2002. Vulnerability to flooding: Health and social dimensions. *Philosophical Transactions of the Royal Society of London A* 360: 1511–1525.

Tercek, Mark. 2017. *Insuring nature, ensuring resilience.* https://blog.nature.org/conservancy/2017/07/21/insuring-nature-ensuring-resilience/.

The Nature Conservancy. 2013. Joint Industry whitepaper. *The Case for Green Infrastructure*.

The Nature Conservancy. 2014. *Reducing climate risks with natural infrastructure.* San Francisco: The Nature Conservancy California Program (28pp). https://www.nature.org/ourinitiatives/regions/northamerica/unitedstates/california/ca-green-vs-gray-report-2.pdf?redirect=https-301.

The Nature Conservancy. 2017. *Investing in nature: Improving the US economy and protecting our communities by investing in natural infrastructure*.

Wamsley, T.V., M.A. Cialone, J.M. Smith, J.H. Atkinson, J.D. Rosati. 2010. The potential of wetlands in reducing storm surge. *Ocean Engineering* 37: 59–68. https://doi.org/10.1016/j.oceaneng.2009.07.018.

Websites

https://www.nature.org/ourinitiatives/habitats/riverslakes/natural-solutions-for-reducing-flood-risk-factsheet.pdf.

https://www.nature.org/ourinitiatives/regions/northamerica/unitedstates/california/ca-green-vs-gray-report-2.pdf?redirect=https-301.

http://nrcsolutions.org/coastal-marshes/.
http://nrcsolutions.org/wp-content/uploads/2017/03/NRC_Solutions_Dunes.pdf.
http://nrcsolutions.org/oyster-reefs/.
http://nrcsolutions.org/coral-reefs/.
https://www.nature.org/ourinitiatives/regions/northamerica/areas/gulfofmexico/economics-of-oyster-reef-restoration-in-the-gulf-of-mexico.pdf?redirect=https-301.

Part III
Case Studies of Threatened and Vulnerable Coasts

Chapter 12
Pearl River Delta and Guangzhou (Canton) China

Lynn Donelson Wright and Wei Wu

> *If it be said that your object is to exercise control over your country's trade, your nationals have had full liberty to trade at Canton for many a year, and have received the greatest consideration at our hands.*
>
> —Emperor Qian Long's Letter to King George III, 1793

12.1 Historical Center of Trade Is Sinking and at Risk

The city of Guangzhou (Figs. 12.1 and 12.2), formerly called Canton by the British, in China's Guangdong Province was established in the Qin Dynasty ~220 BCE as a port city within the Pearl River Delta (PRD) and near the coast of the South China Sea or Nan Hai (Fairbank and Goldman 1998). Following roughly two millennia as a hub of trade, Canton became the focus of international conflict in the "Opium Wars" with Britain in 1841. China lost and one of the concessions it was forced to make was to grant Britain the rights to the nearby island of Hong Kong, which, at the time, was desolate and malaria infested (Schell and Delury 2013). Today, Guangzhou is one of the world's coastal megacities (discussed in Chap. 6), with a population of 13 million people. However, considering the "agglomeration" of the eight urban centers within the Pearl River Delta including Shenzhen, Hong Kong and Macau the total regional population rose to 66 million by 2017 (The Economist 2017). By 2015, the Pearl River Delta urban agglomeration had overtaken Tokyo as the world's largest megacity (van Mead 2015; Fig. 12.2).

Hallegatte et al. (2013) estimate that by the year 2050, the annual cost of flooding of coastal cities worldwide could be on the order of \$60–\$63 billion (USD) and Guangzhou and the PRD are among the cities most at risk. Westphal

L. D. Wright (✉)
Southeastern Universities Research Association (SURA), Washington, DC, USA
e-mail: ldwright@bellsouth.net

W. Wu
Department of Coastal Sciences, University of Southern Mississippi, Ocean Springs, MS, USA

L. D. Wright and C. R. Nichols (eds.), *Tomorrow's Coasts: Complex and Impermanent*, Coastal Research Library 27,
https://doi.org/10.1007/978-3-319-75453-6_12

Fig. 12.1 View of a small part of the modern city of Guangzhou, Guangdong Province, Peoples Republic of China (PRC). *Photo credit* Pixabay free images

Fig. 12.2 Map of Guangdong Province, PRC showing the location of Guangzhou and 8 other cities. The Pearl River Delta is highlighted in Orange. (From HKTDC Research 2016, *PRD Economic Profile*)

et al. (2013) conclude that the Peoples Republic of China (PRC) is the most threatened East Asian nation in terms of the number of people who may be forced, by sea level rise, to relocate. By 2050, millions of PRC coastal residents may be displaced at a total cost of about $150 billion (USD) per million people; the greatest threat in the PRC is to Guangzhou and the Pearl River Delta. The actual number of people displaced will depend on which sea level rise scenario prevails. Like Miami, Florida, Guangzhou has been built at an elevation of only about 1 m (3.28 ft) above sea level. The Pearl River Delta has a mean elevation less than 2 m (6.56 ft) above mean sea level (Chen et al. 2014) with a mean elevation of 0.3–0.4 m (0.98 to 1.31 ft) above mean sea level in the southern delta (Tracy et al. 2007). With its many high-rise buildings it is "very heavy" and it rests on a deltaic surface. The Pearl River Delta is sinking at an average rate of 2.5 mm/year (0.1 in./year) (Wang et al. 2012). This subsidence, combined with rising sea level, increased intensity of impact from typhoons and storm surge (Du et al. 2013) and projections for increased urban expansion have placed Guangzhou at, or near the top of the list of the world's most threatened coastal cities. In monetary terms, the cost of threatened assets in Guangzhou over the coming decades is $268 billion, roughly equivalent to Miami's jeopardy (Parker 2015). This does not include the potential costs of relocating people displaced by urban flooding. Westphal et al. (2013) suggest that, with appropriate adaptation strategies, some mitigation of future climate change impacts is feasible. However, these strategies will also be expensive and undoubtedly complex.

12.2 Rising Wealth, Rising Water

In 1980, Deng Xiaoping established the Shenzhen and Zhuhai Special Economic Zone in the Pearl River Delta (PRD) region of Guangdong Province. Since then, the economy of the PRD has grown at an explosive rate. For the past decade, the annual rate of growth has averaged 12% and in 2017 the PRD supported one of the world's most successful economies with a GDP that exceeds $1.2 trillion (USD) which is about 10% of China's total GDP (The Economist 2017). The PRD is now considered to be China's most innovative economic zone thanks to globalization and the promotion of free international trade. It is a major manufacturing center for automobiles as well as high-tech products. The American computer company Apple is currently building a research and development center in the PRD. Although the PRD occupies only 1% of China's land area and supports only 5% of its population, it accounts for 25% of China's exports (The Economist 2017). The economy and population of the PRD are expected to continue to grow over the decades ahead although there is uncertainty as what the growth rates might be.

Among various climate change impacts, rising sea level is of great concern for the Pearl River Delta due to its flat terrain and urban development (Tracy et al. 2007). The frequency and height of flood events that regularly affect the PRD are also expected to grow over the course of this century. AVISO altimetry data since 1990s has shown a mean sea-level rise of 3.72 mm/year (0.15 in./year) on the Pearl River Delta coast (He et al. 2014). Longer tide gage records from Hong Kong's Victoria Harbor (Hong Kong Observatory 2017) show that over the period 1954–2016, the average rate of sea level rise was 3.1 mm/year (0.12 in./year), which is identical to the global rate as described in Chap. 3 (Fig. 3.3). However, as concluded by IPCC (2013), models predict that, depending on which warming scenario prevails, the global average rate of sea level rise over the coming decades is likely to be somewhere between 8 and 16 mm/year. Since Hong Kong's current rate of rise corresponds almost exactly to the global rate, we may assume that the future rate of rise will also conform to the global rate. This means that by the end of the century, mean sea levels in the Pearl River Delta may be up to 1.3 m (4.3 ft) higher than at present. Figure 12.3 shows observed and predicted sea level rises at Hong Kong and the Pearl River Estuary from 1948 to 2100 for the three IPCC global warming and sea level rise scenarios discussed in Chaps. 2 and 3 (RCP 2.6, RCP 4.5 and RCP 8.5). To the rises in mean sea level we must add the rate of PRD subsidence, which varies locally averaging about 2.5 mm/year (0.1 in./year) (Wang et al. 2012) but reaching maxima of up to 15 mm/year (0.59 in./year) in some locations (Ao et al. 2015). The total relative rise in sea level could exceed 2 m (6.6 ft) in some places by 2100. Another analysis based on a 72-year tidal record from Hong Kong (1920–1992) and other factors including backwater effects and long-term geological subsidence showed a possible 30 cm (11.81 in.) rise in relative sea-level rise at the mouth of the Pearl River (Huang et al. 2003, 2004). Hence, without protective engineering structures such as levees and floodgates much of Guangzhou and the PRD will be flooded (Du et al. 2013). Figure 12.4 shows the land configuration of the PRD today (Fig. 12.4a) and as it might be with a 1.4 m (4.6 ft) rise in relative sea level (RSL) without protective engineering structures (Fig. 12.4b; from Kindel 2016). Most dykes on Pearl River were built for 1 in 20 year flood events ignoring the effect of sea-level rise (Woodroffe et al. 2006). In order to build new protective infrastructure and raise the standards of old infrastructures to meet sea-level rise, a multi-million dollar investment is required. However, this investment is necessary to reduce the recurring costs of flooding. Huang et al (2003) showed that the total engineering work will require 175 million m^3 of soil and stone to meet the design standard of 30 cm (11.81 in.) elevation rise by 2030 for each of the 95 defenses distributed across the Pearl River delta, which amounts $263 million (1998 USD), comparable to the direct economic loss attributed to the 1994 flood in the region. Without infrastructure protection, the potential future flooded area will be much larger (Huang et al. 2004).

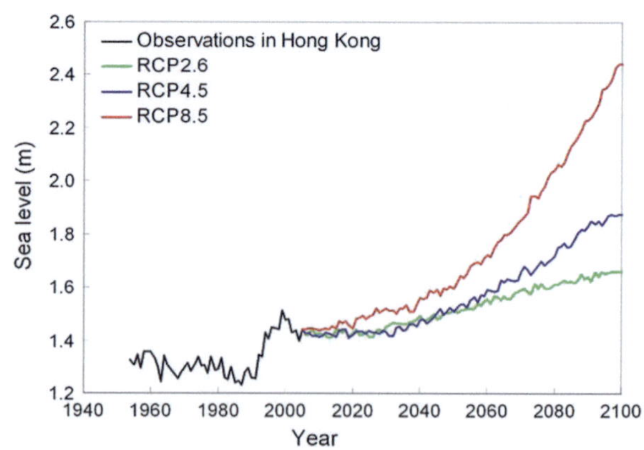

Fig. 12.3 Observed and predicted mean sea level at Hong Kong and the Pearl River Estuary from 1948 to 2100. Figure from Yang et al. (2014). The three scenarios are explained in Chap. 2, Sect. 2.4

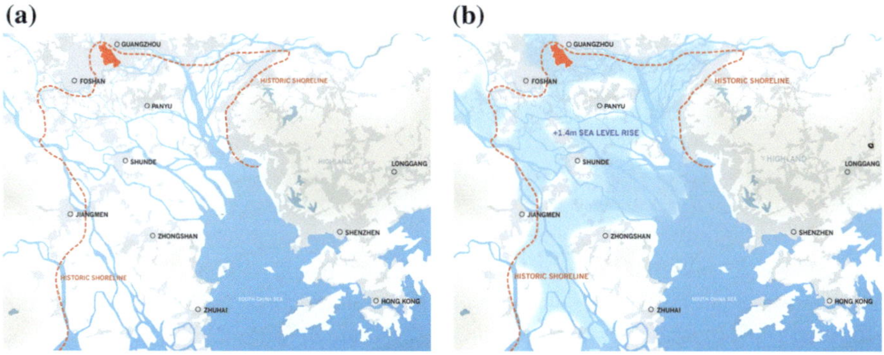

Fig. 12.4 **a** The PRD land surface today. **b** The PRD with 1.4 m RSL rise. Figures are from Kindel (2016). *Lessons from the World's Largest Megacity: How to confront the challenges facing China's Pearl River Delta*

Flooding at recurrent high tides will be common before mid century. The tides that affect the Pearl River Estuary and Delta are mixed-semidiurnal with a large diurnal inequality in range (Mao et al. 2004). The average tide range is less than 2 m (6.56 ft) but the spring tide range (full and new moons) can be up to 2.3 m (7.54 ft) and perigean "King Tides" can have a range up to 2.7 m (8.86 ft) and cause street flooding under normal "fairweather" conditions (Hansen 2017). Added to these water levels, more intense and frequent rainstorms now cause episodic flooding (Yang et al. 2014). In the 19th century, Guangzhou, like Venice, had numerous intertidal canals, which quickly drained away floodwaters. But those have long since been paved over and drainage is now impeded and Du et al. (2013) report that urban drainage problems are worsening. Today, when storm surges are superimposed on these recurrent inundation events, the outcomes can be tragic and extremely costly.

12.3 Typhoons and Storm Surges

The Pearl River estuary faces the largest composite risk of typhoon-induced disaster in China's coastal region (Yin et al. 2012). Long-term records from China are extensive and detailed and they go back thousands of years. Liu et al. (2001) analyzed written records of typhoons striking the PRD dating back to AD 975. Their analysis suggests that typhoon landfalls in Guangzhou were infrequent prior to about 1400 (possibly reflecting incomplete records) but were most frequent during certain multi-decadal periods, specifically: AD 1660–1680; 1850–1880; 1900–1940 and 1970–1980. Typhoon frequency during these periods was around 25–30 typhoons per decade. Based on these results, the authors conjecture that the next period of increased typhoon frequency could be in the 2020s or 2030s but there is no model-based support for this. Since 1900, instrumented records have been maintained and these records indicate higher frequencies of typhoon occurrences specifically 40–60 per decade. Yang et al. (2014) show data indicating that since the 1950s, the PRD has been affected by 3–11 typhoons per year, with the highest frequency occurring in the 1970s. The frequency has been diminishing since the 1990s and over the past decade or so has been on the order of 4 or 5 per year. However, Yang et al. (2014) emphasize that the intensity of these storms has increased significantly in recent years, consistent with expectations from rising sea surface temperatures (e.g. Chap. 2, Sect. 2.6). Yin et al. (2012) point out that socioeconomic vulnerability to these more intense storms is high. They developed a risk analysis system from which they estimated that economic loss to Guangdong Province from a single severe typhoon to be about $2 billion (USD).

One recent major typhoon to strike the PRD was Typhoon Nida in August, 2016 which had wind speeds of 151 km/h (93.83 mph) when it made landfall near Shenzhen (Xinhua news Aug 2 2016; Fig. 12.5). The most recent Typhoon Hato that directly hit the PRD region on August 23 of 2017 was the strongest typhoon within the last fifty years in southeastern China. There is not yet a detailed assessment of the impact but it was estimated that the hurricane could have caused Hong Kong $8 billion in economic losses due to wind and flooding damages and paralysis of business, the stock market and traffic (http://www.scmp.com/news/hong-kong/economy/article/2107994/typhoon-hato-could-cause-hk8-billion-losses-after-no-10, last accessed on September 27, 2017).

According to Li and Li (2010, 2011) Guangdong Province experiences the most severe storm surges in China and most of these are in the PRD. Zhao et al. (2016)

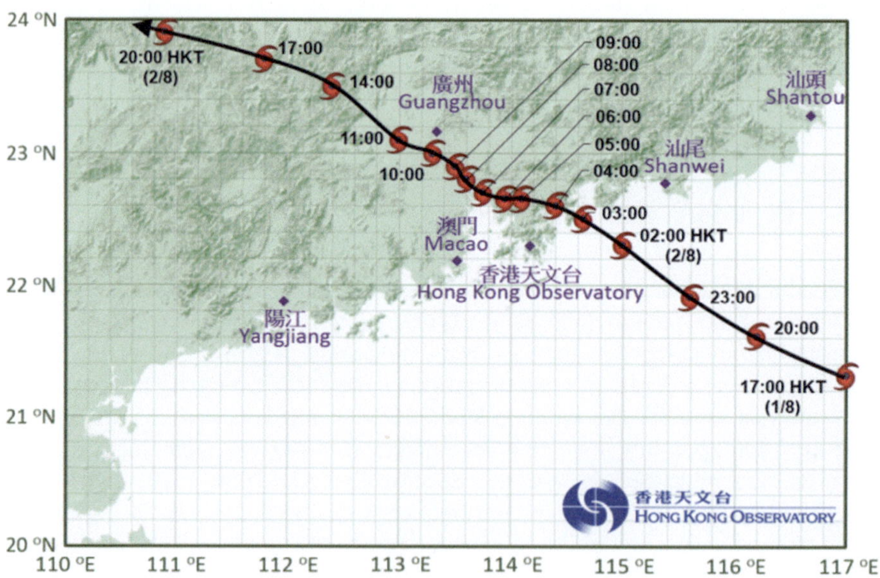

Fig. 12.5 Track of Typhoon Nida near Hong Kong and over the Pearl River Delta, August 1–2, 2016. Image from Hong Kong Observatory 2017

analyzed the storm surges affecting the western region of Guangdong Province that were generated by Typhoons Rammasun (TR) and Kalmaegi (TK) in 2014. The results showed that although the two storms were of similar strength, TR generated a storm surge that was less than 2 m (6.56 ft) high while TK generated a 4.5 m (14.76 ft) storm surge. The difference was attributed to the fact that TK was moving across the shelf at a much higher speed than TR and that speed roughly matched the speed over the shelf of a long wave (which an advancing storm surge is). Storm surges propagating up the funnel-shaped Pearl River Estuary are typically amplified more than the surges examined by Zhao et al. (2016) and often exceed 5 m (16.4 ft) or more in height. Du et al. (2013) state that storm surges in the PRE are the most severe in China and have intensified over the past 20 years or so. Those affecting Guangzhou are now typically 0.76 m (2.49 ft) higher than in the past. The storm surges not only cause flooding but also cause significant upstream intrusions of salt water. Although storm surges are a major cause of damaging floods in Guangzhou, severe rainstorms, which may or may not accompany typhoons are worse. According to He and Yang (2011), rainstorms are responsible for 63% of the economic losses in Shenzhen while typhoons directly account for 33%.

12.4 Environmental Impacts of Climate Change in the PRD

Beyond the extensive built-up urban areas are agricultural lands, coastal wetlands, intertidal mud flats and various marine ecosystems, all of which are already being degraded and are under serious threats from both climate change and urbanization. The main causes of environmental degradation are coastal erosion and land loss, saltwater intrusion, and reduced water quality related to pollution and sewerage overflows (Du et al. 2013; Tracy et al. 2007). Nutrient and sewerage runoff into estuarine waters can lead to low dissolved oxygen (hypoxia) impacting fish and aquaculture. Saltwater penetration into ground water, agricultural areas and city water supplies occurs often during dry conditions. Wetlands that support numerous bird species are diminishing because of saltwater intrusion and erosion combined with the fact that the ability of the wetlands to adapt by migrating inland is precluded by human construction of such structures as paved roads and levees (Tracy et al. 2007). Mangroves and coral reefs have been severely impacted since the 1950s. According to Du et al. (2013) the area covered by mangrove forests has diminished by 65 and 80% of the coral reefs surrounding nearby Hainan Island have disappeared.

The multiple stressors related to sea-level rise, climate change, and anthropogenic activities interact, often in non-linear ways, to affect natural resources to degrees that are greater than the sum of individual effects. Based on an ecosystem health index, Pearl River estuary has deteriorated from healthy to unhealthy status over the last three decades (Chen et al. 2013). Over-reclamation and exploitation have resulted in dramatic reduction of coastal wetlands in the past decades in China (Han et al. 2006). From 1970s to 1990s, tidal marshes in the Pearl River estuary declined from ~7000 ha to ~3000 ha. The mangroves in Guangdong province, where Guangzhou is located, declined from 21,289 ha in 1950s to 9411 ha in 2002. The degradation and loss of coastal wetlands will be even worse under sea-level rise. Sea-level rise leads to more frequent, longer, and deeper flooding in mangroves. The ecosystems need to accumulate sediments at a rate larger than sea-level rise rate in order to be maintained. However, higher nutrient availability due to environmental pollution, interacted with the predicted more severe drought due to climate change, will likely increase mangroves' mortality or vulnerability to sea-level rise by favoring the growth of shoots relative to roots (Lovelock et al. 2009, 2015). The below-ground biomass contributes organic matter to sedimentation and is important to maintain mangroves under sea-level rise.

One of the most immediate threats is to the security of urban and rural water supplies in the PRD and surrounding areas. Although Guangzhou does not yet suffer from water scarcity, 6 of the other PRD cities are considered to be water impoverished (Hu 2017). The combination of salinity intrusion and pollution is increasingly causing water supplies to be impacted. Cheng and Hu (2012) emphasize the need for improved water resource management throughout China. Proposed strategies include improved efficiency of water utilization, reductions in man-made pollution, more effective regulations and enforcement and improved reservoirs.

12.5 Impacts of Climate Change and Flooding on Human Health

Since 1970, average annual temperatures in the PRD have been rising at rates between 0.3 °C (0.54 °F) and 0.4 °C (0.72 °F) per decade depending on specific location; in 2007 the average summer temperature was 28.2 °C (83 °F) and rising (Guangdong Meteorological Administration 2007). During heat waves, however, temperatures are much higher than this. The PRD is hot and getting hotter and high heat can be very harmful to cardio-vascular and respiratory health. For example, IPCC models predict that by 2050, there will be 3.6–7.1 times more heat-related deaths in Shanghai than at present (Tracy et al. 2007). But warmer temperatures and wetter surroundings bring other threats including mosquitos, which spread vector-borne diseases. Bai et al. (2013) conclude from an analysis of numerous health studies that incidences of mosquito-transmitted diseases are increasing throughout China but particularly in South China. The most prominent of these diseases are malaria, Dengue fever and Japanese encephalitis. Those authors also report that Guangzhou has recently experienced an outbreak of Dengue fever.

Lv et al. 2013 have described the prevalence of ten water-borne parasitic diseases in China. They report that socioeconomic development, accompanied by a decline in sanitation improvement is contributing to increases in parasitic diseases. These diseases are contracted through drinking water, direct skin penetrations and consumption of undercooked aquatic plants and shellfish. Among these ten parasites are *Giardia and Cryptosporidium.* Other water-borne diseases include *Vibrio cholerae* and *Vibrio parahaemolyticus* (Li et al. 2014), which causes infectious diarrhea.

China's hazy (smoggy) days have been increasing within the last few decades. For example, the hazy days increased significantly from 70 days in 2001 to 144 days in 2004 in Guangzhou (Liu et al. 2013; Leng et al. 2016). Air pollution due to economic development is mainly responsible for this. In addition, recent studies showed that climate change is likely to cause air circulation changes. This could interact with air pollution to intensify China's severe haze episodes (Cai et al. 2017). The concentration of fine particles with a diameter of 2.5 μm or smaller ($PM_{2.5}$) increases significantly during severe haze. This reduces visibility, disrupts economic activities, and harms human health (Zhang et al. 2014). The toxic substances in the fine particles could severely affect the respiratory, cardiovascular, immune, and nervous systems, and increase morbidity and mortality, especially premature mortality (Bai et al. 2006; Kan et al. 2007; Pope and Dockery 2006; Wang et al. 2006; Xu et al. 2013).

12.6 Socioeconomic Impacts of Climate Change and Flooding on the PRD

Yang et al. (2014) argue that socioeconomic vulnerability of the population of the PRD to floods and storms has been increased in recent years by increased urban population density and by the increase in flood frequency and severity. In addition, the flood-protection dykes are near their limits of effectiveness and would likely be overwhelmed with moderately increased sea level combined with more intense storm surge. Yang et al. (2014) also point out that the most vulnerable segments of the population are children, the elderly, the poor and, from an economic perspective, small businesses. The elevation of most of the PRD is only 1 m (3.28 ft) or less above sea level and according to Du et al. (2013), 13% of the land is below sea level. A 5 m (16.40 ft) storm surge on top of a 1 m sea level rise would inundate 15 counties in the PRD, ruin crop lands and displace many villagers and urban residents. The GDP of Guangzhou, Foshan and Zhongshan together would be reduced by 52.3% (Du et al. 2013). In the case of Foshan, 67% of the population would be forced to relocate. Extensive, and potentially prolonged power outages would have unknown socioeconomic outcomes. Notably, the design criteria of one nuclear power plant have already been exceeded by storm surge (Du et al. 2013). Tracy et al. (2007) reviewed data that suggest that a normal storm surge and high tide superimposed on a 65 cm (2.13 ft) sea level rise would cause economic losses in the

PRD of about $ 22 billion (USD). In May 2014, a single downpour inundated 100 factories and shops, damaged tens of thousands of homes and cost $100 million for repairs (Hansen 2017).

Yang et al. (2014), among others, have developed quantitative indices of vulnerability for identifying the specific areas of the PRD that are at greatest risk. These indices consider three primary determinants: (1) relative exposure to the threats such as elevation, exposure to storm surge, likelihood of failure of dykes etc.; (2) sensitivity of the population which depends on age, health, income and education; and (3) adaptive capacity which is function of flood prevention infrastructure, GDP, education and general mobility. In combination, these factors determine overall vulnerability. Note, for example that Hong Kong has a very high degree of exposure and also a high sensitivity, in part because of the high population density. But Hong Kong also has a high index of adaptive capacity because the population is relatively affluent, well educated and healthy and the city's emergency management infrastructure is good. Therefore, Hong Kong's vulnerability index is low. Another national assessment on China's coastal vulnerability to sea-level rise based on predicted sea-level rise, coastal geomorphology, elevation, slope, shoreline erosion, land use, mean tide range, and wave height showed that the Pearl River delta is in a highly vulnerable region (Yin et al. 2012).

12.7 Adaptive Strategies for the Future

Future resilience of the PRD will depend critically on advancements in resource and emergency management strategies, on investments in flood protection infrastructure including nature and living shorelines (O'Donnell 2017), on improvements in water resource management and evolving predictive models of storms, storm surges and hydrology for both long- and short-range forecasting. In addition, continuing education of PRD residents about natural hazards and how to respond to them is important. Cheng and Hu (2012) have emphasized the urgent need for new and sophisticated approaches to water management in order to cope with water scarcity and with degraded water quality caused by pollution and salt-water intrusion. Fresh water is needed not only for drinking and sanitation but also for agricultural irrigation and power (Hu 2017). Wu et al. (2014) stress the importance of improving flood control protections for reservoirs.

Du et al. (2013) outline four main categories of strategies that must be implemented in order to ensure future resilience of the Pearl River Delta cities and inhabitants. First among these, is the need to improve monitoring, forecasting and early warning of impending floods, storm surges or other damaging events. Advanced numerical models will undoubtedly be essential in this strategy. The second set of adaptation strategies primarily involves innovative coastal engineering including a range of flood protection structures, submerged breakwaters, strengthening and water proofing the walls of buildings and protections of sewers. Thirdly, Du et al. (2013) advocate restoration of coastal ecosystems to strengthen natural protection. Finally, protection against saltwater penetration is essential to the security of water resources. Accomplishing this will require complex management of reservoirs throughout the Pearl River basin and well-orchestrated release of freshwater to suppress seawater penetration during times of drought or storm surges. Implementation of these strategies will have to consider the interdependence of all four categories as well as potentially impacted and constantly changing socioeconomic circumstances.

Mitigating climate change impacts by controlling emission of greenhouse gas and other pollutants is essential. Ecological thresholds can be useful to guide policy making. For example, critical air pollutant loadings based on ecological thresholds have been implemented in European countries to control atmospheric acidic deposition (Porter et al. 2005). China is actively supporting basic research to understand these ecological thresholds, but the research results must be effectively disseminated to policy makers to help them make more-informed decisions.

References

Ao, M., et al. 2015. Monitoring the land subsidence with persistent scatterer interferometry in Nansha District, Guangdong, China. *Natural Hazards* 75 (3): 2947–2964.

Bai, L., L.C. Morton, and Q. Liu. 2013. Climate change and mosquito-borne diseases in China: A review. *Globalization and Health* 9: 10. https://doi.org/10.1186/1744-8603-9-10.

Bai, Z.P., B.B. Cai, H.Y. Dong, and H. Bian. 2006. Adverse health effects caused by dust haze—a review. *Environmental Pollution and Control* 28 (3): 198–201. (in Chinese).

Cai, W., et al. 2017. Weather conditions conducive to Beijing severe haze more frequent under climate change. *Nature Climate Change* 7 (4): 257–262. https://doi.org/10.1038/nclimate3249.

Chen, Y.M., W.R. Huang, and S.D. Xu. 2014. Frequency analysis of extreme water levels affected by sea-level rise in East and Southeast Coasts of China. *The Journal of Coastal Research* 68: 105–112.

Chen, Z., J. Wang, G. Ma, and Y. Zhang. 2013. China tackles the health effects of air pollution. *The Lancet* 382: 1959–1960.

Cheng, H., and Y. Hu. 2012. Improving China's water resources management for better adaptation to climate change. *Climatic Change* 112: 253–282. https://doi.org/10.1007/s10584-011-0042-8.

Du, Y.-D., X.-H. Cheng, X.-W. Wang, H. Ai, H.-L. Duan, J. He, and X.-X. Wu. 2013. A review of assessment and adaptation strategy to climate change impacts on the coastal areas in South China. *Advances in Climate Change Research* 4 (4): 201–207.

Fairbank, J.K., M. Goldman, 1998. *China: A new history.* Cambridge, MA: Harvard University Press 546p.

Guangdong Meteorological Administration. 2007. In *Assessment report on climate change in Guangdong*. www.gdemo.gov.cn.

Hallegatte, S., C. Green, R.J. Nicholls, and J. Corfee-Morlot. 2013. Future flood losses in major coastal cities. *Nature Climate Change* 3: 802–806.

Han, M., X. Guo, and Y. Zhang. 2006. The human capital loss of air pollution in cities, China. *China Environmental Science* 26: 509–512. (in Chinese).

Hansen, K. 2017. Guangzhou rising—canton sinking. *Watts Up With That?* (April 12, 2017).

He, C., L. Yang. 2011. Urban development and climate change in China's Pearl River Delta *Land Lines*. Lincoln Institute of Land Policy, 7p.

He, L., G. Li, K. Li, et al. 2014. Estimation of regional sea level change in the Pearl River delta from tide gauge and satellite altimetry data. *Estuarine, Coastal and Shelf Science* 141 (141): 69–77.

Hong Kong Observatory, 2017. *Climate change in Hong Kong: Mean sea level.* http://www.hko.gov.hk/climate_change/obs_hk_sea_level_e.htm.

HKTDC Research. 2016. *PRD Economic profile.* http://china-trade-research.hktdc.com/business-news/article/Facts-and-Figures/PRD-Economic-Profile/ff/en/1/1X000000/1X06BW84.htm.

Hu, F. 2017. Pearl River Delta: 5 water must-knows. *China Water Risk.*.

Huang, Z., W.Q. Zhang, Q.L. Chen, et al. 2003. Impacts of sea level rise on design parameters of coastal engineering in Guangdong province. *Scientia Geographica Sinica* 23 (10): 39–41. (in Chinese).

Huang, Z., Zong, Y., and Zhang, W. 2004. Coastal inundation due to sea level rise in the Pearl River Delta, China. *Natural Hazards* 33 (2): 247–264.

Intergovernmental Panel on Climate Change (IPCC). 2013. *Climate Change 2013: The Physical Science Basis. Contribution of Working Group I to the Fifth Assessment Report of the Intergovernmental Panel on Climate Change* [Stocker, T.F., D. Qin, G.-K. Plattner, M. Tignor, S.K. Allen, J. Boschung, A. Nauels, Y. Xia, V. Bex and P.M. Midgley (eds.)]. Cambridge University Press, Cambridge.

Kan, H., S.J. London, G. Chen, Y. Zhang, G. Song, N. Zhao, L. Jiang, and B. Chen. 2007. Differentiating the effects of fine and coarse particles on daily mortality in Shanghai, China. *Environment International* 33: 376–384.

Kindel, P. 2016. *Lessons from the World's Largest Megacity: How to confront the challenges facing China's Pearl River Delta.* Owings and Merrill (SOM): Skidmore.

Leng, C., et al. 2016. Insights into a historic severe haze event in Shanghai: synoptic situation, boundary layer and pollutants. *Atmospheric Chemistry and Physics* 16: 9221–9234. https://doi.org/10.5194/acp-16-9221-2016.

Li, K., and G.S. Li. 2010. Calculation of return period for storm surge in the Pearl River Delta Region. *Progress in Geography* 2010 (29): 433–438.

Li, K., and G.S. Li. 2011. Vulnerability assessment of storm surges in the coastal area of Guangdong Province. *Natural Hazards and Earth System Sciences* 11: 2003–2010.

Li, Y., X. Xie, X. Shi, Y. Lin, Y. Qiu, J. Mou, Q. Chen, Y. Lu, L. Zhou, M. Jiang, H. Sun, H. Ma, J. Cheng, Q. Hu. 2014. *Vibrio parahaemolyticus*, Southern Coastal Region of China, 2007–2012. *Emerging Infectious Diseases* 20(4):685–688

Liu, K.-B., C. Shen, and K.-S. Louie. 2001. A 1000-year history of Typhoon landfalls in Guangdong, Southern China, reconstructed from Chinese historical documentary records. *Annals of the Association of American Geographers* 91 (3): 453–464.

Liu, X.G., et al. 2013. Formation and evolution mechanism of regional haze: A case study in the megacity Beijing, China. *Atmospheric Chemistry and Physics* 12: 16259–16292. https://doi.org/10.5194/acpd-12-16259-2012.

Lovelock, C.E., M.C. Ball, K.C. Martin, and I.C. Feller. 2009. Nutrient enrichment increases mortality of mangroves. *PLoS ONE* 4: e5600. https://doi.org/10.1371/journal.pone.0005600.

Lovelock, C.E., et al. 2015. The vulnerability of Indo-Pacific mangrove forests to sea-level rise. *Nature* 526: 559–563.

Lv, S., L.-G. Tian, Q. Liu, M. Qian, Q. Fu, P. Steinmann, J.-X. Chen, G.-J. Yang, K. Yang, and X.-N. Zhou. 2013. Water-related parasitic diseases in China. *International Journal of Environmental Research and Public Health* 10 (5): 1977–2016. https://doi.org/10.3390/ijerph10051977.

Mao, Q., P. Shi, K. Yin, J. Gan, and Y. Qi. 2004. Tides and tidal currents in the Pearl River Estuary. *Continental Shelf Research* 24: 1797–1808.

Parker, L. 2015. Treading water. *National Geographic* 226 (February 2015).

O'Donnell, J.E.D. 2017. Living shorelines: A review of literature relevant to New England Coasts. *Journal of Coastal Research* 33 (2): 435–451.

Pope, C.A., and D.W. Dockery. 2006. Health effects of fine particulate air pollution: Lines that connect. *Journal of the Air and Waste Management Association* 56 (6): 709–742.

Porter, E., T. Blett, D.U. Potter, and C. Huber. 2005. Protecting resources on federal lands: Implications of critical loads for atmospheric deposition of nitrogen and sulfur. *BioScience* 55 (7): 603–612.

The Economist. 2017. Jewel in the crown: Special report on the Pearl River Delta. *The Economist Magazine,* April 8 2017.

Tracy, A., K. Trumbull, C. Loh. 2007. The impact of climate change in Hong Kong and the Pearl River Delta., China perspectives, URL: http://chinaperspectives.revues.org/1173.

Schell, O., and J. Delury. 2013. *Wealth and power: China's long March to the twenty-first century*, 478. New York: Random House.

van Mead, N. 2015. China's Pearl River Delta overtakes Tokyo as world's largest megacity. *The Guardian* 28 January, 2015.

Wang, H., T.J. Wright, Y. Yongping Yu, H. Hong Lin, L. Jiang, C. Changhui Li, and G. Qiu. 2012. InSAR reveals coastal subsidence in the Pearl River Delta. *China. Geophysical Journal International* 191 (3): 1119–1128. https://doi.org/10.1111/j.1365-246X.2012.05687.x.

Wang, Y., G. Zhuang, Y. Sun, and Z. An. 2006. The variation of characteristics and formation mechanisms of aerosols in dust, haze, and clear days in Beijing. *Atmospheric Environment* 2006 (40): 6579.

Westphal, M.I., G.A. Hughes, and J. Brömmelhörster. 2013. *Economics of climate change in East Asia*, 196. Mandaluyong City, Philippines: Asian Development Bank.

Woodroffe, C.D., R.J. Nicholls, Y. Saito, Z. Chen, and S.L. Goodbred. 2006. Landscape variability and the response of Asian megadeltas to environmental change. In *Global change and integrated coastal management*, vol. 10, ed. N. Harvey, 277–314. Dordrecht, The Netherlands: Springer.

Wu, S., F. Deng, Y. Hao, et al. 2014. Fine particulate matter, temperature, and lung function in healthy adults: Findings from the HVNR study. *Chemosphere* 108: 168–174.

Xinhua News. 2016. Typhoon Nida lands in south China's Guangdong. *Xinhua News*, 2 Aug 2016.

Xu, J., X. Wang, and S. Zhang. 2013. Risk-based air pollutants management at regional levels. *Environmental Science & Policy* 25: 167–175.

Yang, L., J. Scheffran, H. Qin, Q. You. 2014. *Climate-related flood risks and urban responses in the Pearl River Delta, China*. Working paper **CLISEC-27**, Research Group Climate Change and Security, University of Hamburg. 27p.

Yin, J., S. Wu, and E. Dai. 2012. Assessment of Economic Damage Risks from Typhoon Disasters in Guangdong. *China. Journal of Resources and Ecology* 3 (2): 144–150. https://doi.org/10.5814/j.issn.1674-764x.2012.02.006.

Zhang, Z., J. Wang, L. Chen, et al. 2014. Impact of haze and air pollution-related hazards on hospital admissions in Guangzhou. *China. Environ Sci Pollut Res* 21: 4236. https://doi.org/10.1007/s11356-013-2374-6.

Zhao, X., Q. Xie, B. Hong, H. Xu, L. Chen, J. Hu, M. Min Zhang, B. Huang, and Z. Che. 2016. Comparison of two typhoon-induced storm surges at the Zhanjiang Coast. *Earth and Environmental Science* 38 (2016): 012003. https://doi.org/10.1088/1755-1315/38/1/012003.

Chapter 13
Coastal Louisiana

Lynn Donelson Wright and Christopher F. D'Elia

> (The upper course of the Mississippi River)... *is held in place ... by the gorge in the Commerce hills. Its mouth in the Gulf of Mexico is fixed by the works of man. Between these points it writhes like an imprisoned snake constantly seeking ... equilibrium.*
> —D. O. Elliot, U.S. Army Engineer quoted by Barry (1998)

13.1 A Unique and Imperiled Coastal System

Just as "Egypt is the Gift of the Nile" (Chap. 4), Coastal Louisiana owes its existence entirely to sediments supplied to the Gulf of Mexico coast by the Mississippi River (Fig. 13.1) from a catchment that covers 3.2 million square kilometers (1.2 million mi^2) of North America including roughly 41% of the area of the contiguous 48 states of the USA (EPA 2017). Over the past 10,000 years, these sediments have yielded vertical accumulations of 0.1–0.5 km (0.06–0.3 mi) with the highest accumulations occurring nearest to river mouths (Coleman and Roberts 1988). As sediments have accumulated, the land surface has undergone progressive subsidence, partly because of large-scale tectonic subsidence of the structural Gulf Coast Geosyncline and partly because of compaction of the fine grained silts and clays. According to Allison et al. (2016), the rates of Mississippi Delta subsidence are up to 18 mm/year. When this subsidence is added to the projected rates of global sea level rise of between 8 and 16 mm/year, the total relative rate of sea level rise in coastal Louisiana will conceivably be between 26 and 34 mm/year or roughly up to 1 foot per decade (Chap. 3; Figs. 3.6 and 3.7). Prior to European settlement in Coastal Louisiana over the past two or three centuries, deposition of

L. D. Wright (✉)
Southeastern Universities Research Association (SURA), Washington, DC, USA
e-mail: ldwright@bellsouth.net

C. F. D'Elia
School of the Coast and Environment, Louisiana State University,
Baton Rouge, LA, USA

L. D. Wright and C. R. Nichols (eds.), *Tomorrow's Coasts: Complex and Impermanent*, Coastal Research Library 27,
https://doi.org/10.1007/978-3-319-75453-6_13

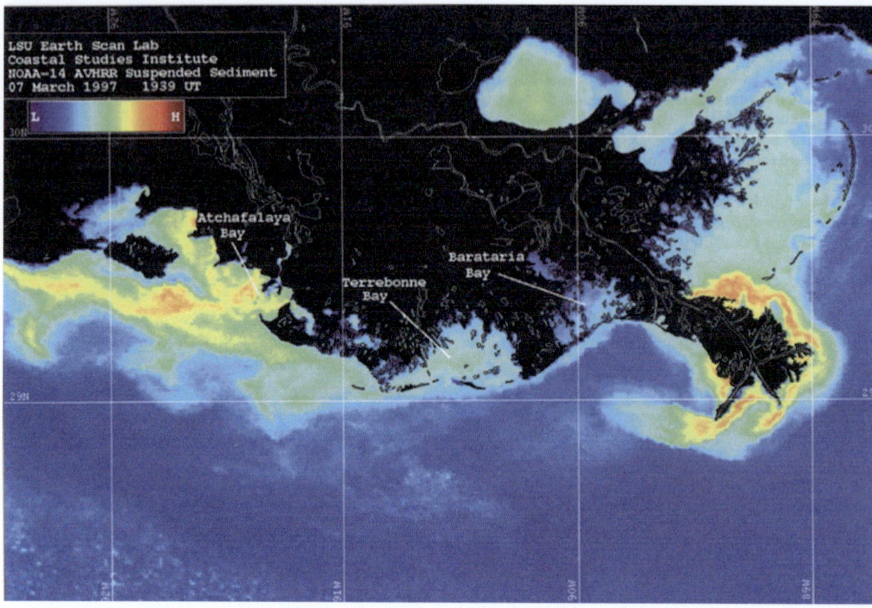

Fig. 13.1 The Mississippi Delta showing sediment plumes issuing from the mouths of the modern "bird's foot" delta lobe in the east and the Atchafalaya River in the west. NOAA AVHRR image from LSU's "Earth Scan" Lab, March 1997

new sediment was able to keep pace with subsidence by a combination of overbank discharge, crevasse splays and occasional avulsions of the river and its tributaries. However, since the early 1800 s, increasing engineering constraints have confined the river, reduced the amount of sediment reaching the delta, and greatly limited the natural dispersal of sediment over the deltaic surface. Today, land loss greatly exceeds land creation. As the area of open water increases, so does exposure to erosive wave action. At the present time, the rate of land loss is averaging 41 km^2 (16 mi^2) per year (or, as popularly stated: "one football field per hour").

Coastal Louisiana includes the largest wetland in the U.S.A. and one of the largest in the world. This unique and diverse ecosystem is regarded as one of the world's natural treasures. The cultural landscape is equally unique and diverse and includes, among its population of roughly 2 million people, long-time Native American residents, extensive African American communities, descendants of French and German immigrants, mixed race Creoles and Acadians ("Cajuns") who were originally displaced from Nova Scotia. In economic terms, up to $138 billion in business, residential, and infrastructure assets are at risk and could be lost by 2050; a single severe storm could cause disruption of $53 billion in economic activity (Barnes and Virgets 2017). Beyond conventional economic activity, Shepard et al. (2013) report that "In 2010, revenues from provisioning ecosystem goods and services generated by the five U.S. states bordering the Gulf of Mexico contributed over $2 trillion per year to the nation's gross domestic product, including $660 billion from the coastal county revenues and $110 billion from ocean revenues" (Shepard et al. 2013, p. 10). It is, perhaps, ironic that the extensive engineering works that were intended to protect people and assets from floods and support navigability of the lower Mississippi River are now profoundly implicated in the disappearance of the wetlands that once provided natural protections. Numerous dams on the upper reaches of the river have significantly reduced the amount of sediment that ultimately makes it to the coast. The levees that protect Louisiana communities from floods and storm surges also prevent sediments and nutrients from nourishing wetlands and agricultural lands. Prevention of the river and its distributaries from switching course, as is natural, does not allow sediments to be deposited in the western parts of the delta. The sediments remain confined to the river channels until they reach the river's mouths where man-made jetties maintain navigable depths by funneling the outflows and causing the much-needed sediments to be lost to the deep Gulf of Mexico waters. The dissolved nutrients that were denied to farmlands and wetlands upstream are now dispersed over the continental shelf where they contribute to eutrophication and oxygen depletion (Chap. 6, Fig. 6.3).

13.2 The Physical Setting of the Mississippi Delta

Today, the Mississippi Delta covers an area of 29,000 km^2 (11,200 mi^2). The rate at which the Mississippi River discharges fresh water into the Gulf of Mexico averages 15,360 m^3/s (552,960 ft^3/s) over the course of a year but reaches a maximum rate of 57,400 m^3/s (2,066,400 ft^3/s) when the river is at flood stage during spring. Annually, the river delivers 210 million tons of suspended sediment to the Gulf of Mexico (Milliman and Meade 1983). Compared to many major river systems, this is a fairly small sediment load; the Ganges-Brahmaputra system has an annual sediment load of 1.67 billion tons per year and the Amazon discharges 900 million tons per year. Although most of the Mississippi sediment bypasses the coastal wetlands that need it, much of the material that has not spilled over the edge of the continental shelf has accumulated on the shelf yielding an extraordinarily low gradient (slope of only 0°.02′) and soft inner shelf profile (Chap. 5, Fig. 5.2). For coastal Louisiana, this is both good and bad. It is good because the flat, muddy seabed causes wind-generated waves to be substantially dissipated before they reach the shore (Sheremet and Stone 2003). It is bad because low gradient shelves favor amplification of long-wave storm surges as was exemplified by Hurricane Katrina in 2005. The average wave height in deep water off the coast is about 1 m (3.3 ft) but is typically less than 20–30 cm (<1 ft) near shore. The maximum wave heights recorded by buoys offshore during storms have been up to 10 m ($\sim$33 ft) but less than $\sim$5 m ($\sim$15 ft) inshore. Tides and tidal currents are small. The tides are diurnal (1 tide per day) and the range is only 40 cm (1.3 ft) and since the tide only occurs once per day, the resulting currents are very weak. Consequently, vertical mixing between the low salinity river plumes and the higher salinity Gulf waters is weak allowing the mud-laden freshwater plumes to be carried westward

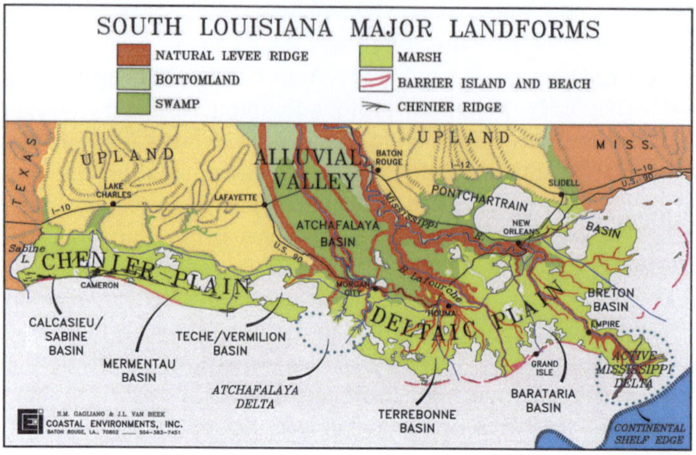

Fig. 13.2 The major geomorphologic regions of coastal Louisiana. Figure is from Gagliano and van Beek 1993 with permission from Coastal Environments Inc. Baton Rouge, LA

by the prevailing coastal currents which normally flow from east to west (Dinnel and Wiseman 1986).

Since the last post-glacial sea level rise reached its current position about 7000 years ago, the lower Mississippi River has switched its course at least 6 times. Following each avulsion, the river produced a major deltaic lobe (Draut et al. 2005). These overlapping lobes have left behind what now constitutes the rapidly receding deltaic plain of the Mississippi. The result has been the existing landscape (Fig. 13.2), which includes the remaining deltaic plain comprised mostly of disappearing wetlands and shallow bays. The western part of the deltaic plain includes the newly forming deltaic deposits from the Atchafalaya River, an active distributary of the Mississippi River. Further to west, is a plain of low beach ridges, locally referred to as "cheniers" because they are surmounted by oak trees (chene, in French). The Chenier Plain is composed of sediments transported westward from the active delta and pushed onshore from the inner continental shelf during episodic high wave energy events. Inland from the deltaic plain is the alluvial valley of the lower Mississippi River, which is incised within "uplands" consisting of older Pleistocene deposits.

13.3 The Cultural and Socioeconomic Landscape of Coastal Louisiana

In the broadest sense, the population of coastal Louisiana can be separated into two very general categories: those who live in and around New Orleans and those who do not. New Orleans proper covers an area of 170 square miles (440 km^2) and has roughly 400,000 permanent residents (U.S. Census Bureau 2016). African Americans make up 60% of the population, whites account for 33% and Hispanics a little over 5%. In the mid 1800 s, New Orleans was the wealthiest and the third largest city in the U.S.A. Today, the average per capita income of the city is $27,700 (USD) and 27% of the population is living at or below the poverty level making them highly vulnerable to floods and severe storms. The loss of protective wetlands has greatly increased this vulnerability (Tibbetts 2006). This historical and architecturally unique city attracts large numbers of tourists and tourism, including hotel accommodations and food services, generates $2.8 billion (USD) in annual revenue. It is largely for this reason as well as the vulnerability of the population that the U.S. Army Corps of Engineers has invested $14.5 billion in surrounding this sinking city, which is already below sea level, with an elaborate flood protection system consisting of hardened levees, floodgates, storm surge barriers and high-volume pumps. It should be noted that, when one considers the urban "agglomeration" immediately surrounding New Orleans proper, including the eight contiguous parishes, the total population swells to 1.3 million (in year 2000; U.S. Census Bureau) and the demographic composition becomes more similar to that of the state as a whole.

Beyond the "Big Easy," Coastal Louisiana includes extensive rural lands, wetlands, bayous, and several other cities including Lake Charles, Abbeville, Lafayette, New Iberia, Morgan City, Houma, Thibodaux and Cameron as well as barrier island communities like Grand Isle (Fig. 13.2). About 63% of the inhabitants are white and 32% are African American. French speaking Cajuns (Acadians), Creoles and direct descendants of French and Haitian immigrants make up a high percentage of the population and many of these residents are involved in commercial fishing (which supplies 26% by weight of U.S. commercial fisheries landings) or agriculture. Many others are employed by the oil and gas industry, which supplies 18% of the nation's oil (CPRA 2017). Although their populations are relatively small, there are also unique and threatened Native American tribal communities, primarily of the Choctaw and Muskogee Nations, throughout the lower Mississippi River Delta. Many of these communities have lived in the delta for up to 300 years, existing on a subsistence economy. Some notable examples include Grande Bayou Village in Plaquemines Parish and the Pointe-au-Chien Tribal Community in Terrebonne Parish (Peterson 2012).

The dominant economies to the west of New Orleans are oil and gas, commercial and recreational fisheries and agriculture. Shipbuilding and boat building are also significant in specific localities. The oil and gas industry accounts for over 25% of the total revenues collected by the state of Louisiana (NRC 2006). In addition to being a major producer of oil and gas, much of it from offshore production platforms, Louisiana is among the nation's top 3 importers of crude oil (U.S. Energy Information Administration 2017). The "Henry Hub" is a natural gas delivery point where 13 major pipelines intersect. The Louisiana Offshore Oil Port (LOOP) is the only U.S. port facility capable of berthing Ultra Large Crude Carriers. Much of the oil imported via such facilities is processed by the 18 operating Louisiana petroleum refineries (U.S. EIA 2017). In the aggregate,Louisiana supplies 18% of the

Fig. 13.3 Port Fourchon, LA. A center of oil and gas activity near Houma and highly vulnerable to rising sea levels and future storm surges. From Barnes and Virgets (2017)

US's oil. The elaborate, multi-billion dollar infrastructure that supports this industry, mostly in the vicinity of Houma, LA (Fig. 14.3; Barnes and Virgets 2017) is now at serious risk of damage or destruction when future storms are superimposed on rising seas and sinking lands (Fig. 13.3).

Agriculture and commercial fisheries are also very important sources of income for residents of rural Coastal Louisiana. In 2005, agriculture in Louisiana's coastal parishes contributed $410 million to the state's economy (NRC 2006). The major crops are sugarcane, rice and soybeans. Freshwater pumped from bayous, such as Bayou La Fourche, provides irrigation for fields but progressive upstream penetration of sea water threatens the continuation of this practice. The smaller, family farms of the past are rapidly being replaced by large "agribusiness" farms. Louisiana is regarded by many as the U.S. seafood capital: second only to Alaska, it accounts for the nation's highest commercial fish landings and 37% of the nation's oysters. The wetlands serve as nurseries for these fisheries and their disappearance will have a very negative impact on commercial and recreational fisheries. Ironically, however, some in the seafood industry have expressed concerns that plans for reclaiming land from open bays could have a negative impact on some fisheries including oysters (NRC 2006).

13.4 Hurricanes and Storm Surges

From an analysis of tropical storm and hurricane landfalls in the Northern Gulf of Mexico, Doyle (2009) concluded that storm frequency is greatest along the Florida Panhandle and decreases toward the west becoming half as frequent along the Texas coast as in West Florida. Nevertheless, over the past century or so Louisiana has taken direct hits by some very intense and destructive hurricanes. According to Roth (2010) in a technical report of the National Weather Service, some of the deadliest and most destructive hurricanes to hit the U.S. have made landfall on the Louisiana coast. Notable examples were Audrey in 1957, Betsy in 1965, Camille in

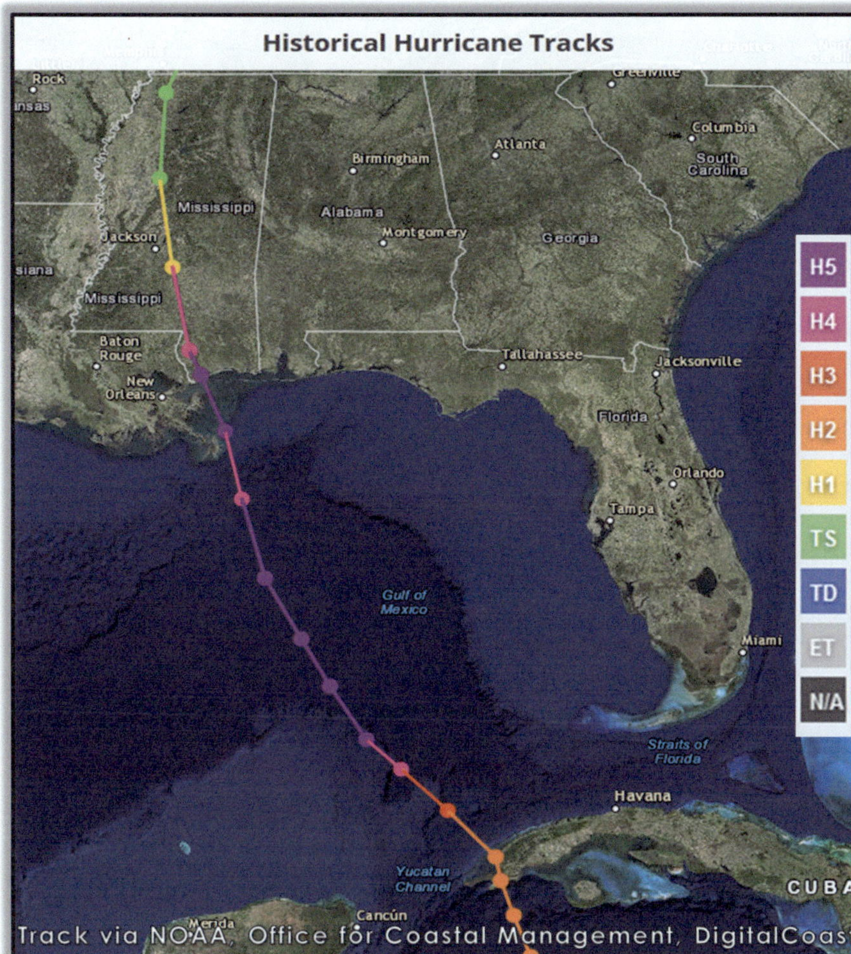

Fig. 13.4 Northward track of Hurricane Camille up the Gulf of Mexico to landfall in the Mississippi Delta in mid August, 1969. Image from NOAA Office of Coastal Management, Digital Coast

1969, Andrew in 1992 and Katrina in 2005. Between 1850 and 2010, a total of 106 tropical cyclones made landfall on Louisiana's shores (Roth 2010). The most intense storm of the 20th Century and early 21st Century was Category 5 Hurricane Camille (Fig. 13.4) in August 1969, which struck Plaquemines Parish along the narrow strip of the modern delta lobe with 160 mile per hour (256 km/h) winds and completely leveled the towns of Venice and Buras. Camille also severely damaged coastal counties of Mississippi and Alabama. The warm waters of the Gulf of Mexico contributed to the intensification of Camille as it progressed northward.

Although it was less intense than Camille, Hurricane Katrina caused far more damage, loss of life and general devastation in eastern Louisiana than Camille, primarily because of losses of protective wetlands over the 36-year period separating the two storms. Less than 4 weeks after Katrina, Hurricane Rita made landfall near the Louisiana-Texas state line inflicting damage on the western part of Louisiana's coast.

Although it is the direct impact of winds that cause damage to, and destruction of buildings, it is coastal flooding that causes the most fatalities as well as much property damage and substantial human health hazards (Chap. 10). The two most common causes of coastal flooding are storm surges and heavy rainfall (as exemplified by hurricane/tropical storm Harvey in August 2017). In some cases, river floods that coincide with storms can add to the effects of surges. In the case of the Pearl River Delta (Chap. 12), heavy rainfall can cause more severe flooding than storm surges. But, in the case of Coastal Louisiana it is storm surges that are most devastating and this was very much true of the hurricane that drowned 6000 people in Galveston in 1900 (Chap. 10). Needham and Keim (2011) compiled a data base on storm surges that have affected the northern Gulf of Mexico and point out that the most severe storm surges in the U.S. occur along the Gulf Coast. The torrential rainfall flooding of Texas and Western Louisiana by Harvey in 2017 may well represent the onset of a new regime.

The low gradient inner continental shelf combined with the trapping or funneling effects of convergent bays such as the Bay St. Louis area immediately east of the active "Birds-foot" delta causes the long-wave surge to experience amplification as it moves ashore. The results of Needham and Keim's (2011) analysis show that surges affecting the Florida panhandle are much less severe than is the case for the crenulated and embayed Louisiana coast. Although Hurricane Andrew that made landfall just south of Miami in 1992 was a Category 5, it caused a maximum storm surge that was typically less than 2 m high (4–6 ft.) throughout most of Biscayne Bay reaching a local maximum of 5 m (16.5 ft; Chap. 15). The weaker Hurricane Katrina over the shallow shelf of the northern Gulf of Mexico generated a surge of 8.5 m (28 ft). The surge caused by Camille was 7.5 m (25 ft) high and it completely inundated Plaquemines Parish. The Mississippi River Gulf Outlet (MRGO), a dredged and artificially maintained shipping channel to the east of the river proper, was reportedly responsible for enhancing the Katrina storm surge that ultimately flooded New Orleans' Ninth Ward (NRC 2006). Chen et al. (2008) have hypothesized that if Katrina had tracked farther to the east and made landfall on the Alabama coast, the storm surge would been 4 m (13 ft) lower. As sea surface temperatures in the Gulf of Mexico increase in the decades ahead, it is likely that tropical cyclones will become more intense, but, there is no existing evidence that they will become more frequent.

13.5 Disappearing Wetlands

From uplands to the Gulf of Mexico, Louisiana's coastal vegetation types include (1) forests and freshwater swamps with cypress and tupelo gums, (2) floating marsh, (3) freshwater marsh, (4) intermediate marsh, (5) brackish marsh and (6) salt marsh. Figure 13.5 shows the distribution of these coastal vegetation types as they exist in 2017 (CPRA 2017). As pointed out earlier, the present rate of wetlands loss is around 41 km^2 (16 mi^2) per year and accelerating. The Louisiana Coastal Protection and Reclamation Authority (CPRA 2017) projects that by 2050, without reclamation, most of the wetlands will have been replaced by open water as portrayed in Fig. 13.6. Houma will be inundated and the narrow coastal barrier islands, such as Grand Isle, will be gone.

There are multiple contributors to Louisiana's wetlands loss. Inundation is the most obvious of these. As explained in Chap. 9, Sect. 9.6, the maximum rates of vertical accretion of salt marshes as well as deltaic surfaces in general are about 5 mm/year. Relative sea level rise rates in excess of these "tipping points" are likely to cause wetlands or deltaic lands to be replaced by open water (Morris et al. 2016; Turner et al. 2017). We pointed out earlier in this chapter that subsidence rates alone are causing inundation of as much as 18 mm/year and, when the modeled rates of future global sea level rise of between 8 and 16 mm/year are added to

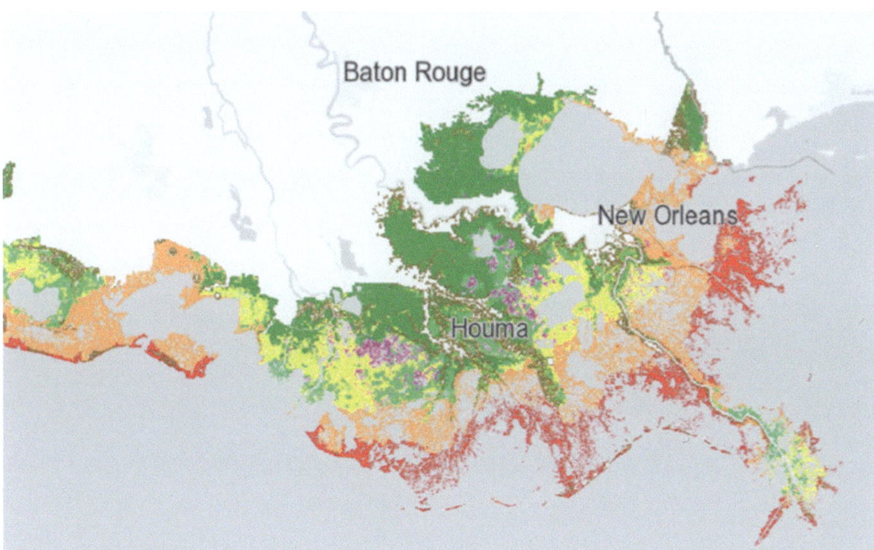

Fig. 13.5 Distribution of vegetation types in Coastal Louisiana in 2017. Figure from Louisiana Coastal Protection and Reclamation Authority (CPRA) on line Master Plan Data Viewer (2017) (https://cims.coastal.louisiana.gov/masterplan). *Key* Dark green—forest and swamp; purple—floating marsh; light green—freshwater marsh; yellow—intermediate marsh; orange—brackish marsh; red—salt marsh

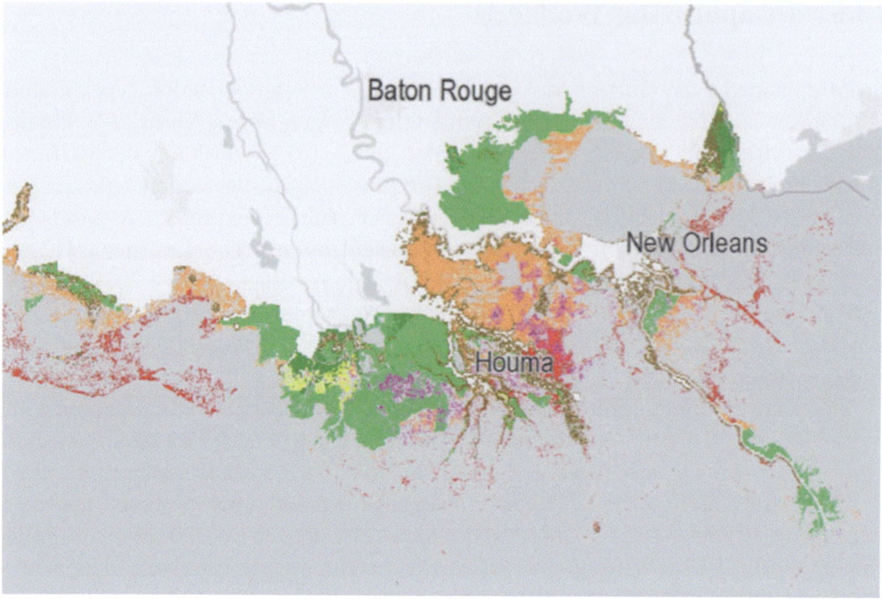

Fig. 13.6 CPRA projections for wetlands distribution in 2050 without reclamation

Fig. 13.7 Dredged access channels in Louisiana coastal wetlands. *Photo source* U.S. geological survey Louisiana coastal wetlands: a resource at risk. https://pubs.usgs.gov/fs/la-wetlands/

subsidence, the "tipping points" are exceeded by 5–6 fold. As open water bays grow in size, wave action becomes more energetic and progressively nibbles away at the marsh edges (Chap. 9, Fig. 9.3) creating a "positive feedback" loop. Landward intrusion of salt water causes dieback of freshwater swamps and marshes. Another important contributor to wetlands disappearance is the denial by levees and other engineering works of the supply of necessary sediments to nourish delta accretion. Finally, perhaps one of the most egregious and highly visible detriments to wetlands health is man-made destruction by the dredging of access canals and channels for oil and gas development (e.g. Dardis 2010; Fig. 13.7).

13.6 Engineering Impacts

Flood control levees on the Mississippi had been constructed by the latter part of the 19th Century and the impact of the levee system on the Mississippi Delta was clearly understood. E.L. Corthell wrote the following in 1897: "*The conditions are very different now from those existing prior to the construction of levees. There are at present no annual accretions of sedimentary matters from the periodical overflows of the river. These accretions formerly were a little more than equal to the annual subsidence of the lands... No doubt the great benefit to the present and two or three following generations accruing from a complete system of absolutely protective levees, excluding the waters entirely from the great areas of the lower delta country, far outweighs the disadvantages to future generations from the subsidence of the Gulf delta lands below the level of the sea and their gradual abandonment due to this cause.*" While this ominous prediction of land loss has come true, the levees of that time were unfortunately not designed to be high enough or strong enough to be effective. The Great Mississippi River Flood of 1927 motivated Congress to mandate the construction of levees high enough to keep the river completely contained in future such events (Barry 1998).

Today, the Mississippi's artificial levees serve the purpose for which they were intended: protecting lives and property along the course of the river. This protection has the added effect of increasing property values and supporting economic growth. However, these levees have also prevented overbank flows from supplying sediments to the deltaic plain and coastal wetlands and nutrients to farms and wetlands. Kesel (2003) compared the situation that existed in 1850 prior to levee construction to that of the late 20th century with levees and other controls. Notably, in 1850, Bayou Lafourch carried about 30% of the river's water and sediment load to Terrebonne and Barataria Bays to the west of the active Birdsfoot and overbank dispersal of sediment prevailed along the main and secondary channels. In contrast, by 1990, overbank flows had been eliminated and sediment was discharged directly into the Gulf of Mexico. In addition, river control structures upstream have completely cut off flows down Bayou Lafourch and directed 30% of the discharge down the Atchafalaya River to the west. Atchafalaya sediments discharged into

Atchafalaya Bay are at presently contributing the progradation of the Wax Lake Delta, the only actively accreting delta lobe on the Louisiana Coast (Xing 2015).

While protection of human health and wealth was a prime driver of engineering investment and construction, maintaining Mississippi River navigation for commerce and trade has always been close behind. Crucial to the navigability goal has been maintaining relatively deep-water channels from the river mouths to as far upstream within the river as ships may need travel. The three main mouths of the river in the active "birdsfoot" are Southwest Pass, South Pass and Pass a'lOutre. Pass a'lOutre the eastern most pass is not navigable by ships but the other two passes are. For ships entering from the east the Mississippi River Gulf Outlet was designed to provide a navigable short cut to New Orleans and the main channel of the Mississippi. To ensure navigable depths across the river mouth bars at the Southwest Pass and South Pass distributary mouths, the U.S. Army Corps of Engineers designed and constructed converging river mouth jetties. By concentrating the outflow, these jetties prevent rapid river mouth deposition and maintain depths sufficient to allow conventional shipping traffic to enter the passes. Unfortunately, they also cause the sediments that are needed to nourish the delta plain to be shunted into deep water. The third important type of impactful structures are oil and gas platforms and pipelines. These are extensive and are highly disruptive to wetlands ecology.

13.7 Adapting to Change: Complex Plans for the Future

The loss of wetlands, barrier islands and other lands in coastal Louisiana is severe and becoming worse. This has been known for several decades and, in 1990 Congress passed the federal Coastal Wetlands Planning Protection and Restoration Act (CWPPRA). With state and federal collaboration, an initial plan was developed in 1993. In 1998 the Louisiana Coastal Wetlands Conservation and Restoration Task Force and the Wetlands Conservation and Restoration Authority (1998) prepared a long-range plan entitled *Coast 2050: Toward a Sustainable Coastal Louisiana.* In 2004, the U.S. Army Corps of Engineers published a report entitled *Louisiana Coastal Area (LCA), Louisiana Ecosystem Restoration Study* (USACOE 2004). In late 2005, in the wake of Hurricanes Katrina and Rita, Louisiana restructured its Wetlands Conservation and Restoration Authority to form the Coastal Protection and Restoration Authority (CPRA). In 2017, CPRA released its most recent comprehensive master plan for a sustainable coast.

The National Research Council was asked to assess the USACOE LCA report and their report was published in 2006. In their assessment, NRC (2006) emphasized that returning coastal Louisiana to earlier conditions or even stopping future land loss on a state-wide scale is not really possible. What is more feasible, however, is to limit or slow future damage by targeting specific vulnerable or economically important localities for "triaged" protection or restoration. The NRC report specifically states that: "The challenge of slowing the loss of coastal wetlands and adjacent barrier islands and levees is unprecedented. The geographic extent of these wetlands and the range of natural and human forces that cause wetland degradation contribute to what would be one of the largest civil works projects in U. S. history" (NRC 2006, p. 17). They explain that the reasons for this challenge have much to do with the conflicts among nature, politics, regulations, economics and social traditions. Reconciliation among these conflicts is a challenge that must be, and surely can be, over come.

The CPRA has a plan and it is laid out in detail in the 2017 report. An interactive version can also be accessed on line at http://cims.coastal.louisiana.gov/masterplan/. Plans for future protections, and some reclamation, include a suite of site-specific strategies. One of these, involves a significant diversion of the lower Mississippi River to discharge sediment into Terrebonne and Barataria Bays to create a new delta lobe. Other strategies involve construction of new levees and elevating old ones as well as installation of floodgates. Non-structural strategies include flood mitigation by relieving or redirecting river flows during flood season, elevating and flood proofing commercial, public and residential structures, removing dangerous or environmentally harmful works, cessation of canal dredging, and, in some cases, relocating residents to upland areas. Several reclamation projects also rely on multiple structural and non-structural methodologies. One traditional approach is barrier island reclamation using dredged material or sediment pumped from offshore. Despite the ambitious plans and the likelihood of a large investment in implementing the envisioned projects, the prospects for the future of coastal Louisiana remain grim —the same human actions that have been taken to protect life and property have, in the long term, jeopardized them as well.

References

Allison, M., B. Yuill, T. Törnqvist, F. Amelung, T.H. Dixon, G. Erkens, R. Stuurman, C. Jones, G. Milne, M. Steckler, J. Syvitski, and P. Teatini. 2016. Global risks and research priorities for coastal subsidence. *EOS* 97 (19): 22–27.

Barnes, S.R., S. Virgets. 2017. *Regional impacts of coastal land loss and Louisiana's opportunity for growth.* LSU E. J. Ourso College of Business Economics & Policy Research Group. 39 pp.

Barry, J.M. 1998. *Rising tide: The Great Mississippi River flood of 1927 and how it changed America.* New York, Simon and Schuster 524 pp.

Chen, Q., L. Wang, and R. Tawes. 2008. Hydrodynamic response of northeastern Gulf of Mexico to hurricanes. *Estuaries and Coasts* 31: 1098–1116.

Coastal Protection and Restoration Authority of Louisiana (CPRA). 2017. *Louisiana's Comprehensive Master Plan for a Sustainable Coast.* Coastal Protection and Restoration Authority of Louisiana. Baton Rouge, LA. 172 pp.

Coleman, J.M., and H.H. Roberts. 1988. Sedimentary development of the Louisiana continental shelf related to sea level cycles, I. Sedimentary sequences. *Geo-Marine Letters* 8: 109.

Dardis, D. 2010. *Louisiana's oil: Understanding the environmental and economic impacts.* Website.

Dinnel, S.P., and W.J. Wiseman. 1986. Freshwater on the Louisiana and Texas shelf. *Continental Shelf Research* 6: 765.

Doyle, T. 2009. Hurricane frequency and landfall distribution for coastal wetlands of the Gulf Coast, USA. *Wetlands* 29 (1): 35–43.

Draut, A.E., G.C. Kineke, D.W. Velasco, M.A. Allison, and R.J. Prime. 2005. Influence of the Atchafalaya River on recent evolution of the Chenier Plain inner continental shelf, northern Gulf of Mexico. *Continental Shelf Research* 25: 91–112.

Environmental Protection Agency (EPA). 2017. *Mississippi River/Gulf of Mexico Watershed Nutrient Task Force 2017 Report to Congress*. Washington, DC, 121 pp.

Gagliano, S.M., J.L. van Beek. 1993. *A long term plan for Louisiana's Coastal wetlands*. Office of Coastal Restoration and Management, Louisiana Department of Natural Resources, Prepared by Coastal Environments, Inc., Baton Rouge, Louisiana.

Kesel, R.H. 2003. Human modification of the sediment regime of the lower Mississippi River flood plain. *Geomorphology* 56: 325–334.

Louisiana Coastal Wetlands Conservation and Restoration Task Force and the Wetlands Conservation and Restoration Authority. 1998. *Coast 2050: Toward a sustainable Coastal Louisiana*. Baton Rouge, LA: Louisiana Department of Natural Resources.

Milliman, J.D., R.H. Meade. 1983. World-wide delivery of river sediment to the oceans. *Journal of Geology* 91: 1–21.

Morris, J.T., et al. 2016. Contributions of organic and inorganic matter to sediment volume and accretion in tidal wetlands at steady state, *Earth's Future,* 4. https://doi.org/10.1002/2015ef000334.

National Research Council. 2006. *Drawing Louisiana's new map: Addressing land loss in coastal Louisiana*, 190. Washington, DC: The National Academies Press.

Needham, H., B.D. Keim. 2011. Storm surge: Physical processes and an impact scale. In *Recent hurricane research-climate, dynamics, and societal impacts*.

Peterson, K., 2012. Vulnerability of coastal Louisiana tribes in a climate change context. *In Tribal Climate Change Profile: Coastal Louisiana Tribes*. University of New Orleans—Center for Hazards Assessment, Response and Technology 10 pp.

Roth, D. 2010. *Louisiana hurricane history*. National Weather Service Technical Report 70 pp.

Shepard, A.N., J.F. Valentine, C.F. D'Elia, D.W. Yoskowitz, E. David, and D.E. Dismukes. 2013. Economic impact of Gulf of Mexico ecosystem goods and services and integration into restoration decision-making. *Gulf of Mexico Science* 2013 (1–2): 10–27.

Sheremet, A., and G.W. Stone. 2003. Observations of nearshore wave dissipation over muddy sea beds. *Journal of Geophysical Research* 108 (C11): 3357.

Tibbetts, J. 2006. Louisiana's wetlands: A lesson in nature appreciation. *Environmental Health Perspectives* 114 (1): A40–A43.

Turner, R.E., M.S. Kearney, R.W. Parkinson, 2017. Sea-level rise tipping point of delta survival. *Journal of Coastal Research,* published online October 2017. https://doi.org/10.2112/jcoastres-d-17-00068.1 (in Press).

U.S. Army Corps of Engineers. 2004. *Louisiana coastal area (LCA), Louisiana ecosystem restoration study*. New Orleans District: USACOE.

U.S. Census Bureau. 2016. QuickFacts New Orleans city, Louisiana. https://www.census.gov/quickfacts/fact/table/neworleanscitylouisiana/PST045216.

U.S. Energy Information Administration. 2017. Louisiana: State profile and energy estimates. https://www.eia.gov/state/?sid=LA.

Xing, F. 2015. *Deltas' responses to fluvial and marine forces*. PhD dissertation Submitted to the Graduate School of the University of Colorado 165 pp.

Chapter 14
Florida

Lynn Donelson Wright, C. Reid Nichols and Gary Zarillo

For nineteen years my vision was bounded by forests, but to-day, emerging from a multitude of tropical plants, I beheld the Gulf of Mexico stretching away unbounded, except by the sky.
—John Muir, upon reaching Cedar Key on the West Coast of Florida (from *A Thousand-Mile Walk to the Gulf,* published 1916)

14.1 "Nuisance Floods" in Paradise

In a February 2015 article in *National Geographic* Magazine, Laura Parker describes the plight of coastal Florida as rising seas add to the frequently recurrent effects of perigean "King tides", non-tidal effects related to fluctuations in the Gulf Stream, and the occasional storm surge. The article points out that, in financial (but not humanitarian) terms, Southeast Florida is more threatened than any urban region in the world: in Miami alone, assets worth $22.6–$300 billion are in jeopardy. These assets are primarily luxury high-rise condominiums, hotels and resorts and are the major source of tax revenue for the state, county and city. More recently, in 2017, Jeff Goodell discussed the threats facing Miami in more detail in his popular book *The Water Will Come* (Goodell 2017). Ironically, in the year 2017, the official position of the Florida State Government is to deny the existence of climate change and sea level rise. In addition to episodic storm surges, there is a recurring and growing inundation threat in the form of "nuisance flooding" (Fig. 14.1; see explanation in Chap. 7). Across the Florida peninsula, on the state's

L. D. Wright (✉)
Southeastern Universities Research Association (SURA), Washington, DC, USA
e-mail: ldwright@bellsouth.net

C. R. Nichols
Marine Information Resources Corporation (MIRC), Ellicott City, MD, USA
e-mail: rnichols@mirc-us.com

G. Zarillo
Ocean Engineering and Sciences, Florida Institute of Technology, Melbourne, FL, USA

L. D. Wright and C. R. Nichols (eds.), *Tomorrow's Coasts: Complex and Impermanent*, Coastal Research Library 27,
https://doi.org/10.1007/978-3-319-75453-6_14

Gulf of Mexico coast, assets are also at risk but they are measured less in monetary terms and more in terms of unique and endangered ecosystems. In this chapter, we describe two dramatically different Florida realms: the highly urbanized Miami-Dade and Broward Counties of Southeast Florida and the "Nature Coast" of Citrus and Levy Counties. Both realms are at serious risk and both are distinguished from most coastal systems by Florida's uncommon natural and demographic circumstances.

14.2 Florida's Physical Setting

Florida is composed mostly of porous limestone, the remains of relict corals. The emergent subaerial land surface of the state of Florida rests on the eastern side of the much larger calcium carbonate Florida Platform (Schmidt 1997), which now constitutes not only the substrate of the state, but also provides the surrounding continental shelf. Because the emergent lands are on the platform's eastern margin, the continental shelf off Florida's south east coast is very narrow, whereas that off the west coast is exceptionally wide. Figure 14.2, from Google maps, shows Florida in its regional setting with its continental shelf. The red colors in the figure represent relatively shallow water. Florida's karst landscape is characterized by extensive underground rivers, springs, caves (Fig. 14.3) and sinkholes as well as by a high degree of permeability. This permeability precludes the feasibility of building protective levees or dykes because rising seas will simply come up through the ground. The Floridan aquifer, which underlies most of the state (Miller 1997), is at or near the surface of the northwestern region and lies at depth beneath southeast Florida where it is overlain by the Biscayne aquifer and a confining layer of fine sedimentary rock. These limestone aquifers are critical components of Florida's

Fig. 14.1 "Nuisance flooding" of the historic Stranahan House, Ft. Lauderdale, October 2015 (Photo-C. R. Nichols)

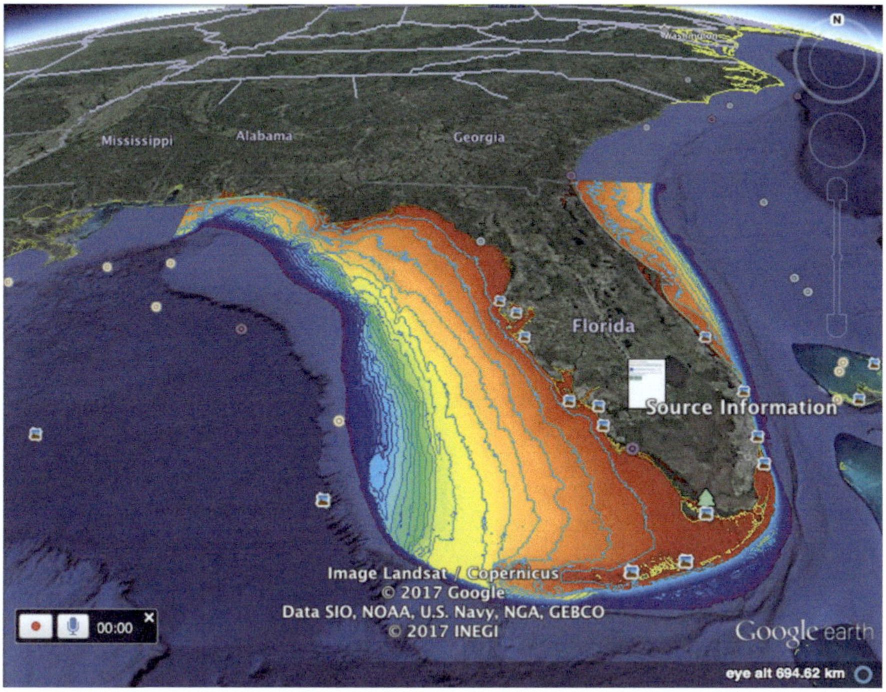

Fig. 14.2 The karst land surface of Florida rests on the eastern side of the much wider calcium carbonate Florida platform. The continental shelf (red and yellow) is narrow off the Atlantic coast and exceptionally wide and gently sloping off the Gulf (west) coast. Landsat image from Google based on NOAA and U.S. Navy data

complex groundwater hydrology network and are subject to saltwater penetration when freshwater flow is significantly reduced.

The Florida Current (Schott and Zantopp 1988; Gyory et al. 2013; Fig. 14.4), the southern component of the Gulf Stream system (or AMOC; see Sect. 2.4), is the most prominent feature of the ocean circulation affecting Florida's coast and commonly attains speeds of up to 2 m/sec (4.5 mi/h). Off southeast Florida this current impinges directly on the inner continental shelf within a few miles of Miami and Ft. Lauderdale. Coastal currents off the coast of west Florida are much weaker and more variable. The dominant ocean circulation feature in the Gulf of Mexico is the Loop Current (Schmitz et al. 2005; National Academies of Sciences, Engineering, and Medicine 2018). This swift Atlantic current, transports warm water from the Caribbean Sea into the Gulf of Mexico at speeds of about 0.8 m/s (1.8 mi/h). Sea surface temperatures in the summer months in the Gulf Stream and within the Gulf of Mexico are hot and getting hotter; in late June 2017, the sea surface temperatures are between 29.44 and 31.67 °C (85 and 89 °F). As pointed out in Chap. 2, Sect. 2.5, warm sea surface temperatures can fuel rapid intensification of tropical cyclones as was the case for Hurricane Matthew in 2016.

Fig. 14.3 Entrance to an underwater cave system in the Floridan Aquifer. Ginny Springs, Gilchrist County Florida (photo-L. D. Wright)

For Florida as a whole, the rise in mean sea level over the past century has been around 21 cm (8 in.; Maul 2015), but this rate is likely accelerating due the steric effects of warming seas. The U.S. Army Corps of Engineers (2011, 2013) estimates that by 2030 sea level will be roughly 18 cm (7.1 in.) higher than at present while Boon and Mitchell (2015) conclude from statistical analyses that by 2050 mean sea level in South Florida could be on the order of 50 cm (1.6 ft) higher than at present. Model projections, however, suggest that by 2050 Miami sea levels could be up to 1 m (3.3 ft) higher than at present (Sweet 2017; Chap. 3, Fig. 3.7).

14.3 Tropical Storms and Hurricanes

Florida, especially south Florida, lies directly in the path of tropical cyclonic systems spawned over the warm waters of the tropical Atlantic Ocean, Caribbean Sea and Gulf of Mexico. Many of these storms originate when hot Saharan winds carry "easterly waves" off the west coast of Africa, across the Cape Verde Islands and out over the Atlantic Ocean in late summer or early autumn. The most intense hurricane to affect Florida in the past 25 years was Category-5 Hurricane Andrew (Rappaport 1993; National Hurricane Center et al. 2015), which made landfall near Homestead in south Florida in 1992. Prior to Hurricane Irma in 2017, Andrew was considered to be the most destructive storm to ever hit Florida where it killed 85 people

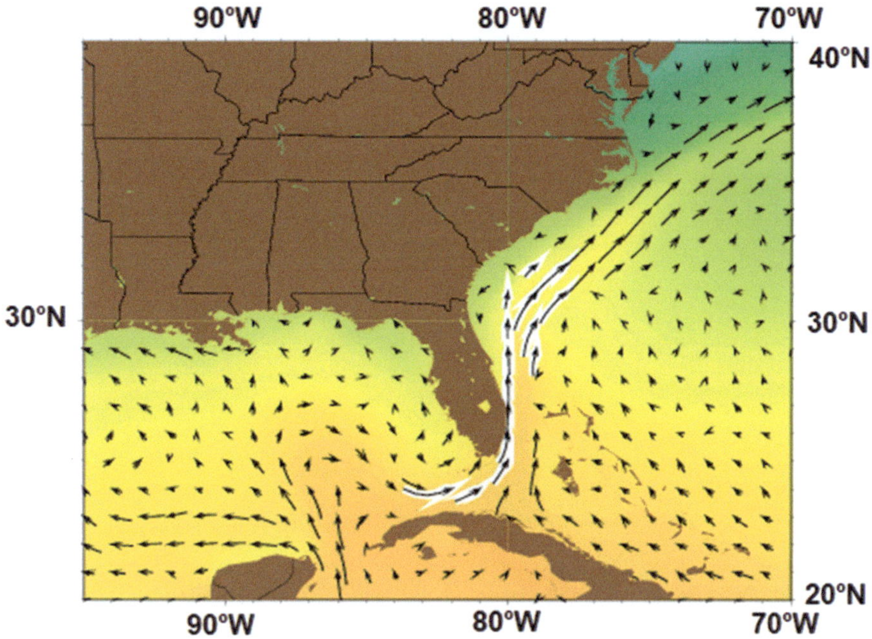

Fig. 14.4 Generalized diagram of the ocean circulation affecting the Florida Coast. The arrows indicate the directions and relative strengths (proportional to arrow length) of currents. The strong Florida current is highlighted in white from Gyory et al. (2013)

and caused $26.5 billion (1992 US$) in damage. An August 23, 2017 article in *USA Today* pointed out that if a hurricane like Andrew hit south Florida today, the damage would be on the order of $300 billion. Since 1992, 8 major hurricanes (Category 3 or higher) and 1 minor hurricane have made landfall on the Florida coast. In addition, since 2000, Florida's coast has been directly affected by 66 tropical storms or hurricanes, including Hurricane Sandy that passed to the east of Florida (but did not make landfall) in 2012 and generated large waves and moderate storm surge along the state's Atlantic coast. In 2005, Tropical Storm Katrina, crossed the southern tip of Florida before entering the Gulf of Mexico where it intensified into the hurricane that devastated New Orleans. In September, 2017, Hurricane Irma, the largest and most intense Atlantic hurricane ever recorded made landfall in south Florida as a Category 4 storm after devastating parts of the Caribbean as a Category 5. Because of the immense size of the storm (Fig. 14.5), it brought hurricane force winds and large storm surges to both the east and west coasts of the state. The unusually warm waters of the tropical Atlantic and Caribbean fueled the growth and rapid intensification of Irma. The impacts of this storm are still being assessed.

Fig. 14.5 NOAA satellite GOES-16 geocolor image of Hurricane Irma as it passed over Cuba on its way to eventual landfall in South Florida. *Image Credit NOAA/CIRA Updated: Sept 10, 2017 Editor: Sarah Loff*

14.4 Invasive Species: A Growing Threat

Numerous plants, animals, and organisms such as the mosquito-transmitted West Nile Virus have been introduced to Florida. While these newcomers do not all present a threat to native species, some cause harm, pose a threat to human health and safety, or cause economic damage. These invasive species are dispersed during hurricanes and floods or may be illegally released into the wild by pet owners. There are reports that invasive lionfish (*Pterois volitans* and *Pterois miles*) are now abundant on reefs and in ship wrecks in south Florida. The lionfish spawn and their larvae are dispersed up the east coast of the United States and into the Caribbean through currents with water temperatures that range from 10 to 35 °C (50–95 °F). Agriculture and mariculture producers are responsible for prevention of the escape of non-native species and may utilize a combination of culture and holding systems located inside buildings, greenhouses, or sea cages. Introduced species that establish their presence and spread over a non-native environment are considered to be invasive. As an example impact, native species such as the endangered Key Largo woodrat (*Neotoma floridana*) or the American alligator (*Alligator mississippiensis*) that may have a low population or rely on isolated or limited habitats such as the Everglades are imperiled by the invasive Burmese pythons (*Python bivittatus*). A hurricane that devastates broad expanses of critical habitat and introduces new and hardier species, could push vulnerable species over the brink. The invasive species in Florida are thriving in or around the suburbs, Everglades and even the coastal ocean. Table 14.1 lists some of the key invasive species challenging Florida.

14.5 Miami-Dade and Broward Counties: Endangered Wealth and Poverty

Miami-Dade and Broward Counties respectively cover 5200 km^2 (2000 mi^2 and 3186 km^2 (1230 mi^2). Both counties are bounded on the east by the Atlantic Ocean and on the west by the Everglades. The major cities are Miami, Miami Beach and Ft. Lauderdale along with numerous smaller cities. In 2012, the population of Miami-Dade County was 2.5 million (Miami-Dade County, Dept. of Regulatory and Economic Planning 2014) and in 2015 Broward County had roughly 2.0 million residents (Broward County Planning Services 2015). In Miami-Dade County, 64% of residents are Hispanic, 17% are African-American and 16% are non-Hispanic whites. Non-Hispanic whites account for 44% of Broward's residents; 25% are Hispanic and 26% are African-American. The average per capita income in 2012 in Miami-Dade County was $38,000; however, 31% of households had combined incomes of less than $25,000. Attesting to the income gap, the mean annual income of the top 5% was $338,000 and in 2012 the gross regional products of Miami-Dade and Broward Counties were $124 billion and $85.4 billion respectively, reflecting a 3% growth from the previous year (Miami-Dade County, Dept. of Regulatory and Economic Planning 2014).

The contrast between wealth and poverty in southeast Florida is dramatic. The abundance of multi-million dollar mansions and yachts in both counties would seem to belie the fact that the average owner-occupied home in Miami Dade was valued at $181,000 in 2012, 46% of houses are rented and most are valued at under $100,000. In fact, only 3% of homes are valued in excess of $1 million. In Broward County, 202,000 residents are living below the poverty line but in Miami-Dade the situation is worse. Cruz and Hesler (2012) identify 15 neighborhoods which they classify as "Targeted Urban Areas" (TUAs) that are in serious need of economic improvement. The residents of the TUAs are mostly poor and their socio-economic conditions seriously lag the rest of the county. The median income of TUA residents is 44% less than the countywide level; the average TUA per capita income in 2011 was $14,561. African Americans account for 58% of the 362,000 TUA residents and Hispanics make up 35%. Notably, the TUA population is growing, as is its vulnerability to future disasters. Vulnerability is exacerbated by poor education: 31% of TUA residents over 25 have not completed high school and 65% have no education beyond high school. It was pointed out in the opening paragraph of this chapter that $22.6 billion worth of assets are in jeopardy in Miami. But these assets are largely luxury buildings and infrastructure that would probably not be occupied by people during times of an impending disaster. Although damage to these assets would be costly, insurers would probably bear most of the costs. The real tragedy could take the form of human fatalities and suffering: the most vulnerable residents of Southeast Florida are those over 65-years old and the nearly half-a-million impoverished people living in TUAs or low-lying neighborhoods. Other factors that may slow reaction to impending disasters such as hurricanes are physical, mental, emotional, or cognitive status; religion; language; and citizenship.

Table 14.1 Selected list of invasive species in Florida. These species are non-native to their ecosystems and are likely to cause economic or environmental harm or harm to human health

Common name	Scientific name	Issue
Burmese python	*Python bivittatus*	A large non-venomous snake native to tropical South and Southeast Asia. Probably released by pet owners and responsible for eating many small mammals. Their bites and sharp teeth can cause severe lacerations
Brazilian peppertree	*Schinus terebinthifolia*	A sprawling shrub native to subtropical and tropical South America, which has been introduced outside its range as an ornamental plant. Because it produces abundant seeds that are dispersed by birds and ants and basal shoots if the trunk is cut, it is hard to control. The Brazilian Peppertree shades native plants and has an aromatic sap that can cause skin reactions in some sensitive people
Cuban tree frog	*Osteopilus septentrionalis*	A large amphibian native to the Caribbean that was probably introduced as a hitchhiker on cargo containers. They eat snails, insects, and the prey of Florida's native frogs. They have a sticky skin secretion that is extremely irritating to the eyes and nose in some people
Giant African snail	*Lissachatina fulica*	Considered one of the world's most invasive species, this land snail was introduced either as a commercial food source (for humans, fish and livestock) or as a novelty pet. Native to East Africa, this fast-growing polyphagous plant pest may also carry the parasitic nematode *Angiostrongylus cantonensis*, which can cause a very serious meningitis in people
Green iguana	*Iguana iguana*	A large herbaceous lizard native to the rain forests of northern Mexico, Central America, the Caribbean Islands, and southern Brazil. They are a popular pet and have probably escaped from or have been released by their owners. Their feces carry salmonella bacteria. Salmonella infection in humans can lead to severe nausea and dehydration
Red lionfish	*Pterois volitans*	A venomous coral reef fish native to the Indo-Pacific and Red Sea region. Their introduction into the Atlantic Ocean and Caribbean Sea was caused by inadvertent escapes from aquaria and purposeful releases. Their venomous spines deter most potential predators. A sting from a lionfish is extremely painful to humans and can cause nausea and breathing difficulties, but is rarely fatal

There are many more invasive species in Florida. To learn more visit the Florida Fish and Wildlife Conservation Commission online at http://myfwc.com/wildlifehabitats/nonnatives/

14.6 Miami-Dade and Broward Counties: Water Threats and Challenges

In Miami-Dade and Broward Counties, flooding of streets, residential neighborhoods and low-elevation homes now occurs several times a year during high tides. In addition, salt-water intrusion into the Biscayne aquifer frequently accompanies high water events creating episodic shortages of fresh water to homes. The frequency of these flood events is increasing. As a recent report by the Union of Concerned Scientists (2017) puts it: "*In recent years tidal flooding in Miami-Dade County has grown from occasional to chronic—a visible sign of rising sea levels—and is causing disruption to the local economy and infrastructure*". This flooding is not solely due to rising mean sea level but to the additive effects of tides, non-tidal water level fluctuations and sea level rise. Tides in southeast Florida are semi-diurnal (2 tides per day) and the maximum tidal range (elevation distance from low to high) varies over the course of a month with the maximum range occurring at times new and full moons (called "spring" tides). The range also varies somewhat over the course of a year and is greatest during perigean tides when the earth and moon are closest (e.g. at times of "harvest moons" in October). Perigean spring tides are locally referred to as "King Tides" because the range is above normal. The mean tide range at Miami is 0.62 m (2 ft) and increases to approximately 1.2 m (3.9 ft) during "King Tides" (Park and Sweet 2015). Simple coincidence of moderately strong onshore winds and high tide can add measurably to these levels.

The maximum heights of King Tides are increasing annually because of superimposition of other non-tidal effects that contribute to the coastal flooding (e.g., Ezer 2013). Recurring flooding of streets, historic sites, and homes occurs during perigean high tides (e.g., Wright et al. 2016). These effects are superimposed on mean sea levels, which are rising. As explained earlier (Sect. 14.2), sea levels by 2050 could be between 50 cm (1.6 ft) and 1 m (3.3 ft) higher than at present. These estimates do not take account of any unexpected glacial melting or calving in Antarctica or Greenland. Regional contributions to non-tidal water levels include atmospheric-pressure and wind induced changes, variations in offshore Ekman transport caused by fluctuations in Gulf Stream transport intensity (Ezer 2013; Ezer and Atkinson 2014; see Chap. 3, Sect. 3.5), storm surge, wave-induced set up, and land sinking. Annual fluctuations in coastal sea level of up to 1 m (3.28 ft) are attributed to aperiodic variations in Gulf Stream transport with higher sea levels corresponding to times of Gulf Stream slackening (Fig. 14.6). Also important are the tidal currents that are created by astronomical forces, which may combine with the waves to severely erode beaches and scour coastal structures.

Hurricanes can cause high storm surges and if these surges coincide with "King Tides", the outcomes can be catastrophic. Fortunately, because the continental shelf fronting southeast Florida is extremely steep, surges do not experience the same degree of amplification that take place over much gentler, shallow shelves (see Part 2, Chap. 7). The continental shelf extends from approximately 100 km (62 mi) in width off of St. Augustine to less than 2 km (1.2 mi) off of West Palm Beach. In

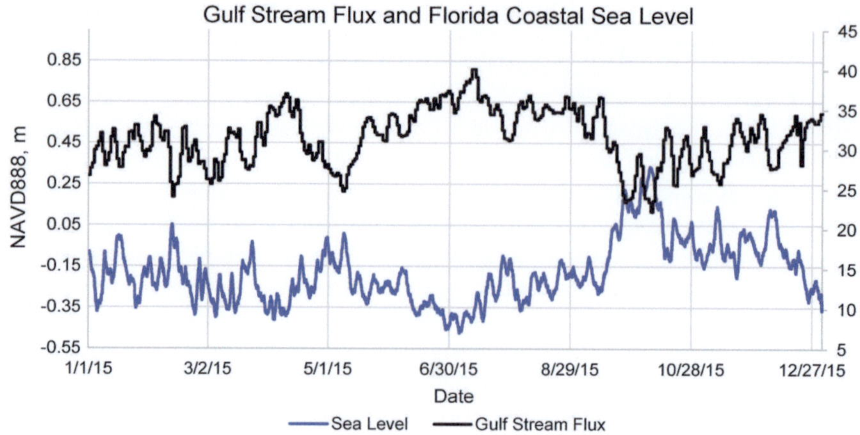

Fig. 14.6 Time series showing Gulf Stream transport and mean sea surface elevation off the east central Florida Coast in 2015. Left vertical axis and blue line indicates coastal sea surface elevation (in meters) relative to a standard datum. Right vertical axis and black line indicates Gulf Stream transport expressed in Sverdrup units (SV). One Sverdrup is equivalent to 1 million cubic meters of water per second. Water level Data are from Gary Zarillo, Florida Institute of Technology. Gulf Stream data are from NOAA

general, wide gently sloping shelves have more storm surge and relatively smaller waves than narrow shelves with a steeper slope. For example, the storm surge generated by the Category-5 Hurricane Andrew was typically less than 2 m high (4–6 ft) throughout most of Biscayne Bay reaching a local maximum of 5 m (16.5 ft) in the western corner of the bay (Rappaport 1993). Though non-trivial, these surge heights were significantly lower than that of 8.5 m (28 ft) generated by the weaker Hurricane Katrina over the shallow shelf of the northern Gulf of Mexico in 2005. Other factors impacting total surge depend on (1) storm characteristics such as angle of approach to the coast, wind intensity, forward speed, size of the windfield and (2) local features such as coastline shape, coastal inlets, lagoons, oyster reefs, and mangrove forests.

A particularly complex aspect of the challenges facing Southeast Florida pertains to the interplay of surface hydrology, ground water hydrology, fluctuations in sea levels at different time scales and spatial and temporal variations in demand for fresh water by county residents. In the years around 1900 the Everglades was roughly twice as large as today and provided a much greater volume of fresh water to recharge the Biscayne Aquifer that immediately underlies Broward County. Reductions in that recharge have allowed the penetration of seawater into the aquifer. Today, the South Florida Water Management District maintains a complex engineered system involving levees, drainage canals, containment ponds and pumping stations (Fig. 14.7). This system serves the multiple functions of mitigating flooding during heavy rainfall events, ensuring adequate distribution of freshwater to local communities and limiting seawater intrusion into the aquifer.

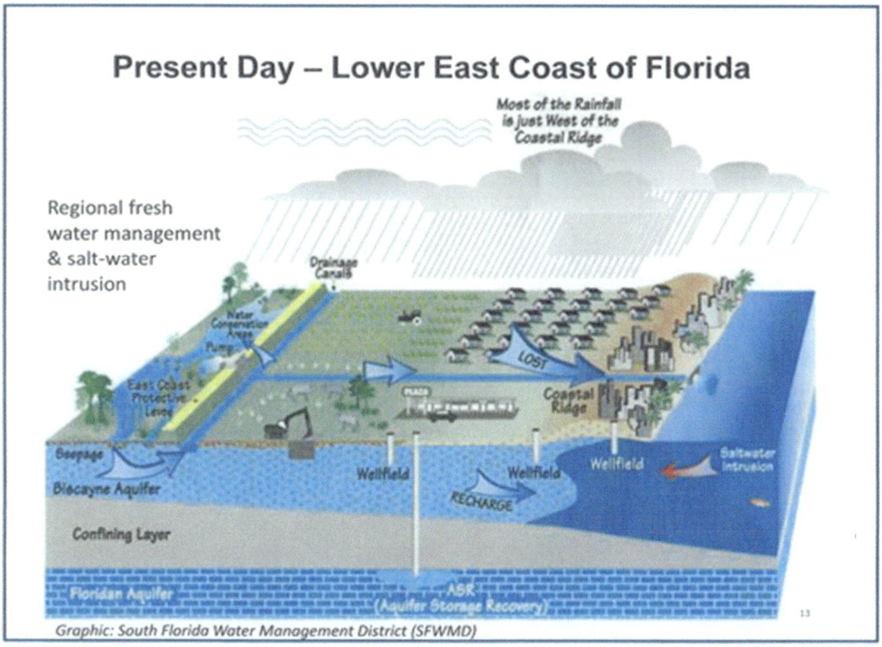

Fig. 14.7 Water management in Broward County, Florida. *Source* South Florida Water Management District

Future modeling efforts should involve linking surface hydrologic, ground water, ocean inundation and sociologic models to optimize and prioritize solutions to this complex suite of problems.

14.7 Florida's Nature Coast: Natural, Rural and Vulnerable

Florida's "Nature Coast" (Fig. 14.8) covers 4000 km^2 (1500 mi^2) and embraces 8 counties on Florida's Gulf Coast but we focus here on Citrus and Levy Counties, which comprise the central part of the region. Much of this area is water, small islands or intertidal wetlands. Environmentally and socio-economically the contrast between southeast Florida and the Nature Coast is truly extreme. This diverse region which includes vital sea-grass ecosystems for blue crabs (*Callinectes sapidus*) and bay scallops (*Argopecten irradians*) and warm spring and river aquatic ecosystems that support species such as the Florida manatee (*Trichechus manatus latirostris*), which is protected under both the Endangered Species Act and the Marine Mammal Protection Act. The Nature Coast is accessible via a coastal road, the north-south trending U. S. Route 19. There are no expensive resorts or high-rise

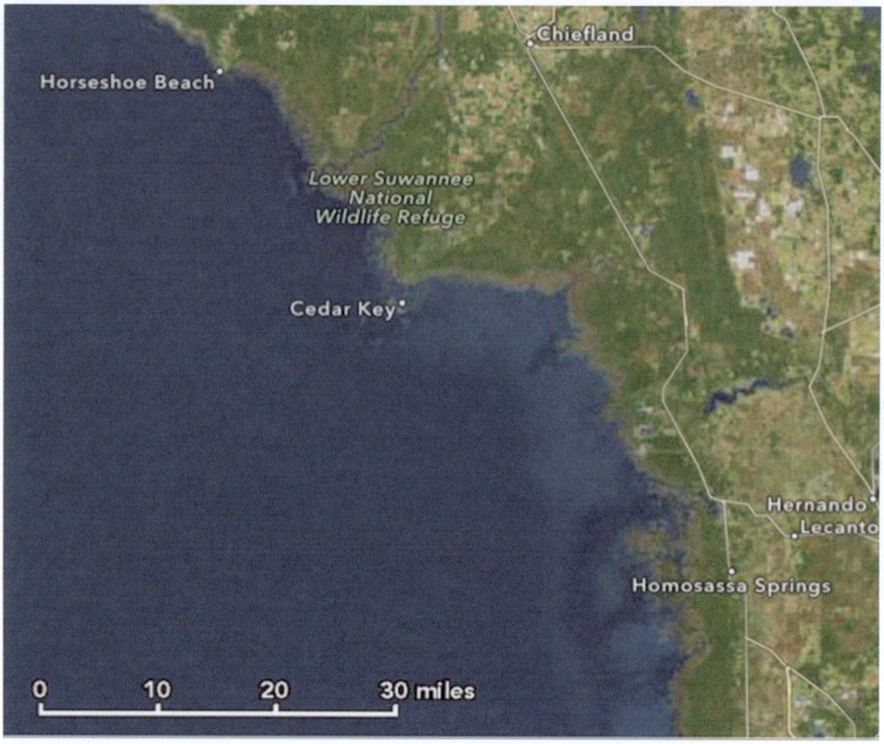

Fig. 14.8 Google earth image of the central part of Florida's "Nature Coast"

buildings on the Nature Coast. There is, however, a large coal-fired and recently nuclear-fueled power plant in Citrus County. Even though the reactors have been shut down, the fuel has not yet been relocated and the plant is situated in the intertidal zone. Demographically, Citrus and Levy Counties (Florida Legislature, Office of Economic and Demographic Research 2017a, b) are totally opposite to Southeast Florida: there are only 183,000 residents and most are white, non-Hispanic (95% in Citrus County and 80% in Levy County). African Americans constitute 2.3 of the Citrus County residents and 9.1% of Levy County residents. Hispanics are 2.7 and 8.0% of the populations in Citrus and Levy respectively. The median annual income averaged across the two counties is roughly $33,000. However, the threats to the Nature Coast cannot be adequately expressed in monetary terms or even in humanitarian terms. So what is special about this coast? The natural ecosystem is beautiful, diverse and unique in the world- and it is receding and may soon disappear forever.

Unlike the shores of southeast Florida, the Nature Coast is not fringed by sandy beaches, but by brackish marshlands that grade almost imperceptibly into shallow open water bays and, ultimately, the Gulf of Mexico (Fig. 14.9). From sea to land the coastal environments progress from open water to tidal flats to salt marsh to

Fig. 14.9 Coastal salt marsh on the shores of a shallow bay near the mouth of the Withlacoochee river on Florida's Nature Coast. The trees on the higher ground "hammock" in the background are typical of the coastal forest (photo-L. D. Wright)

transitional salt marsh to coastal forest (aka "swamp" or "hammock") yielding an extraordinarily diverse ecosystem in the aggregate (Geselbracht et al. 2011; Fig. 14.10). Among the dozen or so plant species are several rare and endangered species. There are at least 25 animal species including Florida black bear (*Ursus americanus floridanus*). These wetlands have long been undergoing gradual flooding by the recent sea level transgression and are slowly receding eastward, often leaving behind relict shorelines beneath the shallow waters (Davis 1997). Although the extremely wide and shallow shelf fronting this coast amplifies storm surges, it dissipates waves approaching from the Gulf of Mexico and the waves that reach the marshy shore are very low energy and only a few centimeters high. This is one of the reasons why there are no beaches. The other reason is that there is no supply of silica sand since the eastward transgression has taken place across a karst limestone surface and the rivers that enter the sea along this coast carry minimal sediment loads (Davis 1997). Low sediment input has favored the growth of fairly extensive oyster reefs on the shallow limestone surface beneath the waters of the shallow bays. Oyster and clam farming are now a major source of livelihood for the residents of Cedar Key.

Despite the deficiency of sediment, it is the rivers that contribute one of the most unique environmental attributes of the Nature Coast: all are fed by freshwater

Fig. 14.10 The Yellow-Crowned Night Heron (*Nyctanassa violacea*) inhabits marshes, mangrove swamps and eats mainly crustaceans as well as insects, some fish, and worms (photograph from the Waccassassa River mouth, Florida courtesy of Gordon Hart)

springs from the Floridan aquifer. The waters issuing from these numerous springs maintain a constant temperature of 22.22 °C (72 °C) year-round and this allows the coastal rivers to support manatees (Fig. 14.11) with refuges from the cold winter waters of the Gulf of Mexico. Manatee watching is the main industry of the city of Crystal River in Citrus County. The seven Nature Coast rivers, from north to south, are Suwanee, Waccasassa, Withlacoochee, Crystal, Homosassa, Chassahowitzka and Weeki Wachee. In addition to the input of freshwater to the coast from rivers, the Floridan aquifer also has numerous subsurface connections with the Gulf and sub-sea springs are common on the inner shelf. In some cases, seawater penetrates upstream within the underground rivers and when this occurs the water from some wells can become brackish.

The tides affecting the Nature Coast are mixed, which means that the secondary diurnal contribution from the solar component of the tide interacts with the primary semidiurnal lunar component in such a way as to cause two high and two low tides a day but with the range of one tide being significantly greater than the other (CO-OPS 2000). The normal maximum spring tide (i.e. new or full moon) tidal range is 1.5 m (5.0 ft). However, when high tides are added to water levels raised by strong onshore winds or winds from the south, flooding of low areas and coastal

Fig. 14.11 Florida Manatee (*Trichechus manatus latirostris*), in Manatee Springs Florida adjacent to the Suwanee River (photo courtesy of L. D. Wright)

roads can occur. And, as is the case for southeast Florida, the frequency of "nuisance flooding" is increasing in communities like Crystal River and Cedar Key as sea level rises. Sinkholes may form by erosion caused by frequent exposure to water that is able to dissolve the limestone bedrock.

Adding to the nuisance flooding, the low gradient continental shelf fronting this coast (Fig. 14.1) amplifies storm surges. The threats were dramatically illustrated by severe inundation caused by the "minor" Category 1 Hurricane Hermine on September 1, 2016, which flooded at least 2000 homes, caused a massive sewage spill in Tampa Bay, and severely impacted the clam aquaculture industry of Cedar Key. Notably, the approximately 2 m (6.5 ft) storm surge that prevailed was comparable to that generated in Biscayne Bay by the much more intense Hurricane Andrew in 1992. Fortunately, Nature Coast flooding caused by the much larger

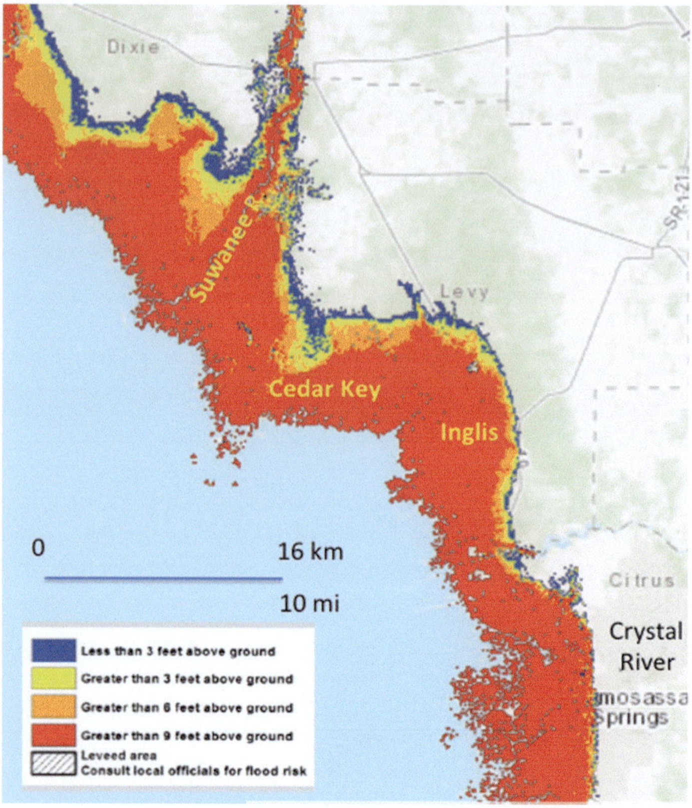

Fig. 14.12 Predicted maxima of maximum envelope of water (MOMs) for a category 3 Hurricane affecting Florida's Nature Coast. Red = greater than 9 ft (3 m) above ground; Orange = greater than 6 ft (2 m) above ground; yellow = greater than 3 ft (1 m) above ground; blue = less than 3 ft (1 m) above ground. *Source* NOAA/NWS/NHC/Storm surge unit, NOAA/NOS/Office for coastal management

Hurricane Irma in September 2017 was minimal because the dominant winds on the storm's advancing margin were easterly and pushed water offshore. As explained in Chap. 7, NOAA's National Hurricane Center uses the SLOSH model to generate maps of worse case scenarios of flooding expressed as *Maximum Envelope of Water (MEOW)* and the *Maximum of the Maximum Envelope of High Water (MOM).* Figure 14.12 shows corresponding MOMS map for the Nature Coast in the event of a hypothetical Category 3 Hurricane. Note that inundation in excess of 9 ft ($\sim$3 M) above the level of normally dry ground is predicted to extend as far as 5 mi. landward of the present coast. Figure 14.13 illustrates the high social vulnerability of communities on Florida's West Coast to 1.83 m (6-ft) storm surges. As sea levels rise, the vulnerability can be expected to increase. Vulnerable Nature Coast communities include Crystal River, Homosassa and Cedar Key.

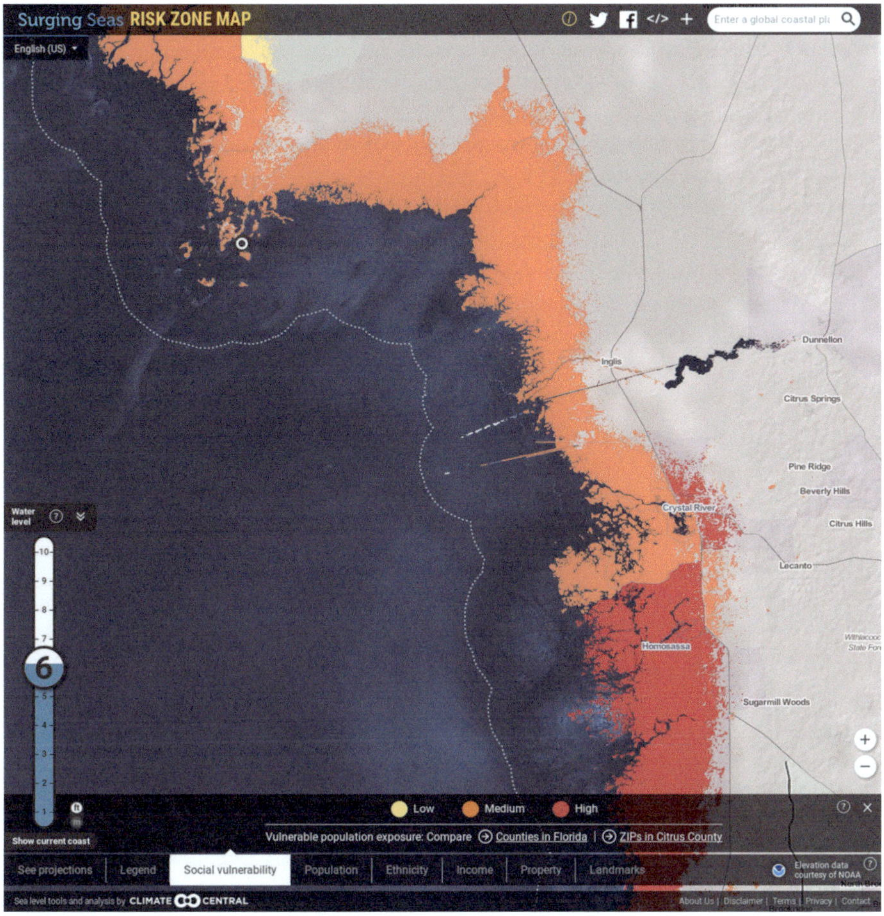

Fig. 14.13 Social vulnerability of Florida Nature Coast residents to a 1.83 m (6 ft) storm surge. From "Surging Seas" risk zone maps produced by *Climate Central*. Orange = medium vulnerability; red = high vulnerability

The future prospects for the Nature Coast are dire and they are made more so, not just by the global changes that are already in play, but also by state and local policies and commercial activities that are impacting the Floridan aquifer and the coastal environment. Among these are recently approved permits for drawing down water from the aquifer for horse and cattle agriculture and commercial water bottling. This will measurably reduce freshwater discharge at the coast and affect the salinity regime. Mining of limestone from the substrate for road aggregate has been ongoing for several years. To assess the likely future impacts of sea level rise on the Nature Coast wetlands, the numerical model *Sea Level Affecting Marshes Model* (SLAMM; Clough et al. 2010) was applied by Geselbracht et al. (2011) for three sea level rise scenarios: 0.64 m (2.1 ft), 1 m (3.3 ft) and 2 m (6.6 ft). The model results suggest that for the 0.64 m (25.2 in) sea level rise scenario, the area of the saltmarsh and transitional saltmarsh would increase by displacing 69% of the coastal forest but without significant recession of the seaward margin of the marsh, which would presumably be maintained by accretion. A 1 m (3.28 ft) S.L. rise is predicted to enable continued eastward expansion of saltmarsh at the expense of 83% of the coastal forest and 25% of the currently developed dry land. With a sea level rise of 2 m (6.5 ft) by the year 2100, open water would encroach inland 7–8 km (approximately 5 mi.) displacing 48% of the salt marsh, 99% of the coastal forest and 79% of developed dry lands (i.e. towns and residential areas). The loss of critical habitat, particularly transitional marsh, would seriously reduce biodiversity and several bird species would disappear (Geselbracht et al. 2011).

References

Boon, J.D., and M. Mitchell. 2015. Nonlinear change in sea level observed at North American tide stations. *Journal of Coastal Research* 31 (6): 1295–1305.

Broward County Planning Services Division. 2015. *Broward-by-the-numbers, Broward 100 No. 62 October 2015 Special Edition.* Ft. Lauderdale: Broward County Environmental Protection and Growth Management Department, Broward County Planning Services Divisio (3pp).

Clough, J., R. Park, R. Fuller. (2010). SLAMM 6 beta technical document, release 6.0. Warren Pinnacle Consulting, Inc. (41pp).

CO-OPS (Center for Operational Oceanographic Products and Services). 2000. *Tide and current glossary.* Silver Spring, MD: National Ocean Service, National Oceanic and Atmospheric Administration. Available online https://tidesandcurrents.noaa.gov/publications/glossary2.pdf. Accessed 29 Aug 2017.

Cruz, R.D., and R. Hesler. 2012. *Socio-economic conditions in miami-dade's targeted urban areas 2007–2011.*

Davis, R.A. 1997. Geology of the Florida Coast. In *The geology of Florida*, ed. A.F. Randazzo, and D.S. Jones, 155–168. Florida: University of Florida Press.

Ezer, T. 2013. Sea level rise, spatially uneven and temporally unsteady: Why the U.S East Coast, the global tide gauge record, and the global altimeter data show different trends. *Geophysical Research Letters* 40: 5439–5444.

Ezer, T., and L.P. Atkinson. 2014. Accelerated flooding along the U.S. East Coast: On the impact of sea-level rise, tides, storms, the Gulf Stream, and the North Atlantic Oscillations. *Earth's Future 2* https://doi.org/10.1002/2014ef000252.

Florida Legislature, Office of Economic and Demographic Research. 2017a. *Citrus County Statistics.*

Florida Legislature, Office of Economic and Demographic Research. 2017b. *Levy County Statistics.*

Geselbracht, L., K. Freeman, E. Kelly, D.R. Gordon, and F.E. Putz. 2011. Retrospective and prospective model simulations of sea level rise impacts on Gulf of Mexico coastal marshes and forests in Waccasassa Bay, Florida. *Climatic Change* 107: 35–57. https://doi.org/10.1007/s10584-011-0084-y.

Goodell, J. 2017. *The water will come: Rising seas, sinking cities, and the remaking of the civilized world.* Little Brown (340pp).

Gyory, J., E. Rowe, A. Mariano, and E. Ryan. 2013. The Florida current. Ocean surface currents. Available online http://oceancurrents.rsmas.miami.edu/atlantic/florida.html. Accessed on 29 Aug 2017.

Maul, G.A. 2015. Florida's rising seas: A report in feet per century for coastal interests. *Florida Scientist* 78 (2): 64–87.

Miami-Dade County, Deptartment of Regulatory and Economic Planning. 2014. *Miami-Dade county economic & demographic profile 2014* (16pp).

Miller, J.A. 1997. Hydrogeology of Florida. In *The geology of Florida*, ed. A.F. Randazzo, and D. S. Jones, 69–88. Florida: University of Florida Press.

National Academies of Sciences, Engineering, and Medicine. 2018. *Understanding and predicting the gulf of mexico loop current: Critical gaps and recommendations*. Washington, DC: The National Academies Press (110pp). https://doi.org/10.17226/24823.

National Hurricane Center, Hurricane Research Division, Atlantic Oceanographic and Meteorological Laboratory. (Feb 2015). Continental United States Hurricanes (detailed description). Miami: United States National Oceanic and Atmospheric Administration's Office of Oceanic & Atmospheric Research. www.aoml.noaa.gov. Retrieved 28 June 2017.

Park, J., and W. Sweet. 2015. Accelerated sea level rise and Florida current transport. *Ocean Science* 11 (4): 607–615.

Parker, L. 2015. Treading water. *National Geographic* 226 (Feb 2015).

Rappaport, E. 1993. Hurricane Andrew. *National hurricane center* (preliminary report). Miami, Florida: United States National Oceanic and Atmospheric Administration's National Weather.

Schott, F., T.N. Lee, and R. Zantopp. 1988. Variability of structure and transport of the Florida current in the period range of days to seasonal. *Journal of Physical Oceanography* 18: 1209–1230.

Schmidt, W. 1997. Geomorphology and physiography of Florida. In *The Geology of Florida*, ed. A.F. Randazzo, and D.S. Jones, 1–12. Florida: University of Florida Press.

Schmitz, W.J., D.C. Biggs, A. Lugo-Fernandez, L.-Y. Oey, and W. Sturges. 2005. A synopsis of the circulation in the Gulf of Mexico and on its continental margins. In *Circulation in the Gulf of Mexico: Observations and models*, eds. W. Sturges, and A. Lugo-Fernandez, 11–29. American Geophysical Union, Geophysical Monograph 161.

Sweet, W.V., R.E. Kopp, C.P. Weaver, J. Obeysekera, R. Horton, E.R. Thieler, and C. Zervas. 2017 *Global and regional sea level rise scenarios for the United States*. NOAA Technical Report NOS CO-OPS 083 (64pp).

Union of Concerned Scientists. 2017. *Encroaching tides in Miami-Dade county, Florida* (10pp). www.ucsusa.org/EncroachingTidesMiamiDade.

U.S. Army Corps of Engineers. 2011. USACE Climate Change Adaptation Policy Statement. 3 June 2011. http://www.corpsclimate.us/docs/USACEAdaptationPolicy3June2011.pdf.

U.S. Army Corps of Engineers. 2013. Incorporating sea level change in civil works programs. USACE ER 1100-2-8162. http://www.publications.usace.army.mil/Portals/76/Publications/EngineerRegulations/ER_1100-2-8162.pdf.

Wright, L. Donelson, C. Reid Nichols, Arthur G. Cosby, Christopher F. D'Elia. 2016. Collaboration to enhance coastal resilience. *Eos, Transactions AmericanGeophysical Union* 97. Available online http://bit.ly/2bwBwrU.

Chapter 15
Mid Atlantic Bight and Chesapeake Bay

C. Reid Nichols, Gary Zarillo and Christopher F. D'Elia

This chapter is dedicated to research asst. professor Arthur Schwarzschild (1966–2017) who is best known for his leadership as the site director at Anheuser-Busch Coastal Research Center in Oyster, Va., which is the facility that hosts the Virginia Coastal Reserve Long Term Ecological Research program.

15.1 Migratory Lighthouses

The Mid-Atlantic Bight (MAB) off the East Coast of the United States covers the continental shelf region from Nantucket Shoals in Massachusetts to Cape Lookout in North Carolina. This indentation or concave stretch of changing coastline includes the New York Bight and important estuaries such as Long Island Sound, Great South Bay, Delaware Bay, Chesapeake Bay, and Pamlico Sound. The geography of the MAB is depicted in Fig. 15.1. The Chesapeake Bay is the largest estuary in the United States, extending nearly 321.9 km (200 miles) from the Susquehanna River in the north to its entrance between the Virginia Capes. The Pamlico Sound is the largest lagoon along the North American East Coast, extending 129 km (80 miles) in length and having widths from 24 to 48 km (15–30 miles). Storms resulting in coastal erosion along the MAB include wave developments preceding cyclogenesis, strong cold fronts with associated squall lines, intense cyclones moving northeastward, and hurricanes. The MAB rising sea level,

C. R. Nichols (✉)
Marine Information Resources Corporation (MIRC), Ellicott City, MD, USA
e-mail: rnichols@mirc-us.com

G. Zarillo
Ocean Engineering and Sciences, Florida Institute of Technology,
Melbourne, FL, USA

C. F. D'Elia
School of the Coast and Environment, Louisiana State University,
Baton Rouge, LA, USA

L. D. Wright and C. R. Nichols (eds.), *Tomorrow's Coasts: Complex and Impermanent*, Coastal Research Library 27,
https://doi.org/10.1007/978-3-319-75453-6_15

which has accelerated during the past few decades in some locations, not only inundates low-lying coastal regions but also contributes to the redistribution of sediment along sandy coasts. Barrier islands respond by migrating landward (Part 2, Chap. 9). The impact is illustrated along the MAB by (1) the erosion observed on barrier islands such as Hatteras Island, NC, and glacial headlands, such as Nantucket Island's Sankaty Head and (2) ongoing engineering efforts required to safeguard the historic Cape Hatteras and Sankaty Lighthouses. While these lighthouses are located at either end of the MAB others such as the historic Montauk Point Lighthouse in New Jersey are also threatened by erosion. The rapidly increasing availability of meteorological and oceanographic data and innovative marine technologies since the 1950s has greatly improved the frequency and accuracy of numerical forecast models and other products that can be directly used to protect property or adapt to sea level rise by moving coastal structures and infrastructure away from unsuitable areas.

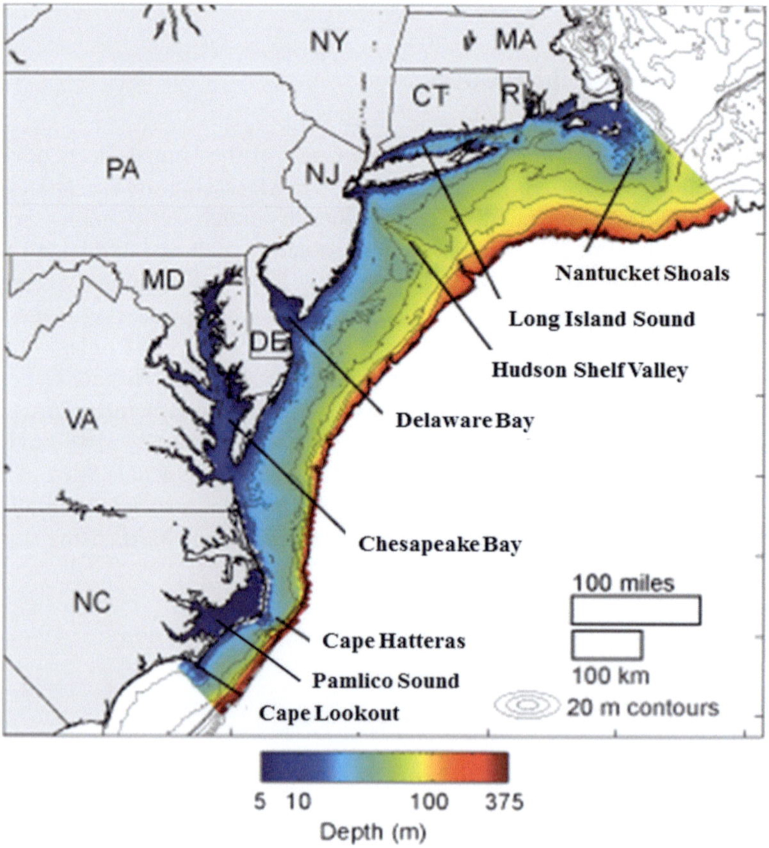

Fig. 15.1 Geography of Mid Atlantic Bight. *Source* U.S. Geological Survey

The first Cape Hatteras Lighthouse was made of sandstone and completed in 1803. Owing to modernization, a new tower was constructed in 1870 and encroachment by the sea required the lighthouse to be replaced by a steel tower in 1935. The beacon was returned to the brick tower during 1950, following considerable accretion of beach material in front of the lighthouse. From the 1960s to the 1980s, efforts were made to stabilize the beach in front of the lighthouse. During March 1980, a winter storm swept away the remains of the 1803 lighthouse and caused significant dune erosion. During September 1999, when the 61-m (200-foot) tall and 5000-ton Cape Hatteras lighthouse was just 4.6 m (15 ft) from the ocean's edge, a well-planned and executed engineering feat resulted in its movement from its original location at the edge of the ocean to a safer location approximately 460 m (1500 ft) from the shoreline. The general contractor International Chimney and Expert House Movers won the 40th Annual Outstanding Civil Engineering Achievement Award from the American Society of Civil Engineers in 1999 for their accomplishment in moving the Cape Hatteras Lighthouse, the tallest masonry structure ever moved.

During the 1850s, Sankaty Light on Nantucket Island was reportedly 85 m (280 ft) from the edge of the bluff. The bluff has reportedly lost an average of 0.9 m (3 ft) a year since 1980, but erosion is episodic. For example, approximately 5.18 m (17 ft) were lost off Sankaty Head after the Perfect Storm, also known as the No-Name Storm, of 1991. This nor'easter absorbed the remnant of Hurricane Grace and ultimately evolved back into a small unnamed hurricane late in its life cycle. With their success in moving the Cape Hatteras Lighthouse, the Buffalo, N.Y.-based company International Chimney was contracted to move the 70-foot-tall, 550-ton Sankaty Lighthouse inland during October 2007, when it was approximately 20 m (66 ft) from the edge of the bluff (Benchley and Felch 2009). The lighthouse was jacked up and slowly moved to its new location approximately 85 m (280 ft) from the cliff along steel rail tracks. In addition to being a historical landmark, Sankaty Light remains an important navigation aid.

15.2 New Jersey to Cape Charles, Virginia

The Intergovernmental Panel on Climate Change, or IPCC (2007), estimates that global sea level will rise from 0.18 to 0.59 m (7–23.23 in.) by the end of the 21st century. As sea level continues to rise along the MAB with the concomitant erosion from tides, waves, and currents, the future of coastal structures such as lighthouses and natural areas such as shorelines is uncertain.

The shoreline is defined as the wider fringe of land that is geologically modified by the action of the ocean past and present. Many coastal residents in places such as Ocean City, New Jersey, and the Eastern Shore of Maryland face recurring flooding, especially during higher high tides or Spring tides and during periods when favorable winds set-up estuarine and coastal waters. The spring tides occur just after a new or full moon, when the gravitational pull of the sun is "added" to the gravitational pull of the moon on Earth. Flooding may also be caused by the combined effects of wind waves, high tides and storm surges in response to fluctuations in local and remote winds and atmospheric pressure. The term 'surge' refers to the pile of water pushed ashore by the strong winds. People living along coastal lagoons and estuaries need to be aware of sea-level rise rates. In the Chesapeake Bay, people living on Smith and Tangier Islands are especially susceptible to the impacts of sea level rise.

Many of the Chesapeake Bay islands have been shrinking in size for centuries, due to a combination of their low elevation, lack of rock substrate, land subsidence, storm erosion, and man-made factors, such as dredging channels through the marsh. Langland and Cronin (2003), report sediment accumulations of 25–30 m (82–98 ft) in some parts of the Bay. Further, land subsidence in some parts of the Chesapeake Bay region is caused by the pumping of ground water from unconsolidated sediment (Eggleston and Pope 2013). In order to be resilient to flooding, some coastal communities spend many millions of dollars on dredging, building new structures, such as elevated roads, and making improvements to drainage systems and pumping stations. Smith Island, located 19.3 km (12 miles) west by water from Crisfield on the Maryland Eastern Shore, was particularly devastated in 2012 by Hurricane Sandy. Residents formed the group Smith Island United to fight efforts by the State of Maryland to buy up and demolish homes that were impacted by hurricane storm surge. Smith Island United, in concert with various private and public partners, has developed plans to be resilient against future storms and eroding land. Other communities are adapting by introducing flood protection schemes that integrate habitat restoration with traditional measures, e.g., developing wetlands that store floodwaters, planting woodlands to slow flash floods, and restoring rivers to hold more water.

15.3 Rising Sea Level Along the Maryland and Virginia Chesapeake Bay Shoreline

Numerous investigators (e.g., Cronin 2005; Curtin et al. 2001; Pritchard 1951, 1952a, b, 1955; Ward et al. 1989; White 1989) have characterized physical features common to the Chesapeake Bay, a semi-enclosed body of water partially protected from the open sea, referred to as a "Type II" (moderately stratified) estuary (Pritchard 1951). Marine scientists classify the Chesapeake Bay as a drowned river valley estuary and have studied its response to notable storms such as Hurricanes Hazel (1954), Agnes (1972), Irene (1999), Isabel (2003), Ivan (2004), Sandy (2012). The area of land that drains into Chesapeake Bay includes parts of Delaware, Maryland, New York, Pennsylvania, Virginia, West Virginia, and Washington D.C. This area, known as the Delmarva Peninsula, is especially affected by cyclones that brush the East Coast. Central and Western Maryland, as well as Washington, D.C., commonly receive rainfall from the remnants of storms that make landfall elsewhere and track northward. These episodic events, which are detailed in Fig. 15.2, directly contribute to flooding and geomorphological changes such as shoreline erosion and shifting channels. As illustrated in Fig. 15.3, the Chesapeake Bay includes major tributaries such as the Susquehanna, Potomac, Rappahannock, York and James rivers, and more than 100,000 streams, creeks and rivers. The bay and its tidal tributaries include about 18,803.6 km (11,684 miles) of shoreline and have depths averaging 6.4 m (21 ft) along with deep holes that are up to 53 m (174 ft) deep.

Physical processes in the MAB, such as sea level rise, greatly influence the life cycles of marshes, submerged aquatic vegetation (SAV), marine birds, shellfish, and fish. Along with natural and anthropogenically influenced processes, sea level rise contributes to coastal erosion, driven by the movement of beach materials. Atmospheric and oceanic weather, such as El Niño, storms, winds, tides, currents, waves, and storm surges, are major factors in coastal erosion, which can be quantified by geophysical surveying and remote sensing. Discharge from rivers and water movement due to the astronomical tides also produces coastal erosion as evidenced by the sediment plumes that are depicted in Fig. 15.3. Solid particles

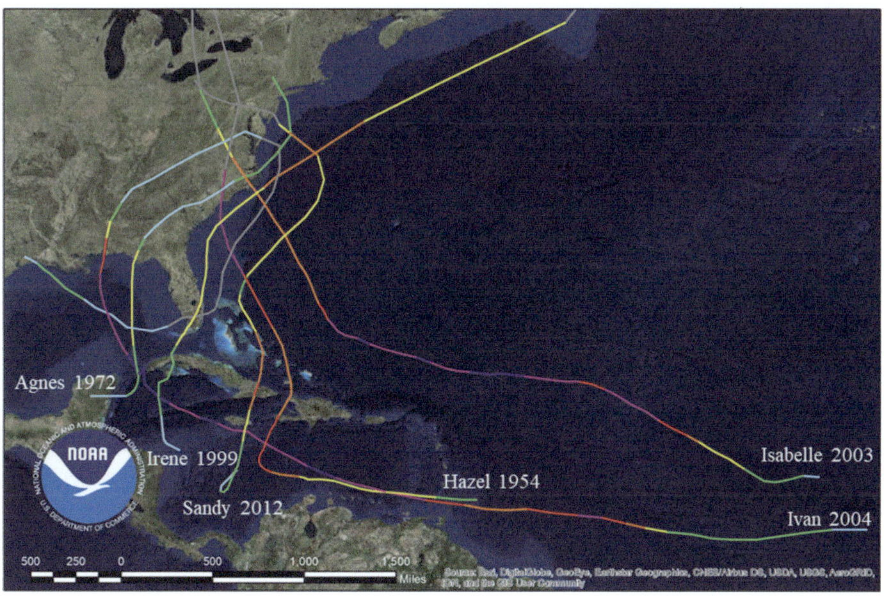

Fig. 15.2 Tracks of hurricanes causing wind damage and severe flooding along the Chesapeake Bay. Numerous investigators (Blankenship 2004; Chesapeake Research Consortium, Inc. 1976; Horowitz et al. 2014; Smith 2014; and Wang et al. 2015) describe the impacts that these extreme events have caused on the Bay's rivers, wetlands, forests, and coastal communities. Tropical cyclones traversing the MAB tend to move toward the west-northwest, which brings the hurricane into the vicinity of the East Coast of the U.S. The hurricane tracks were obtained from the NOAA National Hurricane Center (see http://www.nhc.noaa.gov or https://coast.noaa.gov/hurricanes/)

Fig. 15.3 In Maryland, the National Weather Service reported rainfall totals of 613 mm (24.13 in.) in Largo, 609 mm (23.98 in.) in Forrestville, and 546 mm (21.49 in.) in Forest Heights

from eroding river banks, drainage from construction sites, and agricultural runoff are entrained in rivers such as the Susquehanna and are transported into the Chesapeake Bay. Aside from sea level rise and other natural processes, humans have contributed to coastal erosion by altering coastline transport processes with dams, groins, and jetties, which impact the supplies of sand to the shorelines.

Long-term environmental data are made available through sensors that comprise capabilities such as the National Water Level Observation Network, the U.S. Army Corps of Engineers Field Research Facility at Duck, N.C., Physical Oceanographic Real-Time Systems, and the Chesapeake Bay Interpretive Buoy System. Data from these networks indicate that water levels are rising along the North American coast due to climatic changes that expose more and more of the coastline to wave action (Sweet et al. 2017). The Chesapeake Bay's wetlands safeguard natural services of the coastal ecosystem and the temperate climate offers a fertile and diverse environment for waterfowl such as mallards (*Anas platyrhynchos*), black ducks (*Anas rubripes*), canvasbacks (*Aythya valisineria*), and redhead ducks (*Aythya americana*). Winds and rains can cause high sea states and currents that uproot bay grasses. These same SAV beds that contain species such as eelgrass (*Zostera marina*) and Widgeon grass (*Ruppia maritima*), provide food and habitat for many species of waterfowl on their migration from summer breeding grounds. The beds provide a refuge from predators for finfish such as spot (*Leiostomus xanthurus*), croaker (*Micropogonias undulatus*), and juvenile striped bass (*Morone saxatilis*). The Bay grass beds are also important for grass shrimp (*Palaemonetes pugio*) and blue crabs (*Callinectes sapidus*), especially after molting. These wetlands provide an important natural buffer for storms by temporarily storing floodwaters and filtering pollutants. Shorelines that lack adequate wetlands, beaches, and oyster reefs are highly vulnerable to forces generated by sea level rise, waves, tides, and other shallow water processes. People living in flood prone areas that lack buffers have to contend with recurring floods that enter front yards and garages, overtop docks, clog storm drains, block streets, and damage lawns and trees.

Sea level rise exacerbates the vulnerability of infrastructure to flooding from storm surges along the MAB. In the Chesapeake Bay region, port facilities, coastal roads, pipelines, and other important structures in Maryland and Virginia are especially susceptible to storm surge. Locations of recurring floods include Hampton Roads and the heavily traveled US 17 coastal highway. Flooding caused by hurricanes, tropical storms, and nor'easters occurs as a result of elevated water levels from atmospheric pressure changes and favorable wind directions associated with storm passage. Coastal communities also face flooding when sea level rise is compounded by semi-monthly spring tides, which increases coastal erosion and the weathering of roads. Wave action associated with storm surge has been responsible for the removal of beach and dune sediments in places such as Virginia Beach. One of the often overlooked dangers to motorists driving on previously flooded roads is that the inundation reduces the bearing strength of the pavement by allowing water to get into the underlying layers. This is evidenced by numerous ruts, cracks, and potholes.

There are many physical processes contributing to long and short-term variations in Chesapeake Bay water levels. Short-term variations that occur on a daily basis include waves, tides, or episodic flood events, such as those associated with winter snow melt, hurricanes or coastal storms. The majority of the Chesapeake Bay has a semidiurnal tidal period and harmonically predicted tides are made possible by a long-term water level observation network that includes more than 20 stations, which are maintained by NOAA. Areas that experience tides have two high and low waters and accompanying ebb and flood tidal currents per day in response to astronomical forces. The tidal day is 24 h and 48 min long and is the reason why tides arrive one hour later each solar day. For some species such as horseshoe crabs (*Limulus polyphemus*), spawning activities occur in synchronization with the full moon and spring tides. Currents caused by tides, winds, rains, and differences in water levels are responsible for the transport of passive drifters such as plankton, seeds, and fish larvae. Longer-term processes may occur over time scales ranging from months to several years, and may present as repeatable cycles, gradual trends, or intermittent anomalies. Seasonal weather patterns, variations in the Earth's declination, changes in Gulf Stream circulation (Ezer et al. 2013), anthropogenic influences (such as dredging), vertical land motion, and the El Niño Southern Oscillation are just a few examples of long-term fluctuations. When estimating sea level trends, a minimum of 30 years of data should be used in order to account for long-term sea level variations and to reduce errors in computing sea level trends based on monthly mean sea level. Accounting for repeatable, predictable cycles, such as tidal, seasonal, and interannual variations allows for computation of a more accurate long-term sea level trend. Changing sea levels have been documented by the Coast and Geodetic Survey, as evidenced by Fig. 15.3, which depicts historic seasonal to inter-annual cycles of sea level in the MAB area using a low-pass filter on NOAA water level records.

NOAA tide stations that measure Local Sea Level, which refers to the height of the water as measured along the coast relative to a specific point on land, are critical to understanding long term changes in sea level. According to Cooper (2016), the Eastern Shore is the third most vulnerable area to flooding behind South Florida and Louisiana. As indicated by 20th century water level records in Fig. 15.4 and references such as Cronin (2005), Maryland residents have been coping with rising sea levels and shoreline erosion for many years. The combination of sinking lands and rising sea levels has significantly eroded previously inhabited islands such as Holland Island in the Chesapeake Bay. Holland Island, which once supported a community with approximately 60 homes, a church and other buildings, was abandoned in 1922. The island, which has receded into the Bay, was located in the Holland Strait between South Marsh Island on the south and Bloodsworth Island and other smaller uninhabited low marshy islands on the north. Figure 15.5 is a

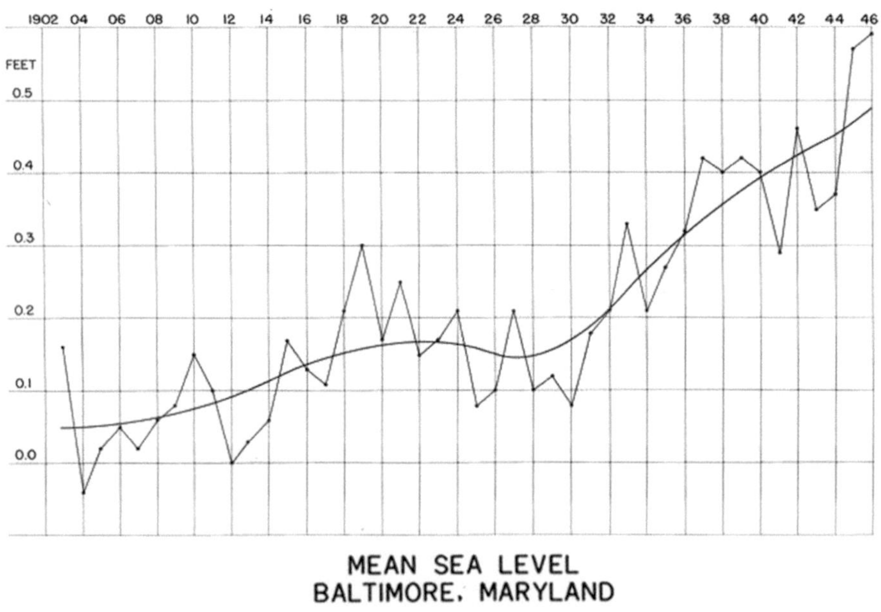

Fig. 15.4 A historic graph showing the rise in sea level in Baltimore, which includes water level fluctuations caused because the underlying land is falling with respect to the sea surface. Data from 1902 through 1946 were obtained through C&GS tide gage observations (courtesy NOAA Photo Library)

Fig. 15.5 The last house on Holland Island, which was built in 1888, collapsed into the Bay in October, 2010. This photograph of the Holland Island, which is barely discernible today, was taken on November 9, 2008 (courtesy of the Chesapeake Bay Foundation, Photo by Chuck Foster)

picture of the remaining house on Holland Island before its total collapse in October 2010. Schulte et al. (2015) hypothesizes that Tangier Island, which is located in Virginia, will have to be abandoned prior to 2100. Neighboring Smith Island residents who have been reported to say, “we live in the water,” highlight the fact that people will fight to defend the places they call home. These islands have a rich history dating back to the earliest European explorers and remain important sites for marine archaeology and anthropology.

15.4 Human Intervention

Effective solutions for shoreline protection rely on a complete understanding of the physical, financial, government permitting, and policy environments. Built-up areas that are subject to coastal erosion are expensive to protect. Most shore protection structures (e.g., jetties and sea walls) are eventually undermined and broken apart by waves. Therefore, when possible, it makes sense to allow coastal erosion to continue and to site new structures far enough back from the shoreline, dunes, or banks to minimize the need for shore protection structures. Climate change will

Fig. 15.6 Following hurricane sandy, the U.S. Army Corps of Engineers replaced roughly 6.1 million m^3 (8 million $yards^3$) of sand to repair and restore engineered beaches along the Atlantic coast of Monmouth County, NJ (courtesy of U.S. Army Corps of Engineers)

accelerate sea-level rise and erosion of beaches and necessitate building rock defenses against coastal erosion. For example, the Monmouth Seawall was heavily damaged by Hurricane Sandy after the "superstorm" made landfall in Brigantine, New Jersey on October 29, 2012. The National Weather Service reported storm surge of up to 3.35 m (11 ft) along the New Jersey Shore with wave heights in excess of 4.27 m (14 ft). Portions of the seawall have been rebuilt and new additions to the wall are being planned as indicated in Fig. 15.6. Data collected during storms such as Hurricane Sandy have been used to improve the accuracy and resolution of numerical modeling of storm effects. The ensuing street-scale forecasts have proven to be especially useful to urban planners. Blumberg et al. (2015) developed a hydrodynamic model to accurately describe Sandy's flooding along the urban coastal waters in Hoboken, New Jersey. Other solutions for shoreline protection include control of ground water pumping and development of surface water to offset the reductions of ground water pumping. Ground water recharge has also been practiced in some locations threatened by subsidence.

15.5 Wetlands Restoration

Human interactions such as heavy land development have been blamed for declines in some of the more than 3600 species of plants, fish, and animals that inhabit the Chesapeake Bay. As an example, runoff from the watershed can pick up pollutants such as nutrients, sediment, and chemical contaminants. The ensuing siltation blocks sunlight from reaching SAV beds and smothers oyster reefs, causing problems for feeding and the settling of new larvae on old shell material. Commercial, residential, and recreational development along the coast has reduced favorable habitat of the piping plover (*Charadrius melodus*), which prefers wide, flat, open, sandy beaches with very little grass or other vegetation. Besides high turbidity from siltation, diseases and invasive species have contributed to declines in sea grasses. Other factors such as the parasites MSX (*Haplosporidium nelsoni*) and Dermo (*Perkinsus marinus*), overharvesting, and loss of habitat are contributing to declines in Eastern Oyster (*Crassostrea virginica*) populations.

Programs have been implemented in the MAB to help improve water quality, prevent overfishing and reopen spawning grounds. Restoration of SAVs benefits through nitrogen removal (e.g., Reynolds et al. 2016) and oyster restoration benefits by helping to stabilize and raise shorelines (e.g., La Peyre et al. 2014). Organizations such as the Maryland Department of Natural Resources, Chesapeake Bay Foundation, and the Oyster Restoration Partnership are working to re-establish once flourishing SAV beds and oyster reefs. The reduced loading of nutrients is now contributing to a marked expansion of SAV beds throughout the Chesapeake Bay. Table 15.1 provides an example list of selected rivers and SAV species that are responding well to restoration efforts. Restoring native oysters, planting underwater grasses, and planting trees and stream buffers helps to restore the Bay's natural filters. For example, eelgrass (*Zostera marina*) with their long, ribbon-like leaves

Approximately 20 species of Chesapeake Bay marine and freshwater SAVs grow into impressive beds and provide important ecosystem services such as habitat and protection for fish, crustaceans, and mollusks. In recent years, SAVs have made an unprecedented recovery owing to restoration efforts. Some selected examples by river and species are listed below

Tributary	Common Name	Scientific Name
James River	eelgrass	*Zostera marina*
Magothy River	widgeon grass	*Ruppia maritima*
Patapsco River	wild celery	*Vallisneria americana*
Patuxent River	eelgrass	*Zostera marina*
Potomac River	eelgrass	*Zostera marina*
Rappahannock River	eelgrass	*Zostera marina*
Severn River	widgeon grass	*Ruppia maritima*
South River	sago pondweed	*Stuckenia pectinata*
Susquehanna River	wild celery	*Vallisneria americana*
York River	eelgrass	*Zostera marina*

that grow in the saltier waters of the middle and lower Chesapeake Bay, absorb CO_2 through photosynthesis. As generations of oysters settle on top of each other, they sequester carbon from the water column as they form calcium carbonate shells. The benefit of restored reefs is structured habitat for many fish species and crabs and a natural water-cleansing system, filtering the estuary's entire water volume of excess nutrients on the order of weeks rather than today's rate of more than a year. Restored SAV beds and oyster reefs help to prevent ocean acidification since they are carbon sinks that absorb carbon dioxide.

15.6 Chesapeake Bay Management

Many government agencies and industries are working to safeguard the Chesapeake Bay. As an example, the Environmental Protection Agency (EPA), the National Oceanic and Atmospheric Administration (NOAA) and the Department of Defense spend millions for restoration projects in the Chesapeake Bay watershed. EPA programs include (1) air and water pollution control, (2) toxic substances, pesticides and drinking water regulation, (3) wetlands protection, (4) hazardous waste management, (5) hazardous waste site cleanup, and (6) some regulation of radioactive materials. Activities include compliance and enforcement, inspection, engineering reviews, ambient monitoring, analysis of environmental trends, environmental planning, pollution prevention, risk assessment, and education and outreach. NOAA provides services ranging from protecting lives and property through the distribution of meteorological, hydrographic, and oceanographic information to evaluating the status and trends in stressed fisheries such as Menhaden (*Brevoortia tyrannus*). Military bases, such as the United States Army's Aberdeen Proving Grounds and the United States Navy's Naval Surface Warfare Center, Dahlgren Division, have exemplified principles of stewardship through compliance with the National Environmental Policy Act of 1969. This Act, which is generally called "NEPA," requires U.S. Federal agencies to consider environmental effects that include, among others, impacts on social, cultural, and economic resources, as well as natural resources.

Since 1983, governors from Maryland, Virginia, and Pennsylvania, along with the mayor from the District of Columbia have sponsored a Chesapeake Bay Program that reflects a shared vision for the restoration and protection of the Chesapeake Bay, one of our nation's most wonderful natural resources. The Chesapeake Bay Program is officially managed by EPA and staff from a number of federal and state agencies, non-profit organizations, and academic institutions that comprise the Chesapeake Bay Program Office. Restoration efforts sponsored by the Chesapeake Bay Program have shown that the development of intertidal oyster reefs near shorelines protects against erosion and could possibly help keep up with sea level rise. These kind of living shoreline techniques are also being applied by others in the MAB region to help stabilize coastlines, improve water quality, and reduce recurring floods and erosion. This type of green infrastructure utilizes natural materials such as plants, sand, or rock and possibly offshore sills or reefs to stabilize the shoreline.

15.7 Virginia Beach to Cape Lookout

Coastal lagoons, such as Long Island and Pamlico Sound that are connected to the sea through inlets, are common along the MAB. Numerous authors (Barnes 1980; Kjerfve and Magill 1989; FitzGerald et al. 2008; Carrasco et al. 2016) have described the formation of these estuaries as the result of rising sea level during the Pleistocene or Holocene epoch and the building of coastal barriers by marine processes. Sea level rise may contribute to the conversion of marshes to open water and increasing flows through coastal inlets. These inlets provide important commercial and military navigation links and may be spanned by bridges or crossed by ferry. These shallow estuaries are productive ecosystems and are a vital nursery area for many species from different taxa. They provide recreational opportunities for the nation and assets for the economic strength of coastal communities. Lagoons are also vulnerable to anthropogenic influences such as increased nutrient loads because of land use change, alterations to freshwater inflows and modification to tidal flushing regimens. On longer time scales, sea level rise causes changes in the exchange of water occurring between the shallow lagoon and the sea because of tides, river flow, wind, and waves. This contributes to sand sequestration in tidal deltas and erosion of adjacent barrier shorelines.

Research sites such as the Virginia Coast Reserve (VCR) provide long-term data to study coastal lagoons that extend for 110 km (63 mi.) along the Atlantic shore of the Delmarva Peninsula (e.g., Callahan 1984; Hayden et al. 1991). Erosion along coastal lagoon shorelines that includes barrier islands along what is now part of the VCR required the decommissioning and demolishment of Hog Island Light in 1948. The site where the lighthouse once stood is now nearly 1.6 km (1 mile) offshore. The VCR Long Term Ecological Research (LTER) program includes an archive of historic information, long-term data sets from sensors that provide access to real-time observations, various types of high-resolution imagery, experimental data on marshes, coastal bays and barrier islands, and numerical model output. Fundamental research at the VCR has included the integration of these data to achieve new levels of understanding on ecological systems such as SAV beds, salt marshes, and barrier islands and processes such as shoreline retreat and resilience to climate change. These data support research that has been focused on succession, disturbance, and system-state change and helps address management questions about what resources should to be protected, saved or conserved. VCR LTER contributions include large seagrass restoration and monitoring projects that can be applied by others (Orth and McGlathery 2012; McGlathery et al. 2012; Moore et al. 2014). The VCR LTER has also demonstrated the importance of educating society on climate changes through outreach and education activities that have included the full spectrum of students from elementary school to university to life-long learners.

The North Carolina Outer Banks consist of low, sandy barrier islands including Cape Hatteras National Seashore, Cape Lookout National Seashore, and Pea Island National Wildlife Refuge. The origin of the Outer Banks is likely to have been from coastal deposition and erosion along the drainage divides between late Pleistocene river valleys. This would be the geological epoch with repeated glaciations which lasted from about 2,588,000 to 11,700 years ago, also commonly called the Ice Age. The barrier island segments along the Outer Banks provide the seaward boundary of a system of large, ecologically important sounds. The Pamlico Sound is connected to the north with Albemarle Sound through passages provided by the Roanoke Sound and Croatan Sound. Core Sound is located at the Pamlico Sound's narrow southern end. Today, it is fed by the Neuse and Pamlico Rivers from the west and by the Atlantic Ocean through Oregon, Hatteras, and Ocracoke Inlets. Inman and Dolan (1989) report shoreline retreat rates along the Outer Banks to be around 1.58 m/year (5.2 ft/year). Sea level rise has caused significant loss of both wetlands and uplands at the estuarine water-land interface. Further, wave climate and longshore transport causes these barrier island inlets to migrate in a southerly direction. The process involves development of distal bar through deposition and erosion of the downdrift shore owing to the longshore current (Nichols and Williams 2009).

The cuspate forelands that comprise the Carolina coast include three major capes, Cape Hatteras to the north, Cape Lookout, and Cape Fear to the south. The formation of these capes has been attributed to the Gulf Stream, wave climate, and longshore currents. The barrier islands along the North Carolina coast are separated by coastal inlets. Most of the tidal inlets form as a result of severe storms and the passages remain open owing to the to-and-fro motion of tidal currents. Tidal inlets have played a major role in the development and maintenance of the Outer Banks. Along the Banks, three to eleven tidal inlets have cut through the barriers islands over the historical record (Stick 1958). These tidal inlets provide a connection between the ocean and bays, lagoons, marshes, and tidal creek systems. Passages such as Oregon Inlet are very important for species that travel from the ocean to estuarine areas and back and navigation by vessels from the sounds to the coastal ocean supports maritime commerce. The North Carolina Department of Transportation runs a ferry system to cross Currituck and Pamlico Sounds, and also the Cape Fear, Neuse, and Pamlico Rivers. When bridges are damaged due to storms such as Hurricanes Irene and Sandy, the ferry system provides a crucial transportation link between the barrier islands and the mainland.

Coastal inlets provide a migration path for fish larvae spawned on the continental shelf that need to return to nursery areas in the coastal lagoons. Wind drift currents facilitate the shoreward transport of fish larvae to the littoral zone, where longshore currents and associated breaking waves should facilitate their movement toward the inlets. Fry then aggregate near the inlet plume front as a result of stimuli such as olfactory cues and high turbidity until they are transported inland on flood tides. Subtidal frequency flood currents are known to jet oceanic water laden with young fish into the Pamlico Sound estuary via Oregon Inlet in related pulses. These exaggerated flood currents are caused by the set-up of water on the ocean side of the barrier islands, which also helps to retain the fry in the sounds. The hydraulic currents resulting from these seaward pressure heads that form during periods of favorable longshore winds drives water from the ocean to the sounds.

Inlets are also regions of mixing for freshwater from land drainage and salty ocean waters. Inlets that close in response to sediment transport may reduce the salt content for organisms living in the back bay. Estuarine organisms have different tolerances and responses to salinity changes. While species such as Eastern Oysters, Bay Scallops (*Argopecten irradians concentricus*) and Blue Crabs (*Callinectes sapidus*), can tolerate some change in salinity, decreased salinities beyond an acceptable range will negatively affect their growth and reproduction, and ultimately, their survival. Further, sea level rise has caused significant loss of wetland and upland habitats at the estuarine water-land interface.

Fishermen from Manteo and Wanchese, two towns on Roanoke Island, navigate through Oregon Inlet to reach fishing grounds on Diamond Shoals, also known as the Graveyard of the Atlantic. The Herbert C. Bonner Bridge, which spans Oregon Inlet, was built in 1963 and is being replaced with a new bridge to better withstand environmental loads (wind, waves, currents) and provide improved options for navigation under the bridge. For many years, watermen and local government officials lobbied for construction of twin jetties that would theoretically stabilize the inlet by blocking sand traveling along the shoreline from entering the inlet. In 1970, Congress approved a $108 million jetty project, but failed to provide construction funds. However, the Departments of the Interior and Commerce both maintained that the twin jetty project should be rejected in favor of dredging alternatives owing to environmental impacts. For example, the National Marine Fisheries Service opposed the jetties because of unresolved issues regarding the blocking of migration routes for marine fish larvae that are carried by natural currents through the inlet to estuarine nursery areas. In 2002, the White House Council on Environmental Quality announced that the jetty project was not warranted. Instead, the federal government promised Dare County and North Carolina that maintenance dredging would be accomplished to maintain a navigable tidal inlet. Navigation through Oregon Inlet, North Carolina is only recommended during good weather and calm seas.

Wave climate contributes to the southward migration of inlets found in North Carolina's Outer Banks, which is a process vital to the health of the barrier islands. Breaching of barriers by tidal inlets allows for the buildup of flood shoals on the bay side of a barrier island that increases the width of the island and provides a platform for further increase in the elevation of the barrier island superstructure and source for sand dune generation. When left to nature, tidal inlets open, migrate, add sediment to the barrier system, and then eventually close as new inlets open at other locations and capture the tidal drainage. When inlets are stabilized, as in the case of Rudee Inlet, Virginia, unintended consequences may occur, including prolonged interruption of littoral sediment supply and erosion of downdrift beaches. Rudee Inlet is located about 122 km (76 miles) to the north of Oregon Inlet.

15.8 With an Eye Toward Marine Policy

The rising population in the coastal zone is a modern phenomenon, which requires the reduction of coastal risks. Today there are many more hotels, restaurants, and shopping centers located along the shoreline than during the 1950s. This development has not always considered the impact of sea level rise and recurring storm events, which may cause major damage. Without doubt, the first step must be for the scientific community to raise awareness of coastal vulnerabilities and the value of science in understanding it. Also important is developing a broad appreciation of the notion of resilience to improve development patterns and principles. Organizations such as the Southeastern Universities Research Association (SURA) have promoted a coastal resilience workshop series to focus on essential elements such as a quantitative understanding of local erosion rates, predictions of shoreline positions for various sea-level rise scenarios, costs to maintain the shoreline in a fixed position, and increased vulnerability of beaches to storms with a rising sea level (Wright et al. 2016). Further, SURA has been instrumental in managing the development of the Coastal and Ocean Modeling Testbed (Luettich et al. 2013). This testbed has been developed to assess state-of-the-art numerical models of storm surge, wave, and coastal flooding which are needed to provide hindcasting of past hurricanes, provide real-time forecasting of current hurricanes and assess future flood risks. One recent advance in numerical modeling that has been demonstrated by this testbed is the coupling of wave models such as Simulating WAves Nearshore or SWAN that computes random, short-crested wind-generated waves with traditional models used by NOAA, such as the Sea, Lake and Overland Surges from Hurricanes (SLOSH) model that was developed by the National Weather Service to estimate storm surge heights resulting from historical, hypothetical, or predicted hurricanes. Luettich et al. (2017) demonstrated that an ensemble mean of multiple models has a higher skill predicting hypoxia than any individual model in the Chesapeake Bay. Since there are numerous numerical models with different model physics and numerics, a testbed helps organizations such as NOAA identify the best approach to ensuring a weather-ready nation.

Dynamic coastal environments can change dramatically across a range of temporal and spatial scales in response to natural processes and human modifications. Rebuilding expensive structures and homes in vulnerable areas such as flood plains located along the MAB is not sustainable. Current flood insurance policies are not in synchronization with environmental factors such as current and projected sea level rise (Dennis 2017). Today's policy makers have competing interests as they work to recover from coastal inundation, and hyperbolic and partisan rhetoric has done little to foster free public discussion of the issues without rancor and finger pointing. Short term economic gains are a critical factor in decisions as to whether a community should restore and rebuild what has been damaged by shoreline retreat rates rather than to cede to the erosion that even faces historic landmarks such as the Sankaty Head and Cape Hatteras lighthouses. However, in the case of these lighthouses, recurring nor'easters and hurricanes put these structures in imminent jeopardy of being beyond the municipality's fiscal and engineering capabilities to save. As climate change causes more intense storms, surges, and higher sea levels in areas such as the Chesapeake Bay (Boesch et al. 2013), coastal communities can respond by (1) mapping floodplains and identifying land areas less than 1.5 m (5 ft) above mean sea level, (2) identifying impacts on structures occurring during spring tides (e.g., how high the water, waves and debris reach), (3) controlling storm-water runoff to prevent erosion (e.g., plant native plants to reduce erosion and flash floods), and (4) ensuring a resilient shoreline that acts as a buffer from storm surge and waves (e.g., living shorelines that include tidal wetland vegetation). Tomorrow's coast will require new flood insurance policies with provisions to buy and destroy or move properties that are associated with repetitive claims.

References

Barnes, R.S.K. 1980. *Coastal lagoons*. Cambridge, UK: Cambridge University Press.

Benchley, R., and R.D. Felch. 2009. *Keeping the light: The epic move and preservation of Nantucket's Sankaty head lighthouse*, 2009. Siasconset, MA: The Sconset Trust.

Blankenship, Karl. 2004, October 1. Floodwaters from hurricane Ivan Trash Upper ChesapeakeBay, *Bay Journal*. Available online. http://www.bayjournal.com/article/floodwaters_from_hurrican_ivan_trash_upper_chesapeake_bay. Accessed July 26, 2017.

Blumberg, Alan F., Nickitas Georgas, Larry Yin, Thomas O. Herrington, and Philip M. Orton. 2015. Street-scale modeling of storm surge inundation along the New Jersey Hudson Waterfront. *Journal of Atmospheric and Oceanic Technology* 32: 1486–1497.

Boesch, Donald F., Larry P. Atkinson, William C. Boicourt, John D. Boon, Donald R. Cahoon, Robert A. Dalrymple, Tal Ezer, Benjamin P. Horton, Zoë P. Johnson, Robert E. Kopp, Ming Li, Richard H. Moss, Adam Parris, Christopher K. Sommerfield. 2013. *Updating Maryland's Sea level rise projections*. Special Report of the Scientific and Technical Working Group to the Maryland Climate Change Commission. Cambridge, MD: University of Maryland Center for Environmental Sciences.

Callahan, James T. 1984. Long-term ecological research. *BioScience* 34 (6): 363–367.

Carrasco, A.Rita, Óscar Ferreira, and Dano J.A. Roelvink. 2016. Coastal lagoons and rising sea level: A review. *Earth-Science Reviews* 154: 356–368.

Chesapeake Research Consortium, Inc. 1976. *The effects of tropical storm Agnes on the Chesapeake Bay estuarine system.* CRC Publication No. 54, Baltimore, MD and London, UK: The Johns Hopkins University Press.

Cooper, Dick. 2016. The Eastern shore needs to plan for rising sea level. *Tidewater Times* 65 (1): 27–28. 30, 32, 34, and 36.

Cronin, William B. 2005. *The disappearing Islands of the Chesapeake*. Baltimore, MD: The Johns Hopkins University Press.

Curtin, Philip D., Grace S. Brush, and George W. Fisher. 2001. *Discovering the Chesapeake*. Baltimore, MD: The Johns Hopkins University Press.

Dennis, Brady (2017, July 16). The country's flood insurance program is sinking. Rescuing it won't be easy. *Washington Post*. Available online. Available online: https://www.washingtonpost.com/national/health-science/the-countrys-flood-insurance-program-is-sinking-rescuing-it-wont-be-easy/2017/07/16/dd766c44-6291-11e7-84a1-a26b75ad39fe_story.html?utm_term=.c8ed0d4976dd. Accessed July 17, 2017.

Eggleston, Jack and Jason Pope, 2013. Land subsidence and relative sea-level rise in the southern Chesapeake Bay region: U.S. Geological Survey Circular 1392, 30pp. http://dx.doi.org/10.3133/cir1392.

Ezer, Tal, Larry P. Atkinson, William B. Corlerr, and Jose L. Blanco. 2013. Gulf stream's induced sea level rise. *Journal of Geophysical Research: Oceans* 118 (2): 685–697.

FitzGerald, Duncan M., Michael S. Fenster, Britt A. Argow, and Ilya V. Buynevich. 2008. Coastal impacts due to sea-level rise. *Annual Review of Earth and Planetary Sciences* 36: 601–647.

Hayden, Bruce P., Raymond D. Dueser, James T. Callahan, and Hank H. Shugart. 1991. Long-term research at the Virginia Coast Reserve: Modeling a highly dynamic environment. *BioScience* 41 (5): 310–318.

Horowitz, Arthur J., Kent A. Elrick, James J. Smith, and Verlin C. Stephens. 2014. The effects of hurricane Irene and tropical storm Lee on the bed sediment geochemistry of U.S. Atlantic Coastal Rivers. *Hydrological Processes* 28: 1250–1259.

Inman, Douglas L., and Robert Dolan. 1989. The outer banks of North Carolina: Budget of sediment and inlet dynamics along a migrating barrier system. *Journal of Coastal Research* 5 (2): 193–237.

IPCC. 2007. Summary for policymakers. In: Climate change 2007: The physical science basis. In *Contribution of Working Group I to the Fourth Assessment Report of the Intergovernmental Panel on Climate Change* [Solomon, Susan, Dahe Qin, Martin Manning, Zhenlin Chen, Melinda Marquis, Kristen B. Averyt, Melinda M. B. Tignor, and Henry L. Miller, Jr. (eds.)]. Cambridge, United Kingdom and New York, NY, USA: Cambridge University Press.

Kjerfve, Björn, and Karen E. Magill. 1989. Geographic and hydrographic characteristics of shallow coastal lagoons. *Marine Geology* 88: 187–199.

La Peyre, Megan K., Austin T. Humphries, Sandra M. Casas, and Jerome F. La Peyre. 2014. Temporal variation in development of ecosystem services from oyster reef restoration. *Ecological Engineering* 63: 34–44.

Langland, Michael, and Thomas Cronin (eds.). 2003. *A summary report of sediment processes in Chesapeake Bay and watershed, water-resources investigations report 0304123*. New Cumberland, PA: U.S. Geological Survey.

Luettich, R.A., Jr., L.D. Wright, C.R. Nichols, R. Baltes, M.A.M. Friedrichs, A. Kurapov, A. van der Westhuysen, K. Fennel, E. Howlett. (2017). A test bed for coastal and ocean modeling, *Eos*, 98. https://doi.org/10.1029/2017EO078243.

Luettich Jr., Richard A., L. Donelson Wright, Richard Signell, Carl Friedrichs, Marjorie A.M. Friedrichs, John Harding, Katja Fennel, Eoin Howlett, Sara Graves, Elizabeth Smith, Gary Crane, and Rebecca Baltes. 2013. Introductions to special section on The U.S. IOOS Coastal and Ocean Modeling Testbed. *Journal of Geophysical Research: Oceans* 118: 6319–6328.

McGlathery, Karen J., Laura K. Reynolds, Luke W. Cole, Robert J. Orth, Scott R. Marion, and Arthur Schwarzshild. 2012. Recovery trajectories during state change from bare sediment to eelgrass dominance. *Marine Ecology-Progress Series* 448: 209–221.

Moore, Kenneth A., Erin C. Shields, and David B. Parrish. 2014. Impacts of varying estuarine temperature and light conditions on Zostera marina (eelgrass) and its interactions with Ruppia maritima (widgeongrass). *Estuaries and Coasts* 37 (1): S20–S30.

Nichols, C. Reid., Robert G. Williams. 2009. *Encyclopedia of marine science* New York: Facts on File, Inc.

Orth, Robert J., and Karen J. McGlathery. 2012. Eelgrass recovery in the coastal bays of the Virginia Coast Reserve. *Marine Ecology-Progress Series* 448: 173–176.

Pritchard, Donald W. 1951. The physical hydrography of Estuaries and some applications to biological problems. In *Transactions of the Sixteenth North American Wildlife Conference*, pp. 368–376.

Pritchard, Donald W. 1952a. *A review of our present knowledge of the dynamics and flushing of Estuaries*. Chesapeake Bay Institute of The Johns Hopkins University, Technical Report 4, Reference 54–7, 45pp.

Pritchard, Donald W. 1952b. Salinity distribution and circulation in the Chesapeake Bay Estuarine system. *Journal of Marine Research* 11 (2): 106–123.

Pritchard, Donald W. 1955. Estuarine circulation patterns. Proceedings, American Society of Civil Engineers, Vol. 81, Separate No. 717, pp. 717–1 to 717–11.

Reynolds, Laura K., Michelle Waycott, Karen J. McGlathery, and Robert J. Orth. 2016. Ecosystem services returned through Seagrass Restoration. *Restoration Ecology* 24 (5): 583–588.

Schulte, David M., Karin M. Dridge, Mark H. Hudgins. 2015. Climate change and the evolution and fate of the Tangier Islands of Chesapeake Bay, USA. *Scientific Reports*, Vol. 5, Article No. 17890; doi: 10.1038.

Stick, D. 1958. *The outer banks of North Carolina, 1584–1958*. Chapel Hill, NC: UNC Press.

Smith, Pam (2014, October 13). Hurricane Hazel 60 years later. *Coastal Review Online*. Available online: https://www.coastalreview.org/2014/10/6041/. Accessed July 26, 2017.

Sweet, William V., Robert E. Koop, Christopher P. Weave, Jayantha Obeysekera, Radley M. Horton, E. Robert Thieler, Chris Zervas. 2017. *Global and regional sea level rise for the United States*. NOAA Technical Report NOAA CO-OPS 083, Silver. Spring, MD: NOAA.

Wang, Harry V., Jon D. Loftis, David Forrest, Wade Smith, and Barry Stamey. 2015. Modeling storm surge and inundation in Washington, DC, during hurricane Isabel and The 1936 Potomac River Great Flood. *Journal of Marine Science and Engineering* 3: 607–629.

Ward, L. G., Peter S. Rosen, P.S. William J. Neal, Orrin H. Pilkey, Jr., Orrin H. Pilkey, Sr., Gary L. Andersohn, Stephen J. Howie. 1989. *Living with Chesapeake Bay and Virginia's Ocean Shores*. Durham, NC: Duke University Press.

White, Christopher P. 1989. *Chesapeake Bay: Nature of the estuary, a field guide*. Centreville, MD: Cornell Maritime Press/Tidewater Publishers.

Wright, L. Donelson, C. Reid Nichols, Arthur G. Cosby, Christopher F. D'Elia. 2016. Collaboration to enhance coastal resilience. *Eos, Transactions American Geophysical Union* 97. Available online: http://bit.ly/2bwBwrU.

Chapter 16
The Alaskan Arctic Coast

Lynn Donelson Wright

There are two kinds of Arctic problems, the imaginary and the real. Of the two, the imaginary are the most real.
—Vilhjalmur Stefansson, *The Arctic in Fact and Fable*

16.1 Melting Ice, Vanishing Ecosystem

In 1845, Captain Sir John Franklin and his crew of 129 sailed westward from England on two ships, the HMS *Erebus* and the HMS *Terror*, in search of a Northwest Passage from Europe to Asia over the top of North America. Franklin and his ships were last seen in Baffin Bay. Some time shortly thereafter, they became icebound and perished near King William Island in the Canadian Arctic. Today, an ice-free Northwest Passage is becoming a reality as the area of summer open water in the Beaufort and Chukchi Seas (Fig. 16.1) continues to grow. As discussed in Chap. 2, Sect. 2.9, the IPCC predicts that for the "worst case" warming scenario, the Arctic Ocean could be entirely ice free in September 2050 (IPCC 2013) or possibly as early as 2030 (AMAP 2017). While this may be good from the perspective of navigation, it is bad from the perspectives of the unique Arctic ecosystem and Native Alaskan subsistence, health and culture. These assets are tightly bound to, and dependent on, a frozen ocean and frozen permafrost on land. The media tends to emphasize the impact of ice melting on Polar bears. The bears are certainly threatened; but the total environmental and human impacts are of equal concern. A comprehensive assessment of the impacts of climate change on all environmental and human aspects of the Arctic realm is offered by ACIA (2005) and most recently by the Arctic Monitoring and Assessment Program (AMAP 2017) in their latest report on *Snow, Water, Ice and Permafrost in the Arctic* (*SWIPA*). The SWIPA report presents a clear and compelling explanation of the

L. D. Wright (✉)
Southeastern Universities Research Association (SURA), Washington, DC, USA
e-mail: ldwright@bellsouth.net

L. D. Wright and C. R. Nichols (eds.), *Tomorrow's Coasts: Complex and Impermanent*, Coastal Research Library 27,
https://doi.org/10.1007/978-3-319-75453-6_16

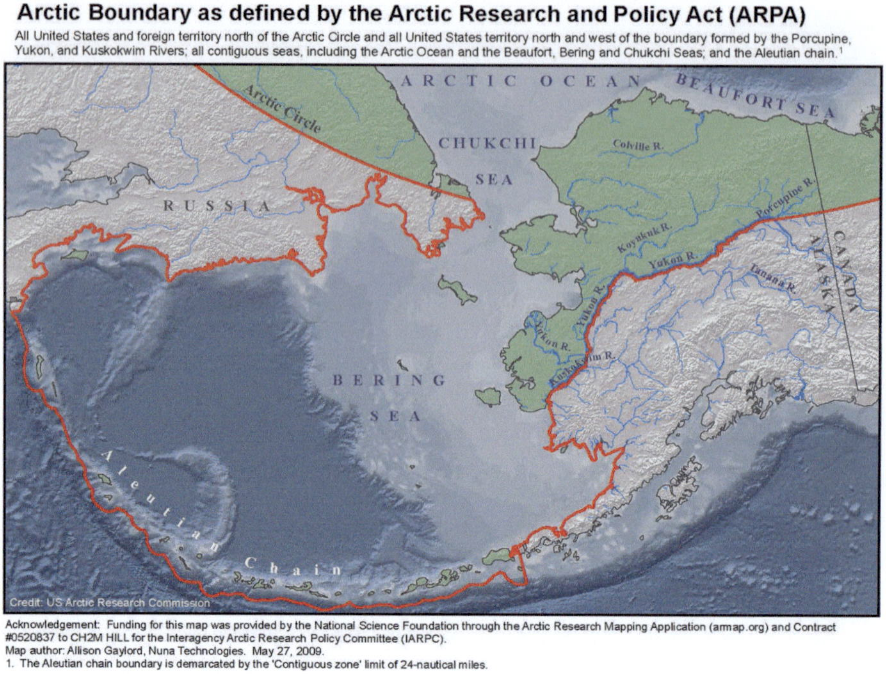

Fig. 16.1 The Alaskan Arctic (red border) as delineated by the Arctic Research and Policy Act (ARPA). *Source* U.S. Arctic research commission

complex feedbacks, triggered by global warming, that are causing rapid and dramatic shifts, not only in the Arctic climate, but in all facets of the Arctic environment including human aspects.

16.2 The Most Rapid Rate of Warming Is in the Arctic

The polar regions of the earth are distinguished from the lower latitudes by two things: (1) the most prominent temporal variations in solar radiation (daylight) are annual rather than diurnal (as in the tropics); and (2) extreme annual temperature variations with very cold winters. The first thing has not, and will not change because it has nothing to do with climate change. The second factor is changing dramatically. Recent studies indicate that near-surface air temperatures in the Arctic are rising 2–3 times faster than are temperatures elsewhere on the earth's surface; this anomalous rate of warming is referred to as the *Arctic amplification* (Wendisch et al. 2017). Sainato and Skojec (2017) conclude in a recent synthesis of results from numerous scientists that the Arctic has been warming twice as fast as the rest of the planet for the past 50 years. Figure 16.2 illustrates the recorded temperature

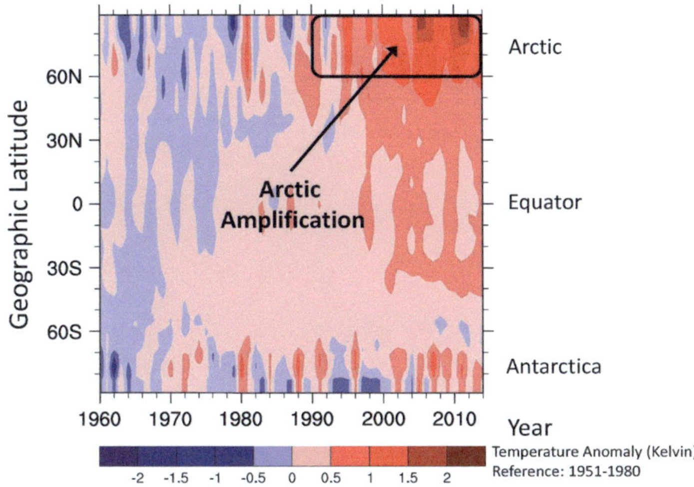

Fig. 16.2 Time series of near surface temperature anomalies as functions of year and latitude. The "Arctic Amplification", which has been growing since the early 1990s is evident in the upper right. From Wendisch et al. (2017)

anomaly since 1960 by year and latitude. Pronounced increases in temperature since the year 2000 at latitudes above 60°N are particularly evident. The corresponding anomalies in the Antarctic are much less dramatic.

The reasons for the Arctic Amplification are understood to be fairly straightforward involving one strong and multiple weaker positive (self reinforcing) feedback loops (Fig. 16.3). The strongest feedback mechanism is triggered by global warming but is accelerated by decreasing albedo (reflection of solar radiation), which in turn causes further melting of snow and ice leading to an even greater decrease in albedo. This processes initiates other feedbacks including increased marine biologic activity, which further increases absorption of solar radiation in ice-free regions of the Arctic Ocean. The impacts of this accelerated warming are profound and are rapidly transforming the Arctic ecosystem and the socioeconomic underpinnings of Native American communities.

16.3 Reduced Sea Ice Area and Thickness

Sea ice does much more than restrict navigation. It provides a "highway" for native peoples and the wildlife that they rely on for subsistence, pack ice limits the fetch over which winds can generate destructive waves and shore-fast ice protects the delicate Arctic shores from erosion. The pronounced melting of Arctic sea ice in the Northern Hemisphere summer (Fig. 16.4) is among the more dramatic manifestations of global warming. Model predictions by the U.S. Geological Survey suggest that by late this century, the Chukchi Sea will be ice free from July through October

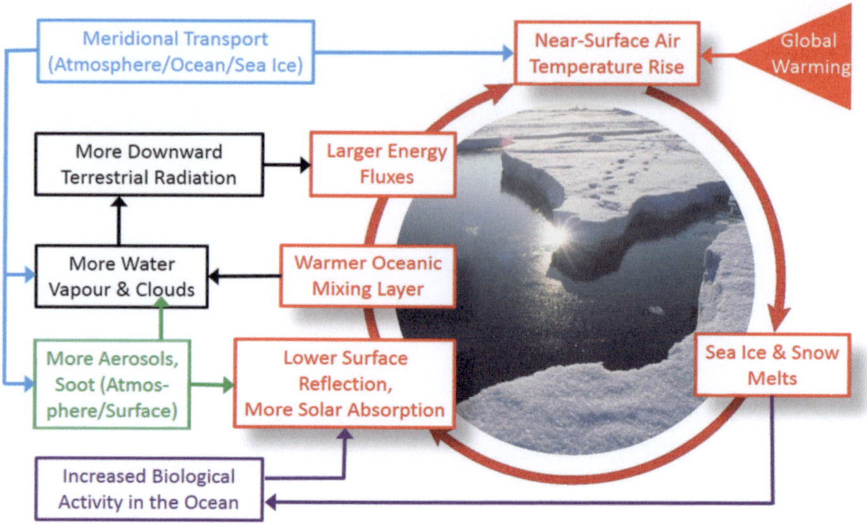

Fig. 16.3 Positive feedback loops that contribute to the Arctic amplification. Reduced albedo and consequent ocean warming leading to snow and ice melting is the most prominent feed back loop. From Wendisch et al. (2017)

and the ice-free period in the lower latitude Bering Sea will increase from today's 5.5 months to 8.5 months (Douglas 2010). In addition to decreases in ice-covered area, the thickness of the ice is also decreasing, as is the age of the ice as described in Chap. 2 (Fig. 2.8). It is worth noting that even though IPCC models are predicting significant reductions in sea ice throughout this century, ongoing observations suggest that the models may be under predicting ice melt (Fig. 2.9). Figure 16.5 shows the extent of summer sea ice in September 2016 in comparison with the September median extent over the period 1981–2010. The significant increase in open water area separating the edge of the ice pack from the Beaufort and Chukchi Sea coasts is apparent.

NOAA, in the U.S. Climate Resilience Toolkit on *Alaska and the Arctic*, states the following: "Arctic change is important to global society not only because of its impacts within the Arctic, but also because of the role of the Arctic in the climate of the entire Earth. The climate and the ocean circulation in all parts of the globe are affected by the temperature and pressure gradients between the poles and the equator. These temperature and pressure gradients are becoming weaker because the Arctic is warming faster than lower latitudes. The resulting changes in climate and weather will affect people everywhere." Locally, within the Arctic region of Alaska, the consequences are immediate, profound and accelerating. Harmful impacts include rapid erosion of shores, ecosystem degradation, wildlife declines, fisheries declines and detrimental effects on the livelihoods and societal traditions of the native Alaskans who must now find new ways to adapt to declining opportunities for subsistence.

Fig. 16.4 Melting sea ice in the Arctic summer. *Image Credit* U.S. Bureau of Ocean Energy Management. Public domain

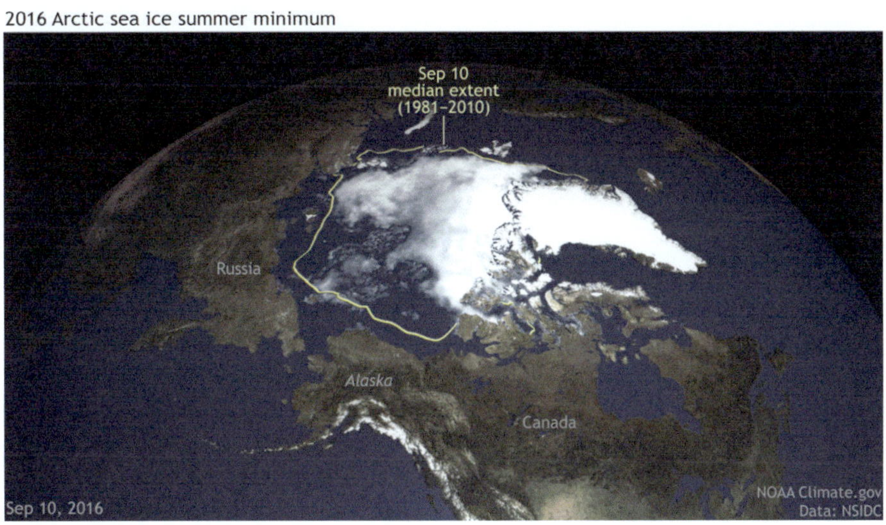

Fig. 16.5 Extent of sea ice in the Arctic Ocean in September 2016 From NOAA Climate.Gov. *Source* URL (modified on 2016-09-15 13:45): https://www.climate.gov/news-features/event-tracker/arctic-seaice—Author Michon Scott National Snow and Ice Data Center Link https://www.climate.gov/news-features/category/climate-change-global-warming

16.4 Physical Process Regime

The Arctic Ocean covers an area of about 14 million km^2 (5.46 million mi^2) and a little more than half of this area is occupied by the continental shelf, a much greater fraction than in any other of the world's oceans (Huntington and Weller 2005). This ocean is surrounded by the landmasses of Eurasia and North America and is almost isolated from the other oceans. Most of the water within the Arctic Ocean enters from the North Atlantic and is relatively warm (Loeng et al. 2005). The prevailing exchange between the North Atlantic and the Arctic Ocean is a crucial element of the thermohaline (i.e. related to density differences caused by temperature and salinity) circulation. Weller et al. (2005) point out that salinity reductions caused by IPCC projections of increased runoff of freshwater could, in the future, decrease or even shut down, the North Atlantic-Arctic Ocean thermohaline exchange. This could actually lead to cooling certain regions of the Arctic Ocean, partially offsetting projected temperature increases. General circulation in the central, polar portion of the Arctic Ocean is characterized by a large-scale, clockwise, gyre, the Beaufort Gyre. A west-to-east flowing counter current dominates the regional transport over the shelf off the Chukchi and Beaufort Sea coast. Tides are negligible: the spring tide range at Utqiaġvik (Pt. Barrow) Alaska is only 15 cm ($\sim$6 in.).

Ice cover prevents storms and high winds from generating waves and storm surges and has protected the sensitive shores of the Arctic coast for millennia. Although strong winds can cause a short wind chop by blowing over a small open water distance, relatively large distances of open water are required for the generation of large waves. The distance over which the wind blows is called "fetch" and increasing fetch allows the winds to generate larger waves. Wind speed is the other factor that controls the height of the waves. Overeem et al. (2011) show that between 1979 and 2009, the duration of the open water period more than doubled from 45 to 95 days per year. Although the fetch distance also increased, the model results of Overeem et al. (2011) suggested that wave heights are more strongly dependent on wind speeds than on fetch. The critical wave-generating wind speeds were considered by Overeem et al. (2011) to be those greater than 5 m s^{-1} (11.2 mi/h). The strongest winds occur in the fall.

Although the locally generated wind waves are only around 0.5–0.6 m (1.6–2.0 ft) high and surges are of similar height, these waves are very destructive in attacking the easily eroded permafrost shores which are susceptible to thermal erosion when waves "slosh" water at temperatures above freezing against them. Rising sea levels exacerbate the thermal erosion problem. Figure 16.6 shows the recent expansion of the open water period since 2001. The red band indicates summer expansion (earlier ice break up) and the blue band indicates autumn expansion (later ice freeze up). Note that the autumn expansion exceeds the summer expansion. The red curve shows the temporal variation in solar radiation from spring to autumn and the black curve shows the corresponding variation in sea

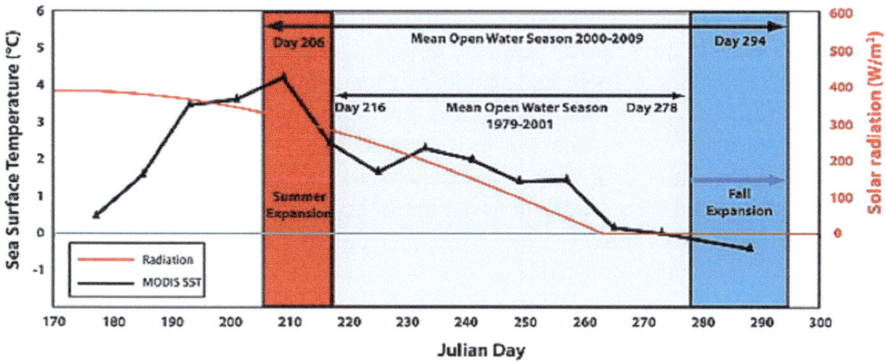

Fig. 16.6 Envelopes of the mean open water seasons in the Arctic Ocean for 1979–2001 and 2000–2009. From Overeem et al. (2011)

surface temperature. The summer expansion allows shores to be exposed to warmer water and the fall expansion allows greater exposure to high winds. Both increase thermal erosion.

16.5 The Marine Ecosystem

Three very prominent characteristics distinguish the marine ecosystems of the Arctic Ocean from those of other seas: (1) variations in solar radiation are seasonal rather than diurnal; (2) relatively shallow water shelf environments are strongly dominant; and (3) surface waters are very cold and are ice covered for much of the year. In the case of the first factor, solar radiation shuts down entirely during winter and prevails for 24 h in the summer. Consequently, primary productivity is light-limited especially during winter and during times of pack ice cover and thus varies seasonally. Increases in the ice-free period are considered likely to increase primary productivity (Loeng et al. 2005). Freshwater runoff from several rivers that flow northward from their catchments in the Brooks Range of Alaska during the thaw seasons of spring and summer provide nutrients, sediments and lower the salinities of shelf waters. As the ice recedes in summer, the edge of the ice pack becomes a zone of high plankton productivity and is particularly important in supporting seabirds, of which there are over 60 species (Huntington and Weller 2005).

Because of the seasonal limitation in solar radiation, winter ice cover and low primary productivity, biodiversity is low and thus the ecosystem is fairly simple in comparison with other marine ecosystems. Marine mammals are the dominant predators and are long lived. Included are several species of whales, seals, walruses and polar bears. These mammals are critical to the traditional subsistence economy and diet of native Alaskans. With the exception of whales, these mammals spend most of their lives on sea ice and are being adversely impacted by the retreat and thinning of the ice. More polar bears are coming ashore during summer and this is reducing their access to food and increasing their mortality (Atwood et al. 2016). Cox et al. (2017) report a dramatic increase in sightings of polar bears on Cooper Island on the Beaufort Sea coast in recent years.

Although the sub-polar Bering and Barents Seas are home to some of the world's most productive fisheries, the Beaufort and Chukchi Seas are much less important for their fisheries. Nevertheless, the Arctic Ocean has 150 species of fish (Murray 1998) of which capelin and polar cod are abundant in the Beaufort and Chukchi Seas.

16.6 The Changing Environment of the Alaskan North Slope

The Alaskan North Slope (Figs. 16.7 and 16.8) extends northwards from the Brooks Mountain Range to the Arctic Ocean and embraces the extensive tundra coastal plain which includes the National Petroleum Reserve-Alaska (NPRA) (centered on Pt. Barrow) and the Arctic National Wildlife Refuge to the east and adjacent to the Canadian Border. North of the Brooks range, the North Slope coastal plain is tundra comprised mostly of small berry plants and a relatively thin active ground layer of lichens and mosses that thaws in the summer and refreezes in winter. The tundra surface is underlain by permafrost, or permanently frozen ground. Today, the permafrost is undergoing summer melting at increasing rates and this is a source of greenhouse gases including carbon dioxide and methane. Numerous rivers, that arise from the Brooks Range, traverse this tundra coastal plain and leave behind river beds of relatively coarse, gravelly deposits and build

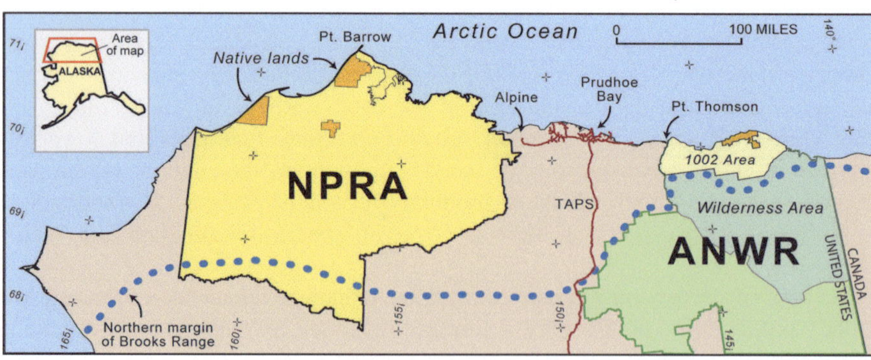

Fig. 16.7 Map of the Alaska North Slope showing the northern limit of the Brooks Range (dotted blue line) and the locations of the National Petroleum Reserve-Alaska (NPRA) and the Arctic National Wildlife Refuge (ANWR). *Source* USGS—http://energy.usgs.gov/images/alaska/NPRA_F1lg.gif

Fig. 16.8 False color satellite image of the Alaskan North Slope prior to summer breakup of ice. The pack ice is seen at the top of the image. The tundra surface is shown in green, the Brooks Range foothills are indicated by the reddish colors and mountain glaciers appear in light blue. Numerous rivers are also evident. Image from USGS Public domain. *Source attribution* NASA-http://earthobservatory.NASA.gov/newsroom

protruding deltas at the coast. Among the largest and most prominent of these rivers is the Colville River, which flows into the Beaufort Sea. The Colville is a rich wildlife habitat and supports large caribou herds, brown bears and polar bears, as well as large concentrations of many birds including peregrine falcons and golden eagles.

The North Slope is currently experiencing numerous rapid and significant changes. As described in Chap. 9, the shores bordering the Chukchi and Beaufort Seas are rapidly receding as a result of thermal erosion of the permafrost that has traditionally "cemented" the low coastal cliffs. Shore recession is only part of the overall degradation of tundra that is resulting from rising temperatures and may lead to the complete destruction of tundra in the decades ahead (Lawrence et al. 2008; Sherwonit 2010). This would represent a severe habitat loss for many species including caribou. Just as the lengthening duration of the ice free period in the Arctic Ocean is altering the marine ecosystem, lengthening of the snow-free period of the land surface is altering the tundra ecology, hydrology and biogeochemistry. Cox et al. (2017) explain how earlier snowmelt and later snowpack onset in autumn are affecting regional phenology, soil temperatures and fluxes of gases from the tundra in addition to the reduced albedo which is contributing to the *Arctic Amplification*. Underwood (2017) describes how melting glaciers are increasing the flows of Arctic rivers and the delivery of fresh water to the coast. Another important change in Arctic river hydrology and ecosystem dynamics involves rapid declines in river icings (Pavelsky and Zanetske 2017). River icings are relatively large and thick ice deposits caused by liquid (unfrozen) ground water flowing into rivers during winter. They cause considerable widening of riverbeds, which are important habitats for many species. Traditionally these features often survived the summer thaw but now they are disappearing in early summer causing serious loss of wildlife habitat.

16.7 Socioeconomic Impacts of Arctic Changes

As pointed out by NOAA (2017) in its *Climate Resilience Toolkit*, oil and gas production provides 80% of Alaska's state revenue and thousands of jobs. This industry may stand to benefit from increased ice-free area and duration, which will allow increased offshore operations. The second and third largest Alaskan industries are respectively mining and fisheries and tourism is close behind. The potential impacts of future changes on mining and tourism are unclear but the impact on fisheries is likely to be negative.

The most serious and damaging socioeconomic impacts of the unfolding changes will be felt by indigenous Alaskan people. Native Inuit, Athabaskan, Métis, and other indigenous inhabitants of Arctic Alaska have long understood that they have a delicate and complex symbiotic relationship with their natural realm and have depended on this relationship for their subsistence livelihood (Stepien et al. 2014). Weller et al. (2005) and NOAA (2017) describe some of the immediate impacts on native peoples who continue to rely heavily on hunting, fishing, trapping, reindeer herding and gathering for sustenance even though some are now also employed by the oil and gas and tourist industries. Reductions in the extent and thickness of the pack ice are making all of these activities more difficult and thinning ice is also making excursions on the ice more hazardous. Decreases in fish stocks and marine mammals are limiting the availability of traditional food sources.

There are also numerous other, but less conspicuous impacts. Among these are direct impacts on housing and village infrastructure. Melting permafrost causes loss of support for structures built on the tundra surface and shore retreat because of thermal erosion and increased wave energy is displacing villages. Some engineering solutions to permafrost thawing utilizing steel pilings are being explored (Poppick 2017) but these methods will probably remain unaffordable for most indigenous people for the near future. Disappearance of protective barrier islands is causing loss of critical wildlife habitat. As pointed out by NOAA (2017) most rural Arctic Alaskan communities have no access to roads or electric grids so limitations on traditional means of transport by boat, sled or overland traffic are quite consequential. Ice cellars traditionally used for storing food are thawing allowing food to spoil and thawing permafrost is contaminating ponds that are sources of drinking water. Climate change is also causing the spread of some diseases not previously present in the Arctic (NOAA 2017). Although resilience has long been a virtue shared by Native Alaskans, the accelerating degradation of their Arctic habitat is forcing many to explore costly relocation (ACIA 2005; Harball 2013). In the recent comprehensive report on the numerous changes taking place in the Arctic, AMAP (2017) stresses the urgency of finding new adaptation strategies supported by new and non-traditional government policies.

16.8 "Unraveling" a Complex Coastal System

In a recent commentary in *Nature* on the AMAP (2017) *SWIPA* report, Tollefson (2017) suggests that the report, to which over 90 scientists contributed, leads one to conclude that the Arctic is "unraveling". This is because of the complex interplay of numerous factors all of which are changing and bringing about mutual changes in the other interconnected factors through positive feedbacks. A few of the more prominent feedbacks as described in Sect. 16.2 of this chapter account for the rapid temperature rise related to the *Arctic Amplification.* There are numerous other feedbacks, however, and all are outcomes, directly or indirectly, of global warming. Some of the key findings of the SWIPA report are summarized as follows:

1. The Arctic Climate is shifting to a new state in which the Arctic Ocean could be entirely free of sea ice in summer by the 2030s and this could affect weather in the mid latitudes.
2. Arctic temperatures are rising faster than anywhere else on earth; the area and duration of snow cover is decreasing; permafrost is warming; and ecosystems are changing.
3. Changes will continue at least to mid century regardless of efforts to reduce greenhouse gases, Arctic temperatures will rise above late 20th century levels by 4–5 °C; and ecosystems will face serious stress and disruptions.
4. Substantial cuts in greenhouse gas emissions can stabilize further impacts after the middle of the century but not in the short term. Compliance with the Paris Agreement will stabilize snow and ice losses but there will still be significant reductions.
5. Effective adaptation policies at regional and local levels can reduce vulnerabilities of Arctic communities.
6. Effective mitigation and adaptation policies will require understanding and acceptance of the scientific realties by government entities and officials at many levels and these understandings will need to be supported by observations and models.

According to Morten Skovgård Olsen, who coordinated the 2017 SWIPA assessment and leads the Danish Ministry of Energy, Utilities and Climate's Arctic Programme: "The Arctic that you will have by mid-century will be very different from the Arctic that we see today".

References

ACIA. 2005. *Arctic climate impact assessment*, 1042. Cambridge University Press.

Arctic Monitoring and Assessment Program (AMAP). 2017. *Snow, water, ice and permafrost in the Arctic (SWIPA) summary for policy makers.*

Atwood, T.C., E. Peacock, M.A. McKinney, K. Lillie, R. Wilson, D.C. Douglas, S. Miller, and P. Terletzky. 2016. Rapid environmental change drives increased land use by an Arctic marine predator. *PLoS ONE* 11: e0155932. https://doi.org/10.1371/journal.pone.1055932.

Cox, C.J., et al. 2017. Drivers and environmental responses to the changing annual snow cycle of northern Alaska. *Bulletin of the American Meteorological Society*. https://doi.org/10.1175/BAMS-D-16-0201.1.

Douglas, D.C. 2010. *Arctic sea ice decline: Projected changes in timing and extent of sea ice in the Bering and Chukchi seas*, 32. U.S. Geological Survey Open-File Report 2010-1176.

Harball, E. 2013. Alaska Natives try to flee climate change impacts but find little help. *Climatewire*, 31 Jan 2013.

Huntington, H., G. Weller, et al. 2005. An introduction to the arctic climate impact assessment. In *ACIA, 2005. Arctic Climate Impact Assessment*, 1–19. Cambridge University Press.

Intergovernmental Panel on Climate Change (IPCC). 2013. *Climate change 2013: The physical science basis. Contribution of working group I to the fifth assessment. Report of the intergovernmental panel on climate change*, ed. Stocker, T.F., D. Qin, G.-K. Plattner, M. Tignor, S.K. Allen, J. Boschung, A. Nauels, Y. Xia, V. Bex, and P.M. Midgley. Cambridge: Cambridge University Press.

Lawrence, D.M., A.G. Slater, R.A. Tomas, M.M. Holland, and C. Deser. 2008. Accelerated Arctic land warming and permafrost degradation during rapid sea ice loss. *Geophysical Reseach Letters* 35: L11506. https://doi.org/10.1029/2008GL033985.

Loeng, H., et al. 2005. Marine systems. *In: ACIA, 2005. Arctic climate impact assessment*, 454–538. Cambridge University Press.

Murray, J.L. 1998. Ecological characteristics of the Arctic. *AMAP assessment report: Arctic pollution issues*, 117–140. Oslo: Arctic Monitoring and Assessment Programme.

NOAA. 2017. *U.S. climate resilience toolkit: Arctic peoples and ecosystems.*

Overeem, I., R.S. Anderson, C.W. Wobus, G.D. Clow, F.E. Urban, and N. Matell. 2011. Sea ice loss enhances wave action at the Arctic coast. *Geophysical Reseach Letters* 38: L17503. https://doi.org/10.1029/2011GL048681.

Pavelsky, T.M., and J.P. Zanetske. 2017. Rapid decline in river icings detected in Arctic Alaska: Implications for a changing hydrologic cycle and river ecosystems. *Geophysical Research Letters* 44 (7): 3228–3235.

Poppick, L. 2017. Engineering new foundations for a thawing Arctic. *Eos*, 98. https://doi.org/10.1029/2017EO080435.

Sainato, M., and C. Skojec. 2017. More than 90 scientists release report that arctic is 'unraveling'. *Observer*, 1 May 2017.

Sherwonit, W. 2010. Arctic tundra is being lost as far north quickly warms. *Yale Environment* 360.

Stepien, A., T. Koivurova, A. Gremsperger, and H. Niemi. 2014. Arctic indigenous peoples and the challenge of climate change. In *Arctic marine governance: Opportunities for transatlantic cooperation*, ed. E. Tedsen, S. Cavalieri, and R. Kraemer. Dordrecht: Springer.

Tollefson, J. 2017. Huge Arctic report ups estimates of sea-level rise. *Nature News*, 28 Apr 2017 https://doi.org/10.1038/nature.2017.21911.
Underwood, E. 2017. Why are Arctic rivers rising in winter? *Eos*, 123. https://doi.org/10.1029/2017EO080803.
Weller, G., et al. 2005. Summary and synthesis of the ACIA. In *ACIA, 2005. Arctic climate impact assessment*, 990–1020. Cambridge University Press.
Wendisch, M., et al. 2017. Understanding causes and effects of rapid warming in the Arctic. *Eos* 98: 2017. https://doi.org/10.1029/2017EO064803. Published on 17 Jan 2017.

Part IV
Collaboration to Enhance Future Coastal Resilience

Chapter 17
Next Generation Numerical Models to Address a Complex Future

Donald T. Resio and C. Reid Nichols

It is far better to foresee even without certainty than not to foresee at all.
—Henri Poincaré, 1913, *The Foundations of Science*

17.1 Numerical Models Are Crucial to Our Future

If we are to adapt to accelerating coastal change and remain resilient despite impermanence, we must be able to anticipate what the future may hold with ever increasing accuracy and decreasing uncertainty. We must improve our ability to predict and prepare for future circumstances on global, regional and local spatial scales and on decadal, annual and short-term time scales. The enabling predictive models should be interdisciplinary and must deal with the interdependence and coupling of physical, ecologic, hydrologic and human factors. Models of global-scale change serve as forcing inputs to regional scale models which, in turn drive more local models of critical processes as illustrated in Fig. 17.1. Examples of global scale models include the NCAR Community Climate Model (CCM3) and large-scale ocean circulation models as described in Chap. 1. The CCM3 model and its more advanced successors will probably provide the long term forcing conditions for predictions of future regional and local scale processes. Regional scale models include the meso-scale atmospheric community models and regional scale ocean circulation models described in Chap. 1. Most of the local scale models currently in use (Chap. 1) are discipline-specific process models. Such models have been used for decades by many professionals, including military planners, engineers, weather forecasters, and natural resource managers. What we now need,

D. T. Resio
School of Engineering, University of North Florida, Jacksonville, FL, USA
e-mail: don.resio@unf.edu

C. R. Nichols (✉)
Marine Information Resources Corporation MIRC), Ellicott City, MD, USA
e-mail: rnichols@mirc-us.com

L. D. Wright and C. R. Nichols (eds.), *Tomorrow's Coasts: Complex and Impermanent*, Coastal Research Library 27,
https://doi.org/10.1007/978-3-319-75453-6_17

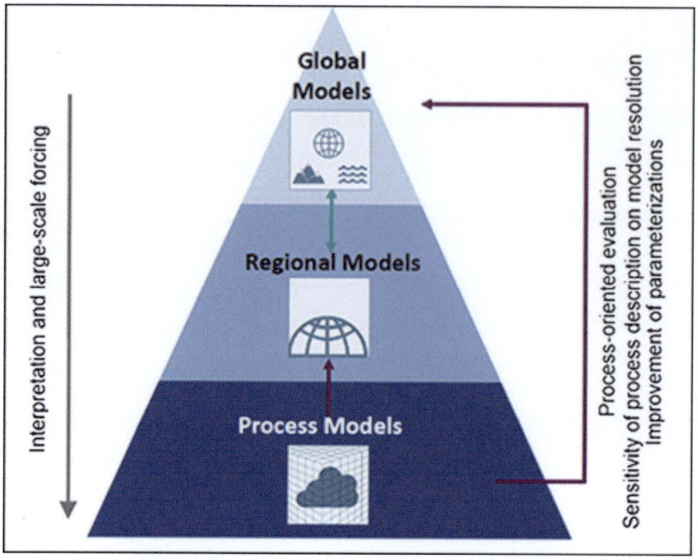

Fig. 17.1 Effective prediction of local conditions at a specific coastal site requires a hierarchy of coupled models from global to local and involving multi-disciplinary models. Figure from Wendisch et al. (2017)

however, is for these models to evolve well beyond being discipline-specific and involve coupling across multiple disciplines including physical oceanography, weather, hydrology, ice, ecosystems, health and socioeconomics. This is now technically possible with rapidly evolving high-performance computing capabilities and the *internet of things*, which connects an ever-growing number of smart devices for up-to-the-minute monitoring and tracking of many physical events. Researchers have deployed thousands of small, low-cost floats to form distributed sensor networks that provide maritime situational awareness over large ocean areas. However, we still need the enabling shared platforms and a new culture of altruistic collaboration.

Models provide representations of reality so that the users can better understand complex processes, causes, and effects to aid in predictions. Prognostic models highlight the evolution of phenomena over time and such models are essential to anticipating a changed future. Researchers test model output with data from ocean observing systems such as the Gulf of Mexico Coastal Ocean Observing System that archive independent data of the actual system being studied. This process is critical to ensure reliable results that meet application requirements. Models are essential to assess management strategies to increase engineering, ecological, and community resilience. Developing authoritative models requires improved understanding of the physical, chemical and ecological processes that occur in large systems, and the impact humans have on natural resources condition and landscape processes. Current and future research is needed to predict fundamental processes affecting the persistence and distribution of plants and animals and people across the land and seascape.

17.2 Status and Future of Physical Oceanographic Models

Numerical modelers construct mathematical models to better understand all relevant physical processes in order to solve physical problems. These models may take the form of differential or algebraic equations or parametric approximations to known natural behavior at scales smaller than that resolved in a model. When designing a model, consideration must be given to the availability of data to drive it. Usually these data are incomplete, duplicative, and often conflicting. In the face of these facts, using a model to obtain a solution requires making specific assumptions. In most real-world applications, the equations in the models do not have exact or "analytical" solutions and must be solved by numerically calculating an approximate solution. Since numerical methods involve discretized, incremental approximations to derive a numerical solution, their accuracy depends critically on careful calibration and validation against pre-existing data to adjust various empirical coefficients within the model. Once a model is implemented, the results which may be displayed as graphics, charts, tables, or other convenient forms, must effectively support users.

Innovations in computational technology have greatly increased the number and types of numerical models and software programs. Numerical modeling has been used extensively in industries for both forward problems such as predicting currents and inverse problems such as wave hindcasting. The numerical models are based on well-developed theoretical frameworks and widely accepted numerical methods. The science of modeling requires that all of the variables that affect important phenomena are identified, reasonable assumptions and approximations are made, and the interdependence of these variables is understood. The art in modeling involves achieving the appropriate level of simplification, distinguishing important phenomena from those that are unimportant, and using acceptable scientific judgment. Successful modeling efforts are able to predict the course of an event such as a current and aspects of the current mathematically such as the speed and direction without actually deploying a mooring with current meters from a research vessel. The samples presented in Table 17.1 provide an abbreviated list of some models that have been shown to provide reasonably accurate results to meaningful practical problems such as wave heights, inundation, and circulation. The use of physical models for understanding the impact of waves, tides, and currents can also provide valuable information for the design of marine structures and potential damages in coastal areas due to natural causes as well as human intervention. However, the preparation of such models requires an adequate knowledge of the natural phenomena and relevant laws, as well as sound judgment. For these reasons,

Table 17.1 Alphabetical list of model acronyms, full names and references to primary developers or users

Model name	Description	Reference
ADCIRC	Advanced Circulation	Luettich and Westerink (1991)
COAMPS	Coupled Ocean/Atmosphere Mesoscale Prediction System	Chen et al. (2003)
Delft3D	Hydrodynamic model, Structured grid	Deltares (2014)
Delft3D FM	Hydrodynamic model, Unstructured grid	Deltares (2015)
HYCOM	Hybrid Coordinate Ocean Model	Chassignet and Verron (2006)
MIKE 21 FM	Hydrodynamic model, Unstructured grid	DHI (2014)
NCOM	Navy Coastal Ocean Model	Martin (2000)
ROMS	Regional Ocean Model System	Shchepetkin and McWilliams (2005)
SLOSH	Sea, Lake, and Overland Surges from Hurricanes	Glahn et al. (2009)
SWAN	Simulating WAves Nearshore	Huang et al. (2013)

fully-integrated observatories such as the NOAA U.S. Integrated Ocean Observing System have contributed and continue to contribute significantly to modeling advances.

Today's computational and networking capabilities provide scientists and engineers with increasing capabilities to study complex processes that are temporally and spatially variable. One example is the Defense Advanced Research Projects Agency's Ocean of Things program, which is demonstrating persistent maritime situational awareness through deployments of thousands of smart floats to collect ocean temperature, sea state, and location data using a cloud network for storage and real-time data analytics. Another example is the Advanced Scientific Computing Research program which operates high-performance computing (HPC) centers and related facilities, and maintains a high-speed network infrastructure at Lawrence Berkeley National Laboratory (LBNL) to support computational science research activities. The HPC facilities include the Oak Ridge Leadership Computing Facility at Oak Ridge National Laboratory, the Argonne Leadership Computing Facility at Argonne National Laboratory, and the National Energy Research Scientific Computing Center at LBNL.

Oceanographic research is advancing through improved instrumentation for measuring properties of the ocean coupled with models that provide a continuous picture of its ever-evolving state. A grand challenge is the operation of coupled ocean and atmospheric models aboard a ship at sea as a means to provide real-time information on environmental phenomena while improving navigation and other maritime operations. Already, researchers have used suites of instrumented autonomous vehicles that relay information to shore-based data-assimilating numerical models to improve the performance of the model's representation of the environment and its predictions.

17.3 Status and Future of Wave and Surf Models

Wind waves impact human activities and have great geophysical significance. The design and installation of coastal structures would benefit from a clear understanding of wave-tide interactions and their impacts on coastal landforms, in particular their effects on bottom friction and sediment transport. In order to help protect property and save lives, most countries run a third-generation wave model that integrates the basic transport equation describing the evolution of a two-dimensional ocean wave spectrum in time and space to provide the products listed in Table 17.2. These models parameterize physical processes explicitly, with minimal parametric constraints on spectral shape or energy levels. Several open source examples include WAve Model (WAM), WAVEWATCH III®, and Simulating WAves Nearshore (SWAN) (see The WISE Group et al. 2007). Comparable commercial models include Oceanweather's third generation wave model (OWI3G) and the spectral wave model developed by DHI known as MIKE 21 SW. Data from wave buoys are critical for off-line model development and quality assessment of operational models (e.g., Allard et al. 2014). A potential extension of these capabilities can be found in Smit et al. (2017), who describe the use of data from 18 lowcost, compact, solar-powered directional wave buoys to drive a regional wave model and produce a data constrained nowcast of the regional wave conditions near Point Sal, California.

The deep-water models WAM and WAVEWATCH III have been used to forecast waves around the globe. Modeling deep water waves accurately is essential to serve as coupled boundary conditions to coastal models where the wave physics becomes more complicated. In shallow water, one has to account for wave shoaling, refraction, diffraction, reflection, and the complexity of wave breaking transitioning into a depth-limited form as surf. For this reason, SWAN, which can be readily nested in WAM and WAVEWATCH III, uses local water depth (bathymetry) and simulates the transformation of off-shore wave conditions into shallow water. It can incorporate winds to simulate wave growth or decay and includes the effects of currents and tides. In the United States, NOAA employs a Global WAVEWATCH

Table 17.2 Wave model output is derived from the spectral description of the wavefield at each grid point

Significant wave height	Peak wave period
Wind sea wave height	Wind sea period
Primary swell wave height	Primary swell period
Secondary swell wave height	Secondary swell period

III model and adaptations such as the Alaskan Waters, Western North Atlantic, and the Eastern North Pacific regional wave models. All regional models obtain hourly boundary data from global models. These models are typically run on the 00z, 06z, 12z and 18z model cycles, and start with 6-h hindcasts to assure continuity of swell. All models provide 120-h forecasts.

Although today's models have made substantial progress toward improving the accuracy of integrated wave parameters (e.g. significant wave height, mean wave period, mean direction, etc.), little progress has been made in model performance related to spectral shape, especially as related to directional characteristics of wave spectra (Resio et al. 2017). Since the coast acts as a directional antenna with respect to incoming waves, this remains a serious problem for nearshore wave models. Although promising progress has been made in wave modeling toward evolving to a fourth generation modeling framework, the inherent inertia in large modeling systems makes it difficult to shift to this new class of model. It is likely that such improved models will be essential to the wave physics necessary to address problems such as the inability of nearshore geomorphic models to capture any of the natural recovery of shorelines between storms. Thus, it is inevitable that future models will eventually supplant today's 3G models. As pointed out by Cavaleri (2006), wave model results from the last decade show an unexplained scatter. Future efforts will need to improve the basic modeling bases for extreme events such as the Halloween Storm (Cardone et al. 1996), rogue waves (Onorato et al. 2013), and shallow water effects which are key to understanding the long-term evolution of coasts during the expected sea level rise occurring over the next several decades. Researchers such as Baschek and Imai (2011) and Cousins and Sapsis (2016) have described some progress in understanding and forecasting extreme events such as rogue waves, and many theories exist for each of the problem areas noted here; however, a modeling consensus is far from being agreed upon in these areas.

Given the importance of waves to many other ocean processes, for example: transport processes and mixing, nearshore inundation, oceanic interaction with the atmosphere, and ocean acoustic, it remains important that researchers do not fall into the trap of accepting a model that performs reasonable in terms of simple wave parameters as the final answer for most purposes, but does not perform adequately in certain critical situations. Thus, an important aspect of future models must be to insist on scenario-based testing, rather than measures of performance adjudicated on a global averaged basis. In this context, measurements are more important than ever, since they are needed to identify problems and quantify the need for improvement in, often rare but important, situations. An important part of this work must improve our understanding and modeling capabilities for wave breaking. The distribution of wave breaking on the water surface is complex and discontinuous. Increased understanding of ocean wave breaking will improve our understanding in other modeling efforts that account for the exchange of momentum, mass and heat with the atmosphere and ocean and provide extremely valuable information for applications in remote sensing, engineering, and navigation.

17.4 Status and Future of Ecosystem Models

The construction and analysis of mathematical models of ecological processes, including both purely biological and combined biophysical models is often called, ecosystem modeling. Such models need to adequately characterize the ecological relationships between system components (individual species or functional groups) and how these components co-vary under different ecosystem drivers and pressures (e.g. harvesting, climate, and management interventions). Models can be analytic or simulation-based and are used to understand complex ecological processes and predict how real ecosystems might change. Analytical models consist of a set of mathematical equations whose behavior is well-known. Simulation models solve problems for which analytic solutions are impractical or impossible. Temporal and spatial considerations are similarly playing an increasingly important role in the development of ecosystem modelling approaches. For this reason, many ecosystem models are coupled to a model which describes physical process, e.g., a one or three-dimensional hydrodynamic model (Werner et al. 2007).

There is a global increase in efforts to develop ecosystem models. This ranges from increasing one's understanding of the ecosystem as a whole or to a subset only, recognizing that both these aims are important in different contexts. Numerous researchers (e.g., Baskett et al. 2007; Fulton et al. 2003, 2004a, b; Hollowed et al. 2000) are extending single-species assessment models to include additional important prey or predator species, and ecosystem models to evaluate policy options for management. The goal involves predicting emergent ecosystem properties such as total system productivity and the mean trophic level of the community, which are frequently proposed as indicators of ecosystem status.

When ecosystem models are used as tools to support decision-making, determining model accuracy or "skill" is essential for decision-makers to consider when weighing forecasts and the possible outcomes of alternative actions.

There is a growing realization that substantial progress towards implementing reliable ecosystem models is still some way off given the need in most regions for considerable data collection and complex analysis. On the other hand, progress in this field has likely developed faster than anticipated given research successes (Fulton et al. 2005), strategies described in NOAA's Ecological Roadmap, and the availability of funding for ecosystem research and development on topics such as hypoxia. Table 17.3 provides a list of ecosystem models and selected references related to ecosystem investigations and models.

Table 17.3 Alphabetical list of model acronyms, full names and references to primary developers or users

Model name	Description	Reference
ATLANTIS	ATLANTIS—considers all parts of marine ecosystems—biophysical, economic and social	Fulton et al. (2004a, b, 2005) and Link et al. (2010)
Ecopath	Mass balance fluxes; linear trophic interactions	Christensen and Pauly (1992) and Pauly et al. (2000)
ECOSIM	Time dynamic simulation	Heymans et al. (2016)
ECOSPACE	Spatial and temporal dynamics	Walters et al. (1999)
EPOC	Ecosystem Productivity Ocean Climate Model	Constable (2005, 2006)
ERGOM	Ecological ReGional Ocean Model	Neumann (2000) and Neumann et al. (2002)
ERSEM	European Regional Seas Ecosystem Model	Baretta et al. (1995)
GADGET	Globally Applicable Area Disaggregated General Ecosystem Toolbox	Begley and Howell (2004)
GEEM	General Equilibrium Ecosystem Model	Tschirhart (2000)
IGBEM	Integrated Generic Bay Ecosystem Model	Fulton et al. (2004a, b)
InVitro	Agent-based ecosystem-level management strategy evaluation modelling framework	McDonald et al. (2006)
OSMOSE	Object-Oriented Simulator of Marine Ecosystem Exploitation	Shin and Cury (2001, 2004)
SEASTAR	Stock Estimation with Adjustable Survey Observation Model and TAg-Return Data	Tjelmeland and Ulf Lindstrøm (2005)
SSEM	Shallow Sea Ecosystem Model	Sekine et al. (1991)

17.5 Status and Future of Hydrology Models for Coastal Applications

Water movement through the biosphere is forced by winds and ocean currents, the flow of rivers, precipitation, the movement of glaciers, evaporation, and transpiration through the porous outer barrier of certain organisms. Hydrology models are important tools since they simulate these movements, distributions, and the quality of water through rivers, soils, and aquifers. Devia et al. (2015) discuss hydrology model characteristics for large complex basins and inputs used by different models such as rainfall, air temperature, soil characteristics, topography, vegetation, hydrogeology, etc. The development of hydrological models is challenged by a lack of data due to short historical observations and spatially insufficient observations. Owing to severe flooding from heavy downpours, which has been increasing over the last three to five decades, there is movement to enhance meteorological and hydrological station networks. These infrastructure advances will improve the model development process and validation efforts. Hydrology models will increasingly be used for predictions, or as inputs to other models. Modeling research efforts will focus on hydrological modeling in un-gaged regions. Key advances will evaluate changing water resources based on distributed hydrological simulations under future climatic situations.

17.6 Status and Future of Sea Ice Models

National ice services or governmental organizations responsible for operational ice services of their countries rely on sea ice models in order to provide forecasts which are extremely important to local communities and industries. Sea ice is an important component of the climate system because it regulates the transfer of heat and momentum between the atmosphere and the ocean. While melting sea ice has no impact on sea level rise, it does contribute to climate change. The runoff from melting glaciers will eventually run into the ocean, which does contribute to sea level rise. Information about sea ice processes can come from sensors installed at field camps (e.g., the 2011 Applied Physics Laboratory Ice Station, on Arctic ice north of Prudhoe Bay, Alaska and the central Transantarctic Mountains field camp, which is periodically set up by U.S. National Science Foundation), or aircraft and satellites. These limited data sources have been instrumental in the development of sea ice models, which provide valuable information on how sea ice evolves and how it will be affected by changing climate. The equations in a sea ice model describe the relevant dynamics (e.g., winds, waves, and currents) and thermodynamics (e.g., air temperatures, ocean temperatures, and albedo) that influence the evolution of sea ice. Short-term operational forecasts (one to five days) by organizations such as the Arctic and Antarctic Research Institute in Russia, Canadian Ice Service, National Ice Center (NIC) in the United States, and Norwegian Ice Service are important aids to navigation for ocean vessels in sea ice-covered regions, as well as seasonal forecasts (one to three months) to aid in planning. At the NIC, the U.S. Navy runs an operational sea ice model, the Arctic Cap Nowcast Forecast System (ACNFS), to provide short-term forecasts of Arctic sea ice. The ACNFS replaced the Polar Ice Prediction System during 2012.

Sea ice models are limited internally by the model physics, and externally by the boundary conditions, or forcing. Models carry assumptions, and forecasts from the models have inherent uncertainties. Further, many sea ice models are coupled to ocean and atmospheric models. The output describes several dynamic and thermodynamic properties such as concentration, extent, thickness, temperature, and velocity. Model skill is evaluated using sea ice observations from satellites, aircraft, and buoys.

17.7 Coupling Oceanographic and Hydrology Models: Progress and Prognosis

As discussed in Chap. 7, compound coastal flooding, involving the superimposition of storm surge along with fluvial and pluvial flooding, as was exemplified by Hurricane Harvey in Houston during 2017, is likely to become more frequent during the years ahead. Mutual interactions between hydrometeorological and ocean models must be advanced in the near future. Those models operate on different scales and simulations represent different sets of interdependent processes. The coupling of hydrology models to coastal ocean models is an area of active research and considered by some agencies as a Grand Challenge. Research questions relate to the determination of necessary resolutions, frequency and spatial locations for the hydrology-ocean model coupling. Objectives include improved forecasts of freshwater-driven events in the coastal ocean environment, especially inundation caused by extreme precipitation events. van der Wiel et al. (2017), describe the historic freshwater flooding that occurred in southern Louisiana during August 2016. Meteorological phenomena such as downpours are small in scale and are simulated using very high-resolution models. These types of events need to be considered as coastal ocean models are pushed into the land-sea margin. Simulations need to account for hydrometeorological, oceanographic, and human-related factors. Major weather factors include heavy or prolonged precipitation, snowmelt, thunderstorms, storm surges from hurricanes, and ice or debris jams. Human factors include structural failures of dams and levees, altered drainage, and land-cover alterations (such as pavement). Strategies may include "offline" coupling, where output from a hydrology model is passed to the ocean model for computation of a variable such as water level. The preference is "online" coupling, where the feedbacks are allowed to pass between the two models. This research is especially important to help determine how climate change affects coastal flooding through sea level rise and storm surge, and increases in heavy rainfall.

17.8 Coupling Ecosystem and Physical Process Models: Progress and Prognosis

Coupled ecosystem and physical models may be applied to better examine the role of physical processes on the ecosystem. The simulated physical processes include meteorological factors such as winds, hydrological factors such as river discharge, and hydrodynamic factors such as water level fluctuations. In the coastal zone, the potential sources of nitrogen to an estuary such as Corpus Christi Bay are through river inputs and wet deposition flux. An ecosystem model might separate the detritus compartment into pelagic and benthic components, based on cohesive sediment processes. The physical model might be tested against observed tidal amplitude and phase, as well as tidal currents from an observational network such as the Texas Coastal Ocean Observation Network. The biological model would be tested to ensure reasonable spatial and temporal patterns in lower trophic level characteristics of the ecosystem in some areas of the Corpus Christi Bay. Model results might also show that the physical processes associated with evaporation produce an intensification of hurricanes owing to surface moisture availability. Hurricanes derive their energy from water vapor which is evaporated from the ocean surface. The water vapor releases latent heat when it condenses to form clouds and rain, warming the surrounding air. Inclusion of physical oceanographic processes such as sea surface temperatures and evaporation is important for a hurricane model to properly forecast the storm's intensity.

17.9 Complex Trans Disciplinary Community Models of the Future

In order to model tomorrow's coasts, scientists from natural and social sciences will need to collaborate to solve important integrated research questions, especially to simulate measures that will ensure sustainability. This type of collaboration will require trans-disciplinary and integrated modeling efforts. The U.S. Army Corps of engineers is already developing an integrated approach to reducing coastal risks and increasing human and ecosystem community resilience through a combination of natural, nature-based, nonstructural and structural measures that have been studied at locations such as Jamaica Bay in New York and Mobile Bay in Alabama (e.g., Rosati et al. 2015; Fox-Lent et al. 2015). The Army has defined resilience in terms of "prepare," "resist," "recover," and "adapt" and metrics associated with engineering, ecological, and community processes. Models in the future will be from different disciplines and will facilitate joint work to create new conceptual, theoretical, methodological, and translational innovations that integrate and move beyond discipline-specific approaches to address a coastal change. The next generation of complex systems models is likely to employ hybrid approaches such as those recently discussed by Vincenot et al. (2016). According to those authors (p. 6): "The non-linear, self-organized dynamics of complex systems, including environmental ones, makes their behavioral trends extremely difficult to predict. Hybrid models, which typically represent the dynamics of a system over a wide parameter space, can be used to explore the envelope of possible, probable and plausible system trajectories". Future efforts can be co-designed with output that meets information requirements for trans-disciplinary users. Modeling systems and analytics will be developed to evaluate impacts of climatic change. They will assume that the coast is complex and impermanent rather than a static infrastructure, economy or demography. Trans-disciplinary community models will enable socio-economic systems to react and to adapt to climatic changes. They will integrate engineering, ecological, and community factors.

17.10 A Cyber Network to Support a Virtual Modeling Community

The pillars of modern science are often considered to be theory, experimentation, and computational science. Despite the great potential of modeling to increase understanding of a variety of important scientific and engineering challenges, High Performance Computing (HPC) has been underutilized to date. Applied computational scientists are developing new mathematical algorithms, tools, and libraries to model complex physical and biological systems. Supercomputing programs are playing increasingly important roles in the scientific discovery process by allowing scientists to create more accurate models of complex systems and simulate problems once thought to be impossible while analyzing an increasing amount of data generated by modeling experiments.

High-performance Computing is moving toward scalable systems software and programming models, that enable computational scientists to effectively utilize petascale computers to advance science in areas important to society. Scalable systems are capable of improving under an increased load when resources such as hardware are added. In computing, petascale refers to a computer system capable of reaching performance in excess of one petaflops, i.e. one quadrillion floating point operations per second. Petascale computing could make the development of a fully-coupled Earth system model feasible, one that provides the most accurate environmental information, including uncertainty and probabilities, to the right people at the right time, and in the right form for optimal understanding and decision-making.

A distributed network environment will involve integrated software tools and advanced network services to enable large-scale scientific collaboration and make effective use of distributed computing and science facilities to support large science. Examples include ocean observatories that provide distributed data and model products for environmental monitoring.

The science potential of emerging computing systems and other novel computing architectures will require numerous significant modifications to today's tools and techniques. Collaborative programs have achieved breakthroughs in scientific advances and many modeling technologies are impossible to complete without these kinds of interdisciplinary efforts. Future modeling will increasingly adopt HPC. Extreme-scale computing will enable the solution of vastly more accurate predictive models and the analysis of massive quantities of data, producing quantum advances in areas of science and technology that are essential to understanding climate changes such as sea level rise, drought and flooding, and severe weather patterns. Researchers are moving from petascale to exascale, which represents a thousand-fold increase over the first petascale computer that came into operation in 2008. Exascale computing refers to computing systems capable of at least one exaFLOPS, or a billion calculations per second. The expertise required to install, utilize and run these assets poses a significant barrier to many organizations due to the levels of complexity built into them to facilitate scientific discovery and research.

Modeling with HPC will be key to dealing with coastal resilience and other complex phenomena that show intricate time evolutionary behavior, mostly on multiple dimensions. Some dynamically complex issues such as inundation are relatively well-know and largely predictable, but have not been forecasted due to the fact that local occurrences are hard to understand. Other phenomena–especially potential future issues such as increasing storm events and grand challenges such as the coupling of hydrological and ocean models—are still largely matters of basic research. Modeling develops and applies computational methods to study these complex issues such as sea level rise and associated problems in engineering, ecology, and urban planning. In fact, as people continue to develop coastal zones, future planners will need to integrate engineering, environmental, and community resilience.

Other exciting technologies that complement modeling include techniques from data science and artificial intelligence. Data analytics will apply cloud-based software and analysis techniques to process the ever growing amounts of *in situ* observations to improve oceanographic and meteorological models. While some of these technologies are already available, collaboration will require a lot of experimentation and will lead to some promising innovations. In the past century design prototypes for manufacturing were scale models and today, we now develop them using CAD software, and produce a replica in minutes using a 3D printer or a CDC router. Developers are designing intelligent sensors to house a passive sensor suite that can survive in harsh maritime environments. Expendable sensors may report information from its surroundings for at least one year before safely scuttling itself in the deep ocean. Data from animal-borne sensors, including marine mammal tags, can help scientists produce analyses and forecasts of ocean temperature and salinity. We can already simulate the interaction of a simulated object with an authoritative ocean environment.

Though every effort is made to insure accuracy in modeling, errors are inevitable. It is essential that models be tested to establish their accuracy and appropriateness to address problems such as sea level rise. The ensuing skill assessments are also known as verification, validation, and accreditation. Most modelers assess their model results against observations. Verification, validation, and accreditation is the process where an objective third party determines whether or not the software product is an accurate implementation of the conceptual model, the extent to which model results provide an accurate representation of the real world, and official determination that output is acceptable to users. No model should be expected to provide universal solutions to all problems in a domain. The value in a model will always be based entirely upon the degree to which it solves a user's real world problem.

References

Allard, Richard, Erick Rogers, Paul Martin, Tommy Jensen, Philip Chu, Tim Campbell, James Dykes, Travis Smith, Jeikook Chu, and Uriah Gravois. 2014. The US navy coupled ocean-wave prediction system. *Oceanography* 27 (3): 92–103.

Baretta, J.W., W. Ebenhöh, and P. Ruardij. 1995. The European regional seas ecosytem model, a complex ecosystem model. *Netherlands Journal of Sea Research* 33 (3–4): 233–246.

Baschek, Burkard, and Jennifer Imai. 2011. Rogue wave observations off the US west coast. *Oceanography* 24 (2): 158–165.

Baskett, Marissa L., Fiorenza Micheli, and Simon A. Levin. 2007. Designing marine reserves for interacting species: Insights from theory. *Biological Conservation* 137: 163–179.

Begley, J., and D. Howell. 2004. *An overview of gadget, the globally applicable area-disaggregated general ecosystem toolbox*, 15. International Council for the Exploration of the Sea (ICES) Document CM 2004/FF: 13.

Cardone, V.J., R.E. Jenson, D.T. Resio, V.R. Swail, and A.T. Cox. 1996. Evaluation of contemporary ocean wave models in rare extreme events: The "Halloween storm" of October 1991 and the "storm of the century" of March 1993. *Journal of Atmospheric and Oceanic Technology* 13: 198–230.

Cavaleri, Luigi. 2006. Wave modeling: Where to go in the future. *Bulletin of the American Meteorological Society* 87 (2): 207–214.

Chassignet, Eric P., and Jacques Verron, (eds.). 2006. *Ocean weather forecasting: An integrated view of oceanography*, 577. Dordrecht, Netherlands: Springer.

Chen, S., J. Cummings, J. Doyle, R.H. Hodur, T. Holt, C. Liou, M. Liu, A. Mirin, J. Ridout, J.M. Schmidt, G. Sugiyama, and W.T. Thompson. 2003. *COAMPS™ version 3 model description—General theory and equations*, 148. NRL Publication NRL/PU/7500–03-448, May 2003, Monterey, CA: Naval Research Laboratory.

Christensen, V., and D. Pauly. 1992. ECOPATH II—A software for balancing steady-state ecosystem models and calculating network characteristics. *Ecolological Modelling* 61 (3–4): 169–185.

Constable, A.J. 2005. *Implementing plausible ecosystem models for the Southern Ocean an ecosystem, productivity, ocean, climate (EPOC) model*, 20. Workshop document presented to WG-EMM subgroup of CCAMLR (Commission for the Conservation of Antarctic Marine Living Resources), WG-EMM-05/33.

Constable, A.J. 2006. *Using the EPOC modelling framework to assess management procedures for Antarctic krill in Statistical Area 48: Evaluating spatial differences in productivity of Antarctic krill.* Workshop document presented to WG-EMM subgroup of CCAMLR (Commission for the Conservation of Antarctic Marine Living Resources), WG-EMM-06/38.

Cousins, Will, and Themistoklis P. Sapsis. 2016. Reduced-order precursors of rare events in unidirectional nonlinear water waves. *Journal of Fluid Mechanics* 790: 368–388.

Deltares. 2014. *Delft3D-flow, simulation of multi-dimensional hydrodynamic flows and transport phenomena, including sediments, User manual*, 684. Version 3.15.34158, May 2014.

Deltares. 2015. *Delft3D flexible mesh suite, D-flow FM in delta shell, user manual*, 376. Version 1.1.148, November 2015.

Devia, Gayathri K., B.P. Ganasri, and G.S. Dwarakish. 2015. A review of hydrological models. *Aquatic Procedia* 4: 1001–1007.

DHI. 2014. *MIKE 21 flow model FM hydrodynamic module, user manual*, 134. August 2014.

Fox-Lent, Cate, Matthew E. Bates, and Igor Linkov. 2015. A matrix approach to community resilience assessment: An illustrative case at rockaway peninsula. *Environment Systems and Decisions.* 35 (2): 209–218.

Fulton, Elizabeth A., Anthony D.M. Smith, and Craig R. Johnson. 2004a. Biogeochemical marine ecosystem models I: IGBEM—A model of marine bay ecosystems. *Ecological Modelling* 174 (3): 267–307.

Fulton, Elizabeth. A., Anthony D.M. Smith, and André E. Punt. 2004b. *Ecological indicators of the ecosystem effects of fishing: Final report*. Report No. R99/1546, Canberra: Australian Fisheries Management Authority.

Fulton, Elizabeth A., Anthony D.M. Smith, and André E. Punt. 2005. Which ecological indicators can robustly detect effects of fishing? *ICES Journal of Marine Science* 62 (3): 540–551.

Glahn, Bob, Arthur Taylor, Nicole Kurkowski, and Wilson A. Shaffer. 2009. The role of the SLOSH model in national weather service storm surge forecasting. *National Weather Digest* 33 (1): 3–14.

Heymans, Johanna Jacomina, Marta Coll, Jason S. Link, Steven Mackinson, Jeroen Steenbeek, Carl Walters, and Villy Christensen. 2016. Best practice in Ecopath with Ecosim food-web models for ecosystem-based management. *Ecological Modelling* 331: 173–184.

Hollowed, Anne B., Nicholas Bax, Richard Beamish, Jeremy Collie, Michael Fogarty, Patricia Livingston, John Pope, and Jake C. Rice. 2000. Are multispecies models an improvement on single-species models for measuring fishing impacts on marine ecosystems? *ICES Journal of Marine Science* 57: 707–719.

Huang, Yong, Robert H. Weisberg, Lianyuan Zheng, and Marcel Zijlema. 2013. Gulf of Mexico hurricane wave simulations using SWAN: Bulk formula-based drag coefficient sensitivity for Hurricane Ike. *Journal of Geophysical Research: Oceans* 118 (8): 3916–3938.

Link, J.S., E.A. Fulton, and R.J. Gamble. 2010. The northeast US application of ATLANTIS: A full system model exploring marine ecosystem dynamics in a living marine resource management context. *Progress in Oceanography* 87 (1–4): 214–234.

Luettich, R.A., and J.J. Westerink. 1991. A solution for the vertical variation of stress, rather than velocity, in a three-dimensional circulation model. *International Journal for Numerical Methods in Fluids* 12: 911–928.

Martin P.J. 2000. *A description of the navy coastal ocean model version 1.0*, 42. NRL Report NRL/FR/7322-00-9962, Stennis Space Center, MS: Naval Research Laboratory.

McDonald, A.D., L. Richard Little, R. Gray, Elizabth Fulton, K.J. Sainsbury, and Vincent D. Lyne. 2006. Multiple use management strategy evaluation for coastal marine ecosytems using invitro. In *Complex science for a complex world*, ed. Pascal Perez, and David F. Batten, 2588–2594. Canberra: ANU E Press.

Neumann, T. 2000. Towards a 3D-ecosystem model of the Baltic Sea. *Journal of Marine Systems* 25 (3–4): 405–419.

Neumann, T., W. Fennel, and C. Kremp. 2002. Experimental simulations with an ecosystem model of the Baltic Sea: A nutrient load reduction experiment. *Global Biogeochemical Cycles* 16 (3): 7–1–7-19.

Onorato, M., S. Residori, U. Bortolozzo, A. Montina, and F. Arecchi. 2013. Rogue waves and their generating mechanisms in different physical contexts. *Physics Reports* 528 (2): 47–89.

Pauly, D., V. Christensen, and C. Walters. 2000. Ecopath, ecosim, and ecospace as tools for evaluating ecosystem impact of fisheries. *ICES Journal of Marine Science* 57 (3): 697–706.

Resio, D.T., L. Vincent, and D. Ardag. 2017. Characteristics of directional wave spectra and implications for detailed-balance wave modeling. *Ocean Modelling* 103: 38–52. https://doi.org/10.1016/j.ocemod.2015.09.009.

Rosati, J.D., K.F. Touzinsky, and W.J. Lillycrop. 2015. Quantifying coastal system resilience for the USACE. *Environment Systems and Decisions* 35 (2): 196–208.

Sekine, Masahiko, Hiroshi Nakanishi, and Masao Ukita. 1991. A shallow-sea ecological model using an object-oriented programming language. *Ecological Modelling* 57 (3–4): 221–236.

Shchepetkin, Alexander F., and James C. McWilliams. 2005. The regional oceanic modeling system (ROMS): A split-explicit, free-surface, topography-following-coordinate oceanic model. *Ocean Modelling* 9 (4): 347–404.

Shin, Yunne-Jai, and Philippe Cury. 2001. Exploring fish community dynamics through size-dependent trophic interactions using a spatialized individual-based model. *Aquatic Living Resources* 14 (2): 65–80.

Shin, Yunne-Jae, and Philippe Cury. 2004. Using an individual-based model of fish assemblages to study the response of size spectra to changes in fishing. *Canadian Journal of Fisheries and Aquatic Science* 61 (3): 414–431.

Smit, Pieter, Tim Janssen, Cameron Dunning, and Wheeler Gans. 2017. Real-time wave assimilation using low-cost sensor arrays. In *Proceedings of the ocean waves workshop*, ed. C.R. Nichols, 7 Dec 2017. New Orleans, LA: University of New Orleans. Available online. https://scholarworks.uno.edu/cgi/viewcontent.cgi?article=1063&context=oceanwaves. Accessed 23 Jan 2018.

The WISE Group, L. Cavaleria, J.-H. G.M. Alves, F. Ardhuin, A. Babanin, M. Banner, K. Belibassakis, M. Benoit, M. Donelan, J. Groeneweg, T.H.C. Herbers, P. Hwang, P.A.E.M. Janssen, T. Janssen, I.V. Lavrenov, R. Magne, J. Monbaliu, M. Onorato, V. Polnikov, D. Resio, W.E. Rogers, A. Sheremet, J. McKee Smith, H.L. Tolman, G. van Vledder, J. Wolf, and I. Young. 2007. *Progress in Oceanography* 75 (4): 603–674.

Tjelmeland, Sigurd, and U. Ulf Lindstrøm. 2005. An ecosystem element added to the assessment of Norwegian spring spawning herring: Implementing predation by Minke Whales. *ICES Journal of Marine Science* 62 (2): 285–294.

Tschirhart, John. 2000. General equilibrium of an ecosystem. *Journal of Theoretical Biology* 203 (1): 13–32.

van der Wiel, Karin, Sarah B. Kapnick, Geert Jan, Kirien van Oldenborgh, Sjoukje Philip Whan, Gabriel A. Vecchi, Roop K. Singh, Julie Arrighi, and Heidi Cullen. 2017. Rapid attribution of the August 2016 flood-inducing extreme precipitation in South Louisiana to climate change. *Hydrology and Earth System Science* 21: 897–921.

Vincenot, C.E., S. Mazzoleni, and L. Parrott. 2016. Editorial: Hybrid solutions for the modeling of complex environmental systems. *Frontiers in Environmental Science* 4: 53. https://doi.org/10.3389/fenvs.2016.00053.

Walters, C., D. Pauly, and V. Christensen. 1999. Ecospace: Prediction of mesoscale spatial patterns in trophic relationships of exploited ecosystems, with emphasis on the impacts of marine protected areas. *Ecosystems* 2 (6): 539–554.

Wendisch, M., et al. 2017. Understanding causes and effects of rapid warming in the Arctic. *Eos* 98. https://doi.org/10.1029/2017EO064803.

Werner, F.E., R.K. Cowen, and C.B. Paris. 2007. Coupled biological and physical models: Present capabilities and necessary developments for future studies of population connectivity. *Oceanography* 20 (3): 54–69.

Chapter 18
Future Societal Vulnerability, Risk and Adaptability

Lynn Donelson Wright

How high's the water Mama? Five feet high and risin'.

—Johnny Cash

18.1 The Most Vulnerable People Must Have Priority

It is imperative for humanity to anticipate and plan better for the future impacts of climate change and coastal flooding on low-income, elderly and infirm communities living in flood-prone areas. The tragedy that unfolded in 2005 when Hurricane Katrina made landfall along the Louisiana and Mississippi Gulf Coast, was most acute in the flooded low lying and low-income neighborhoods of New Orleans, particularly the Ninth Ward. Nearly 2000 people died and hundreds of thousands were displaced. The most severely affected African-American population has still not fully recovered more than a decade later. Elliott and Pais (2006) have articulated the tragedy of the inadequate concerns for the African American community in relief efforts following Hurricane Katrina. At the time of this writing, many Puerto Ricans still lack electricity and adequate drinking water, food and medicine in the aftermath of Hurricane Maria, which hit the island territory in September 2017.

As sea levels rise, low-lying vulnerable urban areas throughout the world will be more frequently flooded by storms. Wealthy populations will migrate to higher ground and the value of these higher elevation properties will escalate. Low-income families will be forced to move into higher density areas or to low-lying, flood-prone areas. Frequent street flooding of low-lying neighborhoods can paralyze traffic, sewers can be flooded, drinking water may be contaminated and water-borne pathogens may be spread throughout neighborhoods. And, as was the case in New Orleans in the days following Katrina, in Texas after Hurricane Harvey

L. D. Wright (✉)
Southeastern Universities Research Association (SURA), Washington, DC, USA
e-mail: ldwright@bellsouth.net

L. D. Wright and C. R. Nichols (eds.), *Tomorrow's Coasts: Complex and Impermanent*, Coastal Research Library 27,
https://doi.org/10.1007/978-3-319-75453-6_18

in 2017 and in Puerto Rico after Hurricane Maria in 2017, extensive inundation of neighborhoods can impede rescue operations, distribution of food, water and medicine and the restoration of power following disasters.

18.2 Indexing Vulnerability and Risk

The past few years have seen significant advances in understanding and modeling societal factors and changes that can impact community resilience (e.g. Gunderson and Holling 2002). Van Zandt et al. (2012) consider neighborhood resilience in relation to social vulnerability and housing. Norris et al. (2008) offer a treatise on the psychology of community resilience as it impacts disaster readiness. More recently, Cutter et al. (2010, 2014) have evolved the concept of *Baseline Resilience Indicators for Communities* (*BRIC*) as empirical metrics for gaging the resilience of communities to disasters. Berkes et al. (2003) offer in depth analyses of social-ecological complexity in assessing community resilience. Guillard-Gonçalves et al. (2014) have developed a "Social Vulnerability Index" (SoVI) which can be readily applied to most regionally specific communities. Balica et al. (2012) developed a multi-term SoVI for coastal cities specifically focused on flooding related to climate change. Flanagan et al. (2011) have developed a social vulnerability index (SVI) for specific application to disaster management. This index considers 15 different factors obtained from census data, most notably income and socioeconomic status, age and disability, minority status and language, type and quality of housing and access to transportation. Figure 18.1, from Flanagan et al. (2011) shows the distribution of drowning deaths caused by Katrina in New Orleans in relation to the Social Vulnerability of elderly residents.

In their analysis, Flanagan et al. (2011) consider *risk* to depend on the probability of occurrence of hazards (such as flooding), social vulnerability and the availability of mitigating resources to counter vulnerability. Risk increases with both hazard and vulnerability but vulnerability decreases with increasing resources, such as emergency management facilities and flood protection infrastructure. Yan et al. (2016) devised an index to evaluate the social vulnerability of the city of Shanghai to rising sea level and storm surge in terms of per capita income and flooding risk, assuming continuation of the existing flood protection infrastructure. They concluded that vulnerability will be minimal until around 2030 but that 23.9% of the population will be highly vulnerable by 2050. As described in Part 3, Chap. 12, Yang et al. (2014) proposed indices of relative exposure, sensitivity, adaptive capacity and total vulnerability of sub regions in the Pearl River Delta. Their results are summarized in Chap. 12, Fig. 12.6. From all of these analyses the take away message is simply this: the vulnerability of coastal populations, now and in the future, depends not only on the environmental threat and the fragility of the people, but also on the investments that local, state and federal governments are willing and able to make in order to protect the most vulnerable.

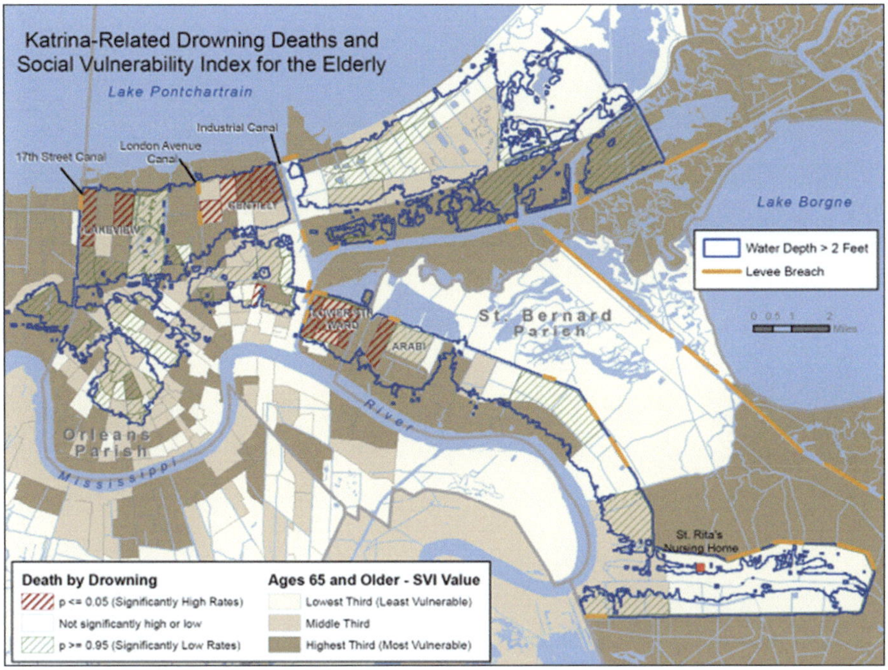

Fig. 18.1 Social vulnerability index for the elderly in New Orleans and locations of drowning deaths caused by Hurricane Katrina in 2005. From Flanagan et al. (2011)

18.3 Linked Ecosystem, Resource and Socioeconomic Vulnerability

Human vulnerability to rising seas, flood events and climate change is not limited to loss of life and health but also includes more complex and non-fatal manifestations involving impacted livelihood and quality of life. Some of the regionally specific impacts were discussed in the case studies presented in Part 3 of this book. To varying degrees, the socioeconomic circumstances of coastal communities are linked to coastal ecosystems. Degradation of those ecosystems has negative consequences for the ability of communities to flourish and adapt. To address such interconnections, Salgado et al. (2013) developed a model that attempts to couple ecosystem vulnerability and socioeconomic vulnerability with applications to Norway, Turkey and Chile. This international project was carried out on behalf of the Belgian organization *Ocean-Certain.* The model focuses on the inter-relationships of climate change, marine ecosystems and coastal communities. The coupled system approach of Delgado-Serrano et al. (2014) and Delgado-Serrano and Ramos (2015) considered the connectivity between ecosystem vulnerability and socioeconomic vulnerability to involve environmental

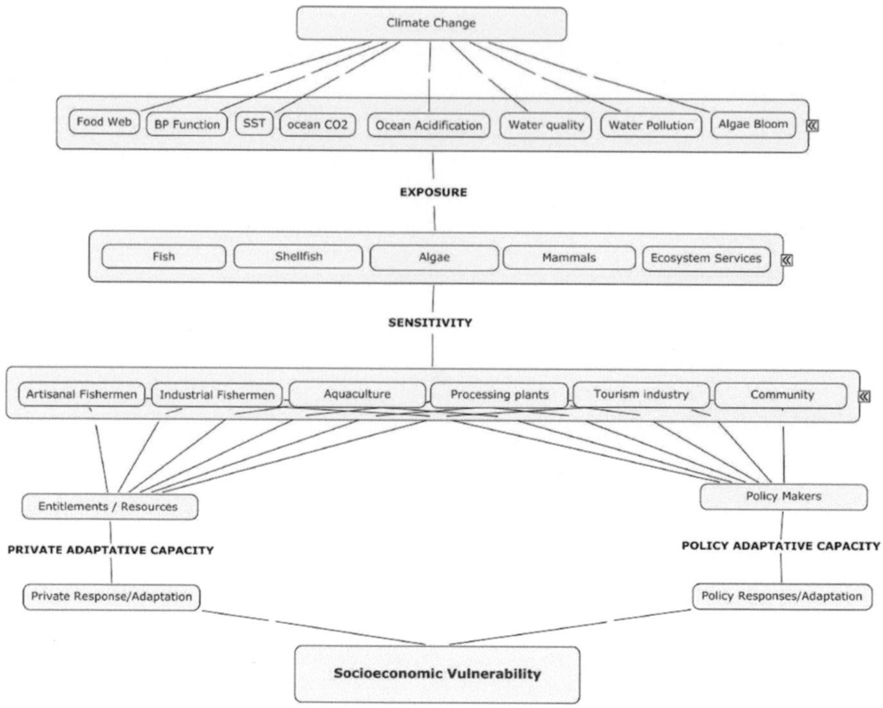

Fig. 18.2 Connected ecosystem and socioeconomic factors affecting socioeconomic vulnerability. From Salgado et al. (2013)

exposure, socioeconomic sensitivity, socioeconomic entitlements and public policy and private responses. Figure 18.2 summarizes the general connectivities identified in the model of Salgado et al. (2013).

In developed, more affluent nations, vulnerability assessments tend to place somewhat greater emphasis on the effects of damaged or destroyed resources and infrastructure since immediate threats to human life are considerably less severe but impacts on livelihood are the primary concerns. Smith et al. (2016) examined socioeconomic vulnerability in the Australian state of Queensland with emphasis on the impacts of climate change and disasters on natural resources and particularly on agriculture. They concluded that the most vulnerable sub regions in their study were ones that were heavily dependent on horticulture. In general terms, they offer a conceptual model for conducting rapid, regionally specific vulnerability assessments. From a general survey of the literature on vulnerability of wealthier nations, there would appear to be somewhat more emphasis on loss of tangible "things" (which are fairly easy to quantify) than on intangibles like human health and wellbeing. However, it is encouraging to note that researchers are increasingly aware of this and are searching for new methodologies to treat the "intangibles" in more rigorous ways.

18.4 Vulnerability and Adaptation of Northern European Coasts

The coastal realms of most European nations bordering the North Sea face the hazards of being low lying, in some cases below sea level or at least below the level of high tide. However, the vulnerabilities of the coastal communities that live there are low by global standards, in part because of the relative affluence of the people and in part because of the investments by the governments in elaborate flood mitigation strategies and protective engineering works. Figure 18.3 shows the relative elevations of these North Sea countries. The Netherlands is, of course, the best known for its extensive flood control system without which the nation would be highly vulnerable to flooding by rising seas and rivers (Dronkers and Stojanovic 2016). The protective system of dykes, sluices, pumps and floodgates is centuries old but now faces numerous increasing challenges including the loss of natural wetlands, salinization of polders during times of low river flow and the insufficient capacity of pumping stations and sluices to accommodate heavy rainfall events (Dronkers and Stojanovic 2016). Other countries such as Belgium, have adopted somewhat "softer" approaches including the nurturing of natural ecosystem services. The UK has fairly advanced adaptation strategies that include managed retreat (De la Vega-Leinert and Nicholls 2008).

Future uncertainties lie ahead that could well overwhelm the adaptive strategies currently in effect for European countries. The occurrence of tropical cyclones in

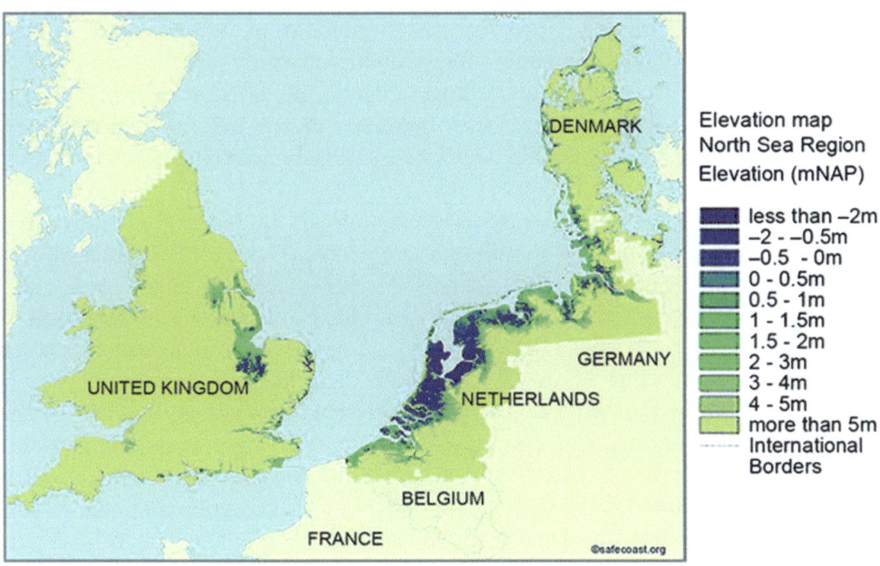

Fig. 18.3 Elevations of the coastal regions of North Sea nations. Note that blue colors indicate elevations below sea level. From Dronkers and Stojanovic (2016)

the northeast Atlantic has been very rare in the past but when former Hurricane Ophelia battered Ireland in October 2017, the potential for future such events was highlighted. As noted by Dronkers and Stojanovic (2016) extreme events that have a low probability of occurrence can have the largest climate-change related impacts on Northern European coasts but uncertainty is one of the biggest obstacles to raising public and political awareness of future threats. Rigid, "blueprint" strategies based on past experience are inadequate. Uncertainties require adaptive approaches that can be changed or reversed as circumstances dictate.

18.5 Bangladesh: Vulnerable to Epic Tragedy

In terms of vulnerability to human tragedy, Bangladesh and India are probably at the opposite extreme from Northern Europe because of the poverty of the people and the general lack of infrastructure and resources. As Fig. 4.6 in Chap. 4 shows, much of Bangladesh is near or only slightly above sea level. Added to this is the fact that summer monsoons bring both torrential rains that cause floods of the Ganges and Brahmaputra Rivers as well as tropical cyclones that generate storm surges. Two hundred million poor people live in this high-risk realm. Brouwer et al. (2007) examined the relationships among environmental risk, poverty, and vulnerability in flood prone regions of Bangladesh. In their study they interviewed 700 floodplain residents and found that residents with the lowest incomes had the highest level of exposure to floods. They also concluded that these most vulnerable low-income residents had the lowest access to flood preparation and post-flood recovery community assets. However, the Brouwer et al. (2007) study did not include projections of future vulnerabilities that might be expected as a consequence of climate change. In recent years, climate change has also brought about other changes in vulnerability. Among these, increases in soil salinity of agricultural lands in the Ganges Brahmaputra Delta are seriously reducing rice production (Goodell 2017).

During the 2007 summer monsoon season, floods in Bangladesh killed 500 people and adversely impacted 20 million more people (Dewan 2015). These floods typically involved heavy rainfall, increased river discharge, and storm surge superimposed on rising sea levels. Accelerated retreat of glaciers is adding to flood waters. According to Dewan (2015) the major impacts of Bangladesh floods are death by drowning, water borne disease, severe diarrhea and snakebites; 75% of Bangladesh flood victims drown. This seemingly contrasts sharply with flood and storm impacts on northern European coasts, which are largely economic and only occasionally involve loss of life. Non-fatal impacts in Bangladesh include complete destruction of housing, loss of income, loss of livestock and grain and general loss of food supply. As noted by Dewan (2015) and others, climate change is already bringing more intense rainfall and higher storm surges and things are predicted to get worse over the coming decades. Effective adaptation to these growing threats is a serious challenge. Although the floods are generally worse than in the past, they are nothing new and the indigenous populations have developed fluid and unique ways of rebounding after devastating events. Fortunately, the Bangladesh Government is taking this as a serious opportunity and incorporating traditional protocols into modern strategies that include a range of engineering approaches.

18.6 Vulnerability of U.S. Atlantic and Gulf of Mexico Coastal Communities

Socioeconomic vulnerabilities of U.S. coastal communities are very much lower than those of Bangladesh, India or most of South and South East Asia but are typically greater than those of northern Europe. The coastal landscapes of the U.S. Atlantic and Gulf of Mexico are generally more vulnerable to sea level rise and storm surge than are the Pacific coastal systems because of the low elevations and low continental shelf gradients for reasons explained in Part 2. The U.S. Geological Survey (Pendleton et al. 2010) has assessed vulnerability of the Gulf of Mexico coast to erosion and land loss as being high. Their analyses took account of geomorphology, recorded rates of coastal recession, significant wave height, tide range and regional coastal slopes but did not involve any socioeconomic indices. When socioeconomic factors such as age, income, education, quality of housing and ethnicity are included, a much higher degree of spatial variability emerges. This is largely the outcome of high socioeconomic diversity and, as described in several of the case studies presented in Part 3, highly vulnerable communities may exist in close proximity to low vulnerability and affluent communities.

The US Army Corps of Engineers (USACE), Engineer Research and Development Center (ERDC) is developing a tiered set of coastal resilience metrics that integrate engineering, environmental and community resilience (Rosati et al. 2015). The USACE's approach considers preparation, resistance, recovery, and adaptation depending on factors such as need, time, space, and available funding. Analyses of U.S. coastal socioeconomic vulnerability now benefit from some fairly elaborate and sophisticated cyber tools and websites that support decision-making and emergency response at local and regional levels. The U.S. Climate Resilience Toolkit (https://toolkit.climate.gov/-tools), an online resource supported by the National Oceanic and Atmospheric Administration (NOAA) outlines the criteria used in regional assessments and provides links to some interactive tools for exploring community vulnerability to different storm and sea level rise scenarios. The local or regionally-specific approach advocated in the toolkit involves: (1) identifying the local climate related threats; (2) assessing risks and vulnerabilities; (3) exploring options; and (4) prioritizing actions with considerations to available resources. The online toolkits for addressing vulnerabilities include the following interactive sites:

- Climate Change Vulnerability Assessment Tool for Coastal Habitats (CCVATCH)
- Coastal County Snapshots
- Coastal Flood Exposure Mapper
- Coastal Change Hazards Portal
- FEMA Flood Map Service Center
- Sea Level Rise and Coastal Flooding Impacts Viewer
- Surging Seas—Sea Level Rise Analysis by Climate Central.

The "Surging Seas" tool is particularly useful for estimating the expected levels of flooding and socioeconomic vulnerability that may prevail in the event of any given height of sea level rise or storm surge inundation. An application to Florida's Nature Coast was presented in Chap. 14, Fig. 14.13. An example of a social vulnerability map of New York City and nearby New Jersey Cities is shown in Fig. 18.4.

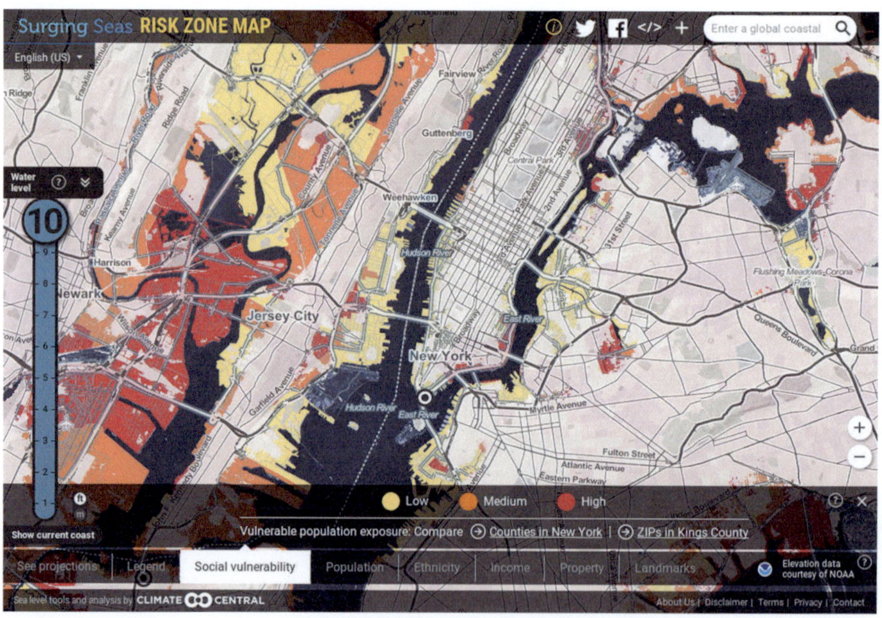

Fig. 18.4 Predicted social vulnerability of communities centered around New York City and Newark, New Jersey to a 10 foot storm surge or sea level rise. From "Surging Seas" risk zone maps produced by *Climate Central*. Orange = medium vulnerability; red = high vulnerability. Note the high level of vulnerability of several New Jersey communities

18.7 Imperiled Urbanites

In his recent book *Extreme Cities,* scholar Ashley Dawson (2017) discusses in great detail the growing vulnerability of impoverished people living in coastal cities, particularly coastal megacities, to the impacts of climate change. Dawson points out that nearly 2 billion people currently live in densely populated, flood prone urban environments that are increasingly subject to storm surges and floods. In addition to the direct threat of drowning, these populations are subject to water-borne infectious diseases such as those described in Chap. 10. The intersections of increases in sea level, tropical storm severity and the growing populations of dreadful urban slums are setting the stage for unprecedented future tragedy.

18.8 Mitigation Strategies to Reduce Vulnerability and Enhance Adaptability

Education and awareness are probably the most effective tragedy preventers, at least in communities that are fortunate enough to have options and a reasonable degree of mobility. Understanding what can happen, or what is about to happen, and having a good idea of what to do about it can save many lives. According to the report of the NOAA-sponsored *Coastal No Adverse Impact Handbook* (CNAI 2007, p. 67) "Perhaps the most effective means of mitigating coastal hazards and protecting sensitive coastal environments is education through outreach". Making coastal residents aware of threats over the long term and helping them to understand what these threats mean in non-technical and simple terms is necessary to prepare these people to respond appropriately in the event of emergencies. For communities where literacy or language is an issue verbal communication via town hall training sessions is required in addition to multi-language brochures. Web-based tools can also be effective as can exploitation of social media to communicate warnings. A few U.S. universities already offer programs focused on helping communities become more disaster resilient through practical education. A prominent example is the *Center for Hazards Assessment, Response and Technology* (*CHART*), an applied social science hazards center at the University of New Orleans (http://scholarworks.uno.edu/chart). Among other activities CHART staff provide risk literacy as a component of more general adult literacy programs. Some coastal counties and a few churches offer similar programs. In order for the general public, particularly in low income or undereducated communities, to trust and believe "academic" messengers, it is often necessary to first reach out in partnership with local community or church leaders, well-known athletes or musicians or trusted organizations such as the NAACP in the U.S. Collaboration across the broadest possible spectrum of the community is required. "Lecturing" by scientists or authorities rarely works and may provoke distrust and resistance in some communities.

Beyond education, the CNAI (2007) Handbook outlines several general strategies for mitigating coastal vulnerabilities. Among these are regionally specific adaptive management policies; these are the subjects of the next Chap. 19 of this book and some have been reviewed in the case studies presented in Part 3. Where feasible, the "preferred" mitigation strategy for residents of low-lying, flood-prone coastal areas is relocation. However, it is understood that this strategy is unrealistic in the majority of cases for various reasons and is in fact impossible for residents of low elevation island nations such as the Maldives. For those people, "relocation" means becoming an environmental refugee. In cases like Bangladesh, the area of inundation threat is vast as is the number of vulnerable residents. Even where feasible relocation options exist or are offered by government entities, they may be resisted for cultural or ideological reasons (e.g. Smith Island, MD in the Chesapeake Bay; Chap. 15).

In the past, attempts to depend on a variety of coastal protection structures such as seawalls, revetments, breakwaters etc. have been found to exacerbate erosion and impede post-storm drainage and are now regarded as the strategies of last resort (CNAI 2007). Preferred approaches involve creating natural buffers, encouraging land reclamation or simply retreating from the shore. The use of natural infrastructure to mitigate land loss and inundation is discussed in Chap. 11. In most cases, there is no single strategy or solution to mitigate vulnerability. Instead, a suite of approaches will be needed as has long been the case for the Netherlands. As described for example in Chap. 12, Du et al. (2013) outline at least four methods that must be employed in concert to make the Pearl River Delta more resilient. These include improved monitoring and forecasting, new and innovative engineering structures, restoration of degraded ecosystems and management of a sophisticated network of reservoirs for impounding and releasing floodwaters.

Other new mitigation and emergency management strategies for the future are likely to take increasing advantage of "big data" and social media to identify and respond to localities in greatest jeopardy in real time. This is the subject of Chap. 20 . Data intensity combined with social media and widespread access to smart phones can enable organic and self-organizing responses to natural disasters (Edwards et al. 2015; Keim and Noji 2011). Collective actions by informed citizens can enhance rapid response and resilience (Colander and Kupers 2014). Community-based response during natural disasters such as hurricanes, tsunamis, or earthquakes are increasingly exploiting social media and are likely to become important aspects of disaster response and community resilience (Landwehr and Carley 2014). Of course, the "Achilles Heel" continues to exist in the form of disabled cell phone service as was prominent after Hurricane Maria made landfall in Puerto Rico. Enterprises such as Uber can also play valuable roles during emergency situations. An adaptive example is the recent Indian based "Uber-like" service Ola that responded to the flooding in Bangladesh by deploying boats to evacuate people during flooding.

References

Balica, S.F., N.G. Wright, and F. Meulen. 2012. A flood vulnerability index for coastal cities and its use in assessing climate change impacts. *Natural Hazards* 64: 73–105.

Berkes, F., J. Colding, and C. Folke. 2003. *Navigating social-ecological systems; Building resilience for complexity and change*. Cambridge: Cambridge University Press.

Brouwer, R.Sonia, S. Akter, L. Brander, and E. Haque. 2007. Socioeconomic vulnerability and adaptation to environmental risk: A case study of climate change and flooding in Bangladesh. *Risk Analysis* 27 (2): 313–326.

CNAI. 2007. *Coastal no adverse impact handbook,* 165. Special Publication by The Coastal Services Center, National Oceanic and Atmospheric Administration and the Association of State Floodplain Managers.

Colander, D., and R. Kupers. 2014. *Complexity and the art of public policy: Solving society's problems from the bottom up*. Princeton, N.J.: Princeton University Press.

Cutter, S.L., C.G. Burton, and C.T. Emrich. 2010. Disaster resilience indicators for benchmarking baseline conditions. *Journal of Homeland Security and Emergency Management* 7(1), Article 51. https://doi.org/10.2202/1547-7355.1732.

Cutter, S.L., K. Ash, and C. Emrich. 2014. The geographies of community disaster resilience. *Global Environmental Change* 29: 65–77.

Dawson, A. 2017. *Extreme cities: The peril and promise of urban life in the age of climate change*, 378. New York: VERSO.

De la Vega-Leinert, A.C., and R.J. Nicholls. 2008. Potential implications of sea-level rise for Great Britain. *Journal of Coastal Research* 24: 342–357.

Delgado-Serrano, M.M., and P. Ramos. 2015. Making Ostrom's framework applicable to characterize social ecological systems at the local level. *International Journal of the Commons* 9(2): 808–830 [Online] http://www.thecommonsjournal.org/index.php/ijc/article/view/567.

Delgado-Serrano, M.M., P. Vanwildemeersch, C. Ortiz, R. Escalante, M. Rojas, J. Berbel, and P. Ambrosio. 2014. *D1.2: Locally-adapted prospective analysis techniques to social-ecological-systems*. COMET-LA, University of Córdoba, Spain. [Online] http://www.comet-la.eu/images/comet_la/deliverebles/Deli_1.2.pdf.

Dewan, T. 2015. Societal impacts and vulnerability to floods in Bangladesh and Nepal. *Weather and Climate Extremes* 7: 36–42.

Dronkers, J., and T. Stojanovic. 2016. Socio-economic impacts—Coastal management and governance. In *North sea region climate change assessment*, ed. M. Quante, and F. Colign, 475–488. Springer.

Du, Y.-D., X.-H. Cheng, X.-W. Wang, H. Ai, H.-L. Duan, J. He, and X.-X. Wu. 2013. A review of assessment and adaptation strategy to climate change impacts on the coastal areas in South China. *Advances in Climate Change Research* 4 (4): 201–207.

Edwards, J.F., S. Mohanty, and P. Fitzpatrick. 2015. *Assessment of social media usage during severe weather events and the development of a twitter-based model for improved communication of storm-related information*. Starkville, MS: Coastal Storm Awareness Program, NOAA.

Elliott, J.R., and J. Pais. 2006. Race, class, and Hurricane Katrina: Social differences in human responses to disaster. *Social Science Research* 35 (2): 295–321.

Flanagan, B.E., E.W. Gregory, E.J. Hallisey, J.L. Heitgerd, and B. Lewis. 2011. A social vulnerability index for disaster management, *Journal of Homeland Security and Emergency Management* 8(1), Article 3. https://doi.org/10.2202/1547-7355.1792, http://www.bepress.com/jhsem/vol8/iss1/3.

Goodell, J. 2017. *The water will come*, 340. New York, NY: Little Brown and Company.

Guillard-Gonçalves, C., S.L. Cutter, C.T. Emrich, and J.L. Zêzere. 2014. Application of social vulnerability index (SoVI) and delineation of natural risk zones in greater lisbon, Portugal. *Journal of Risk Research* 18 (5): 651–674. https://doi.org/10.1080/13669877.2014.910689.

Gunderson, Lance H., and C.S. Holling. 2002. *Panarchy: Understanding transformations in human and natural systems*. Washington: Island Press.

Keim, M.E., and E. Noji. 2011. Emergent use of social media: A new age of opportunity for disaster resilience. *American Journal of Disaster Medicine* 6 (1): 47–54.

Landwehr, P.M., and K.M. Carley. 2014. Social media in disaster relief. In *Data mining and knowledge discovery for big data*, vol. 1, ed. W.W. Chu, 225–257. Springer Berlin Heidelberg.

Norris, F., S. Stevens, B. Pfefferbaum, K. Wyche, and R. Pfefferbaum. 2008. Community resilience as a metaphor, theory, set of capacities, and strategy for disaster readiness. *American Journal of Community Psychology* 41 (1): 127–150.

Pendleton, E.A., J.A. Barras, S.J. Williams, and D.C. Twichell. 2010. *Coastal vulnerability assessment of the Northern Gulf of Mexico to sea-level rise and coastal change*. U.S. Geological Survey Open-File Report 2010-1146.

Rosati, J.D., K.F. Touzinsky, and W.J. Lillycrop. 2015. Quantifying coastal system resilience for the US army corps of engineers. *Environment Systems and Decisions* 35 (2): 196–208.

Salgado, H., J. Bailey, R. Tiller, J.-L De Kok, and R. Richards. 2013. *General model of socio-economic vulnerability*. OCEAN CERTAIN FP7-ENV-2013.6.1-1. Project number 603773.

Smith, E.F., S.N. Lieske, N. Keys, and T.F. Smith. 2016. Rapid regional-scale assessments of socio-economic vulnerability to climate change. *Environmental Research Letters* 11: 034016. https://doi.org/10.1088/1748-9326/11/3/034016.

Van Zandt, S., W.G. Peacock, D.W. Henry, H. Grover, W.E. Highfield, and S.D. Brody. 2012. Mapping social vulnerability to enhance housing and neighborhood resilience. *Housing Policy Debate* 22 (1): 29–55.

Yan, B., S. Li, J. Wang, et al. 2016. Socio-economic vulnerability of the megacity of Shanghai (China) to sealevel rise and associated storm surges. *Regional Environmental Change* 16 (5): 1443–1456. https://doi.org/10.1007/s10113-015-0878-y.

Yang, L., J. Scheffran, H. Qin, and Q. You. 2014.: *Climate-related flood risks and urban responses in the Pearl River Delta, China*, 27. Working paper **CLISEC-27**, Research Group Climate Change and Security, University of Hamburg.

Chapter 19
Future Adaptive Coastal Management

Bruce G. Thom

Scientists may depict the problems that will affect the environment based on available evidence, but their solution is not the responsibility of scientists but of society as a whole.
—Mario Molina—Nobel Laureate

19.1 The Case for Coastal Adaptation Action

In this book there is a clear articulation of coastal systems and how they evolve over time. The forces of climate change are only part of a suite of forces both of natural and human origin that produce changes in coastal conditions. Pressures arise from communities to undertake actions that will address what they see as adverse impacts leading to various scales and types of local, regional or national intervention. The question is whether such intervention can fulfill the needs of both present and future generations given the expanding knowledge base involving different coastal environments and the global array of social-economic-political-legal circumstances. And, as we explain in the final chapter, a key determinant of how effective the intervention is will be critically dependent on the extent to which the planners and policy makers understand and accept the science that must underpin mitigating strategies.

It must be expected that climate change with or without the presence of other contributing forces, will alter coastal conditions around many parts of the world. For most low-lying coastal settlements a significant level of change is unavoidable, regardless of any future reductions in greenhouse gas emissions. So for these areas and societies it is inevitable that steps must be taken to plan to adapt to these changes and as a result set in train actions to invest in reducing adverse impacts. Timing is critical. In some areas there is little time to lose; perhaps already it is too late to save communities from being devastated by erosive storms or long-term

---

B. G. Thom (✉)
School of Geosciences, University of Sydney, Sydney, NSW, Australia
e-mail: bruce.thom@sydney.edu.au

L. D. Wright and C. R. Nichols (eds.), *Tomorrow's Coasts: Complex and Impermanent*, Coastal Research Library 27,
https://doi.org/10.1007/978-3-319-75453-6_19

inundation. But for others the need is less urgent with the greater impacts being expected in the later half of the century. Meanwhile, current climate variability as seen in the historical record of extreme events will require even these communities to act as population in coastal areas continue to grow.

An emphasis in coastal management is the reduction of risk where risk is defined as the product of likelihood and consequences. Adaptive strategies are fundamental in reducing risk. This will involve a clear understanding of two components to future adaptive coastal management: the first is the recognition of impediments to long-term planning at different scales across different nation states; and the second is the requirement of nation states, and other governments (such as constituted states in a federation), to apply a legal system that fully supports the introduction of climate change adaptation policies based on science that will address those barriers and in so doing provide the appropriate legal means to implement and enforce it. Without such legal powers embedded in legislation, any commitments made in the short term can easily be overlooked or discarded with political change. The nature of coastal change is such that the future must not be sacrificed to the whims of the present.

19.2 Impediments to Adaptive Coastal Management

In most settled coastal areas of the world there are differences in opinions as to what are the most appropriate forms of land use. Conflicts often exist between and within the land holder class and government authorities. Pressures to capitalize on property values, investment opportunities, alienation of public land, and construction of major works are some of the outcomes local communities face over the use of land adjoining waterways and the sea. This is why we term coastal lands "contested spaces" subject "to death by a thousand cuts"! Thompson (2007) has eloquently described the various forms of coastal conflict in the U.S.A.

Different interest groups at work in developed lands will continue to operate for their vested benefits and work to thwart longer-term strategic efforts of those who seek to use scientific understanding of coastal change to reduce exposure of future generations to risks. It is vital to know up front who and why those involved in different coastal land uses would want to place barriers in the face of strategic planning. This knowledge would help combat those with vested interests using political and even legal means to allow for developments to occur in places that are or will be in harm's way. The aim should be to prevent or reduce legacies that will require compensation given that such developments create adverse effects to the environment or other land uses.

Six barriers can be recognized that make it difficult to develop adaptive strategies in coastal management:

- It is often very difficult for a community or a government to reach agreement on a long-term vision for the future of any given coastal area. This difficulty is compounded by political opportunism leading to changes in power structures with government officials being forced to change directions at the behest of new leaders. This is manifest in changing priorities for investment or changes in the support for certain stakeholders at the expense of others. All this could and does change after elections as the power base shifts.

- Officials in government agencies are more inclined to follow land use planning procedures, engineering design and management practices that are based on historical evidence, and avoid what may appear to be alarmist predictions from climate science. This innate conservatism can be easily justified to political masters who may otherwise be nervous about incorporating uncertain projections in planning when they know that this will more than likely add to development costs and be readily challenged in courts. Thus short-term thinking is quite common and future risks associated with climate change are seen as beyond decision-making horizons, such as next 4 year budget cycle.
- Many decision-makers are prepared to allow the future to take care of itself; here lies the moral hazard in coastal management. Those with property at risk may be content to accept that future governments or even the insurance industry will support them after a disaster so that there is a disincentive to prepare for future risk of the scale and type forecast under climate change. This results in little local pressure to invest in protective works or other adaptive measures.
- Existing use rights and the pressure on governments to pay market prices for compensating those at risk to relocate often serves to discourage the introduction of policies that stop or force changes to development activity in areas that are and will soon be in harm's way. Long-term benefits are put to one side as land owners can appeal for retention of certain property rights under just terms or similar regulation to continue their existence and then demand help when inevitable disasters occur.
- Across institutions of government there are often jurisdictions with different legislative and budget responsibilities for coastal management. It is the so-called silo effect; agencies or entities within an agency may work to an agenda defined historically or by the political/managerial authority of the day. As noted by Mumford and Harvey (2014) in their discussion of parliamentary inquiries in Australia, a major impediment to achieving integrated coastal zone management (ICZM) is "disintegrated decision making that occurs between local, state, and the commonwealth government". Even at the level of leadership, the role of short-term performance criteria in assessing outcomes within an entity can dominate decision-making so that integrating or coordinating planning for an adaptive future within government takes lower priority, or at worst is treated with disdain. This inhibits the mainstreaming of cost-effective, long-term adaptation approaches.
- In many countries, including the USA, the downsizing of governments within a neo-liberal context has resulted in reduced technical capacity across all levels of government. In many places this has resulted in outsourcing advice and practices to consultants operating within constrained budgets and poorly designed briefs. Lack of qualified staff within government agencies to interpret and

implement actions has also restricted capacity of officials to develop adaptive strategies for the future by being able to assess and reduce climate risk through frank, fearless, and well-informed advice based on best available science.

19.3 A Legal Framework Is Essential for Future Adaptive Coastal Management

In order for societies to assess and reduce future risks associated with impacts of coastal forces on settlements in coastal regions, it will be necessary to embed the potential for harm by those forces in law. James Thornton (in Thornton and Goodman 2017, p. 57) has outlined what he terms "The lifecycle of the law" which provides a 5 phase legally-based framework for governments to approach the management of future risks with a certain level of confidence. From a coastal perspective, the key to this framework is vested in the first step to ensure the legal framework is based on scientific understanding of coastal change. There are several facets to include in the framework and these are discussed in the following subsections.

19.4 Scientific Input

For science to be relevant to decision makers, now and into the future, it must be seen as both credible and relevant. It should be self-evident for coastal scientists to be engaged in the outreach of their findings and understandings of coastal change to the body politic and more broadly into community debates (Oreskes 2017). Scientists must demonstrate that they respect the decision makers and they must strive to communicate with decision makers in non-technical terms and without jargon. They must take care not to be perceived as "talking down". To hide within the confines of the academic or research institutional world deprives society of benefits of their discoveries and intellect. It is apparent that not all scientists are willing or even skilled at communicating their work, but it is important that there are mechanisms to achieve translation of their knowledge into policy and management practices. This book is one such endeavor.

In order to be seen as credible advocates for the science and have the respect of community leaders, public officials, politicians and other professionals, coastal experts should be able to show they have experience in the real world of field studies. To be confronted by science skeptics without being able to demonstrate a background in field observations and experiments quickly allows those on the radical fringe to create doubt. Here is where modeling can be a curse as well as a benefit in articulating the way coastal conditions will change in the future. Modeling tools as discussed in this book are extremely powerful ways to show particular pathways of change. But all too frequently the assumptions on which they are built, or the complexity of processes they are trying to simulate, are ignored and the model output becomes the end and not a means to achieving cost-effective management outcomes for particular areas. Furthermore, communities can be easily stirred to conflict when confronted with cheap, simple models depicting lines on a map of shoreline recession which do not consider geological inheritance, sediment budgets or history of shoreline change, let alone morphodynamic concepts including feedbacks and thresholds (Wright and Thom 1977).

In the development of a legal framework, which incorporates scientific understanding of coastal processes and evolution, it is essential to balance the generic with the specific. Each coastal area has its own unique attributes both in time and space. No policy or legislation can possibly accommodate all the complexities and variables that confront managers and communities in any given jurisdiction. But there are some basic principles and practices involving coastal science that can help decision-makers. An example is the need for states to apply strategies on a regional basis rather than at a local jurisdictional level, for instance how natural processes such as sediment exchanges operate and have impact across administrative boundaries (Psuty and Ofiara 2002; Thom et al. 2018).

Fundamental to the incorporation of biophysical science into policy and law is the need to provide a convincing narrative of history. Communities and law-makers will be more easily convinced to incorporate provisions that reflect the potential impacts of changing climate or the new climate era if what they are being asked to do reflect the history. Here is where we must build on the base of natural variability and how that variability created vulnerability. To divorce climate change from the realities of past is to expose science to criticism. That variability in the future and long term-changes in coastal conditions will in all probability for many coasts fall outside the historical record is accepted, but to press for legal or policy change on those grounds alone will create political and community resistance. There are too many other vested interests of immediate concern to counter such projections. The way forward is to demonstrate that risk is exacerbated by climate change impacts such as warmer seas, higher sea levels and enhanced frequency of inundation from both the sea and land. Social sciences have a key role in this enhanced understanding such as through developing new economic tools to assess risk and determine priorities for action. In addition, there is a need to capture past experiences of living in vulnerable areas and to learn from those experiences.

19.5 Policy Development

One key role of scientists is to inform governments of where there are or will be policy failure. A good example is in the area of fisheries policy. Collapse of many global fisheries was foreseen by fish biologists. But that didn't immediately lead to economic restructuring; too late for some devastated communities. Lessons learnt have resulted in policy change in North America and Europe (Thornton and Goodman 2017).

Climate change is now compelling enough for legislatures to contemplate how best to shift from ignoring potential harmful impacts to a more proactive position. Here is where coastal scientists have been able to engage in policy debates at various administrative and political levels. Some of us who have been embedded within a bureaucracy, or served in some advisory capacity as external experts, have had some successes. However, there are now more and more cases in different countries where coastal and climate scientists in public office have been made redundant or shifted into other positions as political events contrary to climate change advocacy have taken hold. This can be very frustrating and quite upsetting. Yet experience in different jurisdictions does offer some hope as the winds of political change switch direction allowing scope once again for policy development that incorporates climate science in general and coastal science in particular (e.g. New South Wales *Coastal Management Act* 2016, see below).

A key to science–based policy development/revision is the ability of scientists to speak "truth to power". This phrase encapsulates the value of respected, credible scientists serving as public good advocates. Two points here: first, they should not rely on others to advocate their science; lawyers can be very competent and effective, but they usually serve their paymaster; in the field of public good advocacy the science must speak for itself with passion and substance based on procedures that have been the subject of appropriate review. And second, they can demonstrate how emerging policy can best relate to their personal observations, experiments and models as well as the historical/geological record in ways that is hard if not impossible to contest let alone for non-scientists to convey.

In coastal management two of the major stakeholders who wield power are private property owners and property/commercial developers. Wherever coastal land is attractive for housing, resorts, marinas, canal sub-divisions or port expansion, there are pressures on governments to agree to such investments. They are especially attractive to private sector investment interests, but in places developments will occur in partnership with governments seeking to demonstrate their capacity to generate jobs and local taxes. These private land owners or investors can achieve levels of political access and influence for short-term gains far beyond the capacity of others in the broader community that do not share those interests (Thompson 2007). In some cases it leads to a sense of permanent entitlement to secure government support to protect property interests and values even at the expense of natural coastal values. Armoring of the coast by using tax payer funds is one example; in the USA the rules surrounding the role of the US Army Corps of Engineers has been cited as another way to help private beneficiaries at the expense of the public (Pilkey and Dixon 1996).

Science can play a major role in the formulation of public good policy, including coastal flood insurance policy, and in influencing public expectations (McGuire 2015). This can range from a clear articulation of enhanced risk to communities from storm surges and erosion during hurricane events, to the value of ecosystem change as the sea invades coastal wetlands, to how best maintain the physical and biological integrity of coastal dunes. In such cases the public benefit of having natural features, or understanding adverse impacts of human structures, should be explained to policy makers now exposed to climate change forces. This must include those involved in coastal flood and erosion insurance policies. Application of the public trust doctrine has been shown to be effective in this context (Titus 1998), although inconsistent legal interpretation has created problems (Thom 2012).

19.6 Legislative Change

It is not the role of scientists to draft legislation. That is for legal experts working within a given constitutional framework. However, there should be scope for these experts in conjunction with committees of legislatures to receive and debate advice from scientists. This is surely a must in matters as contentious and significant as climate change.

There is now a vast amount of legislation that attempts to integrate environmental and coastal issues within the broader context of coastal management. The US led the way in the early 70s with a range of laws to ensure better protection of environmental values. Many countries have followed and enacted legislation to overcome the rampant acts of land and water degradation that accompanied industrial development. These laws were extended to coastal areas where the spread of urban settlement into areas subject to periodic extreme storm events added to the already set of pressures that were seen by communities as destroying or exposing natural and built assets to increased risk. An example at a national level was the US *Coastal Zone Management Act* 1972 which helped pave the way for state laws such at the *Californian Coastal Act* 1976. But other countries already had legislation, still extant, that pre-dated such initiatives although primarily designed for purposes of coastal protection of coastal land from loss to the sea (e.g. *UK Coastal Protection Act* 1949). But the problem with most of this legislation is their failure to incorporate elements which encompass the uncertainties of climate change.

Opportunities should exist for coastal scientists to engage with legislators or parliamentarians at all stages of the process of drafting or amending legislation. One compelling reason for this is because all politics is local and coastal residents and businesses have every right to push for their vested interest to be considered if not protected. Similarly in a democratic system it is critical that the knowledge base and credibility of science of coastal change is also exposed to the rigours of political debate. When it is time to draft new bills it is not unusual for the lobbying process to be open enough for the voice of science to be included. Many legislatures will involve the use of committees to hear the case for change. Clear explanation of reasons for incorporating new knowledge, or expanding concepts embraced by the bill to include provisions that otherwise would advantage private over public good, can be made at committee stage. It is often frustrating to a scientist if long term cumulative impacts are not included in the legislation. However, in the brave new world of communicating the long term impacts of climate change impacts that should not be seen as a defeat, just a temporary setback. The consequences of inaction or business as usual by legislatures will soon be viewed as short-sighted and engagement by scientists should not cease.

Experience in drafting the new *Coastal Management Act* in New South Wales, Australia, in 2016 demonstrates the value of using coastal science and engineering at all stages in the drafting process. Critical to this engagement was the acceptance of the latest scientific concepts such as those arising from uncertainties with climate change. Members of an Expert Panel were used to explain these concepts to members of NSW Parliament. The concepts included the need to understand sediment compartments and coastal processes that operate across jurisdictional boundaries similar to that advocated by Psuty and Ofiara in New Jersey in 2002. Compartments were spatially defined in a schedule to the Act. The new legislation embraced the need to map a range of different coastal hazards by local governments. Legal experts responsible for drafting the new Bill were able to consult with members of the Expert Panel to ensure the legal and science interpretations were acceptable. All this was time consuming but the process was worth the effort even if not all we set out to achieve proved politically acceptable.

19.7 Implementation Gap

Implementation of policy and law mostly rests with government officials operating within constraints of technical guidelines and budgets approved through the political process. Managers within agencies must assign tasks to ensure the budgets are spent appropriately and can be accounted for according to audit rules. All this seems straightforward except there is often little input from science into technical measures required to ensure adaptation to climate change over long time frames.

Implementation involves many facets presumably with some element of consultation with communities and various vested interests. Different officials work in areas such as asset management, land use planning, protection of environmental values and emergency management. Bringing all this together is often difficult to achieve (that is disintegrated CZM), especially where there is conflict between stakeholders. This leaves governments open to criticism that implementation is not meeting the aspirations of public policy or addressing community values and priorities. Here we have the implementation gap. The gap only gets wider when longer term considerations such as planning for climate change requires consideration and action.

Coastal science with its links to coastal engineering provides opportunities to make officials continuously aware of any deficiencies in the implementation process as it relates to the implications of climate change. Science is in a position to engage in the review of guidelines or manuals that underpin decision-making and actions that individual jurisdictions are obliged in law to follow. In particular, coastal science provides a way to communicate uncertainties in dynamics of change (including “tipping points”) above and beyond what can be expected under natural, historical variability.

One problem that science faces at the implementation stage is that of image. It is so easy to convey gloom and doom. So much of what can be projected with sea level rise or with heightened surge levels and more intense cyclones will require consideration of matters such as new building codes, or more protection works, or even relocation of population. These are highly contentious politically and costly as we saw in the debates following Hurricane Sandy in New Jersey. It becomes a matter of great international significance when nations under threat in the Pacific or Indian oceans or in the Caribbean require massive international aid to survive. Lands currently occupied to feed millions or to house growing populations that may have to migrate will be subject to scrutiny from various perspectives and scientists will be expected to provide expert advice on future change. Here is where more positive outcomes need to be explored as challenges facing jurisdictions at various scales continue to arise.

Science is in a position to contribute to the implementation phase in three ways. First, science plays a major role in the provision of advice on coastal processes and the nature of coastal change under current and future conditions. The advice must aim to reduce contestability of debates around options to mitigate hazard impacts. As noted above, it should not simply be a matter of applying models where assumptions can be disputed. The nature of historical geomorphic change and sediment flows can also provide a basis for extrapolation into the future. In this way the sensitivity of shorelines to change can be identified over different space and time scales. This will assist in the identification of “tipping points” or thresholds that define the need for intervention using a pathways approach to action. It is important to note that some shorelines or estuary basins/entrances may continue to accrete as sea levels rise and these need to be identified as part of an evaluation of the behaviour of discrete sediment compartments (Thom et al. 2018).

Second, guidelines or manuals for coastal managers need not only to be revised according to improvements in our understanding of coastal processes and impacts, but also to be formally incorporated in a coastal adaptation management cycle that aims to deliver a *Coastal Management Plan*. Usually this cycle involves 5 stages: scoping, area study, assess options, implement preferred option, and then monitor outcomes. A respected coastal scientist or engineer should be a member of a committee that is responsible for advising decision-makers on the development and delivery of a plan. As is the case in the UK, such plans cross jurisdictional boundaries where they are linked to sediment cells. Current practices in England and Wales using Shoreline Management Plans (SMPs) designate each stretch of coast according to one of four defense options: hold the line; retreat the line (managed retreat); advance the line; do nothing (Cooper and McKenna 2008). However, there are cases where advice must be revised as more knowledge defines weaknesses in initial understanding of processes and the designation of an option.

Third, it must be continually put that while there are many uncertainties in our understanding of how coasts will evolve in the future, this should never be an excuse for ignoring or discounting coastal research. While that point is self-evident to those involved in coastal science and engineering, it is not necessarily the case where vested interests can argue for business as usual. We include here public officials who see little value in long-term investment in monitoring or experiments that are deemed valuable to the application of, for instance, the pathways approach to coastal management. Experience in Holland and in the Thames estuary in the UK highlights the importance of understanding past natural disasters and how they will be exacerbated in future under climate change. If we are to move towards implementing policies that embrace "sustainable hazard mitigation" (Mileti 1999; Psuty and Ofiara 2002), then more knowledge is required and this knowledge must be shared to help decision-makers reduce social, economic and environmental costs associated with coastal hazards.

19.8 Enforcement

The fifth phase in Thornton's lifecycle of environmental law is enforcement. This phase may be quite challenging to scientists when confronted with two decisions; the first whether to advocate for a legal case against an entity that is apparently in breach of the law; and the second is taking the time and effort to help prosecute the case through the complexities and costs of the legal system.

There are those in government or in NGOs with the responsibility to ensure public policies are enforced. Environment Protection Agencies exist in many countries at both national and state levels (e.g. Coastal Commission in California) to perform this task. In addition, public and private funded NGOs such as ClientEarth are very effective in bringing cases for judicial review.

While most of the hard yards in enforcement must be the province of lawyers, there is certainly scope for coastal scientists to play a role. The courts often rely on expert witnesses to reach a decision. In an adversarial legal system the court must be in a position to hear from experts representing different sides. Unless a mediated solution is reached then the court has to "weigh up" the evidence. This can be a very uncertain business depending on how the evidence is presented and considered in the context of all the legal requirements that are before the court. In many cases this involves interpretations of words and clauses in legislation that may not be consistent with a scientific understanding. Thus it can be frustrating to the expert scientist to hear a ruling that is at odds with science or even best engineering and land use planning practice.

There are many coastal law cases that define the legal framework under which decisions must be taken especially in common law countries. Prior to any consideration of climate change, legal disputes on coastal "rights" reach back centuries flowing from Roman law and the *Magna Carta*. Questions of rights of access, protection of environmental values, and property "takings" have all been subject to litigation including in places where the public trust doctrine prevails (see summary in Titus 1998).

Challenges continue to laws that purport to protect public rights such as the "rolling public beach easement". A recent example was in Texas where the *Open Beaches Act* 1959 came under review in the Severance lawsuit before the Texas Supreme Court (Young 2012). This case has huge implications in ruling that no rolling easement can result from "avulsive events" thus potentially limiting public beach use. Another case with a different result involved a landowner in England seeking to regain lost land to the sea. Along the coast of East Anglia there is a long history of cliff recession and attempts to hold back the advancing sea. At Easton Bavents near Southwold, a Mr Boggis created a giant rampart a kilometer long, 25 m wide, and 5 m high with rocks at the base at his own expense with no permission. Several agencies took exception and a massive fight occurred all the way to the European Court. He lost on two counts: the courts finally ruled against him as the cliffs had been deemed a site of special scientific interest by Natural England; and the sea ate away his rampart (Barkham 2015). Both these cases highlight how coastal management can be affected by enforcement decisions in the context of a dynamic coastal environment.

The issue is how science may influence future enforcement processes with or without lawsuits. Sorell Negro (2013) is one lawyer who has addressed this question. She looked at recent changes in laws that require consideration of sea level rise in planning and development and that tend to favor "living shorelines over armoring" (e.g. Connecticut). These changes may have implications on enforcement of protective works and may affect takings claims. We note already that older UK law (*Coastal Protection Act* 1949) provides discretionary powers to carry out works by local authorities "as may appear to them to be necessary or expedient for protection of land". This has allowed governments in the UK to avoid costly compensation claims and seems fit for the purpose of implementing at minimum public expense. But as seen in the Severance case in Texas, under that state's law, the retreat of the beach as a result of an extreme or avulsion event could constitute a taking if public rights to access a beach are to prevail. This type of judgment in the view of some legal academics requires a more flexible approach to interpretations of coastal law so that public and environmental interests are not compromised under climate change conditions (see Negro 2013 for discussion). Clearly scientific input into the nature of coastal change in the future needs to be embedded in law so that enforcement processes can be clarified. We believe we are heading in this direction with the new *Coastal Management Act* 2016 in NSW, Australia.

19.9 Future Resilience Depends on Effective Adaptive Management

Future adaptive coastal management cannot be seen as a one-size-fits-all. Changing coastal conditions will lead to circumstances where decision-makers will be forced to balance different values. Lawyers, public officials and politicians have to examine priorities and options in addressing both short and long-term issues. Here coastal science has a role to play and should be seen to be active at all phases in the decision process.

There are matters that are very much scale dependent. Regional and local plans must embrace an understanding of coastal evolution, processes and sediment budgets in order to help determine what management strategies will work best in a climate change world. In the U.S., regional and local programs are proving to be more effective than federal programs. It is important to go beyond just biophysical assessment and include social science assessments of demographic, economic and cultural factors that are important to land use planners. In the absence of such engagement, it is too easy for simplified models, or even worse assessments that are designed by those with strong vested interests that will permit developments to grow in increasingly hazardous areas.

Effective adaptive coastal management will require scientists to move beyond the comfort zone of their disciplinary confines. There is a clear need for science input in debates on sustainable use of the coast where risk to built and natural assets is to be minimized, or in the application of all five phases in the lifecycle of environmental law as described above. This can be personally painful and frustrating, but also rewarding. Even when just some of the sound science principles and specifics are incorporated in policy and law and then get implemented and enforced, there are grounds to be satisfied. But this is no justification to stop engaging. There can be no hiding in a climate changing coastal world.

References

Barkham, P. 2015. This sinking isle: the homeowners battling coastal erosion. *The Guardian*, 2 Apr 2015.

Cooper, J.A.G. and J. McKenna. 2008. Concepts of fairness in coastal erosion management. In *Proceedings of the International Pluridisciplinary Conference*, Lille, France, 16–18 Jan 2008.

McGuire, C.J. 2015. U.S. coastal flood insurance, risk perception, and sea-level rise: A perspective. *Coastal Management* 43 (5): 459–464.

Mileti, D.S. 1999. *Disasters by design: A reassessment of natural hazards in the United States*. Washington: D.C., Joseph Henry Press.

Mumford, T., and N. Harvey. 2014. Champions as influences of science uptake into Australian coastal zone policy. *Coastal Management* 42 (6): 495–511.

Negro, S.E. 2013. Preparing, adapting, and rebuilding: Rising sea levels raise new legal issues. *Probate and Property Magazine* 27: 6.

Oreskes, N. 2017. Should scientists serve as sentinels? *Plenary Address*, AAAS Meeting, Boston. Accessed https://www.aaas.org/news/naomi-oreskes-should-scientists-serve-sentinels.

Pilkey, O.H., and K.L. Dixon. 1996. *The corps and the shore*, 272. Washington D.C.: Island Press.

Psuty, N.P., and D.D. Ofiara. 2002. *Coastal hazard management. Lessons and future directions from New Jersey*, 429. New Brunswick: Rutgers University Press.

Thom, B.G. 2012. Climate change, coastal hazards and the public trust doctrine. *Macquarie Journal of International and Comparative Environmental Law* 8 (2): 21–41.

Thom, B.G., I. Eliot, M. Eliot, N. Harvey, D. Rissik, C. Sharples, A.D. Short and C.D. Woodroffe. 2018. National sediment compartment framework for Australian coastal management. *Ocean & Coastal Management* 154: 103–120.

Thompson, R. 2007. Cultural models and shoreline cultural conflict. *Coastal Management* 35: 211–237.

Thornton, J., and M. Goodman. 2017. *Client Earth*, 314. London: Scribe.

Titus, J. 1998. Rising sea, coastal erosion, and the takings clause: How to save wetlands and beaches without hurting property owners. *Maryland Law Review* 57 (4): 1281–1399.

Wright, L.D., and B.G. Thom. 1977. Coastal depositional landforms: A morphodynamic approach. *Progress in Physical Geography* 1 (3): 412–459.

Young, H.S. 2012. Rising to the challenge: Impacts of the Severance lawsuit on Texas open beaches. *ASBPA 2012 National Conference*, San Diego, Oct 2012.

Chapter 20
Data-Intensive Alternatives for Human Adaptation to Coastal Change

Arthur G. Cosby, Gina Rico Mendez and Hasna Khandekar

The measure of intelligence is the ability to change.
—Albert Einstein

20.1 The Nature of Human Utilization of Resources

The history of mankind has demonstrated an incredible capacity of humans to adapt to a variety of physical conditions through the formation of effective social organizations, which have been the most powerful instrument of environmental transformation. For better or worse, the phenomena that encompass the great acceleration (Steffen et al. 2015) are greatly the result of human organizations in action. Crutzen (2002) suggests that during the Anthropocene or human-dominated geological era, the environment has transformed radically as a result of the "rapid expansion of mankind in number and per capita exploitation of Earth's resources" (p. 23). However, human existence is only possible because of the availability of natural resources; that is, the paradox of *sustainable exploitation* frames the relationship between human and earth systems.

Now, considering this connection between humans and the environment, what are the reasons humans use when they decide where to locate and how to transform that place? Due to the need to sustain populations and consolidate power, ancient Asian and middle eastern civilizations succeeded by establishing sedentary societies in geographical places that offered favorable environmental conditions for the development of agricultural systems (e.g., river deltas); this, combined with the creation of institutions capable of implementing technological change (e.g., irrigation and plow systems) became one of the most remarkable advances of human organizations which significantly impacted the Earth (Davis 1955; Jones 1981; Wittfogel 1957). The phenomenon of agriculture brought the emergence of urban areas, by increasing the capacity to feed populations who do not produce their own

A. G. Cosby (✉) · G. Rico Mendez · H. Khandekar
Social Science Research Center, Mississippi State University, Starkville, MS, USA
e-mail: arthur.cosby@ssrc.msstate.edu

L. D. Wright and C. R. Nichols (eds.), *Tomorrow's Coasts: Complex and Impermanent*, Coastal Research Library 27,
https://doi.org/10.1007/978-3-319-75453-6_20

food and allowing a system of specialization and trade (Davis 1955). The development of agricultural systems leveraged an urban-rural divide that radically transformed the foundations of the human-ecological interactions (Rico Méndez 2016) and later boosted what is known as the 'Great Acceleration' (Steffen et al. 2015).

The example of agriculture is important to illustrate human utilization of natural resources. Maslow's hierarchy of needs can be used to explain the foundation of human exploitation of Earth given the requirement of satisfying *basic* needs (Maslow 1943). According to Maslow, physiological needs are those related to the physical survival of the body; if those are considered basic human needs, it can be argued that humans would locate where those can be satisfied. That is, in the case of utilization of natural resources, humans have historically taken risks based on expected pay-offs that, in this case, relate to the satisfaction of physiological and security needs. As limited as this perspective is, it is a useful concept to explain human utilization of natural resources. If that is the case, individual decision to locate in one place or another, is the result of these expected pay-offs.

To make the needs argument more complex, we must add some caveats to the discussion: natural resources are limited, population has grown exponentially in the past century, and incentives move humans to places where they can potentially satisfy needs at some cost (i.e., risk). Let's start by discussing the issue of incentives; in game theoretical approaches, there is an assumption that actors are *rational.* "Rational behavior means choosing the best means to gain a predetermined set of ends" (Morrow 1994, p. 17). Formally speaking, the problem of decision includes: (1) a set of acts; (2) a set of states in the world; (3) a set of outcomes; and (4) a preference ordering over the outcomes. Based on available information, decisions can be made under certain conditions: certainty, risk, and uncertainty. Generally speaking, location decisions occur under conditions of risk or uncertainty, but it is the expected utility which incentivizes the decision, "[u]tility functions capture the risks of less preferred outcomes than an actor will accept to gain more preferred outcomes" (Morrow 1994, p. 29).

The issue of utility and risk can be explained with an example. John decides to move to a coastal area expecting nice weather, amenities, and, in general, a good quality of life (Rappaport 2007); however, John assumes the risk of bad weather too. Certainly, John can do nothing about the occurrence of bad weather, but he can rely on standard harm reduction measures such as building codes, insurance, and emergency management plans, all those built upon availability of information. Institutional economics suggest that availability of and access to information reduces transaction costs (North 1990). Nowadays John can potentially access millions of bits of information that can help him to reduce vulnerability despite living in an area with potentially high risk. The existence of Big Data has the potential to enhance the decision-making of individuals and communities, by combining human and physical data into models that can inform decisions to improve the outcomes of weather-related hazards and the growing uncertainty resulting from climate change. This chapter aims to provide some insights about the emergence of big data and its potential to enhance adaptation efforts given rapid environmental changes, specifically those occurring in coastal areas. This chapter is divided into four sections: (1) coastal areas: arising conditions for disasters; (2) another great acceleration: big data, challenges and opportunities; (3) the power of combing human and physical data; and (4) big data for human adaptation: alternatives to long-term transformations resulting from Climate Change.

20.2 Coastal Areas: Arising Conditions for Disasters?

Disasters are neither natural or a divine punishment, suggests Wilches Chaux (Paredes 2011), a risk management expert, who claims that disasters are the result of poor planning which has created conditions for natural events to become disasters. According to this perspective, a natural event is not a disaster in and of itself, it is the effect on humans, which we interpret to be disastrous. Disasters become disasters because humans are present and loses can be calculated in terms of lives and assets. However, it is not our purpose to disregard the literature that suggests the existence of natural and man-made disasters. Instead, we would like to suggest that the capacity to improve adaptation comes from recognizing that the entangled relationship between human and Earth systems makes us subjects of physical forces and agents of change in the environment. By thinking of disasters as a human condition, we consider the challenges and possibilities that result from the human-environment relationship, given the increasing climate vulnerability. The intensity of climate change effects is even greater in coastal areas due to high population densities and exposure to natural hazards (e.g. hurricanes, tsunamis, seawater rise). When exploring the possibilities of human-Earth connection, this article will suggest that the utilization of big data provides important alternatives for modeling and improving decision-making.

Given the human aspect of disasters, societal challenges from weather-related disasters are continuously increasing. Formally, natural hazards are defined as atmospheric, geologic, hydrologic, seismic, and biologic phenomenon that could lead to an emergency and pose a threat to life, health, property, or the environment (Steinberg 2000). Each year, the world faces a multitude of disasters triggered by natural phenomena that cause thousands of deaths and cost billions of dollars. Worldwide, between 2003 and 2013 the total cost of damage resulting from weather-related disasters was $1.5 trillion, affecting two billion people and killing over a million people (FAO cited by Luxton 2015). In the United States, according to the Center for American Progress, between 2011 and 2013, federal government efforts on disaster relief accounted for more than $136 billion (Weiss and Weidman 2013). This numeric calculation to measure the "cost" of disasters is relevant and accounts for the losses in terms of assets and lives. Importantly, population increase and density make the cost of weather-related disasters even more challenging.

Two of the most relevant socio-economic trends of the great acceleration are world population increase and urban growth. It is well established that the global population is rapidly increasing with population growth estimated to be 8.5 billion by 2030 (United Nations 2014). Significantly, the rapid population increase is not evenly distributed across the landscape. Along with this demographic expansion, there is also an important increase of population density. Population estimates for 2017 indicate that the largest 20 urban areas worldwide have over 15 million people (Cox 2017). Importantly "70% of the world's $90 trillion infrastructure bill centered on urban areas" (Bloomberg and Pope 2017, p. 256). These powerful demographic processes are shaping the future of the coast. In many locations around the globe, coastal regions are experiencing faster population growth. In the United States for example, the coastal population of the Gulf of Mexico region grew more than twice as fast as the total population between 1960 and 2008. This growth is even more pronounced in the number of residential housing structures. In 1960, the U.S. census reported about 1.8 million housing units in the Gulf Coastal region; by 2008, this number has grown to over 4.5 million (Wilson and Fischetti 2010). The attraction of lifestyle commodities associated with coastal living has fueled not only growth in the permanent coastal population, but also the growth of vacation homes of residents whose permanent home is located in non-coastal regions. These temporary vacation residents are not typically accounted for in census population growth estimates. The U.S. census also tells us that the concentration of human populations in coastal regions are frequently denser than other areas. In fact, it is not uncommon for the density to increase as one approaches the coastline (Rappaport 2007). In general terms, this demographic process depicts the rapid grow of coastal areas in terms of population, housing and infrastructure resulting from coastal lifestyle commodities, producing relatively dense populated areas often concentrated close to the shoreline.

The population growth along the coast of the U.S. is clearly transforming coastal environment. Even such dramatic weather events such as hurricanes do not reverse population expansion in most U.S. coastal locations. In 2010, the U.S. Census released a study of coastline population trends that included an analysis of those counties that were frequently hit by intense hurricanes. The ten most intense hurricanes that hit the U.S. coast between 1960 and 2008 impacted nearly 51 million people (Wilson and Fischetti 2010). Because of population growth, if these ten hurricanes hit the same location in 2008, the number of impacted people would increase to 70 million. The growth pattern for the most populous county in Florida, Miami-Dade County, clearly portrays how population growth occurs along the coast in spite of frequent exposure to hurricanes. In a Census study, Miami-Dade County was ranked as one of the most frequently impacted counties in the U.S., with eleven hurricanes occurring between 1960 and 2008 (Wilson and Fischetti 2010). Nevertheless, during this period, there was dramatic population growth; Miami-Dade County's population grew from nearly one million to about 2.5 million (U.S. Census 2016). The expected utility of coastal living in Southern Florida currently exceeds the concern for risk associated with major weather events.

The great acceleration in populations is occurring at the same time as populations are transitioning from rural to urban settings. Initially, the urbanization process along U.S. coasts started by concentrations of population along the coasts leaving sparsely populated areas between them. However, during the last few decades the trend has been for the population to spread through these previous sparsely populated areas. They are now being connected to existing cities and towns creating new metropolitan areas. Currently, about 96% of the coastal population in the U.S. is classified as metropolitan. Later in the chapter we will discuss how the experience of urban expansion is not only a challenge but also an opportunity for adapting to intense climate variability (Wilson and Fischetti 2010).

One must ask what expected utility, beyond the satisfaction of basic needs, drives people to coastal areas. According to Rappaport (2007) there has been an important population shift towards areas with "nice weather". The author suggests that nice weather areas are now seen as a consumption amenity due to rising per capita income, but also to the availability of air conditioning. The draws to scenic coasts, mountains, lakes and national parks are consumptive amenities as are low pollution, low traffic, high quality education and healthcare. Rappaport's work suggests that these might be the key to future local economic development. Coastal areas in the US seem to meet some of those criteria. On the other hand, Chen and Rosenthal (2008) suggest that some cities and regions are likely to thrive in coming years because of their capacity to either create jobs or facilitate consumer amenities. Their research findings indicate that households prefer non-metropolitan areas in warm coastal locations, whereas firms would locate in large growing cities where they have the capacity to attract highly skilled human capital. In contrast, cities with improving consumer amenities would drive retiree populations (Chen and Rosenthal 2008). Coastal areas in Florida for instance, meet the criteria of economic development that has become one of the major recipients of Latin-American migration in the U.S., and has also promoted migration from aging populations who can afford living in places with elevated housing prices (Florida Legislature—Office of Economic and Demographic Research 2012). These demographic and economic trends indicate that there is a great deal of investment and potentially vulnerable populations exposed to natural hazards in coastal Florida. The coastal regions of the world are areas characterized by large and growing concentrations of population, this important population shift has demonstrated to be both beneficial for economic and social development but also increased exposure to natural events to become hazards.

20.3 Another Great Acceleration: Big Data, Challenges and Opportunities

This section will explore another great acceleration: the explosion of data, its implications, challenges, and possibilities given the demographic and economic trends in coastal areas, as detailed in the last section. For the purpose of this chapter, we will refer to the series of societal transformation associated with digital data collection, storage and analytics featured by high volume, complexity, and speed as the data-intensive society or big data. With all new technologies operating, big data provides the potential for a new mode of human adaptation. Its use is a powerful source of scientific and technological innovation. It can be the key to transform the challenges of climate change into opportunities to reduce vulnerabilities resulting from climate variability and provide unparalleled opportunities for significant societal change and advancement. Adapting to the challenges imposed by climate change will require innovations that include the utilization of big data-driven research and applications at the intersection of social-behavioral and natural-physical sciences.

Digital data is exploding at an exponential rate. Many attribute its power to the size, however, beyond volume it is important to take into account its complexity and multi-dimensionality. First, digital data is of many different types: traditional governmental and business data-sets, social media, data from sensors, commerce data, GIS data, satellite imagery, utility data, smart transportation systems, genetic sequencing, monitoring of environmental conditions, mobile data, individually collected images and other sources. Two comprehensive categories of data emerge from these sources: humanly produced and physical data. This is important because it enables the capacity to combine human-behavioral data with environmental information to predict patterns of individual and collective behavior. Due to that possibility of merging human and environmental data, it can create a path for informing decision-making. Second, data-intensity is also driving the development of technologies which collect, store, and process information. The rapid growth of high-performance computing, especially cloud computing, the advent of wearables, the internet of us (IoUs) and the internet of things (IoT) are all examples of technologies and devices that result in the growth of data. Third, we are now able to access and utilize data in real or near-real time, and the demand and utility of real-time data greatly expands our capacity for utilizing data to improve outcomes during and post-disasters.

Bello-Orgaz et al (2016) consider the 5Vs of big data (volume, variety, velocity, value, and veracity) analyzing social media big data sets. They go into detail about how big data can be utilized for social-based applications. Applications of social media data include marketing, crime analysis, epidemic intelligence, and user experience based visualization. However, the authors alert about the challenges imposed by privacy issues, data fusion, data visualization, streaming, and online algorithms (Bello-Orgaz et al. 2016). From the energy sector perspective, Zhou et al. (2016) indicate that big data has started an evolution in our modern era. An evolution that comes with important challenges. The authors suggest that the main challenges of energy big data are in four areas: (1) data collection, storage and management; (2) data mining and analytics; (3) data utilization; and (4) data protection and risk prevention. Their approach suggests a combined modeling approach that merges the 4Vs of big data (Volume, Velocity, Variety, and Value),[1] with the 3Es (Energy, Exchange, and Empathy) of energy models (Zhou et al. 2016).

---

[1]Not 5 V's in the case of the cited work.

From the research standpoint, big data imposes serious challenges to the traditional scientific method. In academia, many aspects of big data approaches are disruptive and represent paradigmatic challenges to existing disciplinary practices and customs. Kitchin (2014) assesses how big data correlates with the scientific paradigm and suggests that big data reveals relationships and patterns that were unrecognizable before its existence. However, unlike science, big data can only be evaluated through a particular lens, which influences how it is interpreted. In this way, big data provides the opportunity to develop more sophisticated, wider-scale but finer grained models of human life. Nevertheless, applying big data to analyze human life poses a risk because people do not act in rational, pre-determined ways, instead they live lives full of contradictions, paradoxes, and unpredictable occurrences (Kitchin 2014). However, many big data advocates find the value of big data to be just its capacity to predict individual and collective behavior. The capacity to store and identify patterns within large unstructured data provides insightful information about human behavior (Stephens-Davidowitz 2017). An important issue is highlighted by Tinati et al. (2014), that is data-driven research must be clear that data are not naturally occurring or unmediated, instead data are socio-technically constructed, produced, and represented in the user interface platforms.

Big data represents not only a technical challenge, but it has also become a societal phenomenon that is radically transforming the social fabric of the contemporary world. Boyd and Crawford (2011) bring about "six provocations for big data": Automating research changes the definition of knowledge, claims to objectivity and accuracy are misleading, bigger data are not always better data, not all data are equivalent, just because it is accessible does not make it ethical, and limited access to big data creates new digital divide (Boyd and Crawford 2011). According to the authors, big data encourages "apophenia" or seeing patterns where none actually exist, due to the sheer volume of incoming data. This issue is relevant because it points out the need to consider what are the analytic assumptions, the methodological frameworks and underlying biases embedded in big data. Although big data allows for the production, sharing, organization, and interaction across fields of study, it is also governed by forces of market, law, social norms, and code. Without disregarding the relevance of big data, it is important to have these issues in mind when conducting big data research.

Nowadays, almost every societal institution is being influenced and changed through the utilization of big data applications. The great acceleration in data about humans and their environment presents ever growing possibilities for data driven approaches to community resilience of environmental threats (McNeill and Engelke 2016). The wealth of data spans almost every aspect of human existence. We now can know patterns of human consumption in great detail from data being collected and analyzed by online retailers such as Amazon, Walmart, etc. (Benkler 2006). We know the type of knowledge that individuals are seeking at any point in time from their utilization of sources such as Google searches (Preis et al. 2013). We are rapidly expanding access to images through sources such as Youtube, GoogleStreet View, and an endless amount of individual photographs with metadata being spread across the internet. Twitter has provided a platform where anyone can publish and share their perspectives globally. Facebook has provided a mechanism that enables the establishment, maintenance, and growth of networks of individuals that share information of topics of interest. Uber and Lyft are essentially data companies that are utilizing the existing stock of privately own vehicles to create a robust, private transportation platform (French et al. 2015). eHarmony and Match.com are among the numerous dating sites that are reshaping the way many individuals meet and form families (Minelli et al. 2012). Of course, these are just a few of the growing number of data driven companies that are reshaping society. All of these are potential ways of tracing human behavior along the coast, and all of these are potential sources of new approaches to improve coastal resilience. Question of course, is how do you access and utilize this information in a modeling framework, and how do you harness the power of these enterprises to improve resilience?

One way to observe the power of big data is by considering networks and information diffusion. A critical aspect of certain types of human-generated big data (i.e. social media), is its relational nature. Indeed, the networked structure of social media makes it possible to think of it as a mechanism to improve dissemination of information in the context of disaster and emergency response (Jones and Faas 2017; Magsino 2009). Social Network Analysis (SNA) has become widely used given the expansion of internet-based social networks. Nowadays, increased computational capacity for data collection and analysis of semi-structured data allows us to obtain useful information from large-size relational data. From the SNA perspective and considering specific social media outlets, research suggests that Twitter is more powerful than Facebook because information can be shared between individuals that have fewer mutual contacts, thus providing more diverse access to information (Bakshy et al. 2012). Schölkopf (2013) details the use of an algorithm created to assess time-varying in network inference, and findings from this work indicate that there is an early relevant increase of information transfer among blogs than among mainstream media for new involving general population and social unrest. Another such way includes, Windhager et al. (2011) who compare visualization paradigms, the process, and network visualization. According to the authors, network analysis could help exploring and controlling actual developments of information diffusion over time. Further findings proved the need to visually unlock specific combinations of structural measures for ongoing organizational self-assessments (Windhager et al. 2011).

As we have detailed, research has used big data experiences to investigate the challenges and possibilities of using it in the field of disaster management and resilience. It is important to understand the role of digital communications in community development. In the work "Applications of Social Network Analysis for Building Community Disaster Resilience. Workshop Summary", the author points out the importance of understanding the dynamic nature of communities as essential for emergency planning and resilience. Communities that had increased their capacity to effectively communicate through social networks are more likely to be more resilient (Magsino 2009). The scenario of effective communication through social networks is critical to resilience efforts. According to the author, in order to create efficient communication within communities, it is critical to delve into the communities to grasp and understand issues of access, feasibility, and reliability of technological resources. The increasing access to technology has allowed for emergency managers to shift focus on what needs to be done for the community to what can be done. Communication is essential for situational awareness because organizational charts cannot explain the importance of relationships and transactions that take place between individuals and within and between organizations (Magsino 2009). The work conducted by Vieweg et al. (2010) analyzes microblog posts generated during two critical concurrent emergency events via Twitter. Findings from this work indicate that it is possible to expand quests for help to communities and/or a region beyond the physical location of the event. The ability to invoke help during emergency situations can be critical for emergency management.

Another important aspect detected in research regarding social media big data use in disaster management is credibility of messages during disasters. Castillo et al. (2011) state that there are signals available in the social media environment that enable users to assess information credibility. Credibility factors include the reactions that certain topics generate and the emotion conveyed by users discussing the topic, the level of certainty of users propagating the information, the external sources cited, and characteristic of the users that propagate the information. A good measure of credibility in Twitter can be the number of "retweets", the more retweets a tweet has the more credible it is considered (Castillo et al. 2011). Experts suggest that social media expands insights to feed into early warning systems. Now, that implies that there is a need to prepare next steps for extracting useful information during emergencies using computational models for information extraction techniques. Gupta and Kumaraguru (2012) state that not all information is trustworthy and/or useful during an event. Using data collected during Hurricane Irene, authors found that only 17% of the content was credible. Computer programs were developed to filter out spam of tweets collected. Data collected illustrated that content and network structure act as prominent features for effective credibility. One program analyzed credibility by whether a tweet contains a URL, the length of tweet, and authority of user. Tweets containing swear words were considered as opinions. Low number of happy emoticons and high number of sad emoticons, as well as number of followers were considered strong predictors of credibility[2] (Gupta and Kumaraguru 2012). As we have discussed in this section, there are many aspects of big data that challenge both research and practice; in the next section we will go hand-on to discover more concrete examples of how big-data can become a useful tool for adaptation.

[2]Keep in mind that this is 2012 study.

20.4 The Power of Combining Human and Physical Data

The phenomenon of digital connectivity is increasingly transforming the way humans interact and deal with problems of collective action. As discussed in the last section, such a paradigmatic change is expanding the frontiers of research in terms of theories, methods and analysis. But overall, big data applications are drivers of paradigmatic change, radically transforming human interaction and its relationship with the Earth system. Here, we will discuss the issues of how to connect human-generated big data with physical big data. Nowadays human activity leaves a digital path. If some argue that digital data are framed by the devices and platforms designed for online interaction, it is also true that these types of interactions are potentially changing individual behavior and collective action. The existence of personal digital devices like computers and cellphones allows for the collection, storage, mining and analysis of human activity. We are aware of the ethical concerns that come with human-generated data (Buchanan and Zimmer 2012; Zimmer 2010), but understanding the profound ethical consequences of utilizing these type of data for identifying individual and collective patterns of behavior is the first necessary step towards understanding the limits between public and private spheres in the context of the data intensive society.

Despite the risks, big data also represents an enormous potential to inform and enhance decision-making in a scenario of intense climate variability, and short-term weather related disasters. Existing literature identifies social capital and human connectivity as critical factors in addressing the challenges of natural disasters. The rise of social media and other digital platforms are *game changers* in the way humans can now organize themselves and respond to vulnerabilities coming from the current climate conditions. Much attention has been drawn to how efficiently NGOs and sometimes governments respond to crises when some sort of big data is available. In today's world, many victims rely on social media to reply their messages/status to friends and family members. Though the use of social media can be a reliable source of information, many flaws come with the use of crowdsourcing. For instance, multiple organizations might respond to an individual request which may lead to a waste of resources. The information provided may also not be accurate, or may not have the geographic location available. With the use of text mining and social computing technologies crowdsourcing may be modified to employ better strategies of obtaining data such as: geo-tag determination, report verification, automated report summarization, spatial-temporal mining for social behavior prediction, and scalability and safety (Gao et al. 2011).

There is growing evidence that data driven technologies are being utilized in creative ways to address human problems that occur in disaster situations. During the 2016 summer floods in Varanasi and Allahabad, India, Ola, India's popular "Uber-like" transportation system developed a near real-time response. They quickly adapted their software to utilize boats as well as cars to assist people in evacuations and other transportation needs in flooded areas. Their online platform was also expanded to include a feature for collecting donations and other material assistance to flood victims. In a similar vein, residents of Louisiana self-organized into a group of boat owners now known as the “Cajun Navy”. They originally came together to respond victims’ requests for help in the aftermath of Hurricane Katrina and the flooding in New Orleans. More recently they used their boats to rescue victims of the 2016 Louisiana flood in the Baton Rouge area. In organizing their rescue efforts, they relied on such online sources as Facebook and Glympse to identify, locate, and rescue victims. The “Cajun Navy” is organized to promote community resilience as reflected in their stated purpose, “We the people of Louisiana refuse to stand by and wait for help in the wake of disasters in our State. We rise up and unite and rescue our neighbors!” They do this in part with the aid of data driven platforms. Their voluntary disaster responses are not limited to their home state of Louisiana. A year later when Hurricane Harvey created record flooding in Southeast Texas, many members of the “Cajun Navy” brought their personal boats to Texas to aid in the rescue and evacuation efforts (“‘Cajun Navy’ answers call in Harvey flood zone,” 2017). Both of these examples reflect how resilience to disasters has increased through the data assisted creative use of private citizens of non-governmental resources.

In an examination of the *Twitterverse* that occurred before, during, and after Hurricane Sandy in the New York area, researchers at the Social Science Research Center at Mississippi State University collected 4.7 million geo-located tweets with thousands of posted photographs of the storm’s impact (Edwards et al. 2015). From this social media data, researchers were able to identify several patterns of individuals and groups using Twitter to cope with and to assist individuals impacted by the storm. There were tweets that were asking for help, tweets offering aid, and tweets organizing groups to assist others. Since many individuals will tweet when their electrical power fails, it was possible to utilize geo-located tweets to map power outages in real time. Captured images of flooding and structural damages were also powerful sources of information about the severity of the storm. Furthermore, analysis of the networks of communication in Twitter revealed three clusters of users: (1) individuals providing support; (2) media outlets; and (3) government agencies with weather-related functions. In analyzing other disasters, these researchers found similar patterns of Twitter usage. Collectively, it appears that internet-based communications during disasters provide additional information that can be used by both the government and private citizens, and also enables a variety of self-organizing behavior that improves disaster resilience. Fortunately, social media and other forms of big data applications that can be utilized during disasters are likely to rapidly increase.

The marketplace is also quite useful providing innovative adaptation efforts for disaster management. From the business perspective Fosso Wamba et al. 2015 found that creating and capturing business value from big data can allow real-time access and sharing of information across local and national government agencies for improved decision making to enhance emergency service response. It allows for improvements of intra and inter transparency within organizations. The study highlights the importance of organizational leadership capable of introducing IT innovations (Fosso Wamba et al. 2015). Disasters create many business opportunities for companies that can provide the needed goods and services to disaster victims in a timely manner. Walmart, Amazon, Lowes, and Home Depot are examples of companies that engage in highly organized disaster entrepreneurship sometimes before, during, and in the recovery phase of disasters. They have a capacity to move massive amounts of essential materials, food, and other supplies to disaster areas on an ongoing basis that dwarfs the capabilities of governments. Typically, these companies have developed highly capable and robust supple chains that utilize numerous big data sources and big data analytics to meet the demands of disaster customers. In this respect, the increasing reliance on big data in commerce is contributing to the resilience by developing systems that can transport goods and services to impacted areas in ways that would not be otherwise. Also, the data that is driving the supply chains for these large companies is also potentially data that could be used to model human consumption patterns in both every-day and disaster conditions, and relate these to the dynamics of natural systems. This is precisely the type of science that the National Science Foundation anticipates in its program of "The Dynamics of Coupled Natural and Human Systems". Accessing this data, however, is a major obstacle in moving such modeling efforts forward. Big data is said to be "the new oil", and the big data underlying the operation of these large influential companies is both very valuable and closely held.

Government agencies with responsibility for coastal environments are beginning to utilize novel data science approaches to addressing coastal challenges. For example, the U.S. Geological Survey (USGS) in 2014 initiated the iCoast program that is based on crowdsourcing. It utilizes volunteers to visually compare time sequence aerial photos of the coast to determine changes in erosion and damage (Greenemeier 2014; USGS 2017). Another valuable example is provided by the National Oceanic and Atmospheric Administration (NOAA) Digital Coast initiative. "This online tool provides managers and citizens with easy-to-understand charts and graphs that describe complex coastal data" (NOAA 2017b). The *Coastal Flood Exposure Mapper* tool provides users critical socio-economic and

Fig. 20.1 Population density in the United States Gulf and Atlantic coastal areas (NOAA 2017a, August 22)

environmental information. This is later combined and used to make planning decisions and coordinate efforts in case of potential emergency situations. These types of initiatives aim at putting together a community of scholars, practitioners, and communities committed to improve access to big data sources through practical applications that allow sharing information, that in turn can be used for decision making in case of disasters or long-term transformations resulting from intense climate variability. One important option, is the availability of socio-economic mapping for coastal areas that details aspects such as population density (Fig. 20.1), poverty rates (Fig. 20.2), employment (Fig. 20.3), and aging population (Fig. 20.4). This socio-economic data can be combined with flooding hazard data (Fig. 20.5) to identify the overlap of highly dense geographic areas with high risk of flooding. Examples of short-term (e.g., preparation for hurricane season in

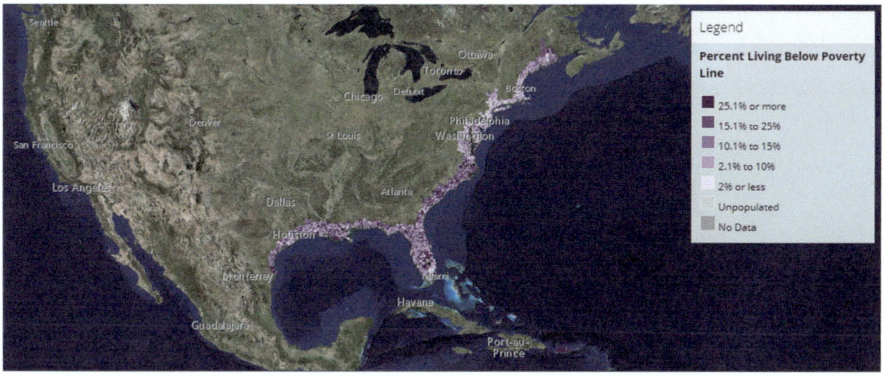

Fig. 20.2 Percentage of population living below poverty line in the United States Gulf and Atlantic coastal areas (NOAA 2017a, August 22)

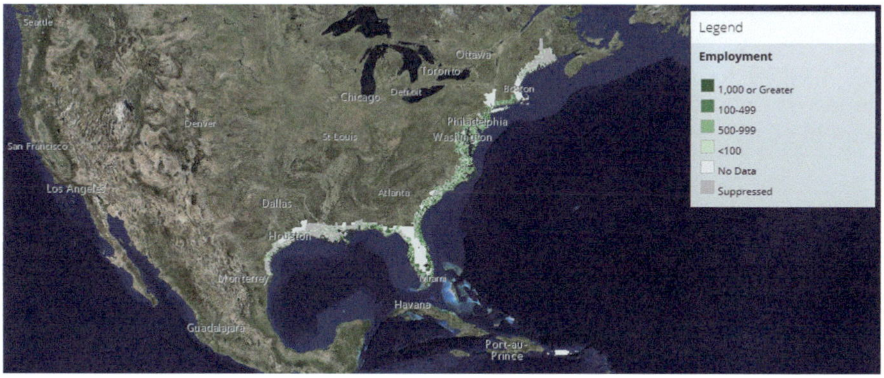

Fig. 20.3 Range in the number of employees for U.S. Census block groups (or geographies) that work in or near coastal flood-prone areas in the United States Gulf and Atlantic coastal areas (NOAA 2017a, August 22)

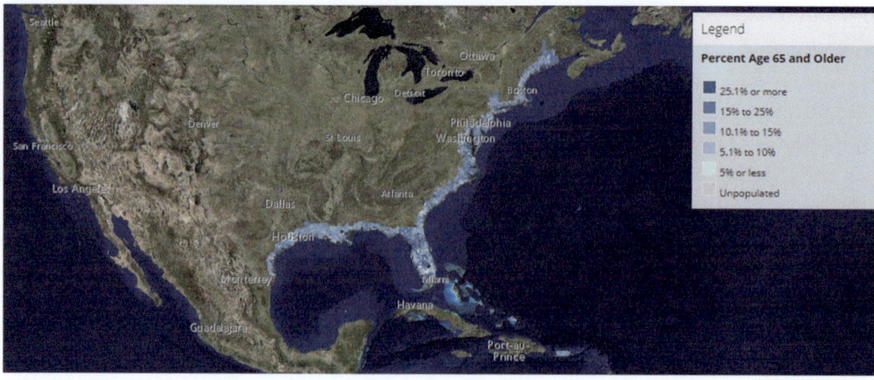

Fig. 20.4 Percent of population age 65 and older in the United States Gulf and Atlantic coastal areas (NOAA 2017a, August 22)

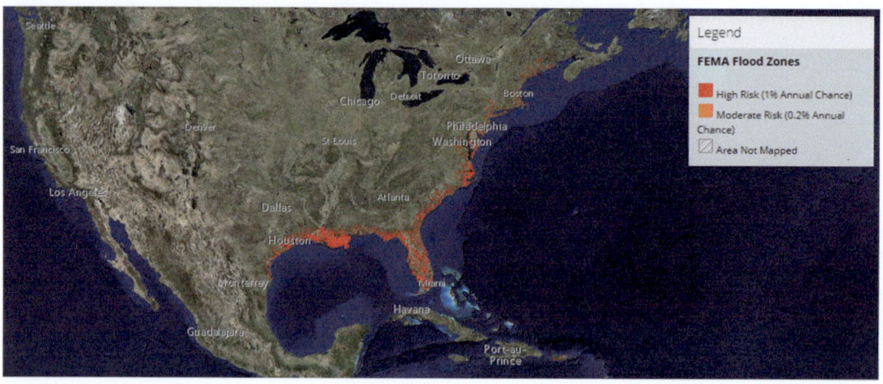

Fig. 20.5 FEMA flood zones in the United States Gulf and Atlantic coastal areas (NOAA 2017a, August 22)

coastal areas) and long-term (e.g., adapting to sea level rise in Miami-Dade County, Florida or Enhancing Resilience to Coastal Hazards in Connecticut) adaptations using big data are highlighted in the NOAA portal (NOAA 2017a).

20.5 Big Data for Human Adaptation: Alternatives to Long-Term Transformations Resulting from Climate Change

Throughout this chapter, we have been discussing issues related to the interactions between human and Earth systems, the capacity of humans to transform their environment, and the contributions of big data to enhance risk and vulnerability management given transformations posed by extreme weather events. In this section we will discuss alternatives for positive change derived from the capacity of humans to self-organize and the ability to innovate which can occur in urban areas.

At the Mississippi State University-Social Science Research Center social media big data research has investigated networks of civic engagement, government accountability, self-organizing for help, and fundraising as examples of social media driven responses occurring in natural disasters such as Hurricane Sandy. These examples most likely are only scratching the surface of the possibilities for community adaptation utilizing data-intensive technologies. From this perspective, the data-intensive society is creating an ever expanding set of technologies that are (or may become) important factors affecting governance. Lessons from this work illustrate the capacity of humans to self-organize and carry out collective action efforts that are almost impossible without the existence of internet-based communications.

Theory has suggested avenues to manage natural resources in a sustainable manner. Gills (2015) studies how under the "World Systems perspective" we have observed large-scale patterns of production and exchange and accumulation of wealth and power resulting in the overuse of natural resources which turns into a global threat unique to this era. Responding to these trends emerge the concept of Sustainable Global Development, defined as "meeting the current needs of our generation without harming future needs". The idea of sustainable development should follow principles of cooperation rather than competition. Contrary to the ideas suggested in the "Tragedy of the Commons" (Hardin 1968)[3] in which a shared resource will be depleted by individuals responding to their self-interest combined in a collective action, Ostrom (1990) suggests that governing common pool resources effectively comes from an ability to identify future *utility* and in further conservation efforts. That is, cooperation is build upon not in the desire to cooperate but in the set of incentives available to cooperate. Analysis of SES depends on resource systems, resource units, governance systems, and users. Sustainability depends on monitoring and enforcement as well as not being overruled by larger governmental policies (Schweitzer et al. 2009).

This perspective about common property regimes suggest that self-organized governance systems are more likely to produce better outcomes for communities and the long-term stock of resources. As a result of the analysis of hundreds of cases from all over the world, Elinor Ostrom and colleagues came to the conclusion that there is a set of design principles of stable local common-pool resource management: (1) Clearly defined boundaries; (2) congruence between appropriation, provision rules and local conditions; (3) collective-choice arrangements allowing for the participation of most of the appropriators in the decision making process; (4) Effective monitoring by individuals who are part of or accountable to

[3]"Each man is locked into a system that compels him to increase his herd without limit—in a world that is limited. Ruin is the destination towards which all men rush, each pursuing his own best interest in a society that believes in freedom of the commons" (Hardin 1968, p. 1244).

the appropriators; (5) Graduated sanctions for appropriators who do not respect community rules; (6) Conflict-resolution mechanisms which are cheap and easy to access; (7) Minimal recognition of rights to organize (e.g., by the government); and (8) In case of larger Common Property Regimes (CPR): Organization in the form of multiple layers of nested enterprises, with small, local CPRs at their bases (Ostrom 1990). At the core of understanding Social-Ecological Systems (SES) and their sustainability is identifying and analyzing relationships through the various levels of these systems on different spatial and temporal scales.

Arguably, the most challenging governing structure that we know of is the city. Cities are nowadays the place where several of the socio-economic trends of the great acceleration are occurring. Understanding large metropolitan areas and their successes, require an examination of the global economy and how those cities are participating in the global network of urban areas (Sassen 2005). The significant demographic shift from rural agrarian settings to large urban/metropolitan regions, has expanded the power of urban areas. This power derived from population size, creativity and potential economic development can be a double-edged sword, but if used appropriately, this power can be harnessed to address costal challenges such as seawater rise, flooding, temperature change, erosion, and increasing power capability. Cities are important because many are located along the coast and their geography often clusters around coastal environmental ecosystems (e.g., Miami ecosystem).

In their 2017 book *Climate of Hope*, Michael Bloomberg, U.S. businessman and former mayor of New York City, and Carl Pope, former head of the Sierra Club, recently teamed to explore "how cities, businesses, and citizens can save the planet". Underlying their insightful work is the reality that the population of the globe is rapidly becoming concentrated in very large and growing urban areas and that the key to adaptations to climate change will reside in the creativity and capabilities of the cities (Bloomberg and Pope 2017). Since U.S. coastal regions are now metropolitan, the coastal cities may become major players in developing the strategies and approaches that can adapt to the myriad of issues brought about by climate change as it intersects with the demographic shift to the coast. There is reason for optimism in the growth of the city. Throughout history, cities have been centers of creativity and innovation. It is apparent that cities are now the location for the rapid expansion of the data intensive society. In fact, there is the concept of "the smart city" that promotes the utilization of data and related technologies to organize and develop a new form of cities (Cocchia 2014). Alex Pentland at MIT's Human Dynamics Laboratory speaks of City Science and Data-Driven Societies that in the context of this discussion present an interesting blueprint of how cities will increase their capacity and become even more effective as centers of innovation in a data intensive world (Pentland 2014). It can be anticipated that innovative activity in coastal cities will be directed towards data driven solutions for coastal challenges brought about by climate change.

Increasing the transparency of climate related risks to the coasts is tied closely to the development of a more capable data-intensive society. The underlying assumption is that better informed individuals, businesses, and government will be better equipped to make decisions and choices that are more adaptable to challenges (Bloomberg and Pope 2017). The explosion of big data, including the development of new techniques such as predictive algorithms, visualization of complex phenomenon, artificial intelligence, and ever-growing number of problem-solving applications, creates a data rich environment that can improve coastal adaptation. For example, data can lead to an improved understanding of risks associated with coastal living and new coastal development. Unsubsidized insurance rates for coastal properties would be a highly useful, short-hand index of the financial risk associated with properties on the coast.

Typically, however, the situation is much more complex and such straight forward risk assessments may not be available. Some recent history of property insurance in Florida illustrates the complexity that currently exist, and the difficulty of having a reliable transparent index of risk. First, critical risk information concerning sea water rise, storm surge, and storm probabilities. In a period of climate change is most likely dynamic and challenging to predict and somewhere rapidly developing like the Florida coast is a region where this information is an essential risk assessment factor (Nyce and Maroney 2011). Second, many private insurers have found Florida to be a high-risk area that is difficult to insure; many have sustained major losses from hurricane-related damages, several companies declared bankruptcy, and others have decided to withdraw from the Florida market. Third, state government supported insurance has been developed to meet this void and now is an important part of property insurance for Florida; the state created the Citizens Property Insurance Corporation that now holds several hundred billion dollars in property risks, including many high-risk properties (Florida Legislature—Office of Economic and Demographic Research 2012). Fourth, the federal government provides insurance for flooding through FEMA that also insures many Florida properties for this risk ("Flood Insurance Reform—Rates and Refunds" 2017). Finally, property value varies over time and may increase dramatically in highly desirable coastal areas which could mean that not only does the property owners' financial risk increase, but also the financial exposure of those who subsidized property insurance. Thus, from the property owner's perspective, the risk associated with ownership is probably underestimated since the true cost of insurance is being subsidized by both the federal and state government. From the perspective of the state, there is a huge potential risk that is difficult to understand since it depends in part on the frequency, severity, and the geographic location of future disasters, possible increases/decreases in the property value and political forces that encourage risk exposure that may not be financially sound. When the aforementioned are taken into account, it is not surprising that taxpayers have a poor basis for understanding the collective risk they are sharing since the combined cost of subsidized insurance is not widely known. While risk knowledge on owning coastal property is complex, big data approaches provide solutions that can more accurately assess, depict and present information both on individual and collective risks that are occurring with coastal property. Since risks for coastal living is often shared broadly, it is essential to have analytical approaches that can estimate the current risks, project future risks, and aggregate the risks at the national, state, city, and property owner levels. This subsidized insurance, while creating value and development through raising property values and incentivizing investment, put people in harm's way by concentrating populations in hazardous locations.

20.6 Disaster Science and Adaptation by a Data-Intensive Society

Elsevier recently released the report, "A Global Outlook on Disaster Science," that illustrates how novel Big Data driven approaches are shaping how science can address global issues. Based on more than 27 thousand disaster-related publications released in the past five years, Elsevier data scientists used complex analytics to generate new insights about the state of disaster science. We now have a better understanding of where the science is being conducted, the impact it is having, and the science's implications for society. Since disaster science has strong implications for the future of the world's coasts, the Global Outlook study is important for our discussions. Because of the Elsevier study, we now know: countries with high death tolls from natural disasters tend to have low disaster scholarly output; while countries with the highest levels of economic loss due to natural disasters, tend to have the largest scholarly output. During the last five years, Elsevier estimates that 0.22% of the world's scholarly output is devoted to disaster studies, and that China and the United States are major contributors to disaster research. Elsevier observes that the focus of disaster science in a country also seems to be driven by the type and occurrence of local disasters. This tendency for a local relevance can lead to a global disconnect between underdeveloped countries with substantial disaster burdens and developed countries with higher levels of disaster science that focuses on their country specific disaster topics (Elsevier 2017). The Elsevier work based on scientific papers and big data analytics creates an information framework for globally accessing, organizing, and planning for disaster science in a manner that would have been impossible prior to the acceleration of data and the emergence of data science analytic techniques. This important report illustrates how we can now rapidly synthesize massive amounts of information that captures the collective knowledge of global science. Such approaches should also have profound implications for linking the Anthropocene to its very broad and complex consequences.

In summary, underlying the perspective of the Anthropocene is the rapid acceleration of human-created influences such as population growth, water usage, fertilizer use, etc. is changing the very physical nature of the globe. However, there is one rapidly accelerating aspect of human behavior, that is not generally recognized as part of the Anthropocene process. The emergence of a data-intensive society, often referred to as Big Data is an important part of the acceleration process. In our mind, this is a profoundly important accelerating influence that could have minimal negative consequences for the environment and possibly provide an approach that would help humans adapt to the challenges and dangers resulted from the Anthropocene. Our optimism for the promise of Big Data should of course be taken with caution. The great acceleration of data is very recent and our enthusiasm for data science solutions might change with more experience. Big Data has great power, and like all things with great power, there is the possibility for great good and great harm. In this respect, it is human intelligence, and perhaps artificial intelligence, that will determine how well we utilize this new powerful resource to address the challenges that occur along the coasts of the world.

References

Bakshy, E., I., Rosenn, C. Marlow, and L. Adamic. 2012. The role of social networks in information diffusion. In *WWW'12 Proceedings of the 21st international conference on World Wide Web*, pp. 519–528. https://doi.org/10.1145/2187836.2187907.

Bello-Orgaz, G., J.J. Jung, and D. Camacho. 2016. Social big data: Recent achievements and new challenges. *Information Fusion* 28 (Supplement C): 45–59. https://doi.org/10.1016/j.inffus.2015.08.005.

Benkler, Y. 2006. *The wealth of networks: How social production transforms markets and freedom*. New Haven, CT: Yale University Press.

Bloomberg, M., and C. Pope. 2017. *Climate of hope*. New York: St. Martin's Press.

Boyd, D., and K. Crawford. 2011. Six provocations for big data. In *A decade in internet time: Symposium on the dynamics of the internet and society*, Oxford, UK: Oxford Internet Institute. https://doi.org/10.2139/ssrn.1926431.

Buchanan, E., and M. Zimmer. 2012, December 20. Internet research ethics. In *The Stanford Encyclopedia of Philosophy*. Stanford University—Center for the Study of Language and Information. Retrieved from http://plato.stanford.edu/archives/win2012/entries/ethics-internet-research/.

"Cajun Navy" Answers Call in Harvey Flood Zone. 2017, August 28. *ABC News,* from. http://abc13.com/2349668/. Retrieved November 16, 2017.

Castillo, C., M. Mendoza, and B. Poblete. 2011. Information credibility on Twitter. In *Proceedings of the 20th international conference on World Wide Web—WWW'11*, pp. 675–675. https://doi.org/10.1145/1963405.1963500.

Chen, Y., and S.S. Rosenthal. 2008. Local amenities and life-cycle migration: Do people move for jobs or fun? *Journal of Urban Economics* 64 (3): 519–537. https://doi.org/10.1016/j.jue.2008.05.005.

Cocchia, A. 2014. Smart and digital city: A systematic literature review. SpringerLink. https://doi.org/10.1007/978-3-319-06160-3_2.

Cox, W. 2017, April 21. The 37 megacitites and largest cities: Demographia world urban areas 2017. Retrieved from http://www.newgeography.com/content/005593-the-largest-cities-demographia-world-urban-areas-2017.

Crutzen, P.J. 2002. Geology of mankind. *Nature* 415: 23.

Davis, K. 1955. The origin and growth of urbanization in the world. *American Journal of Sociology* 60 (5): 429–437.

Edwards, J.F., S. Mohanty, and P. Fitzpatrick. 2015. *Assessment of social media usage during sever weather events and the development of a Twitter-based model for improved communication of storm-related information*. Starkville, MS: Coastal Storm Awareness Program, NOAA.

Elsevier. 2017. *A global outlook on disaster science* (research intelligence). Elsevier. Retrieved from https://www.elsevier.com/__data/assets/pdf_file/0008/538091/ElsevierDisasterScienceReport-PDF.pdf.

Flood Insurance Reform—Rates and Refunds. 2017, March. Retrieved from https://www.fema.gov/flood-insurance-reform-rates-and-refunds.

Florida Legislature—Office of Economic and Demographic Research. 2012. *Florida's economic future & the impact of aging in place*. Tallahassee, FL: Florida Legislature—Office of Economic and Demographic Research. Retrieved from http://edr.state.fl.us/Content/presentations/economic/AginginPlaceHousingSummit.pdf.

Fosso Wamba, S., S. Akter, A. Edwards, G. Chopin, and D. Gnanzou. 2015. How "big data" can make big impact: Findings from a systematic review and a longitudinal case study. *International Journal of Production Economics* 165 (Supplement C): 234–246. https://doi.org/10.1016/j.ijpe.2014.12.031.

French, S., C. Barchers, and W. Zhang. 2015. Moving beyond operations: Leveraging big data for urban planning decisions. In *Proceedings computers in urban planning and urban management 2015: Urban planning and decision support*. Retrieved from http://web.mit.edu/cron/project/CUPUM2015/proceedings/Content/pss/194_french_h.pdf.

Gao, H., G. Barbier, and R. Goolsby. 2011. Harnessing the crowdsourcing power of social media for disaster relief. *IEEE Intelligent Systems* 26 (3): 10–14. https://doi.org/10.1109/MIS.2011.52.

Gills, B.K. 2015. Global development in the Anthropocene. *Globalizations* 12 (5): 824–827. https://doi.org/10.1080/14747731.2015.1062290.

Greenemeier, L. 2014. iCoast: Did the coast change? Help U.S. Geological Survey scientists study storm aftermaths and coastal changes. *Scientific American*. Retrieved from https://www.scientificamerican.com/citizen-science/icoast-did-the-coast-change/.

Gupta, A., and P. Kumaraguru. 2012. Credibility ranking of tweets during high impact events. In *Proceedings of the 1st workshop on privacy and security in online social media*. Lyon, France. Retrieved from https://dl.acm.org/citation.cfm?id=2185356.

Hardin, G. 1968. The tragedy of the commons. *Science* 162: 1243–1248.

Jones, E. 1981. *The European miracle: Environments, economies and geopolitics in the history of Europe and Asia.* (Third). New York: Cambridge University Press.

Jones, E.C., and A.J. Faas (eds.). 2017. *Social network analysis of disaster response, recovery, and adaptation*. Oxford, UK: Elsevier/Butterworth-Heinemann.

Kitchin, R. 2014. Big data, new epistemologies and paradigm shifts. *Big Data & Society* 1 (1): 1–12. https://doi.org/10.1177/2053951714528481.

Luxton, E. 2015, December 16. How much do natural disasters cost the world? Retrieved from https://www.weforum.org/agenda/2015/12/how-much-do-natural-disasters-cost-the-world/.

Magsino, S.L. (ed.). 2009. *Applications of social network analysis for building community disaster resilience. Workshop Summary*. Washington, DC: The National Academy Press. Retrieved from https://www.nap.edu/download/12706.

Maslow, A.H. 1943. A theory of human motivation. *Psychological Review* 50: 370–396.

McNeill, J.R., and P. Engelke. 2016. *The great acceleration*. Cambridge, MA: Harvard University Press.

Minelli, M., M. Chambers, and A. Dhiraj. 2012. *Big data, big analytics: Emerging business intelligence and analytic trends for today's businesses*. Wiley.

Morrow, J.D. 1994. *Game theory for political scientists*. Princeton, NJ: Princeton University Press.

National Oceanic and Atmospheric Administration (NOAA). 2017a, August 22. Coastal flood exposure mapper. Retrieved from https://coast.noaa.gov/floodexposure/#/select.

National Oceanic and Atmospheric Administration (NOAA). 2017b, October 19. Coastal County snapshots. Retrieved from https://coast.noaa.gov/digitalcoast/tools/snapshots.html.

North, D.C. 1990. *Institutions, institutional change and economic performance*. Cambridge: Cambridge University Press.

Nyce, C.M., and P. Maroney. 2011. Are territorial rating models outdated in residential property insurance markets? Evidence from the Florida property insurance market. *Risk Management and Insurance Review* 14 (2): 201–232. https://doi.org/10.1111/j.1540-6296.2011.01204.x.

Ostrom, E. 1990. *Governing the commons: The evolution of institutions for collective action*. New York: Cambridge University Press.

Paredes, C. 2011, January 18. Los Desastres no son ni Naturales ni Castigos de Dios. *Semana*. Retrieved from http://www.semana.com/nacion/articulo/los-desastres-no-ni-naturales-ni-castigos-dios/234261-3.

Pentland, A. 2014. *Social physics: How good ideas spread-the lessons from a new science*. New York: The Penguin Press.

Preis, T., H.S. Moat, and H.E. Stanely. 2013. Quantifying trading behavior in financial markets using Google trends. *Scientific Reports* 3. https://doi.org/10.1038/srep01684.

Rappaport, J. 2007. Moving to nice weather. *Regional Science and Urban Economics* 37 (3): 375–398. https://doi.org/10.1016/j.regsciurbeco.2006.11.004.

Rico Méndez, G. 2016, July. *Governing the rurals: Food security and governance in the 21st century* (Doctoral dissertation). Mississippi State University, Starkville, MS. Retrieved from http://sun.library.msstate.edu/ETD-db/theses/available/etd-07282016-182049/.

Sassen, S. 2005. The global city: Introducing a concept. *Brown Journal of World Affairs* 11 (2): 27–43.

Schölkopf, B. 2013. Structure and dynamics of information pathways in online media. *Acm Wsdm*, 1–10. https://doi.org/10.1145/2433396.2433402.

Schweitzer, F., G. Fagiolo, D. Sornette, F. Vega-Redondo, A. Vespignani, and D.R. White. 2009. Economic networks: The new challenges. *Science* 325 (5939): 422. https://doi.org/10.1126/science.1173644.

Steffen, W., W. Broadgate, L. Deutsch, O. Gaffney, and C. Ludwig. 2015. The trajectory of the Anthropocene: The great acceleration. *The Anthropocene Review* 2 (1): 81–98. https://doi.org/10.1177/2053019614564785.

Steinberg, T. 2000. *Acts of god. The unnatural history of natural disaster in America*. New York, NY: Oxford University Press.

Stephens-Davidowitz, S. 2017. *Everybody lies: Big data, new data, and what the internet can tell us about who we really are (first)*. New York: HarperCollins.

Tinati, R., S. Halford, L. Carr, and C. Pope. 2014. Big data: Methodological challenges and approaches for sociological analysis. *Sociology* 48 (4): 663–681. https://doi.org/10.1177/0038038513511561.

United Nations. 2014. *World urbanization prospects*. New York. Retrieved from https://esa.un.org/unpd/wup/publications/files/wup2014-highlights.pdf.

U.S. Census. 2016. QuickFacts selected: Miami-Dade County, Florida. Retrieved from https://www.census.gov/quickfacts/fact/table/miamidadecountyflorida/POP060210.

U.S. Geological Survey. 2017. USGS iCoast—did the coast change? Retrieved from https://www2.usgs.gov/blogs/features/usgs_top_story/usgs-icoast-did-the-coast-change/.

Vieweg, S., A.L. Hughes, K. Starbird, and L. Palen. 2010. Microblogging during two natural hazards events. In *Proceedings of the 28th international conference on human factors in computing systems—CHI'10*, pp. 1079–1079. https://doi.org/10.1145/1753326.1753486.

Weiss, D.J., and J. Weidman. 2013, April 29. Infographic: Disastrous spending: Federal disaster-relief expenditures rise amid more extreme weather. Retrieved from https://www.americanprogress.org/issues/green/news/2013/04/29/61697/infographic-disastrous-spending-federal-disaster-relief-expenditures-rise-amid-more-extreme-weather/.

Wilson, S.G., and T. Fischetti. 2010. *Coastline population trends in the United States: 1960–2008* (No. P 25-1139). U.S. Department of Commerce, U.S. Census Bureau. Retrieved from https://www.census.gov/prod/2010pubs/p25-1139.pdf.

Windhager, F., L. Zenk, and P. Federico. 2011. Visual enterprise network analytics—visualizing organizational change. *Procedia—Social and Behavioral Sciences* 22 (0): 59–68. https://doi.org/10.1016/j.sbspro.2011.07.056.

Wittfogel, K.A. 1957. *Oriental despotism. A comparative study of total power* (Tenth). New Haven, CT: Yale University Press.

Zhou, K., C. Fu, and S. Yang. 2016. Big data driven smart energy management: From big data to big insights. *Renewable and Sustainable Energy Reviews* 56 (Supplement C): 215–225. https://doi.org/10.1016/j.rser.2015.11.050.

Zimmer, M. 2010. "But the data is already public": On the ethics of research in Facebook. *Ethics and Information Technology* 12 (4): 313–325.

Chapter 21
Promoting Resilience of Tomorrow's Impermanent Coasts

Lynn Donelson Wright and Bruce G. Thom

Get the Black Beauty Kato. Those clowns are in trouble.
—The Green Hornet- c. 1950

21.1 Anticipating and Preparing for the Future

There are three categories of actions that humans need to take in order to minimize the detrimental impacts of global change on tomorrow's coastal systems. The first, of course, is to cause less harm by reducing our carbon footprint and ceasing to do destructive things like polluting, dredging, severing sediment supply, withdrawing groundwater, overdeveloping etc. Much has been written and spoken about this even though we have said relatively little about it in this book. The second category of actions, which has received minimal attention from the popular media but has been the motivating theme of this book, involves promoting deep enough understanding of the myriad complex interconnections of coastal processes to allow long-term predictions of what may lie ahead. Such predictions are essential to evolving effective strategies for adapting and remaining resilient. The third action is to ensure that to the extent possible we embed coastal science, including matters related to future impacts of climate change, into state and federal policies and law. The aim must be to ensure that regional coastal strategies are based on the best available science to reduce risk to built and natural assets from the adverse effects of short-term practices driven by local vested interests.

Regardless of how diligently we may undertake actions of the first category, changes are underway and accelerating so actions of the second and third categories are crucial to our future. However, unless the first category actions are deliberate

L. D. Wright (✉)
Southeastern Universities Research Association (SURA), Washington, DC, USA
e-mail: ldwright@bellsouth.net

B. G. Thom
School of Geosciences, University of Sydney, Sydney, NSW, Australia
e-mail: bruce.thom@sydney.edu.au

L. D. Wright and C. R. Nichols (eds.), *Tomorrow's Coasts: Complex and Impermanent*, Coastal Research Library 27,
https://doi.org/10.1007/978-3-319-75453-6_21

and effective, the challenges of adaptation may be overwhelming in many parts of the world. For example, the likely range of future rates of warming and sea level rise as described in Chaps. 2 and 3 depend strongly on the degrees to which greenhouse gas emissions are reduced beginning in the near future. Regrettably the withdrawal in 2017 of the USA from the Paris Climate Accord significantly dims the likelihood of the "best case" climate change scenario prevailing. This increases the urgency of developing robust adaptive strategies based on reliable numerical models supported by improved observations and monitoring. Hopefully, aggressive and widespread "grassroots" efforts by individuals, cities, businesses, counties and states to reduce carbon emissions as advocated most recently by Bloomberg and Pope (2017) will make a significant difference. But, whether extreme or modest, future changes as described in Parts 1, 2 and 3 of this book are inevitable and we must plan accordingly to ensure the resilience of tomorrow's coasts. In this final chapter, we offer *opening* perspectives on some of the things that the global community might undertake in a collaborative spirit in order to enhance the resilience of tomorrow's coasts. This chapter builds on the contributions of authors to this book, and on the outcomes of two workshops on this subject that the Southeastern Universities Research Association (SURA) led in 2014 and 2015 (SURA 2015; Wright et al. 2016a, b). We intend for future workshops to assist broad based communities of scholars, managers, politicians and stakeholders to further evolve and refine regional strategies for adapting to changing but uncertain futures. While we expect improvements to our understanding and predictive capabilities to take advantage of the global brain trust, resilience must be implemented locally and regionally and the solutions must be fluid and adaptable.

21.2 Summary of Some Future Challenges

In Chap. 1 of this book, we noted that effective solutions to the complex challenges faced by tomorrow's coasts and coastal communities will require not only improved and coupled predictive models along with more accurate, frequent and spatially comprehensive observational data but also, and probably more importantly, finding ways to overcome a range of cultural challenges. Included among these challenges are (1) obstacles to a trans-institutional collaborative environment; (2) attitudes and beliefs of local regulators and policy makers; (3) the extent to which educators understand and communicate environmental science; (4) economic pressures, legal systems and political ideologies; (5) conflicting pressures from environmental and business advocacy groups; and (6) the awareness of the general public. Finding effective approaches to all of these factors represents a suite of serious challenges for the immediate future. Meeting these challenges can only be accomplished by collaborative partnerships that include the general public, educators, planners and managers, politicians and decision makers and the global scientific community. Successes will undoubtedly be incremental and will depend on *communication* and adaptability. In a complex world, models, and even understandings are possibly as impermanent as the coasts themselves. In other words, we still have much to learn from each other and our future resilience will require collective mindfulness. Trans-disciplinary strategies for facilitating collaborations in support of enhanced coastal resilience were outlined by Wright et al. 2016a, b). Collaborations at grass roots levels involving the public and numerous non-governmental organizations have been established by The Nature Conservancy (TNC).

21.3 Resilience as It Relates to Coastal Systems

In the introduction to this book, we stated that: ***A high priority vision for future coastal science should be to enhance resilience of coastal communities by anticipating and mitigating hazards to human health, safety and welfare and reducing economic harm to coastal industries such as tourism, fisheries and shipping.*** As pointed out in Chap. 1, *resilience* is the capacity to change and adapt continually yet remain viable. Humans and nature are interdependent and, through collaboration, natural and social scientists, and stakeholders can improve coastal resilience. According to the Stockholm Resilience Centre, "Resilience thinking embraces learning, diversity and above all the belief that humans and nature are strongly coupled to the point that they should be conceived as one social-ecological system." Low risk is not necessarily requisite for high resilience but risk and resilience should both be considered in planning future mitigation strategies. Coastal risk assessment is considered in detail in a recent NRC report (National Research Council 2014). Considering the complex interdependence of many factors described in the foregoing chapters, community resilience and ecosystem resilience must be considered together, not as separate problems. Furthermore, since the built infrastructure and related services are integral components of communities, infrastructure resilience must be considered in relation to both communities and ecosystems (National Institute of Standards and Technology 2015).

For natural ecosystems, such as wetlands, biodiversity is a source of enhanced resilience. Similarly, economic diversity probably results in increased community diversity. One well-known vulnerability index considers vulnerability to environmental hazards (Cutter 1996). Arkema et al. (2013) discuss the roles that natural habitats can play in enhancing natural resilience of communities. As noted by the National Research Council (2006), the loss of coastal wetlands over the decades preceding Hurricane Katrina, substantially enhanced the vulnerability of New Orleans to that event. The Louisiana Coastal Protection and Restoration Authority (2007) is attempting to address this problem (Chap. 13).

The US Army Corps of Engineers (USACE), Engineer Research and Development Center (ERDC) is developing a tiered set of coastal resilience metrics that integrate engineering, environmental and community resilience (Rosati et al. 2015). Rosati et al. (2015) describe expert elicitation, data driven, and tiered methods to quantify resilience. Expert elicitation is somewhat subjective while the data-driven methods rely on a combination of historical data and numerical modeling. The USACE's approach considers preparation, resistance, recovery, and adaptation depending on factors such as need, time, space, and available funding. The three-tiered approach includes expert elicitation, field data and simple models, and rigorous assessment based on probabilistic analyses (Schultz et al. 2012). An important aspect of any viable long-range resilience program is that it must enable continually evolving adaptive management strategies underpinned by advanced numerical modeling. The USACE has provided extensive guidance for engineering responses to sea level rise including regionally specific estimates of change (US Army Corps of Engineers 2014).

21.4 Global Scientific Collaboration Is Essential and Feasible

According to Marinez-Moyano (2006) "Collaboration is a function of the recursive interaction of knowledge, engagement, results, perceptions of trust, and accumulation of activity over time." If the collaborations we envision are to be successful and persist over a long enough timeframe to make a difference, the collaboration strategies and methodologies that we build must be as rigorous as the models and understanding they are designed to facilitate. The intent is not only to facilitate collaborations within the academic community but, most importantly, among universities and federal, state and local governmental agencies, non-governmental organizations (NGOs) and stakeholder organizations. To accomplish this we must ensure that rigorous and broadly embraced protocols are established and followed. Collaboration involves much more than simply talking to and helping each other. To be effective it must involve mutual acceptance of a common set of goals, critical assessments of diverse approaches, iterative updates and incremental improvements to understanding and predicting, promotion of new paradigms and effective communication with a hierarchy of operational end users. In a true collaborative team or consortium, each member must enable and support, not compete with, the other members. A fairly comprehensive collection of essays on collaboration can be found in a book edited by Schuman (2006). Since 2007, The Nature Conservancy has led the development of **Coastal Resilience, an approach and online decision support tool** to help address the effects of climate change and natural disasters on specific coastal regions around the world (Coastal Resilience.org). For the climate science community, fairly mature global collaborative networks, such as the World Climate Research Programme (WCRP; https://www.wcrp-climate.org/) have played crucial roles in advancing improved climate prediction models for many years (Merryfield et al. 2017; see Chap. 2 for other examples). Coastal sciences can benefit measurably from the creation of comparable global networks of consortia.

Rigorous and generally accepted standards are crucial to collaborative modeling. To date, coastal modeling testbed programs have adopted the standards of the Open Geospatial Consortium (OGC; http://www.OpenGeospatial.org) and, until a more robust successor emerges, this approach should be followed for the near future. The OGC is an established consensus standards organization and an international consortium of 371 companies, government agencies (including NOAA), and universities that develop publicly available interface standards. These services make complex spatial information and data services accessible, interoperable and useful to multiple applications. An ongoing OGC testbed activity is focused on Urban Climate Resilience. Recent collaborative research projects conducted by investigators from different disciplines in the natural sciences have successfully created new conceptual, theoretical, methodological, and translational innovations that address complex coastal problems by integration. With funding from NOAA's U.S. Integrated Ocean Observing System Program (IOOS®), SURA has facilitated strategic collaborations to build and guide the Coastal and Ocean Modeling Testbed or COMT (Luettich et al. 2013, 2017). The COMT has demonstrated considerable success in orchestrating collaboration among more than 20 universities along with agency representation from NOAA, Navy, EPA and the U.S. Army Corp of Engineers. The resulting COMT is now one of 11 official NOAA testbeds with the goal of accelerating the transfer of research results to improve operational modeling skill. Among other things, this has involved maintaining a web site, a data archive, providing high-performance computing resources, and custom code to perform tasks such as skill assessment and format conversions.

Unfortunately, many universities are not yet up to the task of true interdisciplinary research. Part of the problem relates to the accreditation system and its discipline-specific standards. This impedes interdisciplinary work at many traditional universities. Multi-discipline papers with many authors are not really valued and young untenured faculty who engage in too much interdisciplinary work may be denied tenure. The discipline-based distribution of faculty on campuses is also a discouraging factor: social scientists and natural scientists may be based on opposite sides of large campuses or even on different campuses of multi campus state universities. Another part of the problem involves simple competition among universities, which tends to stifle multi-institutional altruism. The world is likely to be very different in 2050, as will the missions of universities that remain relevant. Centers and institutes are one way to promote collaboration among disciplines and are not as constrained as traditional university departments. "Enterprise" entities that promote interdisciplinary synergies but also are designed to evolve as science and needs change may be better models. One model is that of a *Center for Research, Education and Innovation* to facilitate the inclusion of industry and governmental entities along with academics. In Australia, the Wentworth Group of Concerned Scientists performs such a role through the engagement of experienced scientists in public policy (www.wentworthgroup.org.au). An example of the work of this group can be found in the formulation of a method of environmental accounts based on natural asset condition measurements involving a range of different disciplines (Wentworth Group 2015). In Australia, this work is now being incorporated into a national system of environment accounts.

Beyond the confines of individual universities or federal agencies, an independent community-shared consortium can help the global community to take a first step in addressing questions of risk and resilience by facilitating the creation of an open-source base of empirical and numerical model data along with a rigorous set of data standards and an extensible cyber infrastructure for managing, and accessing the necessary information. This will support a combination of discipline-specific and cross-disciplinary numerical modeling, coupling the outputs from physical process models with ecosystem and socioeconomic models, and statistical analyses of socioeconomic factors that might ultimately determine the resilience of communities to expected stressors. In addition, modeling protocols could be extended to enable the potential impacts (positive or negative) of engineering approaches or management decisions to be assessed. Several researchers (e.g., Plag et al. 2015) have stressed the need for international collaboration and virtual research environments to enable knowledge creation in response to societal needs. Cloud computing technologies facilitate the creation of cyber-supported "playing fields" where it is easier to work with others. The supporting cyber services should include successive generations of HPC resources for running and coupling models, platforms for accessing, sharing and archiving data and model outputs as well as for accessing and sharing open-source model codes, and a catalogue of and access to analysis routines and visualization tools. Over the course of the next few years, it should be possible to accommodate most or all of the cyber needs.

21.5 Local and Regional Collaboration Is Essential to Strategic Planning

From the foregoing chapters in this book, it should be clear that natural coastal environments as well as coastal communities are highly diverse and there can never be a "one size fits all" strategy for adapting to future changes. The need for local "bottom up" solutions is discussed by Colander and Kupers (2014) and is dictated in part by the pronounced geographical variability of coastal systems and in part by the simple fact that problems related to coastal flooding, land loss and ecosystem degradation are more immediately apparent- and seemingly urgent- at local and regional scales than at national or global scales. In the year 2017, this is especially true in the U.S.A. where the stated position of the Federal Administration is denial of climate change, sea level rise and attendant environmental changes. Some state governors also share these positions (e.g. Florida). On the other hand, the need for aggressive action is so obvious in threatened localities that some state, county and municipal governments are making significant investments in, and assigning high priorities to, planning in advance for resilient futures. At the state level, prominent examples are Guangdong Province, China (Chap. 12), New South Wales, Australia (Chap. 19), and Louisiana's Coastal Protection and Reclamation Authority (CPRA; Chap. 13). As described in Chap. 14, the coastal problems facing Broward County Florida have mandated the establishment of the county's *Environmental Planning and Community Resilience Division.* In smaller communities with fewer resources, local emergency management officials are taking increasing responsibility for resilience planning- but they need help and guidance from the academic community as well as regional planners, operational agencies (NOAA, FEMA, DHS etc.) and politicians.

Adaptive regional strategies are required to reduce risk of harm to valued coastal assets and to overcoming the numerous barriers to long-term planning at different spatial scales as outlined in Chap. 19. In addition, such strategies must be incorporated into a well-understood public policy and legal framework that has within it climate change adaptation policies that recognize the need for action at certain threshold or trigger points. Communities must understand why there needs to be action at such points when changes in coastal conditions determine the need for action. Without clear and accepted policies embedded in law, supported by an implementation plan and enforcement processes, there will always be the temptation to accept pressures from vested interests driven by ignorance and greed.

To be effective, local and regional collaborations need to involve the broadest possible pool of talent and perspectives; they should be community inclusive. Initially, workshops or a series of "town hall meetings" should assess the concerns, needs and expectations of local residents and stakeholders (e.g. property owners, commercial and recreational fishers, emergency managers, tourist industry representatives etc.). Before academics offer their thoughts or advice, they need to listen carefully to the concerns and perceptions of the citizens. As explained in the previous subsection, true collaboration must involve recursive interaction of knowledge, engagement and perceptions. Community leaders, and, where appropriate, clergy should be part of the collaborative planning process. All members of collaborative teams must be respected as equal partners. In the U.S., representatives of state Sea Grant programs should participate in and, in many cases, lead discussions.

21.6 Finding Common Ground

Formulating a comprehensive plan for an enduring coastal resilience program can begin with determining areas where interdisciplinary synergies can be most readily applied, facilitating the infrastructural advances that are needed to accommodate future modeling and communication and preparing a research plan for moving forward as a community. The scientific community at large can initiate and evolve a network of interdisciplinary scientists and supporting cyber-infrastructure with emphasis on understanding and modeling complex coastal systems and communicating the results to operational end users. A key role for the facilitating consortia will not be to execute models but to provide the virtual environments within which modelers and non-modeling scholars from different disciplines can interconnect. Quite simply, coastal systems science must bring together different components of the system and integrate them.

Some crucial steps in this process summarized by Wright et al. (2016a) include the following:

Step 1: Articulating the interconnections of socio-ecological systems and identifying the societal, legal, biophysical and biogeochemical criteria needed to model resilience in specific coastal regions.

- Refine understanding and articulation of interconnections of human and natural coastal processes.
- Advance understanding of the linkages between regional and ocean systems and scale-dependent inter-connections among societal, biophysical and biogeochemical factors.
- Develop criteria for assessing changes in *ecosystem services* and the impacts that these changes may have on rural and urban socioeconomic systems.
- Following the International Geosphere Biosphere Programme example, develop an analytical framework that is relevant to policy and decision-making at different levels and takes account of legal issues and constraints.

Step 2: Identifying the systems science requirements for future coastal risk and resilience programs.

- Catalyze interdisciplinary collaborations.
- Prioritize coastal threats (by region).
- Identify well-defined, integrated research questions and the required modeling, analysis and visualization products to address these questions.
- Identify and prioritize legal factors that may impact community resilience or vulnerability.
- Assess and refine social resilience indices as they pertain to both urban and rural coastal communities.
- Develop feasible data management structures for trans-disciplinary integration and communication.

Step 3: Creating an accessible and extensible cyber infrastructure for cross-disciplinary communication and collaboration.

- Identify design criteria for a collaborative web portal for cross-disciplinary communication.
- Identify the search tools needed to effectively access existing data sets and model outputs.
- Develop a cyber template(s) to enable social and natural scientists, managers and legal scholars to share information in mutually understandable formats (e.g., utilization of NOAA big data for decision making).
- Define the needs for more effective data and model output visualization.

Step 4: Evolving and validating predictive models to reduce uncertainty and coupling models across disciplines to assess complex interactions among future physical, ecological and socioeconomic processes.

- Couple hydrologic and ocean models to improve predictions of compound flooding such as that caused by Hurricane Harvey.

- Improve decadal time scale models of future coastal environmental and socioeconomic outcomes of multiple climate change scenarios.
- Improve regionally specific predictions of the potential outcomes of alternative strategies to mitigate vulnerability of coastal communities.
- Improve the accuracy and accessibility of coastal observation and monitoring data, especially satellite data and improve techniques for assimilating observational data into numerical models.

21.7 Outreach to Policy Makers, Politicians and the Public

As *Hurricane Katrina* bore down on New Orleans in 2005, numerical models, were predicting high storm surges and waves for Coastal Louisiana (via the now discontinued *Open IOOS* website and NOAA's NWS; e.g. Bogden et al. 2007). Data buoys and integrated observing systems such as Wave-Current-Surge Information System for Coastal Louisiana in the Gulf of Mexico verified wave predictions. Those predictions were readily accessible in real time on the Internet in the form of color-coded animations and numerical data, but they were largely ignored by local, State and Federal leaders as well as by many emergency managers. In 2017, long-range scientific projections of climate-related phenomena and their impacts continue to be widely denied by many politicians and decision makers as well as by the U.S. President and many members of his cabinet. Fortunately, in 2012, short-term wind, storm surge and wave forecasts were heeded as Super Storm Sandy moved up the U.S. East Coast and approached New York Bight and this undoubtedly saved many lives. Similarly, in 2017, predictions of the impending impacts of Hurricanes Harvey, Irma and Maria were trusted and heeded. But substantial improvements in communication and trust are still needed.

The scientific community at large must nurture the enlightenment of politicians and decision makers. This may be the largest challenge of all, but the scientific community must work diligently to persuade emergency managers, land use planners and leaders at all governmental levels to trust science-based model predictions. This will require careful and well-articulated, non-jargonized communication over a prolonged period combined with clear and repeated demonstrations that numerical models really work and are not a hoax. So the question is: How do we do that? On the local level, one obvious way is to get to know the local leaders and gain their trust through one on one visits, open forums involving bi-directional exchanges (but not "lectures") and clear demonstrations of mutual respect among academics, local, state and federal government officials and politicians. Such demonstrations must reflect an understanding by credible scientists of past conditions including observations of geomorphic history and impacts of past extreme events. And, of course, patience and persistence are essential. Consortia and NGOs should be able to greatly broaden the scope of outreach to officials by developing accessible and extensible web sites and cyber tool kits that serve clear graphical and textual explanations of coastal evolution as well as long-and short-term model results.

As pointed out in Chap. 18, a critical, and commonly overlooked, facet of outreach to enhance resilience involves educating the general public, particularly lower income and undereducated vulnerable communities, about hazards and how to respond to them. A few U.S. universities already offer programs focused on helping communities become more disaster resilient through practical education. A prominent example is the *Center for Hazards Assessment, Response and Technology (CHART)*, an applied social science hazards center at the University of New Orleans (http://scholarworks.uno.edu/chart). Among other activities CHART staff provide risk literacy as a component of more general adult literacy programs. A network of collaborating university-based coastal resilience researchers and educators could provide web-accessible resources to other programs that share the goals of UNO's CHART.

21.8 Identifying and Engaging Potential Beneficiaries

As of 2017, one of NOAA's stated goals is that of *Resilient Coastal Communities and Economies*. The collaborative consortia that we envision can contribute to this goal but the effectiveness of the contribution will require a thoughtful, and possibly lengthy, process involving an uncommonly diverse assemblage of social and natural scientists, engineers, legal scholars, health scientists, stakeholders and decision makers. For the ongoing COMT program described in Sect. 21.4, the target beneficiaries have been operational agencies (particularly NOAA) and the main product has been the transfer of methodologies and models from research to operations. For future consortia, the potential stakeholders may include the State Sea Grant Programs, re-insurers, county governments, state governments, health workers, emergency managers, resource managers, FEMA, NGOs such as Nature Conservancy and the Sierra Club, educators, the general public- and operational agencies (particularly NOAA, DHS and USACE). Although the specific needs of each of these stakeholders differ, the universal nature of the most urgent questions should enable the facilitating consortium (or consortia) to focus firstly on problems that are important to a broad range of beneficiaries. In some cases, however, it may be necessary to concentrate on a subset of stakeholders who have a narrow definition of "acceptable benefits" that communities actually value. Risk reduction is one such benefit. County and local government agencies charged with planning for future threats may be among those most willing to engage with the consortium.

21.9 Educating Future Generations

Most of the people who will have to adapt to and live with the coasts of tomorrow have not yet been born and, among those who have, few have so far completed their formal education. There is still time to transmit the essential understandings that future generations will need in order to be resilient in a complex and changing world. The essential education process should begin with preschoolers and continue through higher education and beyond. Appreciation of natural environments and humanity's dependence and impacts on them should be instilled in children at the earliest possible age and reinforced throughout their lives. For children in their formative years, it is important for parents and educators not to perpetuate the dualistic notion that humanity and nature are distinct and that the former must conquer the latter. Instead, the idea of "*intelligent co-operation with nature*" should be promoted as a human virtue. And, of course, if the crucial collaborations that we envision are to succeed, the "me first" mind set must be discouraged, in fact, disdained. Armed with this new, holistic and altruistic set of values, the students of tomorrow should be receptive to a formal education that places greater emphasis on natural complexity and change in the *Anthropocene.*

In future classrooms, environmental science, social sciences and geography as well as arts and humanities need to be elevated to the same priority level as STEM subjects. Environmental ethics should be part of the middle school and high school curriculum. Conceptual understandings of processes and connectivities should be emphasized over facts and data, which are transient. Pragmatists may ask: How do arts and humanities fit in here? The answer is that students should be taught to understand that there is more to humanness than money and material things. They also must learn about and celebrate the diversity of both nature and humanity. They must get to know their local area, its history, its geology, its beauty. At the University level, more students should be attracted to rigorous studies of atmospheric, ocean and environmental sciences and especially interdisciplinary studies that integrate natural and social sciences. For those with technical aptitudes, numerical modeling and coupling of models across disciplines should be encouraged at the graduate level. By whatever educational pathway that works and regardless of the educational level that is ultimately attained, future communities should intuitively realize and embrace the axiom with which we opened Chap. 1: ***"All things are connected. Whatever befalls the earth befalls the children of the earth"***. (Chief Seattle of the Suquamish Native American people, c. 1835).

References

Arkema, K.K., G. Guannel, G. Verutes, S.A. Wood, A. Guerry, M. Ruckelshaus, M.P. Kareiva, M. Lacayo, and J.M. Silver. 2013. Coastal habitats shield people and property from sea-level rise and storms. *Nature Climate Change* 3 (10): 913–918.

Bloomberg, M., and C. Pope. 2017. *Climate of hope*, 284. New York, St: Martin's Press.

Bogden, P., T. Gale, G. Allen, J. MacLaren, G. Almes, G. Creager, J. Bintz, L.D. Wright, H. Graber, N. Williams, S. Graves, H. Conover, K. Gallupp, R. Luettich, W. Perrie, B. Toulany, Y.P. Sheng, J.R. Davis, H. Wang, and D. Forrest. 2007. Architecture of a community infrastructure for predicting and analyzing coastal inundation. *Marine Technology Society Journal* 41 (1): 53–61.

Colander, D., and R. Kupers. 2014. *Complexity and the art of public policy: Solving society's problems from the bottom up*. Princeton, N.J.: Princeton University Press.

Cutter, S.L. 1996. Vulnerability to environmental hazards. *Progress in Human Geography* 20 (4): 529–539.

Jr. Luettich, R.A., L.D. Wright, R. Signell, C. Friedrichs, M. Friedrichs, J. Harding, K. Fennel, E. Howlett, S. Graves, E. Smith, G. Crane, and R. Baltes. 2013. Introduction to special section on the U.S. IOOS coastal and ocean modeling testbed. *Journal of Geophysical Research-Oceans* 118 (12): 6319–6328. https://doi.org/10.1002/2013jc008939.

Louisiana Coastal Protection and Restoration Authority. 2007. *Integrated ecosystem restoration and hurricane protection: Louisiana's comprehensive master plan for a sustainable coast.* Baton Rouge, LA: Louisiana Coastal Protection and Restoration Authority (117pp).

Luettich Jr., R.A., L.D. Wright, C.R. Nichols, R. Baltes, M.A.M. Friedrichs, A. Kurapov, A. van der Westhuysen, K. Fennel, and E. Howlett. 2017. A test bed for coastal and ocean modeling. *EOS* 98 (11): 24–29.

Marinez-Moyano, I.J. 2006. *Exploring the dynamics of collaboration in interorganizational settings* (Ch. 4, p. 83). In *Creating a culture of collaboration: The international association of facilitators handbook*, ed. S. Schuman. San Francisco: Jossey-Bass. ISBN: 0-7879-8116-8.

Merryfield, W.J., F.J. Doblas-Reyes, L. Ferranti, J.-H. Jeong, Y.J. Orsolini, R.I. Saurral, A.A. Scaife, M.A. Tolstykh, and M. Rixen. 2017. Advancing climate forecasting. *Eos* 98. https://doi.org/10.1029/2017EO086891.

National Institute of Standards and Technology. 2015. *Community resilience planning guide for buildings and infrastructure systems.* NIST Special Publication 1190. Gaithersburg, MD: U.S. Department of Commerce.

National Research Council. 2014. *Reducing coastal risk on the east and gulf coasts.* Report of the Committee on U.S. Army Corps of Engineers Water Resources Science, Engineering. Washington, DC: National Academies Press (167pp).

National Research Council, Committee on the Restoration and Protection of Coastal Louisiana. 2006. *Drawing Louisiana's new map: Addressing land loss in coastal Louisiana.* Washington, DC: National Academies Press (206pp).

Plag, H.-P., D. Brocklebank, P. Brosnan, S. Campus, S.J. Cloetingh, S. Plag, and S. Stein. 2015. *Extreme geohazards: Reducing the disaster risk and increasing resilience*. France, European Science Foundation, Group on Earth Observations (67pp). http://bit.ly/1WJf3hi.

Rosati, J.D., K.F. Touzinsky, and W.J. Lillycrop. 2015. Quantifying coastal system resilience for the US Army Corps of Engineers. *Environment Systems and Decisions* 35 (2): 196–208.

Schultz, M.T., S.K. McKay, and L.Z. Hales. 2012. *The quantification and evolution of resilience in integrated coastal systems.* ERDC TR-12-7. Washington, DC: U.S. Army Corps of Engineers (70pp). http://1.usa.gov/1FLvNgo.

Schuman, Sandy (ed.). 2006. *Creating a culture of collaboration.* San Francisco: Jossey-Bass.

Southeastern Universities Research Association (SURA). 2015. *Understanding and modeling risk and resilience in complex coastal systems: Final workshop report.* Washington, DC: Coastal and Environmental Research Committee, Southeastern Universities Research Association (20pp). http://bit.ly/1ScVB67.

US Army Corps of Engineers. 2014. *Procedures to evaluate sea-level change: Impacts, responses, and adaptation.* ETL 1100-2-1. Washington, DC: U.S. Army Corps of Engineers (247pp).

Wentworth Group of Concerned Scientists. 2015. *Accounting for nature, Australian regional environmental accounts trial: Technical analysis*, 235. Sydney: Wentworth Group of Concerned Scientists in Association with NRM Regions Australia.

Wright, L.D., C.R. Nichols, A.G. Cosby, S. Danchuk, C.F. D'Elia, and G. Rico-Mendez. 2016a. *Trans-disciplinary collaboration to enhance coastal resilience: Envisioning a national community modeling initiative*. Washington, DC: Southeastern Universities Research Association (30pp).

Wright, L.D., C.R. Nichols, A.G. Cosby, and C.F. D'Elia. 2016b. Collaboration to enhance coastal resilience. *Eos* 97 (24): 9–10. https://doi.org/10.1029/2016eo057981.

Epilogue

All our science, measured against reality, is primitive and childlike- and yet it is the most precious thing we have.
—Albert Einstein

Throughout all of the scientific understandings that we have discussed in the foregoing chapters, uncertainties prevail and surely always will. The scientific community can never have "all the answers" and no amount of hubris will ever change that. Our theories, models and words are simple abstractions of an outrageously complex nature that involves countless interactions among a vast number of factors. We have described a few of these factors that we know well but there are infinitely more. Nevertheless, we understand much more than we did a decade ago and we can be confident that, unless we consciously regress to another dark age, our understandings will grow and the uncertainties of scientific predictions will shrink. But no single scientist, research team or nation can be relied on to provide the breakthroughs that hopefully lie ahead. The global community must do this together in a spirit of urgent, and selfless collaboration, with or without governmental encouragement.

As we progress, we cannot afford to be dismayed by the lack of perfection but must take serious guidance from the observed trends and model predictions that we have now; they really are pretty good. We can be confident that the future will be notably different from today. Whether or not we have the will to respond to the expected changes is more uncertain than are the predictions. Regrettably, in 2018, the will of some leaders seems to be driven more by ideology and greed than by reason. Uncertainties in predictions are used as excuses for outright rejection of science. We as a people must embrace reason and science over blind ideology. Thomas Jefferson stated: "If a nation expects to be both ignorant and free in a state of civilization, it expects what never was and never will be." The scientific

L. D. Wright and C. R. Nichols (eds.), *Tomorrow's Coasts: Complex and Impermanent*, Coastal Research Library 27,
https://doi.org/10.1007/978-3-319-75453-6

community and humanity at large must oppose the assault on reason and nature or tomorrow may not be very bright for coming generations and the impermanence of tomorrow's coasts and coastal communities will be enhanced.

Don Wright
Inglis, Florida
February, 2018

Appendix
Glossary of Terminology Used in *Tomorrow's Coasts: Complex and Impermanent*

Scholars, I plead with you, Where are your dictionaries of the wind, the grasses?

—Norman MacCaig (1910–1996), from Robert Macfarlane's *Landmarks* (Penguin Books, March 2015)

The multidisciplinary nature of climate studies, coastal oceanography, environmental science and social science creates the need for a glossary to make access to the literature easier by defining some terms that may be unfamiliar to lay readers. Definitions for these terms and others may be found in primary references such as the American Meteorological Society *Glossary of Meteorology*, 2nd Edition, *Encyclopedia of Marine Science*, Second Edition (Nichols and Williams 2017), *Tide and Current Glossary* (CO-OPS 2000), and the *Coastal Engineering Manual* (U.S. Army Corps of Engineers 2002). Some of the following definitions have been modified from the original for ease of use and understanding by the nonscientist or engineer, and the reader is urged to consult a text book if greater detail is required.

Accretion: The accumulation of land by natural causes such as wind, waves, river discharge and longshore currents. There is also an increased potential for accretion when wave energy impacting a shoreline is reduced. In tidal marshes, grasses such as smooth cordgrass (*Spartina alterniflora*) trap sediments and create peat, they may rise with sea level in a process that may be called "wetland accretion." Mariño-Tapia et al. (2014) described accretion or widening of beaches along large segments of Puerto Morelos located on the Caribbean coast of the Yucatán Peninsula to reductions in sediment transport and erosion of dune sands.

Adaptation: Action that helps cope with the effects of natural hazards—for example the use of inundation maps, precipitation projections, and flood models to determine thresholds for inundation and the assessment of options to mitigate losses. Strategies may include construction of defensive barriers to protect against rising sea levels, elevating a building, or moving resources at high risk to secure, safer locations, etc. The National Trust, which was founded during 1895 in the United Kingdom, has embraced adaptive approaches in the management of coastal sites such as Birling Gap in East Sussex by planning to roll back properties in a

L. D. Wright and C. R. Nichols (eds.), *Tomorrow's Coasts: Complex and Impermanent*, Coastal Research Library 27,
https://doi.org/10.1007/978-3-319-75453-6

timely manner as erosion of chalk cliffs along the English Channel such as the "The Seven Sisters" accelerates due to sea level rise.

Adaptive management: An iterative decision making process that accounts for what is uncertain as well as what is known. Through monitoring, managers not only predict how coastal systems are likely to respond to interventions, but also identify what management options are available, what outcomes are desired, how much risk can be tolerated, and how best to choose among a set of alternative actions. This approach enables coastal zone managers to reduce uncertainty and thereby improve management through enhanced understanding of management effects.

Anthropocene: A proposed epoch dating from the commencement of significant human impact on the Earth's geology and ecosystems first popularized in 2002 by Nobel laureate Paul J. Crutzen, who regards the influence of human behavior on Earth's atmosphere in recent centuries as so significant as to constitute a new geological epoch (Zalasiewicz et al. 2008). This proposed epoch would follow the Holocene (11,700 years ago to the present) and begin in the year 1950.

Anthropogenic: Caused or produced by humans as opposed to natural processes. Air and water pollution occur when harmful substances including particulates and biological molecules are introduced into Earth's atmosphere and hydrosphere. These releases may cause diseases, allergies or death of humans; it may also cause harm to other living organisms such as animals and food crops, may damage the natural or built environment, and contribute to climatic changes.

Aquifer: An underground layer of permeable rock (e.g., sand, gravel, and sandstone) that acts as a reservoir for groundwater. Aquifers fill with water that drains into the ground. Wells drilled into aquifers provide water for drinking, agriculture, and industrial uses. Aquifers can be depleted or dry up when people drain them faster than nature can refill them. When too much water is removed from the soil, the soil may collapse, compact, and drop contributing to land subsidence.

Atmosphere: The gaseous envelope surrounding the Earth. The atmosphere consists almost entirely of nitrogen (78.1% volume mixing ratio) and oxygen (20.9% volume mixing ratio), together with a number of trace gases, such as argon (0.93% volume mixing ratio), helium, greenhouse gases such as carbon dioxide (0.035% volume mixing ratio), and ozone. The atmosphere also contains water vapor, clouds, and aerosols.

Atmospheric pressure: The pressure exerted by the earth's atmosphere; it decreases with altitude above sea level. At sea level, the standard atmospheric pressure is equal to 1013.25 mbar (14.7 psi). Researchers such as Gillett et al. (2003) have shown that average sea-level air pressure has risen in recent years over parts of the subtropical North Atlantic Ocean, southern Europe and North Africa while dropping in locations such as the poles and the North Pacific Ocean. Increases in sea-level pressure decrease water level while decreases in pressure increase water level.

Bar: A submerged or emerged embankment of sand, gravel, or other unconsolidated material built on the seafloor in shallow water by waves and currents. Storms tend to carry sand seaward, forming offshore bars and then this sand migrates landward during calm weather. A bar is also a metric unit of pressure equal to the average pressure of the atmosphere at sea level or 106 dyne cm^2 (106 barye), 1000 mbar, 29.53 in. of mercury. Within the ocean water column, the pressure increases by one bar with every 10 m (33 ft) of water depth.

Barrier island: A long, usually thin, sandy stretch of land, oriented parallel to the mainland coast between two inlets. They face the ocean on one side and a lagoon on the other side. Barrier islands provide a natural line of defense against storms that threaten coastal communities by absorbing wind and wave energy.

Beach: The location along a shoreline where the sediment is in motion, being moved by winds, waves, tides, and currents. Loose unconsolidated sediments such as sand that are transported to places suitable for deposition will form a beach. They are dynamic and vary in length, width, composition, and permanence. Sea level rise will affect beaches, but will not occur uniformly worldwide because of complications such as isostatic or postglacial rebound where land masses that were once depressed by ice sheets rise while those that were raised at the edges of the glaciers are now sinking. Beaches will evolve in responses to forces such as sea level rise and coastal development.

Beach material: Granular sediments, usually sand or shingle moved by waves and currents. Dams built for flood control and water catchment along the rivers leading to the coast inhibit the transport of sediments. The lack of sediment from rivers accelerates coastal erosion.

Beach nourishment: The restoration of a beach by placing beach material directly onto the beach. Beach nourishment is expensive and only a temporary solution to counter sediment starvation along severely eroding coasts. During 2017, the towns of Duck, Kitty Hawk, Kill Devil Hills, and Southern Shores along the Outer Banks of North Carolina nourished miles of beach by pumping sand from the ocean floor onto the beach to build up eroded areas. These efforts are intended to create a wider beach area to protect and enhance the shoreline.

Benthic habitat: Physically distinct areas of the seafloor associated with suites of species that consistently occur together. Species of animals and plants that live on or in the bottom are known as the benthos. The creation of benthic habitat maps is essential to the management and analysis of these important ecological resources.

Bight: A geographic term describing a bend or curve in the shoreline such as a large and receding bay. Examples include Bight of Benin, Canterbury Bight, Flemish Bight, Great Australian Bight, Mid-Atlantic Bight, New York Bight, North and South Taranaki Bights, and Southern California Bight.

Biomass: Biological materials including organic material (both living and dead) from above and below, e.g., mangrove trees, marsh grasses, submerged aquatic vegetation, roots, and animals and animal waste.

Breaker: A wave breaking on the shore. As an ocean swell propagates from deep water to shallow water, it will undergo transformations through the effects of refraction, diffraction, and/or shoaling until the wave becomes unstable. Once the wave reaches a critical height, it will overturn on itself and break. Based on the dissipation of energy owing to factors such as beach slope, wind conditions, and nearshore currents, the breaking will result in collapsing, plunging, spilling, and surging surf. These breakers and the ensuing longshore currents tend to pick up and move sand up and down the shoreline, generally offshore during winter storms and back toward land during the summer's fair weather.

Cape: A prominent land area jutting seaward from a continent such as Cape Comorin in Asia or a point on a large island such as Cape Engaño off of Palaui Island in the Philippines. A cape usually represents a marked change in the trend of the coastline. Capes can be formed by glaciers, volcanoes, and changes in sea level and have a relatively short geologic lifespan owing to natural erosion by winds, waves, tides, and currents. Some other cape examples include Cape Cod in Massachusetts and Cape Hatteras in North Carolina and Cape of Good Hope in South Africa.

Carbon dioxide (CO_2): A naturally occurring odorless gas formed during respiration and by the decomposition of organic substances. CO_2 is a needed factor for photosynthesis and is a by-product of the burning of fossil fuels. CO_2 released into the atmosphere is also absorbed in ocean waters. The amount of CO_2 that the ocean can hold depends on the ocean temperatures. Colder waters can absorb more carbon than warmer waters. The uptake of CO_2 from the atmosphere by the ocean may contribute to ocean acidification or a decrease in the pH of the Earth's oceans.

Climate: The prevailing weather conditions for a region. Climatic elements include temperature, air pressure, humidity, precipitation, sunshine, cloudiness, and winds, throughout the year, averaged over a series of decades. Climate differs from weather, in that weather only describes the short-term conditions of these variables in a given region. Example statistics may include labels such as normals, means, and extremes. Normals are generally based on the distribution of all observations over a long-period such as 30 years (e.g., 1981–2010). Means will refer to the average of the maximum and minimum temperatures over a particular period such as a day. These means do not provide any information about how the observations are scattered around the mean, i.e., whether they are tightly grouped or broadly scattered. The extremes are usually observations lying in the most unusual ten percent. Climatic changes take hundreds, thousands, even millions of years to change while the weather can change in just a few hours. Climate change, therefore, is a change in the typical or average weather of a region or city. Earth's climate is always changing as evidenced by climatic records that indicate warmer and cooler periods, each lasting thousands of years. Trends from quality controlled observations currently indicate that the Earth's climate has been warming. The Earth's warming climate is linked to changing rainfall patterns, decreasing snow and ice cover, and rising sea levels. It should also be noted that climatic data can be confounded by various non-climatic effects, such as the relocation of weather stations, land-use changes, changes in instruments, and observational hours.

Coast: A strip of land of indefinite width that extends from the shoreline inland to the first major change in terrain features. In this book, we use the term "coast" to imply the entire region influenced by land-sea interaction, typically from the shelf break to about 200 km (124.3 mile) inland. Coasts are sensitive to sea level rise, changes in the frequency and intensity of storms, increases in precipitation, and water temperatures. Severe storms generate surge, waves and currents that can move large amounts of sediment, destroy roads, buildings and other critical infrastructure as well as alter natural habitats.

Coastal inlet: A passage separating two barrier islands and connecting a lagoon or bay with the ocean. These connections are important for sediment transport and fish migrations. Changes in the morphology and behavior of an inlet and its associated sediment bodies can have far-reaching impacts on adjacent coastlines.

Coastal zone: The dynamic region where land meets water. These regions have become increasingly important because they support large populations and continually change owing to natural (tropical cyclones and tsunami) and man-made (coastal structures) phenomena. Climate change can affect coastal zones through shoreline erosion, coastal flooding, and water quality. Coastal zone management efforts ensure the health and stability of the coast, both environmentally and economically, into the long-term future.

Coastline: The land and water interface, which is also called the shoreline. The coastline is constantly eroding and forming headlands, bays, and cliffs. The weaker or softer rock, such as sandstone, is eroded fastest leaving more resistant rock types, such as granite, which might remain as a sea cave, arch, or stack. Remote sensing surveys of the coast call the land and water interface at time of imaging the waterline. Waterlines can be used to map shoreline changes following severe storms.

Complex systems: A system such as a coastal ecosystem which is composed of many living (biotic) and non-living (abiotic) components. Complexity stems from interactions occurring at the boundary between atmosphere, land, and sea. Abiotic conditions such as winds, waves, tides, and salinity are highly variable as evidenced by observed changes occurring as one moves from land to sea. In many cases it is useful to represent complex systems as a network, where, as in a food web, the nodes represent species in a community and the links map the feeding connections. Another example is the manner in which a coral reef survives wave pounding from severe storms but cannot cope with high siltation rates from sediment-laden runoff.

Compound flooding: Severe flooding related to the simultaneous additive effects of multiple causes, for example river (fluvial) flooding, pluvial flooding by torrential rainfall and storm surge.

Convergence: A new paradigm where disciplines such the physical sciences and engineering, among other disciplines, join forces with the life sciences to solve common problems. Organizations such as the Southeastern Universities Research Association have demonstrated how to move research forward as evidence by development of the Coastal and Ocean Modeling Testbed through collaborations involving multiple types of organizations, such as other universities, industry, government regulators, and community stakeholders.

Coral bleaching: The process in which a coral colony, under environmental stress expels the microscopic algae (*zooxanthellae*) that live in symbiosis with the hard and stony corals. The affected coral colony appears whitened. If the stress-caused bleaching is not severe, corals have been known to recover. If the algae loss is prolonged and the stress continues, the coral will eventually die. Healthy coral reefs support an amazing diversity of marine life and are often called the rainforests of the sea. Coral reefs provide many benefits, including coastal protection, food, jobs, medicine, and recreational activities.

Cyclone: A system of winds that rotate about a low pressure center. Rotation is clockwise in the Southern Hemisphere and anti-clockwise in the Northern Hemisphere. In the Indian Ocean, the term cyclone refers to a tropical cyclone or what would be called a hurricane in the Atlantic Ocean. An example storm was Cyclone Nargis which developed on April 27, 2008 in the Bay of Bengal where it moved eastward and intensified to attain peak winds of approximately 215 km/h (130 mph) before making landfall in the delta region of the Ayeyarwady River (Irrawaddy River) of Myanmar on May 2, 2008, and finally dissipating near the border of Myanmar and Thailand. This cyclone caused catastrophic destruction in Myanmar with a death toll that was estimated to exceed 135,000 people.

Dead zone: A low-oxygen (hypoxic) area in the world's oceans and lakes that is caused by excessive nutrients (e.g., nitrogen and phosphorus) or eutrophication. The enrichment of the water body with nutrients induces growth of phytoplankton and algae and due to the biomass load results in oxygen depletion. The resulting hypoxic areas usually appear near the bottom where the oxygen concentration is so low that benthic and demersal animals suffocate as evidenced by associated fish kills, distressed animals, and sulfur-oxidizing bacteria that cause the anaerobic sediments to turn black. While natural phenomena can cause hypoxia, the improper use of fertilizers and inadequately treated or untreated sewage exacerbates the problem as water high in nutrients drains into streams, rivers, and the ocean. Dead zones have been observed in the northern Gulf of Mexico, North Sea, Black Sea, northern Adriatic, Chesapeake Bay, Kattegat Strait, and more.

Delta: A seaward prograding and fan-shaped sediment body deposited at the mouth of a river. A bird-foot delta (e.g., the Mississippi River delta with its levee-bordered distributaries) extends seaward and resembles the claws of a bird. A cuspate delta such as the Ebro delta in Spain extends into the sea like an arrow. An arcuate delta forms owing to tides and waves as in the case of the rounded convex outer margin of the Nile delta in Egypt. Deltas provide fertile lands, support commercial fishing, and tend to be locations with oil and gas deposits, but these low-lying areas are also associated with severe flooding. For additional information see the World Delta Database (Coleman and Huh 2004; www.geol.lsu.edu/WDD).

El Niño: Coupled ocean-atmospheric interactions across a broad expanse of the equatorial Pacific Ocean with a global impact on weather patterns. The condition is marked by a significant increase in sea surface temperature over the eastern and central equatorial Pacific that occurs at irregular intervals, generally ranging between two and seven years. Warming of the Humboldt current and shifting wind patterns as a consequence of El Niño were determined to also impact the upwelling of nutrient rich waters (off the coast of Peru and Chile) for marine species such as anchoveta (*Engraulis ringens*), which in turn impacts other populations and makes the ecosystem very vulnerable to conventional intensive fishery practices.

Enteric infections: The invasion and multiplication of microorganisms such as amoebas (e.g., *Giardia lamblia*), bacteria, viruses, and parasites in the intestines. Bacteria known to cause enteric infections are *Escherichia coli*, *Vibrio cholerae*, and several species of *Salmonella*, *Shigella*, and anaerobic streptococci. Enteric infections are characterized by diarrhea, abdominal discomfort, nausea and vomiting. Floods may disrupt water sanitation infrastructure and spread enteric infections.

Erosion: The movement of weathered or decomposed rock material or soil by natural forces such as winds, waves, and currents. The combined increased frequency of storms and sea level rise in response to global warming will contribute to coastal erosion, especially where there are cliffs composed of soft rocks such as in Pacifica, located along the Pacific Ocean coast in California, and Hapsburg, located along the North Sea in East Anglia, England. Erosion problems can be worsened by countermeasures that do not consider the effects on adjacent shores. Coastal erosion may also be caused by river damming and diversion, owing to the loss of sediment supply to the coast. Coastal development on Java Island in Indonesia including conversion of mangrove forest to shrimp ponds has been linked to increased rates of erosion.

Estuary: An embayment of the coast such as a river, bay, or lagoon where fresh river water entering at its head mixes with the relatively saline ocean water. A North American example is the St. Lawrence River, which connects the Great Lakes to the Atlantic Ocean. Increasing water levels and salinity changes can be very damaging to an estuary. Plants that are meant to be above the waterline such as least spikerush (*Eleocharis acicularis*) are now being drowned, and there is ever-decreasing light availability to submerged aquatic vegetation such as eelgrass (*Zostera marina*). Sea level rise also increases the amount of flooding and erosion of coastal areas. Inundation and salt water intrusion increases the salinity of water that was once fresh.

Eustatic change: Global fluctuations in sea level due to an alteration in the volume of water in the oceans or, alternatively, a change in the amount of water owing to a change in the shape of an ocean basin. A eustatic sea level rise refers to an increase in the volume of the world's waters caused by the melting of polar ice caps since the last ice age ended 11,700 years ago.

Evaporation: The process by which water changes from a liquid to a gas or vapor. Hypersaline conditions in estuaries such as the Laguna Madre in the United States (e.g., Kolker 2003) or the Coorong Estuary, Leschenault Estuary, and Spencer Gulf in Australia (e.g., Wolanski 2014) occur as salinity levels rise during the dry summer when higher temperatures increase levels of evaporation in the estuary.

Extra-tropical cyclone: A cyclonic-scale storm that primarily gets its energy from the release of potential energy when cold and warm air masses interact. These storms always have one or more fronts connected to them, and can occur over the land or ocean. The tracks of these storms affect weather in mid-latitudes. Extratropical cyclones include blizzards, Nor'easters, and the ordinary low pressure systems that give the continents at mid-latitudes much of their precipitation. Some research meteorologists have provided evidence of an increase in the frequency of winter storms and their intensity since the 1950s, and their tracks have shifted northward over the United States. Extra-tropical cyclones can produce a storm surge and cause coastal erosion.

Fossil fuel: Organic materials formed from decayed plants and animals that have been converted to crude oil, coal, natural gas, or heavy oils by exposure to heat and pressure in the earth's crust over hundreds of millions of years. The burning of fossil fuels contributes carbon dioxide (CO_2) to the atmosphere, which is one of the greenhouse gases that allows radiative forcing and contributes to global warming. High-emissions of CO_2 have been linked to global warming and the melting of Greenland ice sheets (Kintisch 2017).

Glacier: A large, perennial accumulation of ice, snow, rock, sediment and liquid water originating on land and moving down slope under the influence of its own weight and gravity. Glaciers are classified by their size, location, and thermal regime and may terminate on land or in water. One of the largest glaciers in the world is the Lambert-Fisher Glacier in Antarctica. Melting glaciers from global warming cause increased runoff which contributes to sea level rise.

Greenhouse effect: Trapping and build-up of heat in the troposphere. Some of the heat flowing back toward space from the Earth's surface is absorbed by water vapor, carbon dioxide, ozone, and several other gases in the atmosphere and then reradiated back toward the Earth's surface. If the atmospheric concentrations of these greenhouse gases rise, the average temperature of the lower atmosphere will gradually increase.

Greenhouse gases: Any gas that absorbs infrared radiation in the atmosphere. Greenhouse gases include, carbon dioxide, methane, nitrous oxide, ozone, chlorofluorocarbons, hydrochlorofluorocarbons, hydrofluorocarbons, perfluorocarbons, and sulfur hexafluoride.

Hadley cells: Thermally driven and zonally symmetric circulation pattern under the strong influence of the earth's rotation, first proposed in 1735 by George Hadley (1685–1768) as an explanation for the trade winds. It consists of rising air at or near the Equator, poleward flow aloft, and sinking air at or near 30° north or south of the Equator. Global warming will change the structure of Hadley cells.

Harmful algal blooms or HABs: A rapid increase in the population of algae in an aquatic or marine environment. Very dense clouds of these organisms (blooms) can change the color of the water to green, yellowish-brown, or red. Bright green blooms may also occur as a result of blue-green algae, which are actually bacteria (cyanobacteria). Only a few dozen of the many thousands of species of microscopic and macroscopic algae (phytoplankton) are repeatedly associated with toxic or harmful blooms. Species, such as dinoflagellates (e.g., *Alexandrium tamarense*, *Akashiwo sanguinea*, *Karenia brevis*, and *Pfiesteria piscicida*) and the diatom *Pseudo-nitzschia australis* produce potent toxins which are liberated when the algae are eaten. harmful algal blooms can harm or kill marine animals, contaminate shellfish, and threaten human life.

Heat Island: An urban area such as Los Angeles, California where temperatures are higher than those of the surrounding non-urban area. As urban areas develop, buildings, roads, and other infrastructure replace open land and vegetation. These surfaces absorb more solar energy, which can create higher temperatures in urban areas. According to the U.S. Environmental Protection Agency, the annual mean air temperature of a city with 1 million people or more can be 1–3 °C (1.8–5.4 °F) warmer than its surroundings. In the evening, the difference can be as high as 12 °C (22 °F). Heat islands can affect communities by increasing summertime peak energy demand, air conditioning costs, air pollution and greenhouse gas emissions, heat-related illness and mortality, and water quality. Actions to reduce the impact of urban heat islands include increasing tree and vegetative cover.

Holocene: Period of geological time spanning the last 11,700 years from the end of the Pleistocene period (and the end of the last ice age) to the present. Climatic changes during this epoch have contributed to the melting of ice sheets in North America and Europe and the retreat of glaciers. This warming continues to alter landscapes and coastal zones owing to factors such as shifting precipitation patterns, increased intensity of storms, and sea level rise.

Hurricane: An intense tropical cyclone in which winds tend to spiral inward toward a core of low pressure, with maximum wind velocities that are greater than or equal to 33.5 m/s (75 mph) for several minutes or longer. The term hurricane is used in the Atlantic Ocean, Gulf of Mexico, and eastern Pacific Ocean. As an example, tropical depression Harvey moved into the warm waters of the Gulf of Mexico on August 22, 2017, and exploded rapidly into a major hurricane in approximately 40 h. Harvey made landfall as a Category 4 Hurricane with winds of 209 km/h (130 mph) near Rockport, Texas on August 25, 2017. Hurricane Harvey dropped 1–1.5 m (40–61 in.) of rainfall in southeast Texas and southwest Louisiana and created a catastrophic flood disaster in southeast Texas.

Hydrologic cycle: The process of evaporation, vertical and horizontal transport of vapor, condensation, precipitation, and the flow of water from continents to oceans. It is a major factor in determining climate through its influence on surface vegetation, the clouds, snow and ice, and soil moisture. The hydrologic cycle is responsible for 25–30% of the mid-latitudes' heat transport from the equatorial to polar regions.

Hydrosphere: All the waters on the earth's surface including clouds, oceans, seas, rivers, lakes, underground water etc. Like the atmosphere, the hydrosphere is in constant movement through land drainage from rivers and streams and circulation in estuaries and ocean currents. Ocean currents are particularly important in transferring heat from the equator toward the poles.

Hypoxia: Low or depleted oxygen in coastal ocean and freshwater environments that is associated with the overgrowth of algae, which can lead to oxygen depletion when they die, sink to the bottom, and decompose. These areas of low dissolved oxygen concentration are called "dead zones" because animals can suffocate and die. Dead zones caused by agricultural run-off and other pollutants are found all around the world. Examples include Lake Erie, Chesapeake Bay, Lake Tai in China, the Adriatic Sea, Baltic Sea, Black Sea, and Gulf of Mexico.

Impervious surface: Surfaces such as roofs, solid decks, driveways, patios, sidewalks, parking areas, tennis courts, and concrete or asphalt streets that impede the natural infiltration of water into the soil. Since impervious surfaces seal the soil surface, eliminating rainwater infiltration and natural groundwater recharge, they contribute to increased surface runoff and flooding. Elvidge et al. (2004) discuss the impact of constructed areas in the United States on local hydrology, climate, and carbon cycling.

Infrared radiation: Electromagnetic radiation from 700 to 1 mm. Infrared radiation wavelengths are longer than the red (620–750 nm) in the visible part of the spectrum (390–700 nm), but shorter than microwave radiation. Infrared radiation can be perceived as heat. The Earth's surface, the atmosphere, and clouds all emit infrared radiation, which is also known as long-wave radiation. For Earth's temperature to remain stable, the amount of incoming solar radiation (e.g., visible light, ultraviolet, and infrared radiation) should be roughly equal to the amount of infrared leaving the atmosphere.

Inundation: The submergence of land by water, particularly in a coastal setting. The effect, result, or outcome of inundation is potential loss of life, economic losses, and adverse social—environmental impacts. Inundation caused by storms may be measured by water level gages, which provide important information on how high water levels rise and their duration at a particular location. This type of data has proven useful for communities that are interested in preparing for sea level rise. Cities such as Annapolis, Maryland in the United States, located at the confluence of the Severn River and Chesapeake Bay, have flooded in the past and continue to be at risk of flooding. Historic coastal structures in Annapolis such as the Sands House (c. 1740) are considered to be highly vulnerable to future flooding.

Invasive species: Any kind of living organism (e.g., bacteria, fungus, plant, insect, mollusk, fish, reptile, or mammal) that is not native to an ecosystem and which causes harm. Most commercial, agricultural, and recreational activities depend on healthy native ecosystems. For this reason, the impacts of invasive species on natural ecosystems and the economy costs billions of dollars each year. In the United States, costly effects include clogging of water facilities from quagga (*Dreissena bugensis*) and zebra (*Dreissena polymorpha*) mussels and clogging of waterways from aquatic plants such as the weed hydrilla (*Hydrilla verticillata*) and giant salvinia (*Salvinia molesta*), disease transmission, harm to fisheries, and increased fire vulnerability and diminished grazing value.

IPCC: Intergovernmental Panel on Climate Change. This is the extensive group of scientists that have been responsible for publishing the series of assessment reports on climate change.

Isostasy: The balance between changes within the Earth's crust and mantle, where material is displaced in response to an increase (isostatic depression) or decrease (isostatic rebound) in mass at any point on the Earth's surface above. Such changes are frequently caused by advances or retreats of glaciers. Govers (2009) theorized that isostasy sealed off the Mediterranean Sea from the Atlantic Ocean five million years ago.

Isostatic change: Local fluctuations in sea level in response to an increase or decrease in the height of the land. When the height of the land increases in a place such as Richmond Gulf in southeastern Hudson Bay, the sea level falls and when the height of the land decreases in a place such as the Chesapeake Bay the sea level rises. In addition to the sinking of land that was elevated at the edge of a glacier, subsidence may be caused by non-geological processes. For example, land subsidence in the Houston-Galveston, Texas, area and in the Santa Clara Valley, California area were caused by petroleum extraction in Texas and groundwater withdrawals, as well as ground water withdrawals in California. For these reasons, isostatic change refers to local sea level changes while eustatic change refers to global sea level change.

Lagoon: A shallow water body found on all continents which receives little if any fluvial input and is connected to the ocean by coastal inlets. Extreme storms causing the inundation of lagoons by the coastal ocean will change the salinity and possibly alter the ecosystem. Salinity can have a great impact on the type of organisms that live in a body of water such as the Pamlico Sound in North Carolina, the Laguna Madre in Texas, the Laguna Madre de Tamaulipas in Mexico, Lake Nokoué in Benin and Lake Piso in Liberia.

Landfast ice: A type of sea ice that is anchored to shore, grounded icebergs, or the shallow seafloor. It does not move with the winds or currents. Landfast ice is particularly important to polar bears (*Ursus maritimus*) and provides access to prey such as ringed seal (*Pusa hispida*) pups.

Land and sea breezes: Breezes occurring near shorelines that are caused by unequal heating of air over land and water. While the land is warm during the day, air above it rises, and a cool breeze blows in from the sea. As the land cools off at night, air pressure over it increases, and a cool land breeze blows out to the sea. The leading edge of a sea breeze or sea breeze front can provide a trigger to daily coastal thunderstorms.

Land subsidence: The gradual settling or sudden sinking of the Earth's surface owing to subsurface movement of earth materials. Coastal and delta cities around the world are especially susceptible to the sinking of land from the extraction of groundwater, oil, and gas and the drainage of soils (Showstack 2014).

La Niña: Coupled ocean-atmospheric interactions across a broad expanse of the tropical Pacific Ocean with a global impact on weather patterns. The condition refers to the extensive cooling of the central and eastern tropical Pacific Ocean, often accompanied by warmer than normal sea surface temperatures in the western Pacific, and to the north of Australia. La Niña events are sometimes thought of as the opposite of El Niño.

Levee: An artificial bank confining a river or stream channel or limiting adjacent areas subject to flooding. Levees help protect people in the Netherlands from flooding by the North Sea, in Canada from flooding by the Bay of Fundy, and in the United States from flooding by the Mississippi river. Earthen levees must be maintained owing to erosion, especially since failure at one location can result in a catastrophic failure for an entire levee system.

Living shorelines: Shoreline protection that provides erosion control benefits through the strategic placement of plants, stone, sand fill, and organic structural materials such as "biologs" and oyster reefs. Biologs are made of biodegradable (e.g., coconut fiber) materials bound by high strength twisted netting and may be staked to an eroding shoreline to help attenuate wave energy. Oyster larvae typically settle on the shells of other oysters, forming dense, expansive clusters or colonial communities known as oyster reefs, bars or beds. They may also settle on blocks of concrete, limestone, crushed shell and silica.

Longwave radiation: Radiation emitted in the spectral wavelength greater than about 4 μm (4000 nm), corresponding to the radiation emitted from the Earth and atmosphere. Heat resulting from the absorption of incoming shortwave radiation and the emission of longwave radiation warms the lower atmosphere and the Earth's surface.

Macro algae: Large marine and aquatic photosynthetic plants (seaweed) that can be seen without the aid of a microscope. Macro algae are ecologically and economically important primary producers and play key roles in coastal carbon cycles. These sessile organisms are impacted by factors such as land runoff, water pollution, and increasing water temperatures. Macroalgae beds (e.g., *Ascophyllum nodosum*, *Gracilaria pacifica*, *Fucus gardneri*, and *Sargassum muticum*) provide important ecosystem goods and services and the degradation of these systems will have far reaching consequences to society.

Megacities: Cities with populations over 10 million such as Tokyo in Japan, Jakarta in Indonesia, and Seoul in South Korea. Megacities have the potential to contribute significantly to the emissions of greenhouse gases and pollutants.

Meridional Overturning Circulation (MOC): A global-scale circulation cell where surface waters from high latitudes are cooled, become denser than surrounding waters, sink, and then flow towards the equator. The stability and variability of MOC advects warm surface water, forms deep-ocean currents, and affects climate. For example, the North Atlantic Current, which is the upper limb of the MOC in the North Atlantic, moderates the climate of western Europe in comparison to other worldwide regions located at similar latitudes. Researchers such as Liu et al. (2017) have illustrated global warming impacts to the Atlantic MOC which result in the intensification of Atlantic hurricanes.

Monsoon: A thermally driven wind arising from differential heating between a land mass and the adjacent ocean that reverses its direction seasonally. The India monsoon is caused by the asymmetric heating and cooling of the Indian Ocean during summer and winter. It blows from the northeast during cooler months and reverses direction to blow from the southwest during the warmest months of the year. This process brings large amounts of rainfall to India during June and July.

Morphodynamics: The study of landscape and seascape changes due to erosion and sedimentation. Rising sea level, increases in storm intensity, and storm frequency increase the magnitude of forces which impact complex and impermanent coastal systems. Combinations of high-resolution, high-precision spatiotemporal, process-based field observations with numerical models are being applied to predict the evolution of the estuaries, beaches, tidal inlets, and hydrography of the coastal zone.

Mud: Wet clay and silt-rich sediment. Silt and clay sediments hold many answers to climate change. Olajos et al. (2017) have found that deoxyribonucleic acid (DNA) in lake sediment forms a natural archive displaying when various fish species colonized lakes after the glacial period.

Neap tides: Tides of decreased range or tidal currents of decreased speed occurring semimonthly as the result of the Moon being in quadrature. During neap tides, water levels may increase when combined with flood drivers such as storm surge or river discharge.

Northeasters: An extratropical cyclone with winds blowing from the northeast. In parts of the United States such as New England, these storms are called, "nor'easters." A 22 m/s (50 mph) nor'easter during January 1987 combined with a high tide cut a new inlet through North Beach in Cape Cod, Massachusetts forming what the locals call, "South Beach Island." The new inlet has narrowed in recent years as the southern tip of North Beach Island has lengthened, pushing the coastal inlet against an eroding South Beach.

Nutrients: Substances that provide nourishment for growth or metabolism. Plants absorb nutrients and animals obtain nutrients from ingested foods. Too many nutrients entering coastal waters (e.g., nitrogen and phosphorus) will cause an excessive growth or bloom of algae followed by low levels of dissolved oxygen.

Ocean acidification: Increased concentrations of carbon dioxide in sea water causing a measurable increase in acidity (i.e., a reduction in ocean pH). This will impact calcifying organisms such as corals and mollusks.

Perigean spring tide: A tide occurring when the moon is either new or full and closest to Earth (perigee). Perigean spring tides may cause minor coastal flooding in very low-lying areas such as Fort Lauderdale, Florida. It is expected that occurrences of minor "nuisance" flooding at the times of perigean spring tides will increase as sea levels rise relative to the land.

Pleistocene: The epoch of geologic time, informally called the 'The Great Ice Age' that began ~1.8 million years ago and ended ~8000 years ago. During this interval continental glaciers repeatedly formed and covered significant parts of the Earth's surface.

Radiative Forcing (RF): Positive RF leads to surface warming, negative RF leads to surface cooling. RF estimates take account of observed conditions, properties of greenhouse gases, and numerical model characterizations of observed processes (IPCC 2013).

Recession: Landward movement of the shoreline over a specified period of time. Rates of shoreline recession are higher on beaches exposed to high wave energy.

Relative sea level rise: The increase in ocean water levels at a specific location, taking into account both global sea level rise and local factors, such as local subsidence and uplift. Relative sea level rise is measured with respect to a specified vertical datum relative to the land, which may also be changing elevation over time.

Representative Concentration Pathways (RCPs): Four greenhouse gas concentration (not emissions) trajectories adopted by the Intergovernmental Panel on Climate Change (IPCC) (Moss et al. 2008). The pathways are used for climate modeling and research. They describe four possible climate futures, all of which are considered possible depending on how much greenhouse gases are emitted in the years to come. Each different RCP is distinguished by their radiative forcing (in Watts per square meter or W/m^2) in the year 2100 relative estimated 1750 values. The four scenarios are: (1) 2.6 W/m^2 for scenario RCP2.6; (2) 4.5 W/m^2 for scenario RCP 4.5: (3) 6.0 W/m^2 for scenario RCP6.9; and (4) 8.5 W/m^2 for scenario RCP8.5. Model simulations carried out for each scenario assumed CO_2 concentrations by 2100 to be: 421 ppm for RCP2.6; 538 ppm for RCP4.5; 670 ppm for RCP6.0: and 936 ppm for RCP 8.5.

Resilience: A capability to anticipate, prepare for, respond to, and recover from significant multi-hazard threats with minimum damage to social well-being, the economy, and the environment.

Saffir-Simpson scale: A 1–5 rating based on a hurricane's sustained wind speed and potential property damage. Hurricanes reaching Category 3 and higher are considered major hurricanes because of their potential for significant loss of life and damage. Category 1 and 2 storms are still dangerous, however, and require preventative measures. The scale was originally developed by civil engineer Herb Saffir (1917–2007) and meteorologist Bob Simpson (1912–2014). It has been an excellent tool for alerting the public about the possible impacts of various intensity hurricanes.

Salinity: The dissolved salt content of a body of water. Marine salinity levels are influenced by a number of factors including rainfall, evaporation, inflow of river water, wind, and melting of glaciers. Salinity plays a critical role in the water cycle and ocean circulation.

Sand: Loose particles of rock or mineral (sediment) that range in size from 0.0625–2.0 mm in diameter. Sands are common type of marine sediment that are transported alongshore to produce sandy beaches.

Saltwater intrusion: Displacement of fresh or ground water by the advance of salt water due to its greater density, usually in coastal and estuarine areas.

Sea level rise: The upward trend in average sea level height. Causes of sea level rise include the thermal expansion of seawater by warming of the ocean and increased melting of land-based ice, such as glaciers and ice sheets. Higher sea levels allow floods to push farther inland and future rising sea level impacts may be foreshadowed by recurring flooding occurring during spring tides.

Sinkhole: A depression in the surface commonly found in karst landscapes. Sinkholes often form where limestone or some other soluble rock is partially dissolved by groundwater, then collapses to form a depression.

Social vulnerability index: A tool using census variables to help local officials identify communities that may need support in preparing for hazards, or recovering from disaster. Social vulnerability refers to the resilience of communities to stressor related to natural or human-caused disasters, disease outbreaks, etc. Reducing social vulnerability can decrease both human suffering and economic loss.

Spring tide: Tides of increased range or tidal currents of increased speed occurring semimonthly as the result of the Moon being new or full.

Storm surge: An storm-induced rise of water, over and above the predicted astronomical tides.

Subsidence: Land sinking as the result of multiple causes including compaction of coastal sediments, sinking under the shear weight of large urban centers, withdrawal of groundwater or oil, and regional tectonic effects. Subsidence is added to, and exacerbates, sealevel rise. Most of the world's river deltas are experiencing subsidence.

Sustainment: The capacity of a system to endure. Coastal sustainability would relate to the maintenance of rates of renewable resource harvest (fisheries), pollution creation, and non-renewable resource depletion that can be continued indefinitely. Sustainment considers topics from coastal economics to climate change to protecting threatened species. It focuses on maintaining the quality of life and benefits that the coast has provided to humankind while sustaining the integrity of coastal ecosystems.

Symbiosis: Any type of a close and long-term biological interaction (beneficial or negative) between two different biological organisms, be it mutualistic, commensalistic, or parasitic.

Thunderstorm: A rain-bearing cloud that also produces lightning. Every thunderstorm produces lightning. All thunderstorms are dangerous.

Tidal current: The horizontal movement of water resulting from gravitational effects of the Earth, Sun, and Moon, without any atmospheric influences.

Tide: The vertical fluctuations in water level resulting from gravitational effects of the Earth, Sun, and Moon, without any atmospheric influences. As sea levels rise high tides will cause seawater to reach further inland. In some locations, tidal flooding is causing erosion, aquifer and agricultural soil contamination, and lost habitat for fish, birds, and plants.

Tornado: A violently rotating column of air, usually pendant to a cumulonimbus, with circulation reaching the ground. It nearly always starts as a funnel cloud and may be accompanied by a loud roaring noise. On a local scale, it is the most destructive of all atmospheric phenomena.

Tropical cyclone: A warm-core non-frontal synoptic-scale cyclone, originating over tropical or subtropical waters, with organized deep convection and a closed surface wind circulation about a well-defined center. Once formed, a tropical cyclone is maintained by the extraction of heat energy from the ocean at high temperature and heat export at the low temperatures of the upper troposphere. In this they differ from extratropical cyclones, which derive their energy from horizontal temperature contrasts in the atmosphere (baroclinic effects).

Troposphere: The lowest part of the atmosphere from the surface to about 10 km (6.2 mi) in altitude in mid-latitudes (ranging from 9 km (5.6 mi) in high latitudes to 16 km (9.9 mi) in the tropics on average) where clouds and "weather" phenomena occur.

Tsunami: A long-period wave caused by an underwater disturbance such as a volcanic eruption or earthquake. Schnyder et al. (2016) have provided evidence for large submarine landslides on the slopes of the Great Bahama Bank in the Caribbean that have generated tsunamis in the past.

Typhoon: A tropical cyclone in the western Pacific Ocean. Typhoon Tip (aka Typhoon Warling) is an example of a typhoon that rapidly intensified over the open waters of the western Pacific Ocean around October 10, 1979, and then began to weaken steadily after passing east of Okinawa and finally making landfall on the Japanese island of Honshū with winds of about 81 mph (130 km/h) on October 19,

1973. Heavy rainfall from the typhoon caused severe flooding and contributed to a devastating fire at Camp Fuji, a U.S. Marine Corps training facility located in the Shizuoka Prefecture of Japan, at the base of Mount Fuji.

Upwelling: The process where favorable winds blowing across the ocean surface push water away which allows deep, cold, and nutrient rich water to rise up from beneath the surface to replace the water that was pushed away. Upwelling generally occurs in the ocean along coastlines. Two important upwelling centers that vary in strength and frequency occur along the coast from Washington to Southern California in North America and along the coast from Peru to Chile in South America.

Urban sprawl: Uncontrolled expansion of city buildings, houses, and shopping centers from a city to undeveloped land near the city. Urban sprawl increases the need for private automobiles and transportation infrastructure. Urbanization and impervious surfaces decreases the volume of water that percolates into the ground, and increases the volume and decreases the quality of surface water that eventually reaches the coastal ocean.

Visible light: Electromagnetic radiation with wavelengths from 400 to 700 nm. Visible light is located between ultraviolet at the short end and infrared wavelengths at the long end. Light absorption in the sea reduces the amount of visible light rapidly with depth and at a depth of 200 m (656 ft) most of the visible light has been absorbed.

Vulnerability: While vulnerability generally means a potential for harm, the Intergovernmental Panel on Climate Change (IPCC) defined vulnerability as the degree to which a system is susceptible to, and unable to cope with, adverse effects of climate change, including climate variability and extremes. Vulnerability is a function of the character, magnitude, and rate of climate change and variation to which a system is exposed, its sensitivity, and its adaptive capacity (IPCC 2007). Vulnerability may refer to the identification of resources at risk from coastal hazards such tropical storms, which have the potential for causing damage to, or loss of, natural ecosystems, buildings, and infrastructure.

Water level: Elevation of still water level to some datum. Water levels fluctuate as a result of astronomical tides, basin oscillations, climatological effects, evaporation and precipitation, geological effects, storm surge, and tsunami. The water level dictates where waves can reach and attack the coastal zone.

Wave: A ridge, deformation, or undulation of the surface of a liquid. Understanding the interaction of waves with maritime operations such as docking ships, building and maintaining coastal structures, and protecting navigation channels and beaches is critical for coastal resilience.

Weather: The day-to-day state of the atmosphere as regards to sunshine, temperature, humidity, precipitation, wind, cloudiness, moisture, pressure, etc. Weather forecasters apply technology (e.g., meteorological satellites and Doppler radar) and science to predict the state of the atmosphere for a future time and at a given location. This includes collecting as much data as possible and using these observations as input to numerical computer models through a process known as data assimilation to produce outputs of meteorological elements from the surface of the ocean to the top of the atmosphere.

Wind waves: Waves being formed and built by the wind. Wave height is affected by wind speed, wind duration (how long the wind blows), and fetch (the distance over which the wind blows).

Wind stress: The way the faster moving wind transfers energy to the slower moving water. Wind stress drives currents and forces waves and storm surges, which leads to flooding during the passage of storms such as hurricanes.

References

American Meteorological Society. 2017. *Glossary of meteorology*. Available online. http://glossary.ametsoc.org/wiki/Main_Page. Accessed 24 Oct 2017.

Center for Operational Oceanographic Products and Services (CO-OPS). 2000. *Tide and current glossary*. Silver Spring, MD: NOAA.

Coleman, J., and O.K. Huh. 2004. *Major world deltas: A perspective from space*. World Delta Database. Available online. www.geol.lsu.edu/WDD. Accessed on 25 Dec 2017.

Elvidge, Christopher D., Cristina Milesi, John B. Dietz, Benjamin T. Tuttle, Paul C. Sutton, Ramakrishna Nemani, and James E. Volgemann. 2004. U.S. constructed area approached the size of ohio. *Eos, Transactions, American Geophysical Union*, 85 (24), 233–240.

Gillett, Nathan P., Fancis W. Zwiers, Andrew J. Weaver, and P.A. Stott. 2003. Detection of human influence on sea-level pressure. *Nature* 422 (6929): 292–294.

Govers, Rob. 2009. Choking the Mediterranean to dehydration: The Messinian salinity crisis. *Geology* 37: 167–170.

IPCC. 2007. Climate change, 2007. Impacts, adaptation and vulnerability. In *Contribution of working group II to the fourth assessment report of the intergovernmental panel on climate change*, 976, ed. M.L. Parry, O.F. Canziani, J.P. Palutikof, P.J. van der Linden, and C.E. Hanson. Cambridge, UK: Cambridge University Press.

Intergovernmental Panel on Climate Change (IPCC). 2013. Climate change, 2013. The physical science basis. In *Contribution of working group I to the fifth assessment report of the intergovernmental panel on climate change* eds. T.F. Stocker, D. Qin, G.-K. Plattner, M. Tignor, S.K. Allen, J. Boschung, A. Nauels, Y. Xia, V. Bex and P.M. Midgley. Cambridge, Cambridge University Press

Kintisch, Eli. (2017, February 23). The great Greenland meltdown. *Science Magazine*. Available online. http://www.sciencemag.org/news/2017/02/great-greenland-meltdown. Accessed on 28 Oct 2017.

Kolker, Claudia. 2003. The salty lagoon. *Texas Parks and Wildlife Magazine* 61 (7): 50–59.

Liu, Wei, Shang-Ping Xie, Zhengyu Liu, and Jiang Zhu. 2017. Overlooked possibility of a collapsed Atlantic meridional overturning circulation in warming climate. *Science Advances* 3 (1): e1601666. https://doi.org/10.1126/sciadv.1601666.

Mariño-Tapia, Ismael, Cecilia Enriquez, Rodolfo Silva-Casarín, Edgar Mendoza-Baldwin, Edgar Escalante-Mancera, and Francisco Ruiz-Rent. 2014. Comparative morphodynamics between exposed and reef protected beaches under hurricane conditions. In *Proceedings of 34th conference on coastal engineering* (*ICCE 2014*), vol. 1, ed. Patrick J. Lynett, 1705–1713. Seoul, Korea, 15–20 June 2014.

Moss, Richard, Mustafa Babiker, Sander Brinkman, Eduardo Calvo, Tim Carter, Jae Edmonds, Ismail Elgizouli, Seita Emori, Lin Erda, Kathy Hibbard, Roger Jones, Mikiko Kainuma, Jessica Kelleher, Jean Francois Lamarque, Martin Manning, Ben Matthews, Jerry Meehl, Leo Meyer, John Mitchell, Nebojsa Nakicenovic, Brian O'Neill, Ramon Pichs, Keywan Riahi, Steven Rose, Paul Runci, Ron Stouffer, Detlef van Vuuren, John Weyant, Tom Wilbanks, Jean Pascal van Ypersele, and Monika Zurek. 2008. *Towards new scenarios for analysis of emissions, climate change, impacts, and response strategies*, 132. Geneva: Intergovernmental Panel on Climate Change.

Nichols, C. Reid, and Robert G. Williams. 2017. *Encyclopedia of marine science*, 2nd ed. New York: InfoBase Publishing.

Olajos, Fredrik, Folmer Bokma, Pia Bartels, Erik Myrstener, Johan Rydberg, Gunnar Öhlund, Richard Bindler, Xiao-Ru Wang, Rolf Zale, and Göran Englund. 2017. Estimating species colonization dates using DNA in lake sediment. *Methods in Ecology and Evolution*. https://doi.org/10.1111/2041-210X.12890.

Schnyder, Jara S.D., Gregor P. Eberli, James T. Kirby, Fengyan Shi, Babak Tehranirad, Thierry Mulder, Emmanuelle Ducassou, Dierk Hebbeln, and Paul Wintersteller. 2016. Tsunamis caused by submarine slope failures along western Great Bahama Bank. *Scientific Reports* 6: 35925. https://doi.org/10.1038/srep35925.

Showstack, Randy. 2014. Scientists focus on land subsidence impacts on coastal and delta cities. *Eos, Transactions American Geophysical Union* 95 (19): 159.

U.S. Army Corps of Engineers. 2002. *Coastal engineering manual (CEM), engineer manual 1110-2-1100*, vol. 6. Washington, DC: U.S. Army Corps of Engineers.

Wolanski, Eric (ed.). 2014. *Estuaries of Australia in 2050 and beyond*. Dordrecht, The Netherlands: Springer.

Zalasiewicz, J., M. Williams, A. Smith, T.L. Barry, A.L. Coe, P.R. Bown, P. Brenchley, D. Cantrill, A. Gale, P. Gibbard, F.J. Gregory, M.W. Hounslow, A.C. Kerr, P. Pearson, R. Knox, J. Powell, C. Waters, J. Marshall, M. Oates, P. Rawson, and P. Stone. 2008. Are we now living in the Anthropocene? *GSA Today* 18 (2): 4–8.

MIX
Papier aus verantwortungsvollen Quellen
Paper from responsible sources
FSC® C105338

If you have any concerns about our products,
you can contact us on
ProductSafety@springernature.com

In case Publisher is established outside the EU,
the EU authorized representative is:
Springer Nature Customer Service Center GmbH
Europaplatz 3, 69115 Heidelberg, Germany

Printed by Libri Plureos GmbH
in Hamburg, Germany

Mode-Switching for Resilient Security

Michael Riegler

Mode-Switching for Resilient Security

Springer Vieweg

Michael Riegler
LIT Secure and Correct Systems Lab
Johannes Kepler Universität Linz
Linz, Austria

ISBN 978-3-658-49305-9 ISBN 978-3-658-49306-6 (eBook)
https://doi.org/10.1007/978-3-658-49306-6

This book is based on the doctoral thesis submitted to Johannes Kepler University Linz (2024) and was partially supported by the LIT Secure and Correct Systems Lab (funded by the State of Upper Austria), the Austrian Marshall Plan Foundation and ENGEL Austria GmbH.

This Springer Vieweg imprint is published by the registered company Springer Fachmedien Wiesbaden GmbH, part of Springer Nature.
The registered company address is: Abraham-Lincoln-Str. 46, 65189 Wiesbaden, Germany

Foreword

Security has become one of the most pressing concerns in today's digital landscape. As technology advances and connectivity expands, cyberattacks and data breaches are growing at an alarming rate, threatening individuals, organizations, companies, governments, and critical infrastructure. The scale and complexity of these threats have evolved dramatically, requiring continuous adaptation and innovation in cybersecurity strategies.

All stakeholders alike are in a constant battle to safeguard their IT systems and data against evolving threats. The significance of cybersecurity has only intensified in recent years and will continue to do so, driven by the increasing reliance on digital technologies such as cloud computing, mobile devices, *Large Language Models (LLMs)*, and *Artificial Intelligence (AI)*. These technologies offer immense benefits but also introduce new vulnerabilities that cybercriminals may exploit. Regulations like Europe's *General Data Protection Regulation (GDPR)* and the *2nd Directive on Security for Network and Information Systems (NIS2)* underscore the need for stringent security measures. The integrity, confidentiality, and availability of digital assets are no longer optional—they are imperative for the stability of our interconnected world. Recognizing the growing complexity of cyber threats, this book explores mode-switching as a cybersecurity strategy to strengthen system resilience. Traditional perimeter defenses—such as firewalls, encryption, and access controls—are no longer sufficient on their own. As attackers become increasingly sophisticated, systems must adopt more dynamic defense mechanisms. Mode-switching offers a promising solution: By shifting a system into a different operational mode, it can reduce its attack surface and respond to threats in real-time. While perfect security may be unattainable, systems can neutralize and mitigate attacks through smart adaptability. This book introduces a framework in which a system's operational state adjusts according to real-time threat assessments. In this model, the largest mode represents full functionality, while the smallest mode preserves only essential operations—critical for systems like medical devices that must remain operational even under attack. This adaptive mechanism raises key questions about *how these modes are defined, what triggers transitions, and how to implement them without introducing new vulnerabilities.*

This book specifically addresses the cybersecurity challenges of *Internet of Things (IoT)* devices and systems, which are integral to modern life. Embedded hardware/software systems grow increasingly complex and interconnected. They are present in healthcare, industrial automation, telecommunications, defense, and other critical sectors. However, a surge in cyberattacks has accompanied this expansion—whether by rogue private actors or state-sponsored entities—leading to devastating disruptions, financial losses, and even threats to human life. Ransomware attacks on healthcare institutions, large-scale recalls of medical devices, and security breaches in automotive systems illustrate the urgent need for robust security strategies. In response, regulatory agencies and industry leaders are mandating cybersecurity standards, such as those introduced by European automakers, the *U.S. Food and Drug Administration (FDA)*, and international cybersecurity frameworks, ensuring that devices meet rigorous security benchmarks before deployment. While mode-switching is not a new concept, the author innovatively extends it to complex, networked IoT ecosystems, introducing well-defined mechanisms for managing transitions effectively. This book presents a timely and critical discussion on cybersecurity, offering novel insights into securing modern digital systems against ever-evolving cyber threats. Furthermore, the book delves into the technical and practical implications of implementing mode-switching, providing a detailed roadmap for engineers, researchers, and policymakers seeking to enhance the security posture of IT systems.

As cyber threats continue to evolve, a multi-layered, adaptive defense approach will be crucial. Michael Riegler provides a fresh perspective on achieving resilience by leveraging the dynamic properties of mode-switching, contributing valuable knowledge to the cybersecurity landscape.

June, 2025 Johannes Sametinger

Acknowledgments

Over the past four years, many individuals have instructed, advised, and supported me throughout my doctoral journey at *Johannes Kepler University Linz (JKU).* I am deeply grateful for their support that has essentially contributed to my growth as both an academic and a person. This dissertation would be incomplete without acknowledging their many efforts and assistance.

First and foremost, I want to express my deepest gratitude to my first supervisor and mentor, *Prof. Johannes Sametinger*, for his unwavering guidance, encouragement, and trust. Transitioning from industry to academia was a challenge, and your support helped me grow both personally and professionally. Thank you for believing in me, providing opportunities to tackle challenging problems, and for our numerous fruitful discussions. Also, thanks for encouraging me to go abroad and enabling and supporting collaborations with other universities, colleagues, and companies. It has been a great pleasure to work with you. Additionally, I am grateful for the trust I have in teaching the next generation of software engineers and being able to co-supervise master's students.

I would also like to thank my second supervisor, *Prof. René Mayrhofer*, for his critical questions and all the detailed feedback. I appreciate your willingness to take on the role of co-advisor and your critical perspective throughout my work. In addition, I would specifically like to acknowledge the invaluable help of *Prof. Michael Vierhauser* and *Prof. Manuel Wimmer* for the great collaboration on several publications.

My high appreciation goes to *Prof. Jerzy W. Rozenblit*, for letting me be part of his research group at the *University of Arizona.* His guidance significantly contributed to my research on medical device security. My time in Tucson would not have been as enriching my research period abroad would not have been such a great time, both personally and academically, without the great people at the *Model Based Design Lab*: *Nasser Albalawi, Minisk Hong, Sofia Carreón, Vijay Nafria, Olu Sodipe, and Jonathan Biren.* I am very grateful to the *Austrian Marshall Plan Foundation* for awarding me a scholarship for my research stay in the USA. I also appreciate the support from the *State of Upper Austria* with the Internationalization Program for Students.

A significant portion of this dissertation was shaped by a fruitful collaboration with *ENGEL Austria GmbH*, focusing on securing the next generation of injection molding machines and automation systems. Many thanks to *Christoph Schönegger* and *Martin Grubner* for the opportunity to work on impactful, real-world challenges and for our insightful discussions. I am delighted that our collaboration and research on mode-switching continues with my successor, *Marco Stadler*. Thank you for your questions, discussions, and especially your assistance in proofreading this dissertation.

I want to thank all the members of the *Secure and Correct Systems Lab* at the *Linz Institute of Technology (LIT)* for their support and feedback on my work. Special thanks to *Prof. Daniel Große, Prof. Robert Wille, Prof. Josef Küng, Prof. Stefan Rass, Dagmar Auer, Atif Mashkoor, Christoph Mayr-Dorn, Lukas Burgholzer, Daniel Hofer, Barbara Lehner, Gabriela Michelon, Omid Mir, Aya Mohamed, Sibylle Möhle-Rotondi, Daniel Pekarek, Philipp Schwarz, Sebastian Stock, Anja Mayr, Eva Resch, and Katrin Burger*. The community at the *Open Innovation Center* created an inspiring professional environment over the past years.

I would also like to thank the members of the *JKU Institute of Business Informatics—Software Engineering* for their valuable feedback and support, including *Felix Gemeinhardt, Iris Groher, Stefan Klikovits, Daniel Lehner, Wolfgang Narzt, Prof. Reinhold Plösch, Prof. Rainer Weinreich, and Birgit Breitschopf*.

Last but not least, I would like to extend my deep gratitude to my family and friends. Thanks, Mom and Dad, for your tireless support in my life and for enabling my education. Thank you, Anja, for providing me with a strong personal foundation throughout this long and exhausting journey.

June, 2025 Michael Riegler

Abstract

Recent developments in medical and industrial *Internet of Things (IoT)* raised concerns about their security. Vulnerabilities adversely affect essential systems inside an organization, hence introducing risks to several aspects, such as processes, workflows, privacy, and safety. The *Window of Vulnerability (WoV)* varies from a few days to several months, throughout which malicious actors have the opportunity to develop exploits and attack targeted systems. These attacks can cause interruptions, production losses, harm the environment, or even be life-threatening. The manual tracking of vulnerabilities and subsequent implementation of essential measures to mitigate risks, such as system shutdowns, updates, or modifications, can prove to be a laborious and time-intensive process. In addition, vulnerabilities potentially exploit particular versions or combinations of components, impacting critical system elements like servers and libraries.

After analyzing mode-switching in a *systematic literature review*, a *multi-modal architecture* was developed in three design iterations that provide a resilient security mechanism to defend systems while maintaining operations. Complex systems are divided into logical and tangible modes of operation to control them and provide situational awareness. A mode describes a desired system behavior where a specific configuration is active and specific functionality is provided for a period of time. Each mode faces a different security risk and offers more or less attack areas. A *mode-switching framework* was developed to support security engineers and system administrators to mitigate security risks, such as vulnerabilities and attacks. The framework shares security information across distributed devices and provides manual and automated contingency support. Switching modes adapt system behavior, configuration, functionality, and thus the attack surface. The framework employs model-driven techniques and provides a *mode domain specific language* to describe potential mitigation as mode switches.

The evaluation of the approach was conducted with several sample scenarios and case studies, including medical and industrial IoT and edge devices, web servers, and web applications. Simulated attacks and the evaluation of historical vulnerabilities showed that switching to less vulnerable modes provided a smart way to reduce the risk and the attack surface during the WoV. In a specific use case, the risk was reduced in 98.9% of the analyzed time.

Keywords: Mode-Switching · Security · Resilience · Self-Protection · Medical IoT · Industrial IoT

Zusammenfassung

Die Fortschritte im Bereich des *Internets der Dinge (IoT)* für medizinische und industrielle Geräte haben Sicherheitsbedenken aufgeworfen. Sicherheitslücken haben erhebliche Auswirkungen auf Organisationen und Personen. Weiters entstehen dadurch Risiken für Prozesse, Arbeitsabläufe, Datenschutz sowie die Betriebs- und Angriffssicherheit. Das sogenannte *Zeitfenster für Verwundbarkeiten* reicht von Tagen bis Monaten und gibt böswilligen Akteuren die Gelegenheit Schwachstellen auszunutzen. Dies kann zu Produktionsausfällen führen, die Umwelt schädigen oder sogar lebensbedrohlich sein. Die manuelle Verfolgung von Sicherheitslücken und die Umsetzung notwendiger Maßnahmen zur Risikominderung, wie Systemabschaltungen, Updates oder manuelle Änderungen, sind ein mühsamer und zeitintensiver Prozess. Zusätzlich können Sicherheitslücken bestimmte Versionen oder Kombinationen von Komponenten betreffen und kritische Systemelemente wie Server und Bibliotheken beeinträchtigen.

Diese Dissertation präsentiert einen Sicherheitsmechanismus, der Systeme schützt, widerstandsfähiger macht und gleichzeitig den Betrieb aufrechterhält. Durch die Einführung verschiedener Modi kann der Überblick und die Kontrolle über das System verbessert werden. Jeder Modus beschreibt ein gewünschtes Verhalten, hat eine spezifische Konfiguration, stellt bestimmte Funktionen bereit und bietet mehr oder weniger Angriffsfläche. In einer *systematischen Literaturanalyse* wurde der Wechsel von Modi untersucht. Darauf aufbauend wurden *Multi-Modale Architekturen* und *Frameworks für den Moduswechsel* entwickelt, die Sicherheitsingenieuren und Systemadministratoren dabei unterstützen, Angriffe abzuwehren und Risiken zu reduzieren. Modellgesteuerte Techniken und eine *domänenspezifische Sprache* kommen dabei zum Einsatz. Sicherheitsinformationen werden über verteilte Geräte hinweg geteilt, was manuelle und automatisierte Notfallunterstützung ermöglicht. Das Umschalten der Modi passt Verhalten, Systemkonfiguration und Angriffsfläche an.

Die Evaluierung erfolgte anhand mehrerer Beispielszenarien und Fallstudien, darunter medizinische und industrielle IoT- und Edge-Geräte, Webserver und -anwendungen. Historische Schwachstellendaten wurden ausgewertet, Angriffe simuliert und gezeigt, dass der Moduswechsel die Angriffsfläche in 98,9% der analysierten Zeit reduziert und somit eine effektive Lösung zur Überbrückung des Verwundbarkeitsfensters darstellt.

Schlüsselwörter: Moduswechsel · Sicherheit · Widerstandsfähigkeit · Selbstschutz · Medizin · Industrie

Contents

Abbreviations

2FA	Two-Factor Authentication
AADL	Architecture Analysis and Design Language
AAMI	Association for the Advancement of Medical Instrumentation
ABA-IDS	Anomaly Behavior Analysis Intrusion Detection System
ABSP	Architecture-Based Self-Protection
AFT	Attack-Fault-Tree
AI	Artificial Intelligence
AID	Automated Insulin Delivery
AIS	Automated Information System
AMPF	Austrian Marshall Plan Foundation
ANNSIM	Annual Modeling and Simulation Conference
ANSI	American National Standards Institute
ASE	IEEE/ACM International Conference on Automated Software Engineering
ATT & CK	Adversarial Tactics, Techniques, and Common Knowledge
AUTOSAR	AUTomotive Open System ARchitecture
AWS	Amazon Web Services
BLE	Bluetooth Low Energy
BOBE	Break once, break everywhere
BOM	Bill of Materials
BSI	German Federal Office for Information Security
CAPEC	Common Attack Pattern Enumerations and Classifications
CBSE	Component-Based Software Engineering
CCIS	Communications in Computer and Information Science

CEP	Complex Event Processing
CIA	Confidentiality, Integrity, and Availability
CISA	U.S. Cybersecurity and Infrastructure Security Agency
CLSI	Clinical and Laboratory Standards Institute
CMS	Content Management System
CoAP	Constrained Application Protocol
COVID-19	Coronavirus disease 2019
CPE	Common Platform Enumeration
CPS	Cyber-Physical System
CRA	Cyber Resilience Act
CTF	Capture-The-Flag
CVD	Coordinated Vulnerability Disclosure
CVE	Common Vulnerabilities and Exposure
CVSS	Common Vulnerability Scoring System
CWE	Common Weakness Enumeration
DAST	Dynamic Application Security Testing
DBMS	Database Management System
DDoS	Distributed Denial-of-Service
DEXA	Database and Expert Systems Applications
DHS	U.S. Department of Homeland Security
DNS	Domain Name Service
DoD	U.S. Department of Defense
DoS	Denial-of-Service
DSL	Domain-Specific Language
ECA	Event-Condition-Action
ECSA	European Conference on Software Architecture
EDF	Earliest-Deadline-First
EMF	Eclipse Modeling Framework
ENISA	European Union Agency for Cybersecurity
EPL	Event Processing Language
ETSI	European Telecommunications Standards Institute
EU	European Union
F2b	Fail2ban
FAA	U.S. Federal Aviation Administration
FDA	U.S. Food and Drug Administration
FMEA	Failure Mode and Effects Analysis
FMECA	Failure Mode Effects and Criticality Analysis
FTP	Fixed-Task-Priority
GDPR	General Data Protection Regulation

GPL	General-Purpose Language
GUI	Graphical User Interface
HCI	Human–Computer Interaction
HCP	Hierarchical Control Pattern
HDO	Healthcare Delivery Organization
HIPAA	Health Insurance Portability and Accountability Act of 1996
HSCC	Healthcare Sector Coordination Council
HTML	Hypertext Markup Language
HTTP	Hypertext Transfer Protocol
HTTPS	Hypertext Transfer Protocol Secure
IACS	Industrial Automation and Control System
IAS	Industrial Automation System
ICMP	Internet Control Message Protocol
ICS	Industrial Control System
ICSA	IEEE International Conference on Software Architecture
ICSMA	ICS Medical Advisory
IDE	Integrated Development Environment
IDEF4	Integrated DEFinition for Object-Oriented Design
IDPS	Intrusion Detection and Prevention System
IDS	Intrusion Detection System
IEC	International Electrotechnical Commission
IEEE	Institute of Electrical and Electronics Engineers
IIoT	Industrial Internet of Things
IMD	Implantable Medical Device
IMDRF	International Medical Device Regulators Forum
INCOSE	International Council on Systems Engineering
IoMT	Internet of Medical Things
IoT	Internet of Things
IP	Internet Protocol
IPS	Intrusion Prevention System
ISA	International Society of Automation
ISM	International Conference on Industry 4.0 and Smart Manufacturing
ISO	International Organization for Standardization
IT	Information Technology
IWCFS	International Workshop on Cyber-Security and Functional Safety in Cyber-Physical Systems
JKU	Johannes Kepler University Linz
JSS	Journal of Systems and Software

JVM	Java Virtual Machine
KEV	Known Exploited Vulnerabilities
LIT	Linz Institute of Technology
LLM	Large Language Model
LoC	Lines of Code
MAPE	Monitor, Analyze, Plan, and Execute
MAPE-K	Monitor, Analyze, Plan, Execute, and Knowledge
MARTE	Modeling and Analysis of Real-Time and Embedded Systems
MBSE	Model-Based Systems Engineering
MCP	Mode Change Protocol
MCPS	Medical Cyber-Physical System
MCR	Mode Change Request
MCS	Mixed-critical System
MD	Medical Device
MDCG	Medical Device Coordination Group
MDE	Model-driven Engineering
MDM	Medical Device Manufacturer
MDR	European Medical Device Regulation
MDSL	Mode Domain-Specific Language
MitM	Man-in-the-Middle
ML	Machine Learning
MMS	Multi-Mode System
MQTT	Message Queuing Telemetry Transport
ms	milliseconds
MSL	Mode Switch Logic
MSR	Mode Switch Request
MTD	Moving Target Defense
NASA	U.S. National Aeronautics and Space Administration
NEMA	U.S. National Electrical Manufacturers Association
NERC	North American Electric Reliability Corporation
NIS	Network and Information Systems
NIS2	2nd Directive on Security for Network and Information Systems
NIST	U.S. National Institute of Standards and Technology
NSF	U.S. National Science Foundation
NVD	U.S. National Vulnerability Database
OMG	Object Management Group
OOP	Object-oriented Programming
OPC UA	Open Platform Communications Unified Architecture
OS	Operating System

OT	Operational Technology
OTS	off-the-shelf
OWASP	Open Worldwide Application Security Project
PACS	Picture Archiving and Communication System
PHP	PHP: Hypertext Preprocessor
PoC	Proof of Concept
QoS	Quality of Service
Regex	Regular Expression
REST	Representational State Transfer
RIA	Robotic Industries Association
ROS	Robot Operating System
RPi	Raspberry Pi
RQ	Research Question
s	Seconds
SAST	Static Application Security Testing
SBOM	Software Bill of Materials
SEAMS	IEEE/ACM International Symposium on Software Engineering for Adaptive and Self-Managing Systems
SIL	Safety Integrity Level
SLR	Systematic Literature Review
SOC	Security Operation Center
SPL	Software Product Line
SQL	Structured Query Language
SSDLC	Secure Software Development Life Cycle
SSH	Secure Shell
STRIDE	Spoofing, Tampering, Repudiation, Information Disclosure, Denial of Service, and Elevation of Privilege
SysML	Systems Modeling Language
TDL	Timing Definition Language
TEE	Trusted Execution Environment
TPM	Trusted Platform Module
UML	Unified Modeling Language
USA	United States of America
USB	Universal Serial Bus
VPN	Virtual Private Network
WoV	Window of Vulnerability
XML	Extensible Markup Language

List of Figures

List of Tables

List of Listings

1 Introduction

Our modern lives increasingly rely on technology and automation, from the seamless operation of smart factories to the life-saving capabilities of medical devices. According to *Lasi et al.*, the fourth industrial revolution will increase the digitalization and networking of all manufacturing processes and transform industries from product-oriented to service-orientated businesses [LFK+14]. Reducing downtimes, lead times, waste, and costs have long been targets. Industry 4.0 should provide high flexibility and individualization (batch size one) in product development and production and, therefore, decrease time to market. While *Operational Technology (OT)* focuses on industrial processes, *Information Technology (IT)* typically concentrates on digital information, communication, and business operations. However, the worlds of OT and IT are converging. Physical devices like machines, robots, sensors, actuators, and embedded systems are combined into *Cyber-Physical Systems (CPSs)* that communicate with other systems and services over various protocols and are either fully automated, semi-automated, or operated manually. *Internet of Things (IoT)* and *Industrial Internet of Things (IIoT)* are examples of the blurring the boundaries between OT and IT.

Similar situation in the healthcare domain: Concepts like telehealth, telemedicine, and interconnected medical devices, so-called *Internet of Medical Things (IoMT)*, allow remote monitoring with better patient care and reduce in-person visits and costs [VS19]. Therefore, more and more medical devices are connected to the Internet. Especially, the emergence of the *Coronavirus disease 2019 (COVID-19)* pushed this effect. On average, healthcare organizations have more than 26,000 network-connected devices, and the number will continue to rise in the future [Pon22, Rep21].

A problem arises as our world increasingly relies on these systems: the security. In addition to the advantages, the interconnection of these systems can be a host of vulnerabilities that, if left unaddressed, can have serious consequences. Ensuring the security of these systems is not a matter of convenience. It has become a critical

M. Riegler, *Mode-Switching for Resilient Security*,
https://doi.org/10.1007/978-3-658-49306-6_1

requirement as regulations like the *European Medical Device Regulation (MDR)* and standards like IEC 62443 show [MDC19, IEC18].

This dissertation addresses the core of this multifaceted problem and seeks to explore, understand, and ultimately mitigate the security challenges of industrial and medical systems. In a time of increasing cyber threats, this research strives to provide valuable insights into critical infrastructure protection and patient well-being.

1.1 Problem Statement

In the context of medical and industrial CPSs, telehealth, remote monitoring, predictive maintenance, and other techniques aim to reduce costs. However, internet connection increases the attack surface, and the threat of hacks and malicious software is significant. Within minutes, attackers detect and exploit misconfigured or vulnerable services, as a study by *Palo Alto Networks* has shown [Pal21]. They deployed 320 honeypots that appeared as legitimate targets to lure and track hackers. Within 24 hours 80 % of the deceptive systems were compromised. All honeypots were infiltrated within a week [Pal21]. Therefore, it is not a question of whether a device will be attacked, but when. In 2019, two billion IoT devices were affected by zero-day vulnerabilities, which allowed remote code execution [Dof19]. In 2020, not only the lockdown in several countries because of COVID-19 led to production downtimes. The Belgian loom manufacturer *Picanol* and the German car parts company *Gedia* had to send over 1,500 people home after cyberattacks [Goo20, The20]. In 2024, the battery maker *Varta* had to stop production after a cyberattack [Tou24]. Attacked devices can cause interruptions, production losses, and, in the case of insulin pumps or pacemakers, even life-threatening situations. Even if communication is only possible locally, recent research on medical devices identified vulnerabilities that can lead to deadly attacks [HBR+08, HHBF+08, LRJ11, PKK11]. Vulnerabilities are an emerging threat in healthcare [ENI23]. Attackers have also used COVID-19 for scam and phishing emails to distribute malware and exploit the new teleworking infrastructure [CIS20a]. Vulnerabilities in frequently used software like *Log4J*, *Kaseya*, and *Solar Winds* have affected thousands of companies [New11, Tun21, Zet23]. Cyberattacks like the one against the *Colonial Pipeline Company* can lead to loss of productivity, business continuity problems, financial losses, and damage of reputation [Seg22]. According to the Risk Barometer of the corporate insurer *Allianz*, cyber incidents are one of the biggest company threats after business interruptions since 2019 [All19, All20, All21, All22, All23, All24].

IT security is about protecting systems and reducing the risk of intrusions and attacks. Many approaches, like signature-based detection, work only for well-known attacks. What if there is a new vulnerability, a zero-day? Due to safety requirements, developing updates and patches for industrial machines and especially medical devices can take months. Any change needs an impact assessment, complete test evidence, an updated safety case, and formal approval. Workarounds are sometimes provided to mitigate known vulnerabilities. Nevertheless, manual changes are time-consuming and, if performed hastily, involve the danger of disrupting operations. In the meantime, hackers can write exploits to take advantage of the vulnerability. Moreover, the distribution and installation of updates and patches take time. Stopping an industrial machine leads to business interruption and loss of money. Stopping a pacemaker in case of a threat is no option and a danger to life.

1.2 Research Goals

With my dissertation, I want to improve the security and resilience of medical and industrial IoT devices (even in case of zero-day vulnerabilities) by switching modes automatically or manually to reduce the attack surface and attackers' range of activity without stopping a system's core functionality in order to provide mission-critical capabilities for business continuity and even life.

The goal was to develop a security concept with modes that provides a better overview and more transparency about several systems' status and security, especially for non-security experts, manufacturers, and operators. Each mode provides a specific functionality and attack surface, with the ability to change them by switching the mode.

Various mode-switching scenarios and case studies were conducted. This involved analyzing log files, performing integrity checks, and comparing execution times to detect threats, anomalies, and potential intrusions. If an anomaly is detected, a mode switch changes system behavior, adapts functionality, and modifies configurations to reduce the attack surface. During the transition from one mode to another, the system sends notifications to inform the owner and/or vendor. Depending on the risk, software services and containers should be stopped or started, and implementations should be changed during runtime. Attackers IP addresses must be blocked temporarily or permanently, and communication must be restricted to one-way communication. Also, authentication methods must be increased, for instance, with *Two-Factor Authentication (2FA)*. Mode-switching should provide a method to reduce the risk of vulnerabilities in certain software versions until an update is provided. A combined mode based on several devices, for example, a type mode, a factory mode, or a global mode, can lead to a quick overview of what is going on out there. The devices send their mode information to the superior mode. If a certain threshold is reached, the superior mode is changed, which in turn can affect the mode of all other devices, such as preventing the virus from spreading further, similar to the COVID-19 measures.

1.3 Research Questions

This dissertation strives to answer the following *Research Questions (RQs)*:

- **RQ1: What is mode-switching, and how can it be applied to mitigate risks from a security perspective?**
  Considering security risks for mode-switching is relatively new [SRLO15]. Therefore, an in-depth analysis is necessary to provide an overview and study how existing methods and protocols can be applied in a security context. *Multi-Mode Systems (MMSs)* were already used in many devices and applications, so far mainly for safety reasons. For example, pacemakers and airbags use modes for real-time adaptivity. Depending on the environment, self-driving cars switch between levels of automation [CP18]. Efficient usage of resources is also a use-case for MMSs where each mode corresponds to a specific system behavior. What does it take to include mode-switching behavior into existing systems and applications? To what extent can mode-switching reduce the risk of a system being vulnerable and increase its resilience?
- **RQ2: How does mode-switching work on medical IoT devices?**
  To ensure safety and improve security, many regulations, standards, and guidance documents have been published over the years [BJH16, Lec18]. Nevertheless, regulatory processes also lead to longer process design cycles which can take years and cost millions of Euros. Any update needs an impact assessment, an updated safety case, complete test evidence, and formal approval. Regular IT security is not always suitable for these kinds of devices. Medical devices have limited resources and limited functionality and thereby offer special challenges. Even if there are vulnerabilities, life-critical devices must continue to run. This research question aims to propose a way to use mode-switching on medical devices, especially on implantable medical devices like pacemakers and insulin pumps, to reduce the attack surface. Which modes are useful for medical devices?
- **RQ3: How does mode-switching work on industrial IoT devices?**
  In contrast to medical devices, industrial machines have more power and computational resources, but also a larger attack surface through their interfaces, like

USB ports, Ethernet connections, and debug ports. How can modes be applied to industrial systems to ensure more security? What are realistic test scenarios, the effort required for creating and specifying them, and what are the challenges? Which are the modes that are useful for industrial IoT devices? How long does it take to apply countermeasures?

- **RQ4: What are general risk mitigation strategies as compared to strategies tailored to specific domains?**
  Every domain has its own peculiarities. But how different are the risk mitigation strategies for medical devices and industrial automation systems? What works in both domains, and what does not? What are the special challenges? Can frameworks help to facilitate the use of mode-switching? What are the supporting elements? Does a hierarchical structure of devices and their modes and consideration as an overall system provide more overview and transparency for the user, the operator, and the manufacturer? How is communication made about a mode change? Is it useful to derive modes of the hierarchical structure, e.g., for a location, a factory, a company, or globally? Is a global mode switch useful, or is it even a new central attack vector? How fast is a framework in processing configurations, fetching vulnerabilities and patches, comparing mode risks, and executing mode-switching?

1.4 Research Methodology

The structure of this dissertation is based on the design science methodology for information systems and software engineering by *Wieringa* [Wie14]. Figure 1.1 shows the methodological structure. The focus of this work is on medical devices such as pacemakers and insulin pumps, industrial machines such as injection molding machines, and connected IoT and edge devices, and their *stakeholders* like device manufacturers, integrators, asset owners, and operators.

The *knowledge goal* is to analyze existing concepts and to develop a future concept with our own methods. The task was to equip systems with a mode-based architecture, to improve the security and resilience of patients and companies by automatic threat detection and mode-switching. The mode switch should be smooth for the patient/machine and not affect the core functionality. Additionally, there should be a mode switch, kill switch, or emergency stop switch for humans to reduce damage to machines and harm to patients.

The *holistic research goal* is a secure information system architecture and the application of design science research methods. The *aim of the dissertation* is about bridging the time between the discovery of a vulnerability and the patch to fix it. The

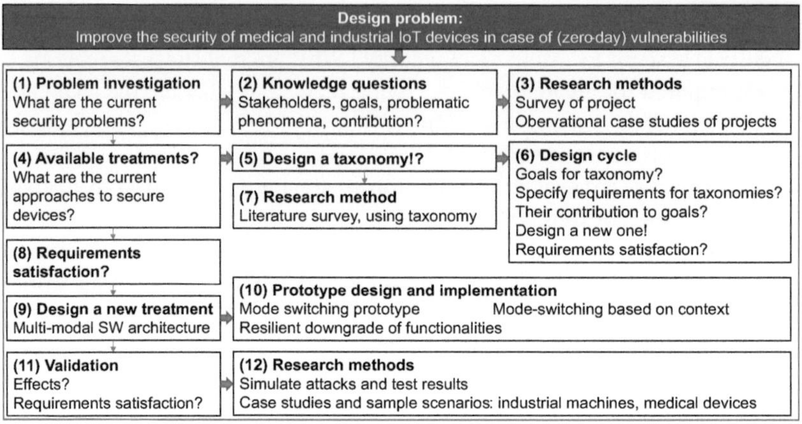

Fig. 1.1 Methodological Structure

contribution is a mode-switching prototype with built-in security mechanisms based on security-by-design principles. Furthermore, case studies on mode-switching in medical devices and industrial systems should show whether and how this can be applied in practice. There exists already a *plenty of existing countermeasures* from the field of computer security, such as authentication, encryption, and firewalls, up to principles such as secure coding and secure by default. In this case, particular challenges are real-time or embedded systems with limited resources that require specific countermeasures. Regular incidents at a wide variety of industrial and medical manufacturers show that fixing vulnerabilities takes too long. Therefore, patients and operators of industrial machines and medical devices are at risk. Sufficient resilience against cyberattacks is rarely given, as current attacks on Cisco switches show [Cis23d]. Attackers can read and modify encrypted traffic. Nevertheless, the manufacturer does not provide a software update or workaround - the only solution is to switch off the function. Due to their range of functions, a further complication is that many systems have become very complex. 100 % security can never be guaranteed.

The *treatment* is based on the considerations of our colleagues at the *University of Arizona* and the work of *Sametinger et al.* [RRLS18, SRLO15]. A multi-modal architecture was designed, and a mode-switching framework was implemented to improve the security of critical systems. Devices should automatically recognize attacks based on the context and ward them off by switching modes. Certain functionalities may be switched back or off to reduce the attack surface. In the worst case, only the core functionality is provided. In three iterations, the architecture and the framework were built, evaluated, extended, and improved; see Fig. 1.2. While the research initially focused on single devices, a broader perspective was adopted over time, as many systems are interconnected. After the multi-modal architecture and the mode-switching framework were extended for multiple systems. Finally, the MAPE-K loop was added. The *validation* was executed with simulated attacks to test how MMSs reacts to them. Furthermore, sample scenarios and case studies showed whether and how these considerations can be used in practice in industrial automation systems and medical devices and whether they can meet the requirements.

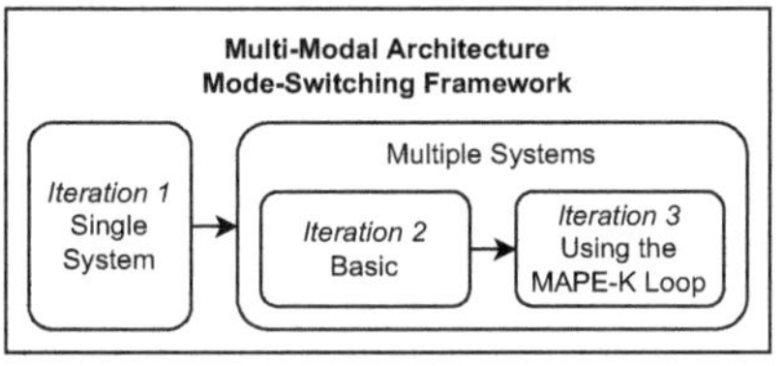

Fig. 1.2 Design Science Iterations

1.5 Research Environment

This dissertation comprises a collection of articles that have been either published or submitted. Additionally, definitions of key terminology, information on state-of-the-art, and unpublished work are part of the dissertation. The upcoming section will provide an overview of the research environment in which these articles were produced and detail my contribution to their development.

This dissertation's theoretical and empirical research was mainly conducted within the *LIT Secure and Correct Systems Lab* funded by the *State of Upper Austria*, and the *Institute of Business Informatics - Software Engineering*. Both belong to the *Johannes Kepler University Linz (JKU)*.

In 2019, I started my research with a *Systematic Literature Review (SLR)*, which resulted in the article *Mode Switching from a Security Perspective* [RS20]. We discussed the importance of security in CPS and proposed using mode-switching to create resilient systems that can continue functioning correctly even under malicious attacks. Using a multi-modal software architecture, companies, operators, and people should be enabled to protect themselves and their customers and minimize attackers' range of activity and the attack surface without waiting for security updates.

Parts of the work were conducted in the ongoing collaboration of the Lab and with the global manufacturer of injection molding machines and associated automation systems *ENGEL Austria GmbH* on *Cybersecurity and Functional Safety in Cyber-Physical Systems.* After penetration tests on their edge device hardware and software, we focused on automated detection and defense against cyber-attacks on machine and edge device software. That included developing software prototypes to demonstrate mode-switching ideas for the next generation of industrial automation systems. Together with *Prof. Johannes Sametinger* (JKU) and *Christoph Schönegger* (ENGEL) the articles *Multi-Mode Systems for Resilient Security in Industry 4.0* and *Mode Switching for Secure Edge Devices* were conducted [RS21b, RSS22].

Additionally, we have an ongoing research collaboration with the *University of Arizona*, especially *Prof. Jerzy W. Rozenblit*, on medical device security. Supported by scholarships of the *Austrian Marshall Plan Foundation (AMPF)* and the *State of Upper Austria*, I had the pleasure of joining the colleagues 2022 in Tucson, Arizona, as a research scholar. During my six-month research stay at the Department of Electrical and Computer Engineering, the articles *Context-Aware Security Modes for Medical Devices* and *Towards Strategies for Secure Data Transfer of IoT Devices with Limited Resources* were conducted [RSR22, ARR22]. Further, I got a scholarship to attend *Black Hat USA 2022* and *DEF CON 30* in Las Vegas, Nevada, the two biggest hacker conferences worldwide. I got valuable insights into cybersecurity companies, emerging threats, and state-of-the-art practices to defend them. Aside from that, the conferences provided an opportunity to engage in meaningful discussions about my dissertation topic with experts in the field. The final research report on *Resilient Middleware for Smart Security in Medical Devices* summarizes my research stay [Rie22].

With the support of three master students that I co-supervised, we are implementing modes more deeply into software and on different layers and investigating how mode-switching can enhance security and resilience. We are analyzing mode-switching for web applications, network communication, and authentication. One result is the article *Mode Switching for Secure Web Applications – A Juice Shop Case Scenario*, which extends a master thesis that focuses on securing the deliberately insecure web application with modes [RS21a]. My tasks for this article included implementing parts of the prototype, simulating attacks, discussing the results, and molding the results into a presentable format. As web applications are part of different systems, we suggest adapting our approach to other scenarios as well. For example, we can switch between different communication protocols to provide more flexibility in case of attacks, errors, or vulnerabilities. Similarly, we deal with authentication. Methods can be switched to make it harder for attackers to guess passwords and break into systems.

As a joint research with *Prof. Johannes Sametinger*, *Prof. Michael Vierhauser*, and *Prof. Manuel Wimmer* from the *JKU Institute of Business Informatics - Software Engineering*, we published the article *A model-based mode-switching framework based on security vulnerability scores* in the *Journal of Systems and Software (JSS)* [RSVW23a]. We proposed an automated approach to reduce the risk of exposure to software vulnerabilities by switching modes. This work relies on model-driven techniques, a multi-modal architecture, and a *Domain-Specific Language (DSL)* to describe mitigation strategies as mode switches. The paper was presented at the *IEEE/ACM International Conference on Automated Software Engineering* (ASE'23).

The article *A Distributed MAPE-K Framework for Self-Protective IoT Devices* builds up on the former results and was presented at the *IEEE/ACM International Symposium on Software Engineering for Adaptive and Self-Managing Systems (SEAMS)* [RSV23]. The goal was to expand the single-device view to a distributed approach. Considering tight resource constraints and disconnections, we propose a self-protection framework for IoT devices. I implemented a software prototype in this study, simulated attacks, and analyzed our framework's detection and automatic reaction.

In the medical domain, we discuss arising security challenges in the article *Architecture for Self-protective Medical Cyber-Physical Systems (MCPSs)* and propose mechanisms to reduce the attack surface and increase patient safety [RSR23]. We highlight the importance of monitoring vulnerabilities and suspicious behavior and switching between context-aware security modes to provide a flexible and secure healthcare system.

I am glad that our research on mode-switching is ongoing, as my successor *Marco Stadler* already started his PhD at the LIT Secure and Correct Systems Lab. The articles *Towards Increasing Safety in Collaborative CPS Environments*, *Cyber-Resilient Edge Computing: A Holistic Approach with Multi-Level MAPE-K Loops*, and *Improve CPSoS Security: An enhanced Anomaly-based Intrusion Detection Architecture* are the first results of our joint efforts [SRS23].

1.6 Contributing Publications

In the following, the main publications contributing to this dissertation are briefly outlined below. One article was published in the *Journal of Systems and Software (JSS)*, one article was published and presented at the *IEEE/ACM International*

Symposium on Software Engineering for Adaptive and Self-Managing Systems (SEAMS'23), one article was published and presented at the *International Conference on Industry 4.0 and Smart Manufacturing* (ISM'20), one article was published and presented at the *Annual Modeling and Simulation Conference* (ANNSIM'22), six articles were published and presented at the *International Workshop on Cyber-Security and Functional Safety in Cyber-Physical Systems* (IWCFS'20, IWCFS'21, two at IWCFS'22, and two at IWCFS'23), and one final research report was published from the *Austrian Marshall Plan Foundation.*

The articles *Towards Strategies for Secure Data Transfer of IoT Devices with Limited Resources, Towards Increasing Safety in Collaborative CPS Environments, Cyber-Resilient Edge Computing: A Holistic Approach with Multi-Level MAPE-K Loops,* and *Improve CPSoS Security: An enhanced Anomaly-based Intrusion Detection Architecture* can be regarded as related to the dissertation but not directly contributing [ARR22, SRS23, SRS24, SRSS24].

The following publications are listed in chronological order and give the authors in the identical order of the original publications:

- **Michael Riegler** and Johannes Sametinger. *Mode Switching from a Security Perspective: First Findings of a Systematic Literature Review.* In: Database and Expert Systems Applications (DEXA), 2020, Bratislava, Slovakia. 4th International Workshop on Cyber- Security and Functional Safety in Cyber-Physical Systems (IWCFS), 2020. Communications in Computer and Information Science (CCIS), vol 1285, pp. 63–73, 10 pages. Springer, Cham. Available at https://doi.org/10.1007/978-3-030-59028-4_6
  This article is used in Chap. 2 and Chap. 3 for the definition of key terminology and introduction on resilient security and modes. Parts were used to discuss safety versus security in Sect. 2.4.
- **Michael Riegler** and Johannes Sametinger. *Multi-Mode Systems for Resilient Security in Industry 4.0.* Proceedings of the 2nd International Conference on Industry 4.0 and Smart Manufacturing (ISM), 2020, Virtual Event. Procedia Computer Science, 2021, vol 180, pp. 301–307, 6 pages. Available at https://doi.org/10.1016/j.procs.2021.01.167
  This article is used in Chap. 2 to define terms like threats for industrial IoT devices, safety, robustness, and resilience. In Chap. 3, parts of the article are used to describe examples and application areas. Also, parts are used in the discussions in Chap. 6.
- **Michael Riegler** and Johannes Sametinger. *Mode Switching for Secure Web Applications – A Juice Shop Case Scenario.* In: DEXA, 2021, Virtual Event. 5th IWCFS, 2021. CCIS, vol 1479, pp. 3–8, 5 pages. Springer, Cham. Available at

https://doi.org/10.1007/978-3-030-87101-7_1
This article is used in Chap. 2 to explain security design principles. Additionally, the OWASP Juice Shop sample scenario is used in the evaluation, see Chap. 5, and the discussion, see Chap. 6.

- **Michael Riegler**, Johannes Sametinger, and Jerzy W. Rozenblit. *Context-Aware Security Modes For Medical Devices*. Proceedings of the Annual Modeling and Simulation Conference (ANNSIM) 2022, San Diego, CA, USA. pp. 372–382, 10 pages. Available at https://doi.org/10.23919/ANNSIM55834.2022.9859283
This article is used in the evaluation in Chap. 5 to provide a sample scenario with a Picture Archiving and Communication System (PACS).
- **Michael Riegler**, Johannes Sametinger, and Christoph Schönegger. *Mode Switching for Secure Edge Devices*. In: DEXA, 2022, Vienna, Austria. 6th IWCFS, 2022. CCIS, vol 1633, pp. 347–356, 9 pages. Springer, Cham. Available at https://doi.org/10.1007/978-3-031-14343-4_32
This article is used in Chap. 2 to define security goals in the context of industrial IoT devices. Additionally, the edge device sample scenario is used for the evaluation in Chap. 5.
- **Michael Riegler**. *Resilient Middleware for Smart Security in Medical Devices.* Final research report of my research stay at the University of Arizona, USA, at the Model Based Design Laboratory, submitted to the Austrian Marshall Plan Foundation, 2022, 29 pages. Available at https://www.marshallplan.at/images/All-Papers/mp-2022/Michael_Riegler_1048.PDF
Parts of the research report are used to define threats and security measures of medical devices in Chap. 2. Additionally, the report is used in the discussion in Chap. 6.
- **Michael Riegler**, Johannes Sametinger, Michael Vierhauser, and Manuel Wimmer. *A model-based mode-switching framework based on security vulnerability scores*. Journal of Systems and Software (JSS), 2023, vol 200, 111633, 16 pages. Available at https://doi.org/10.1016/j.jss.2023.111633
This article is used in Chap. 2 for the definition of key terminology and introduction on resilient security. Main parts of Chap. 4 build up on this work. The web server case study is presented in Chap. 5 and discussed in Chap. 6.
- **Michael Riegler**, Johannes Sametinger, and Michael Vierhauser. *A Distributed MAPE-K Framework for Self-Protective IoT Devices*. 18th IEEE/ACM International Symposium on Software Engineering for Adaptive and Self-Managing Systems (SEAMS), Melbourne, Australia, 2023, pp. 202-208, 7 pages. Available at https://doi.org/10.1109/seams59076.2023.00034

This article is utilized in Sect. 4.9 to adopt a more comprehensive viewpoint. Insights of the IoT case study are presented in Chap. 5 and discussed in Chap. 6.

- **Michael Riegler**, Johannes Sametinger, and Jerzy W. Rozenblit. *Architecture for Self-protective Medical Cyber-Physical Systems*. In: DEXA, 2023, Penang, Malaysia. 7th IWCFS, 2023. CCIS, vol 1872, pp. 57–66, 10 pages. Springer, Cham. Available at https://doi.org/10.1007/978-3-031-39689-2_6
  This article is used in Chap. 4 to present the self-protective system architecture. In Chap. 5, the sample scenario evaluation on insulin pumps is presented, and the results are discussed in Chap. 6.

The following articles are not used in this dissertation but show further joint research:

- Nasser S. Albalawi, **Michael Riegler**, and Jerzy W. Rozenblit. *Towards Strategies for Secure Data Transfer of IoT Devices with Limited Resources*. In: DEXA, 2022, Vienna, Austria. 6th IWCFS, 2022. CCIS, vol 1633, pp. 331–336, 5 pages. Springer, Cham. Available at https://doi.org/10.1016/j.jss.2023.111633
- Marco Stadler, **Michael Riegler**, and Johannes Sametinger. *Towards Increasing Safety in Collaborative CPS Environments*. In: DEXA, 2023, Penang, Malaysia. 7th IWCFS, 2023. CCIS, vol 1872, pp. 79–85, 7 pages. Springer, Cham. Available at https://doi.org/10.1007/978-3-031-39689-2_8
- Marco Stadler, Johannes Sametinger, and **Michael Riegler**. *Cyber-Resilient Edge Computing: A Holistic Approach with Multi-Level MAPE-K Loops*. 21st IEEE International Conference on Software Architecture (ICSA), 2024, Hyderabad, India, pp. 79–83, 5 pages. Available at https://doi.org/10.1109/ICSA-C63560.2024.00020
- Marco Stadler, **Michael Riegler**, Johannes Sametinger, and Christoph Schönegger. *Improved CPSoS Security: An enhanced Anomaly-based Intrusion Detection Architecture*. In: DEXA, 2024, Penang, Malaysia. 8th IWCFS, 2024. CCIS, vol 2169, pp. 3–13, 10 pages. Springer, Cham. Available at https://doi.org/10.1007/978-3-031-68302-2_1
- Marco Stadler, Michael Vierhauser, **Michael Riegler**, Daniel Waghubinger, and Johannes Sametinger. *SAFER-D: A Self-Adaptive Security Framework for Distributed Computing Architectures*. Accepted for publication for 19th *European Conference on Software Architecture (ECSA)*, 2025. Preprint available at https://doi.org/10.48550/arXiv.2506.16545

1.7 Overview

This dissertation is structured as follows. First, Chapter 2 defines key terminologies and introduces the fundamentals of resilient security. Then, Chapter 3 explains the foundations of modes necessary for this dissertation. Next, Chapter 4 presents the findings on various iterations of the multi-modal architecture and mode-switching framework. Chapter 5 shows use cases and sample scenarios where the architectures and frameworks were applied. Chapter 6, discusses the evaluation findings. Finally, Chapter 7 concludes the dissertation by highlighting its contribution and future research prospects.

The definitions of IoT, IIoT, IoMT, CPS, MCPS, *Industrial Automation System (IAS)*, *Industrial Control System (ICS)* and *Industrial Automation and Control System (IACS)* are converging over time [GBWG19]. For conciseness and readability, the term IoT is used throughout.

Abbreviations are initially written out in full with the abbreviation in brackets. To enhance readability, some abbreviations are mentioned in their expanded forms within both the chapter introductions and summaries. A list of ABBREVIATIONS is provided at the beginning of this dissertation. Authors and other important terms are shown in italics.

2 Resilient Security

This chapter defines key terminologies and related works that are relevant to this dissertation, and introduces fundamentals on resilient security.

Parts of this chapter are published in several publication venues, including the *International Workshop on Cyber-Security and Functional Safety in Cyber-Physical Systems (IWCFS)* [RS20, RS21a, RSS22], the *International Conference on Industry 4.0 and Smart Manufacturing (ISM)* [RS21b], the *Journal of Systems and Software (JSS)* [RSVW23a], and my final research report published by the *Austrian Marshall Plan Foundation* [Rie22].

Section 2.1 describes threats, weaknesses, vulnerabilities, and exploits. Referring to Sect. 2.2, the *Window of Vulnerability (WoV)* is a crucial time frame in this context. Section 2.3 then explains fundamental security concepts and how they affect privacy. Additionally, safety is an essential requirement, particularly for devices that can harm people or the environment. Section 2.4 explains safety and security and addresses the difference between them and the influence on mode-switching. Computer systems should detect faults, withstand disruptions, and recover from them. Section 2.5 presents considerations on robustness and resilience. Section 2.6 describes the military strategy *Moving Target Defense (MTD)* and software diversity that makes attacks more difficult. Monitoring and controlling a large number of systems is cumbersome. Section 2.7 shows how self-adaptive mechanisms and reconfiguration can help to increase security and reduce the effort of manual work. Since the focus of this dissertation is on the security of medical and industrial IoT devices, Sects. 2.8 and 2.9 highlight specific threats, previous vulnerabilities, and security and privacy issues. As many medical and industrial IoT devices have limited resources, Sect. 2.10 elaborates on the security impact. Finally, Sect. 2.11 summarizes this chapter.

M. Riegler, *Mode-Switching for Resilient Security*,
https://doi.org/10.1007/978-3-658-49306-6_2

2.1 Threats, Weaknesses, Vulnerabilities, and Exploits

Threats refer to incidents with the potential to harm systems based on exploiting vulnerabilities. They can be caused by attacks, errors, or nature (flood, fire, etc.) [ISO09]. While deleting critical files and misconfiguration is considered unintentional and caused by human errors or accidents, malicious actors intentionally attack systems with phishing scams and malware to gain access or disrupt with *Denial-of-Service (DoS)* attacks. Threat modeling is the process of collecting and thinking about risks and how to make systems more secure [Sho14].

Weaknesses are "flaws, faults, bugs, or other errors in software or hardware implementation, code, design, or architecture that, if left unaddressed, could result in systems, networks, or hardware being vulnerable to attack" [MIT23d]. Examples of such weaknesses include buffer overflows, out-of-bounds read and write, or improper input validation. In that context, Common Weakness Enumeration (CWE) is a classification language describing common software and hardware weaknesses. The MITRE Corporation provides a list of the top 25 most dangerous software weaknesses, demonstrative examples, as well as detection and mitigation methods [MIT23f].

Vulnerabilities in software, hardware, or firmware are concrete instances of these weaknesses [MIT23g]. For example, a SQL injection in software XYZ is a vulnerability related to the weakness of improper input validation. Publicly disclosed vulnerabilities are indexed on the *Common Vulnerabilities and Exposure (CVE)* list with a unique identification number, description, severity score, and references [MIT23e]. Typically, the risk is calculated by the likelihood and impact of exploitation. The numerical severity score, ranging from 0 to 10, is rated in the *Common Vulnerability Scoring System (CVSS)* and can be represented as low, medium, high, and critical, which is used to prioritize vulnerability management and remediation [FIR23]. The score depends on several metrics considering the exploitability, the impact, and the environment. In addition to the CVE list, the *U.S. National Vulnerability Database (NVD)* provides the *Common Platform Enumeration (CPE)* to search for specific systems, software, packages, versions, and other data fields [NVD23a, NVD23b]. Also, information on advisories, solutions, and tools is available. The *U.S. Cybersecurity and Infrastructure Security Agency (CISA)* provides cybersecurity alerts and advisories, including executive summary, risk evaluation, technical details, and possible mitigations [CIS23a]. This information and databases are important for the frameworks in Chap. 4.

Exploits are commands and programs to take advantage of a vulnerability. Attackers use them to gain remote or physical access to systems or networks. For example, *ExploitDB* and *packet storm* provide exploits publicly available, which increase the probability that attackers and script-kiddies will use them [Off23, Pac23]. CISA collects *Known Exploited Vulnerabilities (KEV)*, which are used in the wild [CIS23c]. Typically, the CVSS score of these vulnerabilities is higher.

2.2 Window of Vulnerability

The period of time from the discovery of a vulnerability to the availability of the corresponding patch is commonly referred to as the *Window of Vulnerability (WoV)*. Similar to the situation with COVID-19 and the vaccination between 2019 and 2023. The consequences are different, but there is some overlap I want to present as an introduction. The world is very connected. Who thought before 2020 that bats have such a big impact on the world, especially on people's health? The first reports about the virus came from Wuhan, China [PHM21]. In the beginning, the source of infection was unclear. No treatment was available. The virus spread all over the world, and a local event turned into a pandemic.

That situation is similar to IT vulnerabilities and the recently published attacks on Solarwinds, which became a national security threat in the *United States of America (USA)* [Tur20]. Bloomberg's William Turton describes that a single vulnerability in a monitoring tool had a huge impact. The attackers, undiscovered for months, had access to critical infrastructures like the *U.S. Department of Homeland Security (DHS)*, but also cybersecurity firms like FireEye were affected with their cloud platforms. The NATO, the UK National Health Service, and the European Parliament also use this software. In the beginning, the source of infection was unclear, no treatment was available, and the spreading was still ongoing. [Tur20]

As long as no treatment or patch is available to mitigate a vulnerability—affected people and companies have to wait. Some people, groups, vendors, and governments deny the problem until an exploit is presented, the number of those affected is increasing, and they only understand it when it is in a closer environment, as the rising COVID-19 numbers in neighboring countries showed.

It takes time to develop a vaccine and a patch. It needs to be extensively tested and have formal approval. Even if it is available, much will be unclear, and some people will not use it because they fear negative side effects. What else should you do? During the pandemic, the mode of operation had to be changed: keep distance and wear masks. Additionally, border controls similar to the computer domain were activated: isolate and shield the system and reinforce the firewall. Because some infections are without symptoms, virus scans and tests are made. If something is discovered, contact tracing and further analysis are carried out. However, still then infections can happen, as Stuxnet and COVID-19 numbers have shown. The knowledge grows daily, but some protection measures are still unclear, not known, contradict each other, or are deliberately ignored. A lockdown, stopping the systems, or switching to limp mode seems to be the only solution. But I think there are many variations here. For more details about approaches, see Chap. 4.

The WoV is also referred to as the window of exposure [Sch00, ENI13] or patch gap [Res23]. From an organizational perspective, the WoV is only closed when all the patches have been installed on all their affected systems [ENI13]. While a patch is a small, specific change to fix a software bug or vulnerability, updates commonly combine multiple patches and new features.

Figure 2.1 shows the phases of the WoV with a *coordinated vulnerability disclosure* and events that influence the risk level over time. Based on the web security testing guide of the *Open Worldwide Application Security Project (OWASP)*, the phases are defined as follows [OWA20]: Before a vulnerability is discovered, it exists hidden in a system (t0), but no one can exploit it. Therefore, the risk level is low. When individuals or groups discover a vulnerability (t1), the risk level rises as no countermeasures are known yet. These vulnerabilities are called *zero-day*, as they are unknown to the vendor, and no mitigation is available. Such zero-day vulnerabilities can be an enormous risk if the discoverer is a criminal or foreign intelligence organization. They can secretly use and sell the information and develop so-called zero-day exploits to take advantage of the vulnerability. Initially, only a few people know about a vulnerability, but the risk level rises depending on how the discoverer spreads the information. If security researchers or manufacturers discover the vulnerability, there may be no risk at all. Typically, they consider a coordinated vulnerability disclosure, also called responsible disclosure. In that process, the discoverer informs the vendor (t2) and gives them sufficient time to inform their customers (t2a) and develop a patch. Other attackers may discover the vulnerability independently and develop exploits for follow-on attacks after the zero-day phase. After vulnerability disclosure and a system's initial breach or compromise, follow-on attacks try to further exploit the system by escalating privileges, maintaining access, or achieving additional malicious objectives. If vendors provide advisories, customers can adapt software settings and update security tools like firewalls and antivirus software to mitigate the vulnerability (t3). However, the vendor information can be misused by attackers to target customers who are behind on the countermeasures. After the vendor provides a patch (t4), the customers and the public should be informed about it (t4a) so they can install it shortly (t4b) and reduce the risk level. Until the patch is installed, follow-on attacks are still possible. Finally, the vendor publishes information about the vulnerability and their fix (t4c). If any code to exploit the vulnerability is published after the patches are provided and installed (t5), the risk level is low because, hopefully, many customers have already installed the patch.

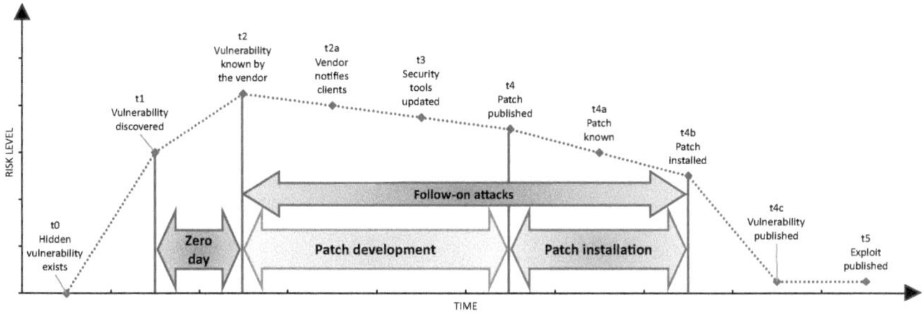

Fig. 2.1 Window of Vulnerability with a Coordinated Disclosure

At *full disclosure*, all vulnerability details are published publicly (t4c) right after discovery (t1) without consultation with the vendor or other responsible authorities [Sch07a]. Figure 2.2 shows an exemplary timeline and indicates a higher risk level for customers affected by the vulnerability because vendors are informed later (t2), which may delay the information of customers (t2a) and the development and distribution of patches (t4–t4b). *Householder et al.* argue that attackers can take advantage of this situation when no mitigation is available yet [HWMK20].

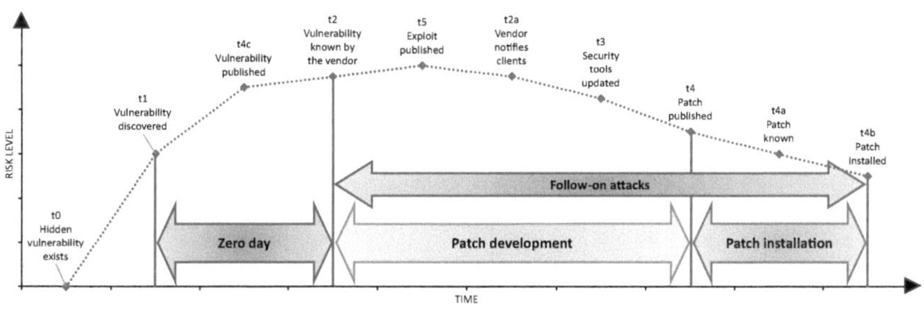

Fig. 2.2 Window of Vulnerability with a Full Disclosure

Readily available and easy-to-use exploits (t5) simplify follow-on attacks and increase the risk level. However, according to *Schneier*, keeping vulnerability information secret is a not good idea [Sch07a]: Hackers can discover vulnerabilities independently, and some vendors may ignore information about discovered vulnerabilities and threaten security researchers with legal action. He argues full disclosure increases public pressure and may push the vendor to provide a patch quickly. *Householder et al.* describe vulnerability disclosure as a "wicked problem" as there is no optimal solution [HWMK20, p. xi]. The time between the vulnerability being published (t4c) and known by the vendor (t2) may be shorter in reality than depicted in the figure. However, if a zero-day exploit is available (t5) before the vendor provides a patch or mitigation, it can have devastating consequences.

2.3 Security & Privacy

Security is about protecting information and information systems from unauthorized access and use. The key concepts of the CIA security triad are important: confidentiality, integrity, and availability. Confidentiality means that only authorized users have access to information. This must be maintained through identity and access management. Sensitive data and device programming must be protected, particularly remote access. Data integrity must be accurate, complete, and protected through hashing algorithms and checksums. Data must be available when it is needed. Availability must be maintained to prevent DoS attacks from overloading IoT devices, causing battery depletion, or disrupting critical functions. According to ISO/IEC 27000:2009, authenticity, accountability, non-repudiation, and reliability are also worth mentioning [ISO09]. Preventive, detective, and corrective security controls will help minimize security risks and help the organization recover faster when something bad happens. There should be multiple levels of protection (defense in depth) for all system parts like a medieval castle: multiple walls and levels of defense for cloud, network, device, application, and data [DHS16]. One compromised part should not affect the entire system.

Several guidelines and frameworks exist to make systems and especially IoT devices more secure [ENI20, NIS22, NIS18, NIS24, IEC18]. They provide general security and privacy controls for information systems and suggest physical controls like *Trusted Platform Modules (TPMs)*, mechanical locks, and minimized external ports. TPM goes hand in hand with a *Trusted Execution Environment (TEE)*, secure boot, and a protected *Operating System (OS)*. For example, *McCormack et al.* provide an architecture for trusted edge IoT security gateways and recommend periodic remote attestation with TPM, to control that there are no fraudulent manipulations [MVL+20]. Common information security defenses include access control, application and network security, authentication, authorization, and encryption. It is state-of-the-art to encrypt all communication with cryptographic protocols to the outside of an IoT device and minimize open ports. Risky legacy protocols should be avoided. Secure and reliable communication prevents *Man-in-the-Middle (MitM)* attacks like packet modification or replay. Each IoT device should have its own client certificate to communicate with the server or cloud services.

Compromised devices can be blocked on the server side if necessary. Even if attackers take over an account, the principle of least privilege reduces the range of activity. Permissions should be as little as possible and as much as absolutely necessary. For example, web servers and other software should run within non-privileged accounts, restrict file and directory access, enable protection mechanisms, disable default behavior like directory browser listing, and remove the server version banner. It is also important to avoid storing hard-coded passwords, keys, and other sensitive information unencrypted on IoT devices, as compromised or stolen devices can threaten other devices.

Default secure settings and fail-secure mechanisms ensure security even if something bad happens. These fail-secure or fail-closed mechanisms are used for fire doors to reduce the spread in case of a fire. When there is an alarm or the power goes out, the door closes automatically. However, manually opening the door is still possible. Applied to the context of this work, this means separating infected devices from others. When an update or patch is eventually available, it should be installed as soon as possible, because hackers can use the information about the fixed vulnerabilities to write malware and attack unpatched systems.

According to *Security by Design* principles, systems should be built secure right from the start. Techniques like STRIDE and kill chains help to identify threats and encourage countermeasures [Sho14]. Following a *Secure Software Development Life Cycle (SSDLC)* with guidelines for secure coding and development will facilitate delivering effective and efficient systems during lifetime [ENI19, IEC18]. Software security is "the idea of engineering software so that it continues to function correctly under malicious attack" [McG04]. For example, OWASP provides a list of the top 10 web application security risks, how to test applications, and how to prevent and mitigate them [OWA23c]. MITRE provides *Common Attack Pattern Enumerations and Classifications (CAPEC)* and *Adversarial Tactics, Techniques, and Common Knowledge (ATT&CK)* [MIT23b, MIT23a]. While CAPEC focuses on application security and should help defenders to understand adversary behavior better, ATT&CK provides examples, detections, and mitigations for enterprise, mobile, and ICS.

Unit testing and code reviews increase code quality. Static and dynamic testing and vulnerability detection tools can be used for analysis. Static application security testing uses white-box testing to find vulnerabilities in the source code. For example, the *NodeJsScan* vulnerability scanner finds 40 issues in the source code of *Juice Shop* used in Sect. 5.1. Dynamic application security testing has a focus on black-box testing. Applications such as *OWASP Zap* assist in conducting vulnerability scans and executing attacks.

Formal methods help to verify the correctness of software updates. Updates and patches should be signed, verified, and should not introduce new vulnerabilities. In addition to technical controls, logical and organizational controls should also be implemented. Logging and audit controls can help monitor and detect suspicious access and changes and provide evidence for the incident response team. *Intrusion Detection Systems (IDSs)* must be in place to detect malicious activities. Even better are *Intrusion Detection and Prevention Systems (IDPSs)* which are able to take countermeasures to prevent further attacks. Moreover, a *Security Operation Center* with active vulnerability management can reduce risks. Backup and recovery plans must be in place to mitigate ransomware attacks.

Over the years, several regulations, standards, and guidance documents were published that provided an attempt to protect systems and their environment. This means to reduce risks and prevent and mitigate threats. The ISO/IEC 27000 family consists of several information security standards. The most commonly used standards and practices include the NIST Cybersecurity Framework, ANSI UL 2900-1, ETSI EN 303 645, ISO 15408, ISO/IEC/IEEE 29119, ISO/IEC 29147, ISO/IEC 30111, and NISTIR 8259. For example, the *NIST Cybersecurity Framework* recommends considering functionality to prevent unauthorized use, loss of confidentiality, integrity, or availability [NIS18, NIS24]. Using the five core functions: identify, protect, detect, respond, and recover, they present guidelines for managing and reducing cybersecurity risks.

Privacy

Another concept related to security is *privacy*. Weak security of systems can affect the privacy of customers, employees, patients, and other parties. When personally identifiable information like the social security number, biometric data, or other sensitive information like health data is processed or kept, privacy issues arise [SRLO15]. The main factor contributing to privacy problems is often inadequate access control. Privacy is violated when attackers can access personally identifiable or sensitive information stored on the cloud or systems. Disclosure of such information can have unfavorable effects. For example, personal identifiable information can be used for identity theft or insurance fraud. Unique device identifiers can allow to identify, track, and locate persons. Collected data about individuals can be used for purposes other than the initially intended one. Data breaches may disclose private information to the public and simplify phishing campaigns and social engineering [AAS17].

In the *European Union (EU)*, the *General Data Protection Regulation (GDPR)* governs how organizations have to handle personal data [Eur16]. The GDPR gives individuals more control over their personal data and imposes substantial fines for non-compliance. Individuals have the right to information, correction, and deletion of data. Organizations have to follow strict rules regarding collecting, processing, and protecting personal data. Principles include data minimization, storage limitation, and protection against unauthorized access. In the USA, no comparable federal data privacy regulation exists. However, some states like California, Virginia, and others issued their own privacy laws, and there exist regulations for specific industries like health insurance [Nay22].

2.4 Safety & Security

Safety aims to protect the environment from a system. For example, medical IoT devices should not harm patients, medical staff, and the environment in any way. The same is true for connected machines and robots. In that context, the fail-safe or fail-open mechanism is often used, e.g., emergency exits should automatically open when there is an alarm or the power goes out. Conversely, the air brakes of trains and trucks automatically close if they lose air or in case of other defects.

From a security perspective, patches and updates should be provided and installed as quickly as possible. Systems should react immediately (fail-secure) if antivirus software detects malicious files, or if an IDS identifies suspicious processes. However, stopping medical and industrial IoT devices in case of potential or real threats leads to a highly secure state (fail-secure), but is rarely an option. For example, stopping a pacemaker in case of known vulnerabilities can be life-threatening. Consider a vacation situation far where medical help is needed. In an emergency, security mechanisms must not endanger patients' lives [MDC19]. Safety-relevant processes must never be interrupted or stopped without human interaction (fail-safe). Therefore, safety and availability are more important for these life-critical systems than security, confidentiality, and integrity. For this reason updates are rare and require long planning. Any change needs an impact assessment, complete test evidence, and formal approval.

In the context of functional safety, the IEC 61508 standard focuses on electrical, electronic, and programmable electronic safety-related systems and their life cycle. In the sector of manufacturing and machinery IEC 62061 extends the standard and provides more details. Also ISO 13849 provides design principles and guidance for safety-related parts of machinery control systems.

The *EU Machinery Directive* primarily focuses on the safety of industrial machinery and equipment and does not explicitly address cybersecurity [Eur06]. However, safety is at stake when security is weak. Security vulnerabilities can undermine safety and open access can be risky and contradict the idea of security by default.

For example, if an attacker gets access to a pacemaker and changes the clock rate or triggers an emergency shock that harms the patient. In this case, a security issue becomes a safety issue. Likewise, a compromised robot can harm people or cause damage to the environment. Critical vulnerability patches may not be available as quickly as needed. Due to safety, certification, and warranty requirements, they must be tested extensively before they can be installed. This takes time.

According to *Witte et al.* it is necessary to consider safety and security together [WGR+22]. Especially for medical and industrial devices attacks often result in actual safety issues due to the tight integration and interplay of hardware and software components. Therefore, the authors propose a modeling approach to generate *Attack-Fault-Trees* as the basis for further analysis. Medical and industrial IoT devices should consider the context, automatically detect critical situations, and adapt the behavior.

Avizienis et al. provide basic concepts and taxonomy of dependable and secure computing [ALRL04]. They compare the attributes of dependability (availability, reliability, safety, integrity, and maintainability) with security (CIA). While dependability "is the ability to deliver service that can justifiably be trusted" [ALRL04, p. 13], security is more focused on preventing unauthorized actions. To improve dependability and security measures, they recommend to prevent, tolerate, remove, and forecast faults [ALRL04]. Although some measures overlap with classic risk management and the functions of the NIST cybersecurity framework, forecasting faults is an interesting approach. By estimating "the present number, the future incidence, and the likely consequences of faults" [ALRL04, p. 28], the authors refer to the *Failure Mode and Effects Analysis (FMEA)* (see Sect. 3.8). Failures can lead to degraded modes like slow service, limited service, or emergency service, which still provide a limited set of functionalities [ALRL04].

2.5 Robustness & Resilience

Robust and resilient IoT devices should be able to detect faults, recover from them, and prevent them in the future. This concept also corresponds with the idea of the *MAPE-K* feedback loop (see Sect. 2.7). Security issues can cause system faults and vice-versa, so tactics and patterns for software robustness are relevant [KBEI22]. Traditionally, robustness is a quality attribute in software engineering and describes the "degree to which a system or component can function correctly in the presence of invalid inputs or stressful environmental conditions" [ISO17a, p. 394]. Tactics like ping/echo, heartbeat, self-tests, and others can be used for monitoring, analyzing, and detecting faults. *Kazman et al.* extended the view on robustness and not only suggest tactics for fault preparation, but also tactics for repair, and re-introduction to recover from faults and adapt the system to achieve the system goals [KBEI22]. For example, graceful degradation and re-configuration can help an attacked IoT device to achieve its goals in the planning and execution activities. That is similar to the concept of modes, see Chap. 4, reducing less critical functionality while providing core functionality during power outages, connection loss, and attacks, which can prevent system outages. The re-introduction tactics can be used for the self-healing aspect of autonomous systems, as they try to bring a failed element back into normal operation. Preventing faults can build on the knowledge base of previous events and the following activities.

In the context of systems engineering, resilience extends robustness and includes also recovering from disruptions and a proactive perspective [INC23]. That can be situational awareness, graceful degradation, and adaptive response. Because traditional IT security is not always suitable for *Operational Technology (OT)* used in medical and industrial IoT systems, Conklin postulates that resilience is the driving factor for security and should be included into the CIA security triad [Con16]. According to *Firesmith*, resilience has a more comprehensive view and includes quality attributes from robustness, safety, cybersecurity, anti-tamper, survivability, capacity, longevity, and interoperability like coding defects, hazards and accidents, vulnerabilities and attacks, excessive loads, long lifespans, and lost communication [Fir20b].

Firesmith provides an extensive overview of commonly-used resilience techniques [Fir20a]. He claims that systems are resilient if they do their work despite adverse circumstances and recover rapidly from any harm [Fir19]. No matter how well systems have been designed, unknown or unresolved vulnerabilities can become a risk in the future. *Resilient Systems* protect their important assets by resisting hardware and software failures, detecting adversities, mitigating them, and recovering. The reaction can include exception handling, degraded modes of operation, and redundancy with voting, e.g., to detect the failed redundant component [Fir19]. As redundant components take over, degraded modes provide limited functionality, and settings are adjusted, modes, redundancy, and re-configuration also aid in the full or partial recovery. To limit the duration of failures, automatic resilience is often preferred [Fir20b]. Human detection and reaction can take longer and are not always possible (offline and autonomous devices).

For resilient security, exposure and authentication measures may be adapted depending on the security context by switching modes [SS17]. Reduced exposure can be a first step, e.g., temporarily blocking communication or reducing the data transfer rate. In addition, modes are useful not only from a security perspective. If a component experiences external or internal disturbances or problems, a restricted mode like vehicles' limp mode can provide at least the core functionality for mission-critical capabilities and, therefore, business continuity.

2.6 Moving Target Defense & Software Diversity

In the context of this dissertation, related concepts include *Moving Target Defense (MTD)* and software diversity. MTD is a military strategy that constantly changes systems, components, configurations, implementations, and versions, making it more difficult for attackers to elaborate system information and successfully execute attacks. For example, *Dsouza et al.* used MTD as a kind of software behavior encryption to build resilient and fault-tolerant cloud services [DRANH13]. They reduce the attack surface by frequently modifying parts of the system execution environment, like switching the OS, the programs written in different languages, and the program versions.

Attackers can cause much damage by exploiting vulnerabilities in widely used frameworks, components, libraries, and plugins, like *jQuery*, *Spring*, and *WordPress*. Software diversification is a method to mitigate *Break once, break everywhere (BOBE)* attacks on monocultures and increase fault tolerance [ABB+15, BM15]. ISO 24765:2017 specify software diversity as "software development technique in which two or more functionally identical variants of a program are developed from the same specification by different programmers or programming teams with the intent of providing error detection, increased reliability, additional documentation, or reduced probability that programming or compiler errors will influence the end results" [ISO17a, p. 421].

Allier et al. used multi-tier diversification in web-based software applications. They transform a monolithic blogging system into a diverse one using different OSs, java virtual machines, frameworks, and regional deployment. *Baudry & Monperrus* surveyed the multiple facets of software diversity which was applied for fault and attack tolerance, testing, and reusability [BM15]. They mention the former terms of N-version programming and recovery blocks that were used to limit the consequences of programming failures. In that context *Software Product Lines (SPLs)* support to control the diversity of a range of varied products for different users and environments [BM15].

2.7 Self-Adaptation, Self-Protection, and Re-Configuration

Self-adaptive systems focus on configuring, re-configuring, optimizing, and updating themselves automatically at runtime [LGM+13]. One prevalent reference model in this context is the *MAPE-K feedback loop* for managing and controlling autonomous and self-adaptive systems in various different domains such as autonomous vehicles, IoT applications, or service-based systems [ARS15, KC03]. Monitoring, analysis, and subsequent adaptation play a fundamental role in these types of systems focusing on architecture adaptation, performance, or response time optimization [MSB+18, KSNW19, BKFL17], as well as ensuring reliability, for example in robotic- and CPSs [CCF+20, VHFST15, PDL+16, MBFD12, GSP+19]. The MAPE-K feedback loop consists of four main phases: *Monitor*, *Analyze*, *Plan*, and *Execute*, followed by a *Knowledge* component, providing continuous feedback to adapt and improve the performance of a system [KC03]. Some approaches in this area have also addressed security concerns. For example, *Tomic et al.* present an approach for detecting attacks on network communication [TCBM18]. Their approach operates on the network level and facilitates data routing adaptations. *Abie et al.* describe a messaging infrastructure facilitating adaptation and targeting security vulnerabilities to improve reliability and robustness [ASB+10]. However, their focus is on message-oriented middleware. In the domain of threat modeling, *Van Landuyt et al.* propose runtime support via reflective threat modeling to support adaptive security [VLPSJ21]. While these approaches facilitate runtime adaptation and—to some extent—incorporate security concerns, they do not provide a flexible mode concept and a *Domain-Specific Language (DSL)* as proposed in the multi-modal architecture (see Chap. 4).

Pacheco & Hariri used an *Anomaly Behavior Analysis Intrusion Detection System (ABA-IDS)* to detect known and unknown attacks in smart cyber infrastructures [PH16, PZBH17]. They trained a baseline model and compared it with real sensor data to distinguish between normal and abnormal behavior. While ABA-IDS mainly deals with attack detection, this work also focuses on the reaction. For example, launch alerts, enforce authentication, reject data, and change configurations in case of attacks.

Jahan et al. propose an approach combining two separate *MAPE-K* loops, each focusing on a different adaptation aspect [JRW+20]. The first *MAPE-K* loop targets functional assurance cases, while the second one is concerned with security assurance cases. As part of this, they adopt the *Regional Planning Pattern* and coordinate the loops in the planning and execution activity [WSG+13]. In comparison, with the extended mode-switching framework, see Sect. 4.9, the focus of this work is on the protective aspect of self-adaptive systems, detecting security issues on multiple levels and executing countermeasures to adapt the systems accordingly.

Khakpour et al. investigated secure architecture-based adaptions based on vulnerability analysis [KSNW19]. Compared to their approach, this work focuses on IoT devices and takes cost-efficient countermeasures in the first step to mitigate attacks at runtime. These measures may involve blocking single IP addresses before taking more complex adaptations, such as mode switches.

Yuan et al. argue for an *Architecture-Based Self-Protection* approach to protect web applications against DoS attacks and provide availability utilizing the Rainbow adaptation framework [YMS+13]. In a later work, they presented a systematic literature review about self-protecting software systems and proposed a taxonomy, patterns, and tactics to protect systems [YEM14].

García-Gordillo et al. proposed a framework using the *MAPE-K* loop to coordinate mode switches at different tiers: the cloud, the edge, and the device [GGVCS24]. They distinguish between local operational modes of the devices and global modes managed by coordinators at the edge and cloud tiers. If cloud services become unavailable, the degraded operation mode is activated on the devices.

2.8 Medical IoT Devices

Medical devices that are actively implanted, such as insulin pumps, brain stimulators, cardiac defibrillators, cochlear implants, and gastric stimulators, are crucial for maintaining patient health. These life-saving devices are usually battery-powered and have a lifespan of several years. Regular checks in hospitals are essential to monitor these devices and their batteries. Remote monitoring and interconnected devices allow better patient care and reduce in-person visits and associated costs [VS19]. Therefore, more and more medical devices are connected to the Internet. Measured values like heart rate, pulse rate, blood pressure, body temperature, and others, can be transferred wirelessly and automatically to healthcare service providers [SRJ20]. If needed, authorized medical personnel can analyze the data and adapt configurations remotely. Additionally, medical IoT devices can automatically alert medical personnel if certain values fall below or exceed a predefined threshold. This is more efficient than traditional personal emergency response systems with manual alerts.

The emergence of COVID-19 has increased the development and adoption of medical IoT devices, especially during times of lockdown. According to *Statista*, the global market value of medical IoT devices will reach over 260 billion U.S. dollars in 2027 [Rep21]. Because of the growing market and the introduction of new mobile communication standards with higher data rates, digital interconnectivity is unstoppable.

Nevertheless, these connections go hand in hand with security vulnerabilities and potential threats to these devices [HLLL17]. According to the technological research and consulting firm *Gartner*, by 2025 "cyber attackers will have weaponized Operational Technology environments to successfully harm or kill humans" [Gar21]. The interfaces of the medical device and remote monitoring can have drawbacks. For example, a pacemaker may deliver a too intensive shock, or an insulin pump may overdose. In 2019, eleven zero-day vulnerabilities affected 2 billion IoT devices, including many medical devices, which among other things, allowed remote code execution [Dof19]. These vulnerabilities resulted from flaws in software development kits used for Wi-Fi, Bluetooth, and Bluetooth Low Energy connections and allowed attackers to cause deadlocks, crashes, buffer overflows, or completely bypass security [GCW20, YAA19].

In 2017, cybersecurity vulnerabilities in implantable cardiac pacemakers affected hundreds of thousands of Americans [FDA17]. These vulnerabilities allowed unauthorized users to modify programming commands, resulting in rapid depletion of the battery or harm to the patient through inappropriate pacing. Fortunately, a firmware update was sufficient to address the issues, and no surgery was required. Although recommended, the update was not mandatory. More than a hundred devices were, and some still are affected by life-critical cyber-attacks [AAS17, HBR+08, Kap11, LRJ11, PKK11, SCL+18, YAA19]. Over the past years, the *U.S. Food and Drug Administration (FDA)* has recalled multiple medical devices due to possible cybersecurity threats. According to the cybersecurity company *Claroty*, there has been a 110% increase in vulnerability disclosures since 2018 [Cla22]. Recent reports by *Palo Alto Networks* have also shown that over 75% of more than 200,000 infusion pumps have vulnerabilities, which can result in serious issues such as battery depletion, malfunctioning, extortion, remote assassination with manipulated settings, fatal doses of medication, and privacy breaches [Pal22].

Some *Medical Device Manufacturers (MDMs)* use *off-the-shelf (OTS)* hardware and software from third-party vendors. These standardized hardware and pre-packaged software address common use cases but make them vulnerable to exploits such as URGENT/11, SweynTooth, Ripple20, AMNESIA:33, BadAlloc, Nucleus:13, Log4Shell, and others. Almost every fourth medical device has KEVs, many of them because of unsupported or end-of-life OSs [Cla24].

Security of medical devices received public attention in 2007 when the former U.S. Vice President Dick Cheney had his pacemaker's remote control deactivated, because a terrorist could hack into and shock his heart [Gup13]. Safety and utility have traditionally been the goals for medical devices [HHBF+08]. Various authors have discussed the challenges in the medical domain, including confidentiality, integrity, availability, reliability, safety, privacy, interoperability, secure communication, software and hardware aspects, malware and intrusion detection and reaction, formal methods, resource constraints, non-technical aspects, and organizational and regulatory issues [AAA+22, AAS17, ETA+21, IMD20, KTGM22, SRLO15, SCL+18, TA21, YNN+20].

According to the chronology of medical device security by *Burns*, considering threats and zero-day vulnerabilities in medical devices are necessary throughout their entire useful life, from design, development, and distribution through maintenance and maintenance decommissioning [BJH16]. Modified data may result in an incorrect diagnosis, treatment, or medication and risk the patient's health. Medical devices must be continuously operational and reliable to deliver clinical functions to patients. Privacy must be protected to prevent violating sensitive individual health and behavior data. Violations may seriously affect patient safety and the reputation of *Healthcare Delivery Organizations (HDOs)* and MDMs, potentially resulting in penalty payments.

Burleson et al. have recommended considering security in early design phases for medical devices, to encrypt traffic and use well-studied crypto, authenticate third-party devices, do industry-standard source-code analysis, and develop realistic threat models [BCRF12]. Additionally, manufacturers should not rely on security by obscurity, e.g., hiding implementation details. Failing to fix vulnerabilities leaves medical devices susceptible to attacks, as attackers can exploit the vulnerabilities [SR16].

There exist several approaches to make authentication more secure, for example:

- Proxy-based [DFK08, GHR+11, Sch10, XQT+11],
- Based on human (biometric) characteristic [HD11, RJK13, SWdH+18],
- Patient passive [GMV06],
- Proximity-based [HBR+08, CZCM11, CVG03, MMV+11, RCHBC09]
- Trusted third party [MSG+16, XWP14a].

To ensure safety and improve security, many regulations, standards, and guidance documents have been published over the years. The following paragraphs provide some insights.

The FDA and Lechner list many recognized consensus industry standards and guidances for medical device security from several organizations [FDA23b, Lec18]. For example, the *Association for the Advancement of Medical Instrumentation (AAMI)*, the *American National Standards Institute (ANSI)*, the *Clinical and Laboratory Standards Institute (CLSI)*, the *Healthcare Sector Coordination 29 Council*

(HSCC), the *International Electrotechnical Commission (IEC)*, the IEEE, the *International Society of Automation (ISA)*, the *International Organization for Standardization (ISO)*, and the *U.S. National Electrical Manufacturers Association (NEMA)* established AAMI TIR57, AAMI TIR97, ANSI/AAMI SW96, ANSI/NEMA HN 1-2019, ANSI UL 2900-2-1, CLSI AUTO11, HSCC Model Contract Language, IEC 60601, IEC 62304, IEC 62443, IEC 81001-5-1, IEC 82304, IEEE/UL 2621, ISO 14971, ISO/IEC 80001, ISO/IEEE 11073, and others.

In the USA, the FDA provides pre- and post-market guidance on building secure medical devices and using a *Bill of Materials (BOM)* to monitor supply chain problems and vulnerabilities of commercial, open-source, and OTS hardware and software [FDA16b, FDA23a]. BOM simplify the monitoring process and reduce response time in case of incidents or vulnerabilities. MITRE provides a playbook to identify and mitigate medical device threats [BCC+21]. Safeguards and other protection methods are needed to limit access and ensure that only trusted and authorized users and devices execute safety-critical commands. The *Health Insurance Portability and Accountability Act of 1996 (HIPAA)* sets the rules in the context of personal health information. In their strategic plan, HSCC argue that "cyber safety is patient safety" [HSC24b, HSC24a, p. 4] and provide a reference guide for secure medical devices and health IT solutions [HSC24b, HSC24a].

In the EU, the *Medical Device Coordination Group* offers guidance for developing and manufacturing medical devices in compliance with state-of-the-art and meeting the *European Medical Device Regulation* [MDC19]. They suggest a defense-in-depth strategy during the product life cycle. MDMs must consider risk management, information security, and IT security measures like protection against unauthorized access and foreseeable misuse across the life cycle. The *International Medical Device Regulators Forum (IMDRF)* provide design principles and best practices, such as secure communication and software maintenance, to assist all stakeholders in their collective duty to ensure medical device security [IMD20]. They mention pre- and post-market considerations for secure architectures, testing, vulnerability remediation, and incident response. The IMDRF provides also principles and practices for legacy medical devices and software BOM [IMD23b, IMD23a]. The regulatory process will lead to better patient safety and medical device security. Nevertheless, it will also lead to a longer process design cycle, which can take years and cost millions of euros.

2.9 Industrial IoT Devices

Industrial IoT devices are based on real-time embedded devices with wired or wireless connections to monitor and control physical systems. They are used in mission- or safety-critical applications like aviation, self-driving cars, unmanned aerial vehicles, medical monitoring, traffic control, power plants, and industrial machines. Such systems require functional correct execution on time. Industrial IoT devices consist of hardware and software interacting with physical components like sensors and actuators and other local and/or remote systems.

Typically, industrial IoT devices have a life cycle from 5 to 20 years. They need to operate 24/7 and thereby master special security challenges. It is essential to take into account a multitude of perspectives and issues: organizational and regulatory challenges, resource constraints, and non-technical aspects like security awareness. Security aims at protecting industrial IoT devices and their components from their environments, e.g., from attacks by hackers. In contrast to security, safety aims to protect the environment, including staff, customers, and assets, from the industrial IoT device and its components. Manufacturers must ensure that their systems do not harm people or damage anything, see further details in Sect. 2.4.

In their survey on CPS security which includes industrial IoT devices, *Humayed et al.* mention criminal, financial, political, privacy, and physical threats [HLLL17]: Disrupting operations, tampering with physical equipment, intercepting communications, or injecting false data can lead to the loss of critical data, financial losses, blackouts, environmental pollution, or even accidents and critical injuries. *Cejka et al.* have compared Amazon Web Services IoT Greengrass, Microsoft Azure IoT Edge, and a self-implemented framework for secure *edge devices* and distributed control of critical infrastructure [CKK19]. They recommend filtering and monitoring communication for anomaly detection and reacting to deviations with countermeasures.

Unlike general IT, where confidentiality is considered the highest priority, availability is the most important goal for OT used in industrial IoT devices [DHS11]. The security of these devices received public attention in 2010 when the computer worm Stuxnet caused widespread damage and even spread through air-gapped systems with removable devices [FOMC10, KRT10]. Security failures in industrial IoT devices can have more severe consequences, like harm to people or environmental damage. The *German Federal Office for Information Security (BSI)* provides a list of the top ten threats and countermeasures for ICS [BSI22]. Examples are malware infections, human error, compromised components, social engineering, phishing, and others. In the context of industrial automation, several standards and guidance, like the ANSI UL 2900-2-2, IEC 62443, the NERC Cyber Security standards, NIST

SP 800–183 for Networks of 'Things', NISTIR 8259, and the ENISA Good Practices for Security of IoT provide support for Smart Manufacturing.

In the EU, the *Cybersecurity Act* established a framework for certification of various products, including IoT devices [Eur19]. The *Network and Information Systems (NIS)* directives consider network and information systems' security, especially critical infrastructures. In 2023, the second version of the NIS directive replaced the former one, introduced new topics like vulnerability management and supply chain security, and affect a larger share of the economy [Eur22]. In 2024, the *European Commission* introduced mandatory cybersecurity requirements for products with digital elements through the *Cyber Resilience Act (CRA)* [Eur23]. By 2026, the regulation will enforce reporting obligations and will become fully applicable by 2027, aiming to ensure cybersecurity throughout the entire product life cycle.

In the USA, specific regulations exist for the health sector, financial services, and others. However, the IoT Cybersecurity Improvement Act of 2020 enforces federal agencies to adopt standards and guidelines [FMB+21]. In that context, the NIST SP 800–213 series provide guidance for federal procurement [FMB+21]. Also, the DHS provides best practices ICS security [DHS11].

2.10 Limited Resources

Due to space and cost limitations, industrial and medical IoT devices are expected to have restricted resources compared to traditional IT systems [PdOG+20]. For example, pacemakers and other battery-driven medical devices are designed to work for years, since device exchange implies surgery that must be associated with enormous risks that may even be life-threatening. These devices have to use resources efficiently. They have limited memory, low processing speed, and restricted network bandwidth. Therefore, such devices are connected to either edge or cloud computing, which offer adequate resources such as computation, memory, and storage. This strategy, known as computation offloading, enables IoT devices to consume less power since most of the computation and storage is performed at the edge or in the cloud [STB+16]. Lightweight communication protocols like *Message Queuing Telemetry Transport (MQTT)* and *Constrained Application Protocol (CoAP)* are used in low-power and lossy networks [EPV17].

Limited resources have a significant impact on security. Standard IT security is often not applicable as IoT devices may not have the ability to implement robust security measures like complex cryptographic methods. This makes them vulnerable to cyber-attacks, data breaches, and other security threats.

2.11 Summary

This chapter introduced essential terminology about resilient security and presented related work relevant to this dissertation. Medical and industrial devices face common and specific threats. As previous incidents have shown, connecting these devices to the Internet increases their attack surface. Weaknesses in hardware and software can lead to vulnerabilities. If left unaddressed, they will be discovered and prone to potential exploitation by attackers. The Window of Vulnerability exists until an effective patch or mitigation addresses them. Therefore, shortening this critical timeframe has always been the goal of denying potential opportunities for attackers. Several standards and guidelines have been presented to improve the design, development, deployment, and maintenance. Manufacturers have to consider security, safety, privacy, and other aspects. It happens that safety and security contradict each other. However, the most important goal is to prevent any harm to people and the environment surrounding them. Some fail-safe mechanisms may offer opportunities for attackers. Security issues can become relevant to safety if guards are bypassed. While robustness considers resisting failures and attacks, resilience extends that functionality to recover from such situations and restore the functionality. The concepts of Moving Target Defense, self-adaptation, self-protection, and re-configuration make possible attacks harder and support the defense reaction. Medical and industrial IoT devices have special challenges concerning longevity and limited resources. The goal of resilient security is to anticipate common attack patterns in system design and development to identify, withstand and respond to attacks during runtime and to recover to normal operation.

3 Modes

As the foundations of this dissertation build up on the idea of modes, the following chapter defines and presents key terms, characteristics, and principles. Based on our SLR on mode-switching from a security perspective [RS20], we extended this chapter with the snowballing approach by reference and citation tracking. Additionally, this chapter includes parts of our work on *Multi-Mode Systems (MMSs)* for resilient security in industry 4.0 [RS21b].

According to the *Oxford English Dictionary* [OED23], several different meanings and definitions of the noun *mode* exist, from music, logic, costumes, to statistics and linguistics. One of them is: "A particular form, manner, or variety in which some quality, phenomenon, or condition occurs or is manifested." The Systems and Software Engineering Vocabulary (ISO/IEC/IEEE 24765:2017) defines a *mode* as a "set of related features or functional capabilities of a product" [ISO17a, p. 278]. In the software requirements specification, *ISO/IEC/IEEE 29148:2018(E)* mentions *system modes* as organizational approaches: "Some systems behave quite differently depending on the mode of operation. For example, a control system may have different sets of functions depending on its mode: training, normal, degraded or emergency" [ISO18, p. 56].

Section 3.1 explains how modes are interpreted in the context of this work. Sect. 3.2 shows examples of MMSs and application areas. Over time, multiple definitions of phases, modes, and states have been established. The difference and relationship of these terms are controversially discussed. The definitions from the literature are presented in Sect. 3.3, while the difference of modes and states and

Supplementary Information The online version contains supplementary material available at https://doi.org/10.1007/978-3-658-49306-6_3.

M. Riegler, *Mode-Switching for Resilient Security*,
https://doi.org/10.1007/978-3-658-49306-6_3

their relationship are explained in Sect. 3.5. Sect. 3.6 presents mode modeling approaches and programming and specification languages to describe and visualize system behavior. In Sect. 3.7, MMSs and common mode-switching protocols are described, along with the conditions and requirements that must be met. Sect. 3.8 present failure, safety, and security modes. Sect. 3.9 discusses the human-computer interaction and why operators need to know the current modes of systems and how to prevent misinterpretation. Finally, Sect. 3.10 summarizes this chapter.

3.1 Overview

One of the simplest mode examples is a light system with the modes "ON" and "OFF", and the transition between them is highlighted in Fig. 3.1. In the *Mode OFF*, the light bulb is not emitting any light because the electrical power supply is disconnected. If the user flips a switch or presses a button, the electrical power to the light bulb is restored, which turns the light on. So, the light system transitions from the mode "OFF" to "ON". To switch the system mode from "ON" to "OFF", the user has to flip a switch again or release a button. Then the electrical power to the light bulb is interrupted, which turns the light off. The user can observe the system behavior by looking at the light bulb.

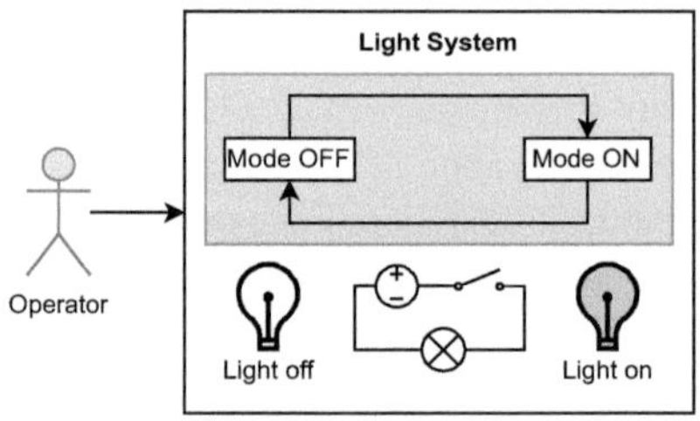

Fig. 3.1 Light System Modes

Industrial and medical IoT devices consist of hardware and software interacting with physical components like sensors, actuators, and other local and/or remote systems. In hazardous conditions, for safety or security reasons, and other circumstances, they must quickly adapt to their environment and behave differently to normal. Additionally, human operators need to monitor, command, and control these devices. To solve such complex problems in computer science, the divide-and-conquer paradigm breaks down problems into smaller pieces, that are easier to solve. Then, the solutions of the pieces are combined to form an overall solution. Likewise, *Wasson* mentions in his book on *System Engineering Analysis, Design, and Development* the important principles of complexity reduction and decomposition [Was15]. In this context, phases, modes, and states of operation are used to increase clarity, manage risks, and reduce complexity [Was15]. According to *Monk*, "almost any system of realistic complexity will have modes of some kind" [Mon86, p. 1]. A common example in the literature is the modes of airplanes [HP87, TBW92, MR98, HHK01, RC04, FR07, Sch07b, YH13, BVEN17]. When a plane flies from A to B, a sequence of modes becomes active, see Fig. 3.2: parking, taxiing, take-off, manual and automatic cruising flight, landing, taxiing, and finally parking.

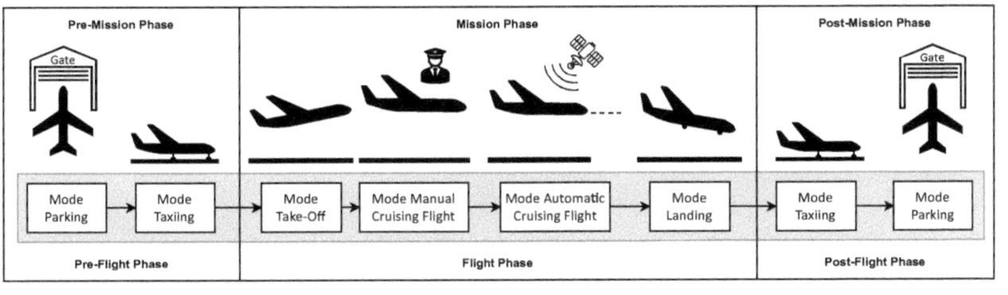

Fig. 3.2 Airplane Modes with Phases. (Adapted from [Was15, p. 169])

Compared to a monolithic implementation, modes support systems with diversified functionalities and make them more re-configurable and adaptive [MMU+16]. *Wasson* splits the airplane modes into pre-flight, flight, and post-flight phases [Was05]. The pre-flight phase includes all modes for ground operations before departure, e.g., load and taxi. The flight phase includes the modes for take-off, cruise, and land. After ground arrival the post-flight phase includes the modes for taxi, park, unload, and maintenance.

However, the reality of airplane operations is more complex and encompasses a broader array of potential mode switches. The schematic representation in Fig. 3.3 illustrates various airplane modes. The flight cycle initiates in the mode parking: the plane parks on the ground, situated at a hangar, gate, or parking area. The switch to the mode taxiing occurs after powering up the plane, conducting system checks, and loading passengers and cargo. During this mode, the plane moves toward the runway, with the pilot communicating with the control tower for guidance. In the event of adverse weather or other issues, the departure may be postponed, prompting a return to the mode parking. If all systems are OK and the pilot receives clearance from the control tower, the plane switches to the mode take-off. It accelerates on the runway until reaching a speed that facilitates lifting off the ground.

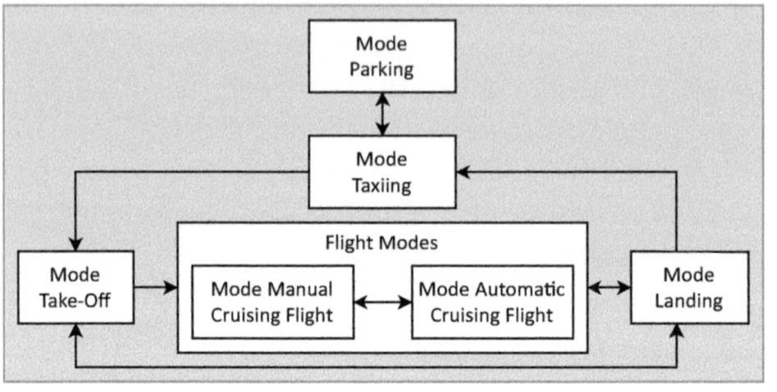

Fig. 3.3 Schematic Representation of the Airplane Modes

When there are engine issues or other failures during take-off, the pilot may opt to abort and switch back to the landing mode. Otherwise, the plane ascends to a specified altitude, and the pilot switches between various flight modes. Depending on the circumstances and regulations, the pilot may engage the mode automatic cruising flight. In this mode, the autopilot maintains predefined settings, including altitude, speed, and direction, with support of satellite systems. In the event of autopilot issues or at the pilot's decision, a switch to the mode manual cruising flight may occur. As the plane nears its destination, the pilot switches to the mode landing, initiates a descent, and prepares for landing, such as lowering the landing gear. The pilot communicates with the control tower to secure landing clearance, aligns the plane with the runway, and descends for landing. After the touchdown, the pilot guides the plane along the runway. Once the plane decelerates sufficiently, it switches to the mode taxiing, and the pilot navigates it to a gate or parking area. Upon arrival, the plane switches to mode parking, the engines are shut down, and ground services may unload passengers and cargo before commencing a new cycle.

When the plane becomes unstable during the mode landing, encounters problems with the landing gear, or experiences other issues related to the plane or the runway, the pilot may execute a go-around, a maneuver involving the aborted landing, and a switch back to various flight modes. Another mode switch commonly occurs during pilot training, known as a touch-and-go landing. During this maneuver, the plane briefly touches the ground, the pilot accelerates the plane and switches back to the mode take-off.

According to *Ryan & Olver*, the concept of phases, modes, and states is also used "to describe requirements that exist outside the normal operating environment" [RO14, p. 2]. Eventually, a failure, emergency, or safe mode exits if the airplane does not act as expected. Switching to any of these modes reduces the risks of catastrophic consequences and provides fault tolerance. Every mode has a set of tasks, and there exist protocols to switch from one mode to another [PL11, CP18]. While some functionalities will be allowed only in specific modes, e.g., cruising flight, some functionalities will also be prohibited, e.g., deployment of the thrust reverser during flight, as it may lead to a crash. Additionally, mode switches are not possible and useful between all modes, e.g., switching from the mode parking to mode landing. For *Paynter* modes are an intuitive way of understanding and naming system functionality and behavior in many real-time systems [Pay96]. Each mode has different goals as well as its unique behavior, and a set of functionalities characterizes it. Such a multi-modal architecture makes a system more flexible to respond quickly to external events. Based on changes in the environment, a mode switch changes the system behavior, e.g., from take-off to cruising or from cruising to landing. The concept of modes has also been motivated by faster development and extensibility, which allows the addition of new system behavior and design and testing in parallel [YH18]. Additionally, modes and states were used to specify functional and performance requirements of systems [Edw03]. For example, during the cruising flight mode, the airplane has stricter performance requirements and needs multiple sensor values than in the maintenance or testing mode, while some functions are only needed in the maintenance mode.

Additionally to the normal mode *Real & Crespo* considered the following modes: initialization, check, emergency, alarm, data dumping, low power, high reliability, and fault recovery [RC04]. For more examples and applications areas, see Sect. 3.2.

3.2 Examples & Application Areas

The idea of *multiple modes* and *mode-switching* is well known in domains like aviation, automobiles, and power supply. This section provides some examples in different application areas, such as aircraft, spacecraft, consumer electronics, operating systems, defense systems, medical devices, motor vehicles, watercraft, trains, robots, power supply, and standards.

Aircrafts & Spacecrafts

Airplanes are a common example where modes are used. Typically, they have a parking mode, a taxiing mode, a take-off mode, manual and automatic flying mode, a landing mode, and perhaps an emergency mode. Some functionalities may only be available in a specific mode, e.g., the navigation subsystem only in flight mode and the wheel control only in taxiing mode [YH13]. *Alspaugh et al.* used modes to describe the alignment, navigation, navigation update, weapon delivery, and test of U.S. Navy's A-7 aircraft [AFB+92]. *Wray* analyzed modes and states to describe space avionic behavior and distinguished between mission and vehicle modes [Wra93]. Mission modes refer to the high-level operational behaviors when interacting with external interfaces, like the tracking and data relay satellite system or the mission control center. Example modes are pre-launch, launch, boost, orbit-coast (space mobile operations, vehicle space proximity operations), deorbit, landing, and post-landing (shutdown, surface operations, etc.). Vehicle modes contain the operational behaviors within the internal interfaces, enabling astronauts and the vehicle's software to utilize its capabilities for specific missions. Examples are off, assembly, initialization, standby, integrated system test, normal operation, training, degraded operation, maintenance, and shutdown.

Degani & Kirlik analyzed modes in the human-automation interaction of pilots with airplanes [DK95]. For example, the vertical navigation mode is active during climbing, so the flight-management computer defines the speed. After reaching the specified height, the altitude hold mode is engaged automatically, and the altitude is maintained. Related to the aircraft automatic pilot, *Micouin* mentions that a system can be divided into exclusive modes based on different purposes [Mic14]. Examples are the heading holding mode, the altitude holding mode, and others. Schoeberl analyzed the starting and stopping of real-time tasks in the development of safety-critical Java applications [Sch07b]. He used mission modes for an airplane example to provide different functionality without a restart.

Consumer Electronics & Operating Systems

Multiple modes are also known from operating systems like Windows and Linux, where a kernel mode, a root mode, and a user mode are present. The latter has limited access to hardware like the CPU and the memory. If something goes wrong, the safe or rescue mode provides limited functionality to solve the problems. In wireless networks, infrastructure and ad-hoc modes have different functionalities like routing and point-to-point connections. Smartphones and other portable devices typically have an airplane mode, which deactivates signal transmission of cellular telephony, Bluetooth, and Wi-Fi to prevent interference during a flight. Additionally, a power-saving mode is automatically turned on when a device's battery is running low, restricting device functionality but increasing the runtime. The smartphone manufacturer Samsung went one step further, providing modes and routines based on current activity or location like sleep, driving, exercise, relax, work, writing, and other modes [Sha23]. For example, when the phone connects to the car via Bluetooth, the phone switches automatically to the driving mode, which in turn activates the do not disturb functionality [Sha23].

The drone manufacturer DJI provides modes for several functionalities of their drones. The remote controller has cine, normal, and sport modes, which enables quickly switching the allowed flight speed, stabilization, and obstacle sensing [DJI23a]. The flight modes provide a mechanism to switch between different drone behaviors: follow me, course lock, waypoints, home lock, and point of interest [DJI23b]. For additional data privacy and security, enterprise users can enable the local data mode, which restricts online data processing, prevents any connection to the Internet, and works completely offline [DJI20].

Defense Systems

Already in the 1980s, the *U.S. Department of Defense (DoD)* considers four security modes for its *Automated Information System* [DoD88]: dedicated, system high, multilevel, and partitioned. These operation modes reflect the restrictions on the user security clearance levels, formal access requirements, need-to-know requirements, and the range of sensitive information permitted:

- In the *dedicated mode*, all users have clearance or authorization and need-to-know.
- In the *system high mode*, all users have clearance or authorization, but not necessarily a need-to-know.
- In the *multilevel mode*, two or more classification levels are allowed, even if not all users have clearance or formal access approval.
- In the *partitioned mode*, all users have the clearance, but not necessarily formal access approval and need-to-know.

But modes also are used to structure the behavior of complex real-time systems in the military domain, like missile defense systems. These systems have modes for searching targets, launching the missile, and guiding the missile to the target [Pay96]. Other mode examples of the DoD system/subsystem specification are idle, ready, active, post-use analysis, training, degraded, emergency, backup, wartime, and peacetime [DoD00]. In their *Unmanned Systems Safety Guide for Acquisition*, the DoD use "modes to identify operational segments within the system mission" [DoD07, p. 40]. This includes modes for "training, maintenance, battlefield weapons-loaded embedded training, initialization, start-up, stop, test, and normal operations" [DoD07, p. 18]. For unmanned ground vehicles, they consider the following control modes [DoD07, pp. 43], see Fig. 3.4:

- *Mode Fully Autonomous*: This mode can accomplish a defined mission without any human intervention.
- *Mode Semi-Autonomous*: In this mode, the human operator, the vehicle itself, or both control the mission.
- *Mode Teleoperation*: Based on video and/or other sensors, the human operator guides the vehicle wired or wireless in this mode.
- *Mode Remote Control*: In this mode, the human operator controls the vehicle without video or other sensors.

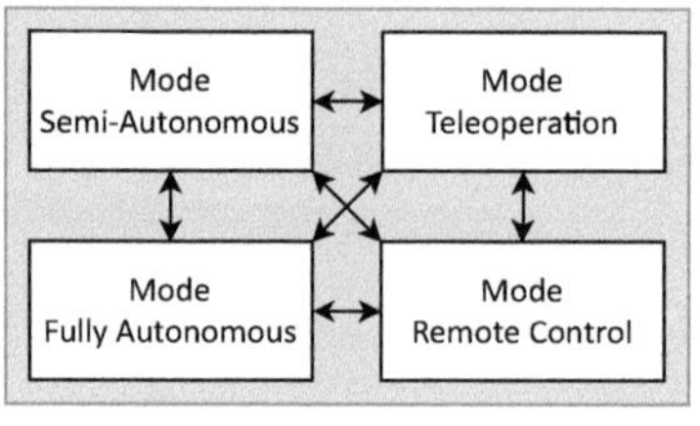

Fig. 3.4 Modes of Unmanned Ground Vehicles

Medical Devices

Medical devices like pacemakers need real-time adaptivity and efficient usage of resources. For example, in an emergency (safety) situation, the device should switch from the normal resource-efficient mode to the emergency mode within seconds. *Yin & Hansson* presented a healthcare monitoring system with multiple modes [YH18]. In case of an emergency like a heart attack, the system triggers an alarm and switches to the attention mode to increase the audio and video transfer to the HDO. In 2018, the BSI has introduced modes for medical operation, device configuration, and technical maintenance and evaluated the patient risk associated with a specific vulnerability with regard to modes [BSI18]. *Rao et al.* propose a trustworthy multi-mode framework for life-critical systems [RRLS18]. With real-time risk assessment and a multi-modal design, they ensure trustworthiness. Similar to our approach, *Easttom & Mei* propose a software shim with a normal and an emergency mode to protect implanted medical devices [EM19]. If an anomaly is detected, the device switches to the emergency mode and only allows data synchronization.

Motor Vehicles, Watercrafts, Trains & Robotics

Multi-Mode System are also used to manage complexity and divide systems into operation modes in the automotive domain. Simple examples are the user-selectable car modes: park, neutral, reverse, and forward [Was14]. Additionally, cars can have normal, sport, and economic modes with different behavior and fuel usage [LB22]. AUTOSAR, the open and standardized software architecture for automotive electronic control units, uses modes for a certain set of states [AUT17]. Example modes for the motor control unit are start, drive, and stop [Gem08]. Another example is self-driving cars, which use a manual mode, an adaptive cruise control mode, a lane-keeping assistant mode, an emergency braking mode, or a parking mode. Most vehicles also have a self-protection feature called limp mode. This mode is activated when abnormal behavior is detected to protect the engine and transmission from major damage. The car's speed and performance are limited, but it can still be used to drive to a nearby service station or a safe location at the roadside to wait for assistance.

Beggs & Tucker used modes to decompose warning zones for autonomous boats and the physical behavior of autonomous racing cars between the start and end mode of a race track [BT18]. *Baduel* applied the concept of states and modes for railway systems and identified modes as the link between train capabilities and train conditions from different perspectives [BBOD18, Bad19]. Examples are modes for non-operation, manual operation, and automatic operation.

Within the Robot Operating System, modes were used to reduce complexity, abstract runtime state information, and for system reconfiguration [NLR21]. For example, if the robot's battery level is low, the system switches from normal to energy-saving mode. For component failures, they suggest a degraded mode to mitigate them. Additionally, a switch from performance to slow mode is possible.

Power Supply

Nuclear power plants also have systems with multiple operating modes like power operation, startup, hot standby, hot shutdown, cold shutdown, and refueling [NRC95]. There are special emergency and accident systems to change modes. If there is a malfunction or parameters exceed or fall below critical values, the reactor is immediately switched off automatically [SFM16]. Such an emergency shutdown, called SCRAM, can also be triggered manually by a kill switch. These emergency stop buttons are used on railways, industrial machines, etc., to abort the current operation mode as quickly as possible in order to prevent damages, injuries, and deaths. A dead man's switch is a similar fail-safe system for vehicles like motorcycles, snowmobiles, and locomotives. It stops or starts a machine if the human operator becomes unconscious or is removed from control.

Modes are also used for secure data transmission and power management, e.g., a power-saving mode to avoid power outages or a mode with lower sampling rates and work offload in case of low battery [AIKR13, STB+19, ARR20].

Standards

In their standard on safety requirements for industrial robots and robot systems R15.06-1999, the *American National Standards Institute (ANSI)* and the *Robotic Industries Association (RIA)* provide guidelines for safe design and protective measures [ANS99]. For example, they suggest a manual and/or automatic safety stop functionality in the automatic mode, to halt the robot and deactivate the power of the actuators. Additionally, they suggest a "slow speed control mode", where robot speed is limited to allow individuals sufficient time to either move away from hazardous motion or stop the robot [ANS99, p. 5]. As an extension, the "teach mode" enables the creation, storage, and replay of positions [ANS99, p. 6].

In CLC/TR 50610:2014, the *European Committee for Electrotechnical Standardisation* provide an overview of train modes and their functional interface specification [CEN14]. A train can be in the modes shutdown, in service, driving, and several sub-modes like parking and emergency.

The *NIST Special Publication 800–160* on *Systems Security Engineering* notes that high-confidence and dedicated-purpose systems like the real-time control of vehicles, industrial or utility processes, weapons, or nuclear may include manual, semi-manual, automated, or autonomous modes [RMCO16]. Failures of such systems may lead to catastrophic consequences. Protection is necessary for all "system execution modes (initialization, operation, maintenance, training, and shutdown), all system states (secure, nonsecure, normal, degraded, and recovery)", and during the transitions [RMCO16, p. 16]. Further operational, contingency, degraded, and alternative modes include: "standby, normal, degraded, reduced capacity, simulation, and test" [RMCO16, p. 140].

According to ISO/IEC/IEEE 29148:2018, software engineers should consider degraded, backup, and emergency modes during the software life cycle [ISO17b]. Examples are "regular, degraded, maintenance, training, emergency, alternate-site, peacetime, wartime, ground-based, flight, active and idle modes" [ISO17b, p. 82].

3.3 History

This section provides a historical overview of the terminology around modes and presents related literature, their citation interconnection, and statistics. Many methodologies and standards describe system behavior using phases, modes, states, configurations, situations, and scenarios. Unfortunately, there is a lack of consistency in the definitions and a controversial discussion. There even exist circular references in the literature. The ELECTRONIC SUPPLEMENTARY MATERIAL provides a comprehensive collection of definitions found in the literature.

Mode-related literature was analyzed, including articles, books, theses, and technical reports from 1965 to 2024. The interconnection of 99 carefully selected contributions is visually represented in Figs. 3.5–3.7, showcasing their referencing relationships. The citation interconnection network is divided into three parts, with highlighted articles in orange connecting to other figures for better presentation and readability. Between 1965 and 1999, a total of 22 articles were identified, followed by the discovery of 28 articles in the 2000s and 42 articles in the 2010s. In the more recent period, spanning 2020 to 2024, a compilation of 7 articles was found. Notably, in 2009 and 2013, there was a concentration of 7 articles. The articles explore various contexts, including security, risks, mode switching, real-time, and critical systems applications. They contribute to the understanding, developing, and analyzing systems incorporating modes, addressing challenges in domains like medical devices, software engineering, real-time systems, and aerospace.

Figure 3.5 covers various aspects of system specification, design, and development, providing insights into strategies, methodologies, and tools for modeling and analyzing system modes and states. Articles in Fig. 3.6 investigate the role of modes in software systems, addressing challenges and proposing techniques across domains like aviation, real-time systems, and *Component-Based Software*

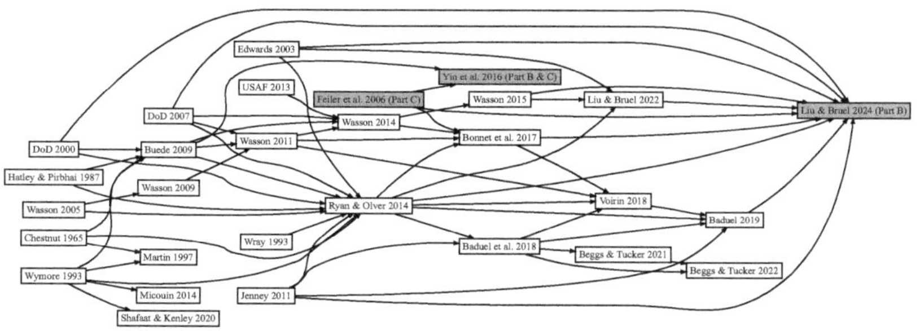

Fig. 3.5 Mode-Related Literature and their Citation Interconnection—Part A

Engineering (CBSE). Common topics include mode-based modeling, formal methods for verification, prevention of mode errors, and considerations for human-machine interaction. Fig. 3.7 presents a wide range of mode-switching topics,

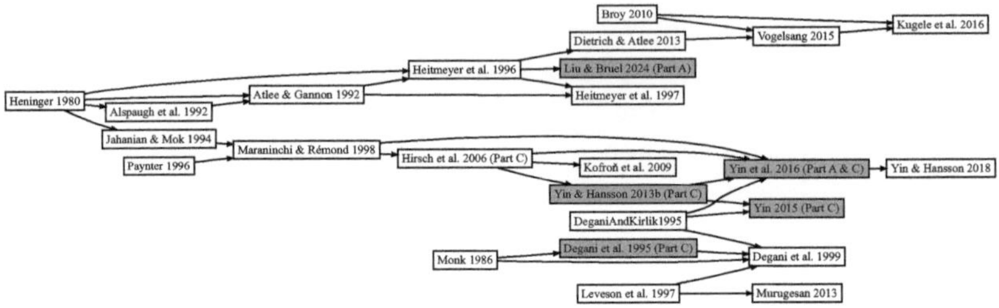

Fig. 3.6 Mode-Related Literature and their Citation Interconnection—Part B

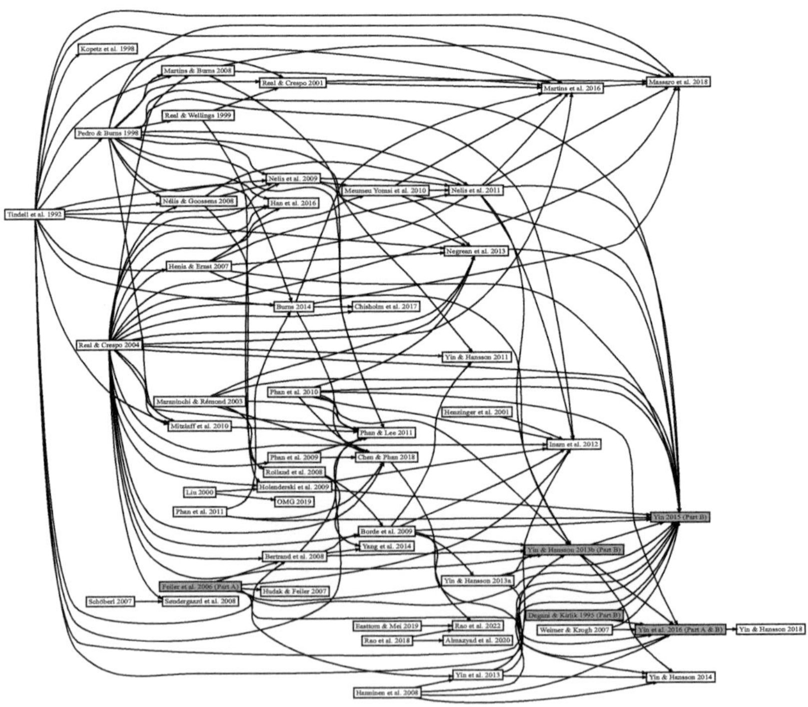

Fig. 3.7 Mode-Related Literature and their Citation Interconnection—Part C

including scheduling, timing analysis, protocols, and methodologies, particularly emphasizing real-time and critical systems. They aid in the development of adaptive systems with different operating modes.

Regarding citations, *Real & Crespo's* survey on *Mode Change Protocols for Real-Time Systems* holds the top spot with a total of 24 citations [RC04]. *Tindell et al.*'s work on *Mode Changes in Priority Pre-emptively Scheduled Systems* closely follows, accumulating 17 citations [TBW92]. In third place with 15 citations are *Pedro & Burns*, who conducted a *Schedulability Analysis for Mode Changes in Flexible Real-Time Systems* [PB98]. When it comes to citing articles, *Yin's* doctoral thesis on introducing mode switching in CBSE cites 19 articles [Yin15], while *Liu & Bruel* reference 11 articles [LB24]. In an extended report of MMSs *Yin et al.* cites ten articles [YHO+16]. *Ryan & Olver* also reference ten articles in the citation interconnection network [RO14]. Particularly worth mentioning are the works of *Wasson. Ryan & Olver*, *Bonnet et al.*, *Baduel et al.*, and *Voirin* [Was14, RO14, BVEN17, BBOD18, Voi18b]. They extensively studied the literature and provide taxonomies of phases, modes, and states in software engineering. The following paragraphs provide further insights in chronological order.

1965–1989

In 1965 *Chestnut* mentioned in his book on *System Engineering Tools* modes of operations in the context of tolerances, variations, and disturbances of systems [Che65]. System missions and modes of operation may have different environmental conditions. He also mentions three operational phases in the context of space vehicles to structure the configuration and design: "the launch phase, the space flight phase, and the return phase" [Che65, pp. 48]. Subphases provide more details.

In 1980 *Heninger* and later *Alspaugh et al.* used modes to organize conditions, simplify function descriptions, and specify software requirements of the flight software for the U.S. Navy's A-7 aircraft [Hen80, AFB+92]. Modes describe "equivalence classes of system states" [Hen80, p. 12] and the "states of many conditions" [AFB+92, p. vi]. Instead of lengthy and complicated expressions for all possible condition values, they use modes as a subset of possible system states [AFB+92]. *Alspaugh et al.* argue, that using modes "simplifies the specification and makes state information manageable by separating concerns" [AFB+92, p. vi]. Typically, a system can be in only one mode at a given time [AFB+92]. However, a system can have several subsystems or components that consist of modes as well. In this context, a mode class describes "a set of modes between which transitions may occur" and "completely and uniquely partitions the system state space" [AFB+92, p. vi]. Transitions between modes are triggered by events like changing conditions, or human mode selection.

In their book on strategies for real-time system specification, *Hatley & Pirbhai* use modes in the context of discrete signals like flight phases (takeoff, cruise, land) or valve conditions (open, closed) [HP87]. They mention that "theoreticians enjoy long philosophical discourses on the exact meaning of state" [HP87, p. 150]. *Hatley & Pirbhai* understand and use the term state equivalent to mode and condition [HP87]. Modes "are represented by digital words with a purely symbolic interpretation" [HP87, p. 144]. Typically, control flows switch the operating mode of the system [HP87]. In their requirements model, they consider two system perspectives: "its information processing (functional) behavior, and its control (state) behavior" [HP87, p. 50].

1990–1999

Tindell et al. describe that modes provide a set of functions and analyzed mode changes in priority pre-emptively scheduled systems [TBW92]. By using modes, they seek to control the system behavior and the timing constraints of tasks during the switch from one mode to another. *Atlee & Gannon* used modes for requirements specification and presented case studies for an automobile cruise control system and a water-level monitoring system [AG93]. They define a mode as "a set of system states that share a common property" [AG93, p. 24]. All combinations of the conditions' values form the system state space. To reduce complexity they abstract the system state space with modes, as not all conditions are always important. Conditions and events trigger mode transitions. For example, by turning on the ignition the cruise control switches from the mode "off" to "inactive" [AG93, p. 30]. From this mode, the system can switch back to off, by turning the ignition off, or switch to cruise, when the engine is running, the brake is released, and the cruise control lever is activated.

In his analysis of modes and states for the *U.S. National Aeronautics and Space Administration (NASA)*, *Wray* defines a mode as "a predefined set of hardware and software configurations, and associated procedures used to organize and manage the conditions of operation for an avionics system's behavior, as planned (in advance) or directed by a human" [Wra93, p. 1–3]. Something similar but different, he defines a state as "a set of hardware and software configurations which identify the intervening or instantaneous organization and constraints on the conditions of operation for an avionics subsystem's behavior, resulting from mode commands, as established by the system. Several states may result from one mode command. The existence of unexpected configurations in states indicates potential operating problems" [Wra93, p. 1–4].

Wymore declares in his book on *Model-Based Systems Engineering (MBSE)* that every system has several "modes of behavior", in the smallest form, a failure mode and an operational mode [Wym93, p. 201]. He mathematically defines the behavior functions in each mode and differentiates among trivial system modes, proper system modes, primary system modes, reachable system modes, isolated system modes, transient system modes, and absorbing system modes [Wym93, pp. 208]:

- "Z1 is a *trivial system mode* of the system Z2 if and only if Z1 is a system mode of Z2 and Z1 is a trivial system".
- "Z1 is a *proper system mode* of the system Z2 if and only if Z1 is a system mode of Z2 and $Z1 \neq Z2$".
- "Z1 is a *primary system mode* of the system Z2 if and only if Z1 is a system mode of Z2 with constant time index 1".
- *Reachable system modes* deal with the fact that some modes are only reachable from specific modes or under specific circumstances. For example, the control software of an airplane is unable to switch from cruising flight to parking mode.
- Similarly, *isolated system modes* influence functionality. *Wymore* mentions the AM/FM radio as an example: The radio display and station choice depends on the initial AM/FM selection.
- *Transient modes* are used for checking functionality during the system startup and can not be reached during restart.
- *Absorbing system modes* represent final modes like deactivated, decommissioned or destroyed.

Degani & Kirlik define modes as "the manner of behavior" of a given system, that can have different ways of behaving [DK95, p. 3]. They classify modes according to their functions and the types of transitions. The input, output, states, and time define the behavior. Different interface formats and control behavior are terms of function. The transition from one mode to another can be manual, automatic, or both. In a later work *Degani et al.* distinguish between three different types of modes [DSK99]:

- *Interface modes* allow the user to change the display format, e.g., switch from view to input mode of a text editor.
- *Functional modes* allow to define different behavior, e.g., the gear shift mechanism of a car: neutral, reverse, and drive.
- *Supervisory modes* specify the level of control: manual, semiautomatic, and fully automatic.

Leveson et al. defines a mode as "mutually exclusive set of system behaviors" [LPS+97, p. 134]. They classify modes based on the control component [LPS+97, p. 135]:

- "*Supervisory modes* determine who or what is controlling the component at any time", e.g., humans or other components.
- "*Component operating modes* control the behavior of the control component itself", e.g., the interface or the process-control behavior.
- "*Controlled-system operating modes* specify sets of related behaviors of the controlled systems an are used to indicate its operational state", e.g., moving, working, waiting.

Pedro & Burns used modes in flexible real-time systems to adapt to changing environments [PB98]. They focused on task schedulability during the switch from the old mode to the new mode and introduced a new mode change model.

Maraninchi & Rémond use modes as "collections of execution states" for reactive systems and synchronous languages [MR98, p. 187]. They consider the normal system behavior as a "sequence of modes", switching from one mode to the other [MR98, p. 186].

2000–2009

Real & Crespo analyzed mode change protocols for real-time systems and argued that many systems consist of multiple modes [RC04]. "Each mode produces a different behavior, characterized by a set of functionalities that are carried out by different task sets" [RC04, p. 161]. Sect. 3.7 provides more details on mode change and mode-switching protocols.

For *Wasson* "each mode of operation represents User selectable options available to perform specific mission operations and tasks" [Was05, p. 189]. He separates common mode patterns by phases [Was05, pp. 199]:

- In the *pre-mission phase*, a system may have modes for off, power-up/initialization, configuration, calibration/alignment, training, and deactivated.
- The *mission phase* includes the normal and abnormal modes.
- In the *post-mission phase*, a system may have a safe mode, an analysis mode, and a maintenance mode.

Each phase is "a label applied to segments of the System/Product Life Cycle or Mission Life Cycle of a System or Product" [Was15, p. 148]. In a later work, he defines a mode as "an abstract label applied to a user selectable option that enables a set of use case-based system capabilities to be employed in conjunction with organizational processes and procedures to Command and Control (C2) a system, product, or service to achieve a specified set of mission objectives, outcomes, and levels of performance" [Was14, p. 4]. In difference to modes, *Wasson* defines states as observable, measurable, and verifiable "physical attributes used to characterize the current configuration, status, or performance-based condition of a System or Entity" [Was15, p. 148]. He differentiate states based on the context [Was14, p. 4]:

- *System states* for the "logistical employment, availability, or status of an organizational asset", e.g., in storage, in operation, and in service.
- *Operational states* for "the operational readiness condition to conduct a mission", e.g., "active or inactive", "operating or non-operating".
- *Dynamical states* for "the time-dependent rate of change", e.g., "initializing, melting, landing, accelerating".
- *Physical configuration states* for "the physical arrangement", e.g., "cabin doors are open or closed and locked" [Was14, p. 8].

In his book on *The Engineering Design of Systems*, *Buede* describes states and modes as "the highest level functional breakout to define a set of functions that system performs at a point in time, e.g., startup, normal operation, recovery operations, shut down" [Bue09, p.67]. He defines a mode as "a distinct operating capability of the system during which some or all of the system's functions may be performed to a full or limited degree" [Bue09, p. 213]. A state of a system is "a static snapshot of the set of metrics or variables needed to describe fully the system's capabilities to perform the system's functions" [Bue09, p. 214]. As examples for a laptop computer *Buede* mentions the states of the battery power level (high, low), the CPU usage (high, low), and the hard disk space (full, empty). The system's functions or tasks are processes that take inputs, transform them to outputs, and eventually change states [Bue09].

2010–2019

Broy considers modes as a valuable approach to structure service behavior and to specify dependencies between services [Bro10]. "Modes are used to discriminate different forms of operations for a service" [Bro10, p. 1204]. Typically, mode sets comprise only a few elements, such as enumerated types, like the operational modes of a car: "moving forward, stopped, and moving backward" [Bro10, p. 1204].

Jenney defines states as an "an exact operating condition of a system" and modes as "the set of capabilities or functions which are valid for the current operating condition" [Jen11a]. In his TV example, he mentions that in the OFF mode, the TV must be able to sense the power-on signal from the remote control. In the ON mode, the TV should support channel changing, volume adjustment, and power off. However, modes do not provide exact information, e.g., what channel is active.

So modes define what functions (on, off, channel up/down, volume up/down) are supported in each condition (state).

In the *Systems Engineering Primer & Handbook* of the *U.S. Air Force Space and Missile Systems Center*, states and modes have nearly the same definition. A state is defined as "the condition of a system or subsystem when specific modes or capabilities (or functions) are valid" [USA13, p. 125]. Very similar a mode is defined as "the condition of a system or subsystem in a certain state when specific capabilities (or functions) are valid" [USA13, p. 125]. They provide different examples for states: "Off, Start-up, Ready On, Deployed, Stored, and In-Flight" and mode examples "within the ready state: Normal, Emergency, Surge, Degraded, and Reset" [USA13, p. 125]. It emphasizes that performance requirements must be specified once states/modes are used.

According to *Yin & Hansson*, it is "common practice to manage and master software complexity to partition system behaviors into different mutually-exclusive operational modes" [YH18, p. 2]. A system can consist of several subsystems with different modes. As an example they mentioned the navigation subsystem, which may only run in cruising flight mode, and the subsystem for wheel control, which may only run in taxiing mode. The modes of the system and the subsystems can influence each other.

Ryan & Olver provide a taxonomy of phases, modes, and states in systems engineering [RO14]. They discuss the need for phases, modes, and states, and compare definitions and existing methods from 27 different sources. Without states and modes the system specification may be incomplete. They argue that it is necessary to define system requirements to allow recovering from failures. Overall, the use of phases, modes, and states can improve customer and supplier communication.

In the NIST SP 800-160 on *Systems Security Engineering*, *Ross et al.* suggest providing "operational and contingency, degraded, alternative, and other modes of operation" as part of the secure operation strategy [RMCO16, p. 140]. According to them, modes and states support the "principle of secure failure and recovery" by maintaining a secure state during errors and attacks until recovery to normal mode.

According to ISO/IEC/IEEE 12207:2017. operational concepts should be identified and defined during the software life cycle process. These include "high level operational modes or states, operational scenarios, potential use cases, or usage within a proposed business strategy" [ISO17b, p. 58]. Modes should be considered for special conditions or contingency operations. In that context, behavioral models describe interrelated scenarios and comprise internal and external functions and interfaces that define how systems or their components behave in various conditions. The model considerations include execution sequencing, synchronization, concurrency, conditions for behavioral changes, and performance. Also behavioral life cycle elements like modes/states, transitions, event triggers, and operational scenarios are part of the modelling process.

Bonnet et al. analyzed modes, states, configurations, and situations in the context of the model-based engineering method Arcadia and its supporting modeling workbench Capella [BVEN17]. They define modes as "the expected behavior of the system (or of its actors, or of its components) in situations foreseen at design-time is captured in the form of system modes. Each mode is mainly characterized by the expected functional content of the system in this mode" [BVEN17, p. 6]. For them, modes can reflect mission phases, system functionality and conditions like connected or autonomous, and system usage like test, training, or maintenance. The system, its users, and external actors can trigger mode switches by a decision to adapt the system to "new needs or new contexts" [BVEN17, p. 6].

In contrast to that, a state "reflects an operating condition or status on structural elements of the system: operational, failed, degraded, absent, etc.", "the physical condition of a component (full or empty fuel tank, charged or discharged battery, etc.)", and "environment constraints (temperature, humidity, etc.)" [BVEN17, p. 6]. The transition between states is typically based on changed physical properties. *Bonnet et al.* use configurations to characterize a system context, which can be a mode, a state, or the combination of both [BVEN17]. Additionally, they introduce situations to reflect simultaneously active modes and states in a system.

In the *NASA Systems Engineering Handbook* states and modes are part of the logical decomposition process during the system design [NAS17]. For the concept of operations, modes and configurations are mentioned to accomplish goals, or are needed for development, testing, training, operating, and decommissioning.

According to *Chen & Phan*, each mode corresponds to a specific configuration or system behavior and "can be characterized by a unique set of tasks" [CP18, p. 105]. These tasks are active when the system is in this mode. They analyzed the design and evaluation of mode-change protocols with a self-driving car example [CP18].

Baduel et al. compared state-of-the-art definitions and use states and modes as general concepts for system design and validation in the context of the train industry [BBOD18, Bad19]. They define a mode as "a characterization of a set of capabilities of a thing under a set of invariant conditions" [BBOD18, p. 6]. These conditions are based on state values, that represent "an information about a thing at a given time inside a context" [BBOD18, p. 5]. *Baduel* differentiates between operational modes and operation modes [Bad19]. While operational modes are expressed from the user perspective, he considers that "operation modes characterize the train capabilities linked to a situation from its own perspective" [Bad19, p. 43].

Voirin presents the concepts of states and modes in his book *Model-Based System and Architecture Engineering with the Arcadia Method* [Voi18b]. He defines a mode as "a behavior expected of the system, a component or also an actor or operational entity, in some chosen conditions" [Voi18b, p. 305]. Compared to that, he defines a state as "a behavior undergone by the system, a component, an actor or an operational entity, in some conditions imposed by the environment" [Voi18b, p. 306]. Both modes and states are characterized by configurations, which specify available functions and components [Voi18b, Voi18a].

Beggs & Tucker decompose and partition complex physical system behavior with modes [BT18]. They suggest using modes for specifying "different physical characteristics, models of behaviour, exceptional circumstances, and assessments of priorities and decisions" [BT18, p. 2]. Thereby complexity should be reduced to ease understanding. Pre- and post-conditions help to know when and how to switch between modes.

The *Object Management Group (OMG)* mention in their UML specification only state machines for modeling behavior [OMG17]. Modes are only referred to in the context of expansion regions and how they proceed: in parallel, iterative, or as a stream. However, in their UML profile for *Modeling and Analysis of Real-Time and Embedded Systems (MARTE)*, operational modes are used for "modal behavior" and provide a mapping to AADL [OMG19, p. 24]. These operational modes can encompass various meanings [OMG19, p. 24]:

- *Operational system (or subsystem) states* with reconfiguration mechanisms, e.g., for fault-tolerance.
- *System operation states* with different QoS levels.
- *System operation phases*, "e.g., starting, stopping, launching, in a mission-critical aerospace system".

"A mode identifies an operational segment within the system execution that is characterized by a given configuration. The system configuration may be defined by a set of active system elements (e.g., application components, platform components, hardware resources), and/or by a set of operation parameters (e.g., QoS parameters or functional parameters)" [OMG19, p. 24]. Only one mode can be active at a time, and events can trigger mode switches.

2020–2024

Shafaat & Kenley examined the use of modes and states from systems engineering in project management and design [SK20]. They suggest a systematic framework for project implementation, verification, and validation and build up on the modal systems notation by *Wymore* [Wym93].

Beggs & Tucker consider a mode as "a high-level categorisation of the operation and behaviour of a system" [BT21, p. 6]. In their work on modes and mode transitions present seven principles that characterize modes [BT22, pp. 3-4]:

- *Completeness*: System modes classify "the operation or behaviour of a system". A system is in one or more modes at any given moment.
- *Localisation*: Each system mode selects and gathers its data "to monitor its behaviour and environment" and to compute "mode's state space and data type".
- *Combination*: When several modes are active in parallel, this combination represents a mode as well.
- *Component*: A system's set of modes comprises a collection of basic and joint modes.
- *Quantification*: It is required to quantify and evaluate the applicability of each mode if a system state has relevance for many modes.
- *Thresholds*: Mode switches are triggered by specified numerical thresholds.
- *Visualisations*: Modes and thresholds must be visualized to support the selection and decisions for mode switches.

Likewise to *Wasson* [Was15], *Liu & Bruel* argue that modes represent system capabilities and states system boundaries [LB22]. While modes affect system behavior, conditions must be met to change states. They propose a DSL for requirements modeling and verification of modes and states [LB22]. In 2024, *Liu & Bruel* extended their work with an Eclipse-based framework for model transformation, symbolic model checking, and validation [LB24].

3.4 Scenarios, Configurations, and Situations

Very similar to the former definitions of operating modes, *Henia & Ernst* call them *scenarios*: "Each scenario is characterized by a different behavior and is associated with a specific set of tasks" [HE07, p. 171]. In their *Systems Modeling Language (SysML)* glossary, the *Object Management Group (OMG)* specify a scenario as "a specific sequence of actions that illustrates behaviors. A scenario may be used to illustrate an interaction or the execution of a use case instance" [OMG06]. *Yin & Hansson* also use the term scenario in the context of modes [YH13]. However, they focus on handling multiple concurrent mode switch scenarios.

To characterize a system context (modes, states, or combinations) *Bonnet et al.* use the term *configuration* to identify active elements like functions and components [BVEN17]. The *NASA* use that term interchangeably with modes during the system life cycle, from development, testing, and operation to decommissioning [NAS17]. *Voirin* used the term *situation* to specify the combination of modes and states as a superposition at a point of time [Voi18b, Voi18a]. He considers scenarios to describe transitions from one situation to another [Voi18a]. The ELECTRONIC SUPPLEMENTARY MATERIAL provide a comprehensive collection of definitions found in the literature. Even though different terms exist, this work focuses on the term "modes".

3.5 Relationship between Modes and States

According to *Liu & Bruel*, “the relationship between states (status of a system) and modes (capabilities of a system) used to describe system requirements is often poorly defined” [LB22, p. 190]. While some authors do not distinguish between the terms, they use states and modes to model natural language requirements and argue that modes directly impact system behavior, while states change based on whether specific conditions are met.

The handbook of the *International Council on Systems Engineering (INCOSE)* mentions states and modes as related concepts to define, model, and manage system functional architectures and behaviors [INC23]. While states represent snapshots of system variables, modes were used to transform systems to new settings.

According to the *U.S. Military Standard Software Development and Documentation (MIL-STD-498B)* and *Data Item Description DI-IPSC-81431A for System/Subsystem Specification*, “the distinction between states and modes is arbitrary” [DoD00, p. 3]. They consider systems with “states only, modes only, states within modes, modes within states, or any other scheme that is useful” [DoD00, p. 3]. However, if modes or states are used, each requirement shall be correlated to them. *Dutcher* mentions that “both states and modes are containers that define control of the functions of the system” [Dut15, p. 3].

Voirin argues “that a states’ machine does not contain modes and vice versa: states and modes are defined separately from one another and potentially cohabit, but always in separate machines” [Voi18a, p. 214]. Nevertheless, several authors like *Edwards*, *Wasson*, and *Ryan & Olver* analyzed the relationship of modes to states [Edw03, Was14, RO14]. For example, *Ryan & Olver* provide an overview of existing methods to present phases, modes, and states and compare the definitions [RO14]. They group the relationship into two categories: states within modes and modes within states, which are examined in more detail in the following subsections.

States within Modes

Many authors consider states within modes [Wra93, Wym93, Pay96, Was05, Bue09, DoD07, Was14, RO14, SK20] but also modes within modes and states within states are common.

Wymore considers modes as high-level system functions with a set of states and differentiates between system mode categories, see Sect. 3.3 [Wym93]. *Wray* used modes and states as interfaces for space vehicles and described their relationship [Wra93]. While modes are directly planned and controlled activities by humans, states represent the resulting machine condition. Mission modes can influence multiple vehicle modes (*modes within modes*), and each vehicle mode can influence multiple subsystem states (*states within modes*). States represent the internal behaviors of flight vehicles or supporting facilities. Likewise, subsystem states influence function states (*states within states*), which are related to configurations of major functions or lower-level subsystems. [Wra93]

Paynter analyzed the relationship between modes and states and argued that mode-machines differ from state-machines in important aspects, "most notably in that the modes of a mode-machine will not necessarily be exhaustive and mutually exclusive, unlike the states of a state-machine" [Pay96, p. 91]. Modes are defined as "collection of states", which are active during a mode [Pay96, p. 92]. Likewise, *Maraninchi & Remond* relate the abstract notion of mode to collections of execution states [MR98].

Wasson considers that each system phase consists of "embedded modes of operation", representing user-selectable options [Was05, p. 189]. He also mentions sub-modes to split modes for desired functionality. States of operation, which are "operational or operating conditions" of a system like the wing flap positions of an aircraft, support these modes and sub-modes [Was05, p. 189]. In a later work, *Wasson* argues that modes contain states and states contain modes [Was14]. As mentioned in Sect. 3.3, he differentiates between states based on the context. During the life cycle, he considers system states like storage, operation, and maintenance. Within the system states, *Wasson* presents vehicle modes of operations like park, neutral, reverse, and forward. Depending on the mode, there are allowed and prohibited actions. For example, moving backward during the mode forward is prohibited. A mode switch, triggered by the user, re-configures the physical configuration state of the vehicle, e.g., the gear train to allow moving backward.

The DoD determine modes within the "system mission", which can be compared to a phase [DoD07, p. 40]. System or subsystem conditions like armed weapons are identified by states and used in various modes. *Ryan & Olver* decided to define a deliberate relationship between the terms and consider that "phases contain modes, and modes contain states" [RO14, p. 12]. For example, they consider that during the cruising phase of an airplane, the landing gear subsystem is not in the mode engaged, meaning the state of the landing gear is stowed. *Shafaat & Kenley* follow the work of *Wymore* but see modes as "time-indexed systems" [SK20, p. 167]. They consider that a system is transitioning through modes from one state to another.

Modes within States

The *U.S. Air Force* claims that the inclusion of modes within states is the most common relationship [USA13]. Similar other authors consider modes within states [Mar97, Edw03, Dut15].

Martin views modes as the "condition of a system or subsystem in a certain state when specific capabilities (or functions) are valid" and mentions the normal, emergency, surge, degraded, and reset mode during the ready state [Mar97, p. 207]. *Edwards* considers states as the "overall condition of a system or a subsystem" and modes as "the way in which functions are performed" [Edw03, p. 13]. Thereby, modes were defined as within states. For example, during the system state 'on' the diagnostic and normal mode allow specific functions. *Dutcher* analyzed the linkage between states, modes, and performance requirements. He mentions that mode switches happen from the system inside while state changes happen externally and regards modes within states [Dut15].

3.6 Mode Modeling & Specification

Several programming and specification languages and models have been developed to describe and visualize system behavior. *Heninger* used *mode transition lists* and *charts*, *condition tables*, and *event tables* for organizing modes, conditions, and their transitions [Hen80]. The mode transition list shows events which cause mode transitions from one mode to another. Each row of the condition table presents one or a group of active modes and "mutually exclusive conditions" [Hen80, p. 9]. Event tables combine events, modes, and actions that are executed when specific events are triggered, and the system is in a specific mode [Hen80].

Similar to *Heningers'* mode transition list, *Hatley & Pirbhai* used *decision tables* with modes to arrange input and output signal combinations [HP87]. Based on their work, Table 3.1 shows an example of a heating system with two modes: idle and auto [HP87, p. 148]. Certain input signal combinations represent "events" and can trigger "actions", such as output signals and state changes [HP87, p. 151]. Depending on the selected mode, temperature, pressure, and the states of the heater and pump are changed.

Table 3.1 Decision Table for a Heating System based on *Hatley & Pirbhai*. [HP87, p. 148]

Inputs			Outputs	
Mode	*Temperature*	*Pressure*	*Heater*	*Pump*
Idle	High	High	Off	Off
		Low	Off	Off
	Low	High	Off	Off
		Low	Off	Off
Auto	High	High	Off	Off
		Low	Off	On
	Low	High	On	Off
		Low	On	On

Alspaugh et al. mention that modes can be represented with "a *finite state machine* for which events are "input" and modes are machine states" [AFB+92, p. vi]. Similarly, *state machines* are used in SysML for modeling discrete behavior [OMG06]. While finite state machines present formal models, *state diagrams* visually represent system behavior.

Harel defines state diagrams as "directed graphs, with nodes denoting states, and arrows (labelled with the triggering events and guarding conditions) denoting transitions" [Har87, p. 232]. For example, Fig. 3.8 shows a simple state diagram of a system with a normal mode and a degraded mode. The degraded mode gets active if there is any failure. When all failures are cleared, the system switches back to the normal mode. *Harel* extended the *state-transition diagrams* with hierarchy, concurrency, and communication elements and introduced *statecharts* to capture the transformational behavior of complex systems. According to *Harel's* definitions, multiple states can be "exclusive, orthogonal or ancestoral" [Har87, p. 265]:

- *Exclusive* means that only one state/mode can be active at the same time, e.g., a light bulb can be on or off, see Fig. 3.1.
- *Orthogonal* means that multiple states/modes can be active simultaneously, e.g., a smart home system's lighting, heating, and cooling.
- *Ancestoral* states, also known as superstates, group multiple states together in a hierarchy, e.g., the home, away, and vacation modes of smart home systems combine lighting, heating, and cooling modes.

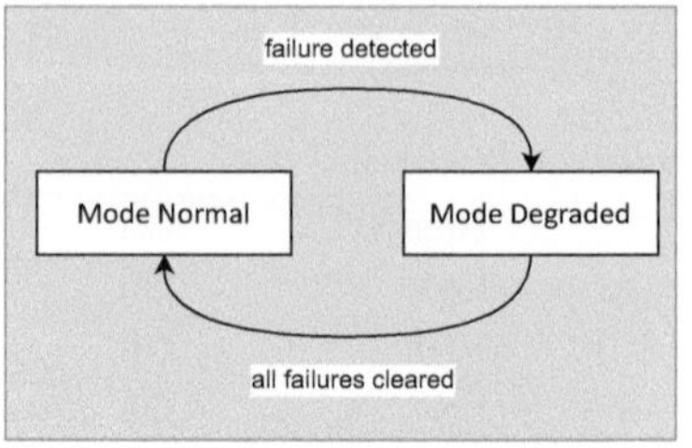

Fig. 3.8 State Diagram of a System With Multiple Modes

Jenney argues that it is necessary to define not only the "in use" mode but also other modes of the systems life cycle [Jen11a]. For that, the *top-level life cycle mode diagram* should include all modes and sub-modes from assembly to end of life and the possible transitions. Then, the allowed sub-mode transitions can be defined in a diagram or a matrix. For verifying functional architecture, the functions to system modes matrix is used [Jen11b]. The matrix helps to check if functions are missing and does the function mode mapping. As an extension *Dutcher* proposes the *state & mode function matrix* [Dut15].

Voirin visualizes sets of modes in so-called *mode machines* and uses the style of state machines [Voi18b]. Fig. 3.9 shows an example of a car with multiple modes. Beginning with the mode off, a driver can turn on the engine for switching to the mode on, and then switch from the parking sub-mode to neutral, forward, backward, and back to park.

In the *UML Profile for MARTE* modal behavior can be described with the "CommonBehavior package" [OMG19, pp. 22]. Fig. 3.10 shows an excerpt of the domain model. "A behavior defines how some system or entity changes over time" [OMG19, p. 23]. During that behavior, actions process inputs and convert them to outputs. Events can trigger a mode transition, which may lead to a changed behavior and a configuration change. The OMG claims that modes can be represented by state machines and state diagrams [OMG19].

Modes and states are also used in programming and specification languages. In *Modechart*, a specification language for real-time systems, modes "represent control information about the behavior of a system" [JM94, p. 937]. The language supports serial and parallel modes and the verification of timing properties [JM94]

In automotive and avionics, the standardized *Architecture Analysis and Design Language (AADL)* presents modes as states, which are integrated into a state machine. In that context, *Feiler & Rugina* distinguish between operational modes and error states [FR07]. "Error states result from occurrences of error events (e.g.,

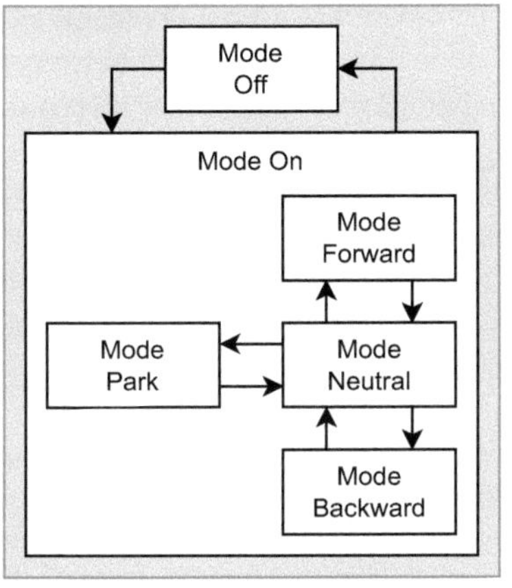

Fig. 3.9 Mode Machine of a Car

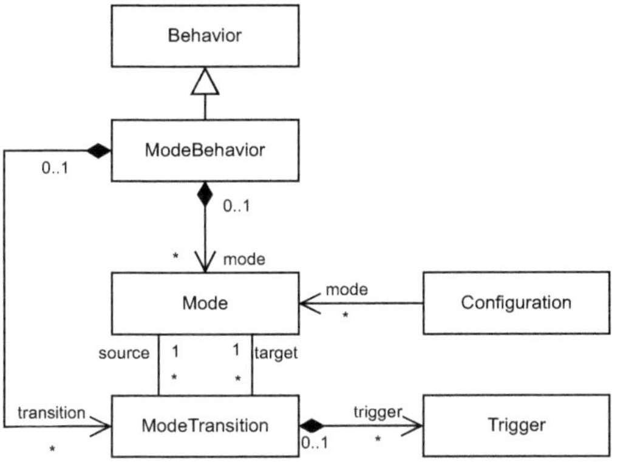

Fig. 3.10 Modal Behavior Excerpt of the UML Profile for MARTE [OMG19]

faults or repair events), while modes represent operational states that may be totally independent of error events" [FR07, p. 48]. They argue that modes are always observable, whereas error states may not be observable.

The time-driven programming language *Giotto* is used for safety-critical applications like fly-by-wire flight control systems with hard real-time constraints and utilizes mode-switching to turn tasks on and off. [HHK01]. The *Timing Definition Language* [RDP20] builds on the concept of Giotto and is used for describing the time behavior of component-based real-time applications. It provides an abstraction for independent execution times of periodic time-triggered computational tasks. The language allows simulations without knowing all physical platform details, like CPU and memory.

Maraninchi & Rémond extended the reactive system programming language *Lustre* with *Mode-Automata* [MR03]. They provide a new construct that supports the direct expression of the mode-structure of complex systems and use modes as "collections of execution states" [MR98, p. 187]. These concrete states are related to the program's memory during execution. A sequence of modes is the normal system behavior.

In the area of service-oriented computing, *Hirsch et al.* propose modes as a new element of architectural descriptions, especially for complex service-oriented systems [HKMU06]. They extend the *Darwin* architectural language to describe and verify complex and adaptable systems in the automotive domain. Thereby, modes help as scenario-based abstraction with service configuration to close the gap between requirements and software architectures.

Another approach from *Tiwari et al.* is to learn modes with machine learning [TDJ+14]. Systems have a specific behavior under normal circumstances. In case of attacks like sensor spoofing or DoS, system behavior changes, and, thereby, abnormality can be detected. This approach is similar to the run-time security detector by *Rao et al.*, which considers run-time changes as security incidents [RRLS18]. Based on expert-defined risk level, the current mode is switched when a certain threshold is reached. Later work on runtime adaptive risk assessment and automated mitigation for life-critical and safety-critical devices was patented [LRS+24]. Through the automatic deactivation of impacted system components, the approach isolates risks.

3.7 Multi-Mode Systems & Mode-Switching

Multi-Mode Systems (MMSs) are used to manage complexity and divide systems into operational modes. These modes change behavior and performance and can be organized hierarchically and by having several sub-modes [MB08]. *Chen & Phan* define a MMS "by a set of modes, the initial mode, and a set of transitions among the modes" [CP18, p. 107]. The functionality of each mode consists of a series of tasks that are active when the system is in that mode. *Tasks* can run periodically or sporadically, some will run independently in the background, and some of them may only be necessary within specific modes.

Burns formalized the description of MMS with six aspects [Bur14]:

- *Type*: Different modes classes of a system, e.g., functional modes, exceptional modes, and degraded modes.
- *Trigger*: What activates a mode switch?
- *Protocol*: How to manage the mode switch?
- *Attributes*: Mode properties like priorities, initial modes, and possible mode switches.
- *Definition*: The software, components and its configurations.
- *Analysis*: The scheduling analysis during a mode switch, e.g., immediately stopping old tasks and starting new tasks or delaying them.

In the literature, a change between modes is often referred to as a "mode switch", "mode transition", or "mode change". In this work, the term mode switch is mainly used. *Real & Crespo* define a mode switch as a "transient stage" between the current operating mode, the old mode, and another mode, the new mode [RC04, p. 162]. Following their work and *Gemuende*, Fig. 3.11 shows a mode switch sequence [Gem08]. A mode switch is triggered by a *timer* or a specific *event* and is initiated by a *request*. For example, smartphones poll mail servers every 15 minutes (timer) and then go into idle or sleeping mode to save battery. When a phone call comes in (event), an active mode takes over from the idle mode. The scheduler has to guarantee task deadlines and prevent an overload of the system [PB98].

New tasks are released during a mode switch, and no-longer-needed tasks are stopped. Depending on the attributes, old-mode tasks can finish their execution or be aborted immediately. For example, the old-mode *task 1* is not used in the new mode and stopped after the mode switch request. A last run of *task 1* was executed because it started simultaneously with the request. In contrast, the old-mode *task 2* is immediately stopped after the mode switch request. *Task 3* needs special actions, such as cleaning up after stopping. Also, special actions like initializing can be needed before starting new tasks. The new-mode *task 4* is started, eventually delayed by a predefined offset. Some modes may execute both old-mode and new-mode tasks like *task 5* but change properties such as the execution frequency. Continuous and mode-independent tasks like *task 6* are not interrupted by the mode switch. For example, monitoring the heart rhythm and pacing on demand will run during all modes of a pacemaker and should not be disturbed by mode-switching. Other tasks, such as sending patient and device status to a remote monitor and receiving configuration changes are only needed in specific modes, e.g., surveillance and configuration. These tasks have a lower priority than vital critical tasks.

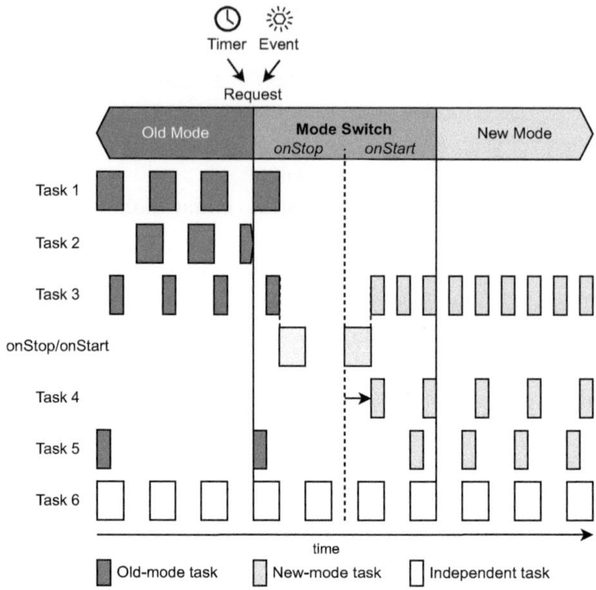

Fig. 3.11 Mode Switch from the Old Mode to the New Mode

Yin & Hansson combine CBSE and MMS and propose a mode switch logic [YH14, Yin15]. Usually, this is contradictory because splitting a system into modes is a top-down approach, and CBSE is a bottom-up approach [YH18]. Nevertheless, they show how it is feasible to combine and integrate them. In the development of Component-Based MMSs, the behavior of the system is divided into several modes with reusable software components to manage complexity. To switch modes quickly in case of an emergency, they propose immediate handling with buffering.

Mode-switching is used in a range of component models:

- Koala [vOvdLKM00]
- SaveCCM [HACT04]
- COMDES-II [KSA07]
- RubusCMv3 [HMTN+08]
- MyCCM-HI [BHP09]

Conditions & Requirements

A mode-switching protocol has to define how a mode switch can be triggered and how this has to be requested. This can be done manually, e.g., by a driver's requesting to take back control of the vehicle from self-driving mode, or automatically by the system itself. Several changing environmental conditions can trigger mode switch events. Thereby a condition is defined as "an attribute of the system at a particular point in time" [Edw03, p. 13]. *Beggs & Tucker* considered also pre- and post-conditions for modes to guarantee behavior [BT18]. For example, the mode giveway has the pre-condition that another vehicle is currently driving through an intersection.

Many IoT devices and real-time systems have limited resources, so it is necessary to prevent resource bottlenecks. Processors are getting faster every year, but the tasks are also growing. Dynamic resource allocation, predictability analysis, and other techniques ensure avoiding bottlenecks [AB01, PL11]. Such mode-switching protocols have to provide quality of service and efficient resource management [AB01, PL11, SYST18]. In this context, a mode switch can be used for energy efficiency. For example, if the batteries of self-driving cars come close to empty, only the core system remains operational. Air conditioning and other convenience features are disabled.

Safety issues may also lead to mode switches. In mixed-critical systems, modes and tasks can have different criticality or safety integrity levels. For example, a car's engine electronics and the emergency braking mode are more critical than the lane-keeping mode and other driver assistance systems. To avoid accidents, vehicles have to adapt their behavior rapidly. The airbag has to deploy quickly in case of a crash to prevent the driver from injuries, e.g., hitting the dashboard. The entertainment system has a lower criticality than airbag deployment, yet all modes coexist in one "mixed critical" system. Mission-critical tasks have a higher priority and must be continued even if errors or system failures occur. Examples of such fault-tolerant designs are proposed by *Burns et al.*, *Capota et al.*, *Chen & Phan*, *Sundar & Easwaran*, and others [BDBB18, CSMC19, CP18, SE19].

Considering *security* risks for a mode switch is quite new [SS17]. A mode switch is necessary if a threat is detected, e.g., with an IDS or IDPS. If a vulnerability becomes known, an operator can manually switch modes to reduce the attack surface. We imagine a middleware as software layer on top of the operating system that provides interoperability, functionality for different software components and applications and last but not least security. Resilience and fault tolerance play a major role here. Especially life-critical systems should not break down by a single error. Possibly, a mode switch to a less risky mode is available, but the system is supposed to continue its mission-critical operation.

Protocols for Single- and Multi-core Platforms

Real & Crespo have examined mode change protocols for single-processor, fixed-priority, preemptively scheduled real-time systems [RC04]. They structured mixed-critical systems in synchronous and asynchronous protocols and classified them based on the requirements for a mode change: schedulability, periodicity, promptness, and consistency.

Real-time systems have to complete tasks within given deadlines. Schedulers have to guarantee these deadlines and, at the same time, have to prevent system overloads. A protocol specifies when new tasks are released and what happens with the unfinished tasks of the old mode. It defines whether new tasks have to wait until the old ones are finished or will run in parallel. It also defines whether the old tasks are completed or stopped immediately. Another requested mode switch during an active mode switch is another challenge. Mechanisms for conflict handling are needed.

Figure 3.12 shows a classification of mode-switching protocols following the work of *Real & Crespo* [RC04]. With *synchronous protocols*, new-mode tasks are only started when the old-mode tasks have been completed, and they are suitable if promptness and periodicity are not critical. *Tindell & Alonso* and *Sondergaard et al.* presented examples of the *Idle Time Protocol*, where old-mode tasks are suspended at the first idle time-instant, and single offset protocols which delay the start of all the new-mode tasks with an offset [SRTS08, TA96]. The *Maximum-Period Offset Protocol* supports periodicity, and the delay of the first activation is equal with the period of the less frequent task in both modes [Bai93]. The *Minimum Single Offset Protocol* finishes the last activation of all old-mode tasks and, after that, releases the new-mode tasks [Rea00]. This protocol is proposed with and without periodicity. In this context, periodicity means the protocol can deal with mode-independent tasks.

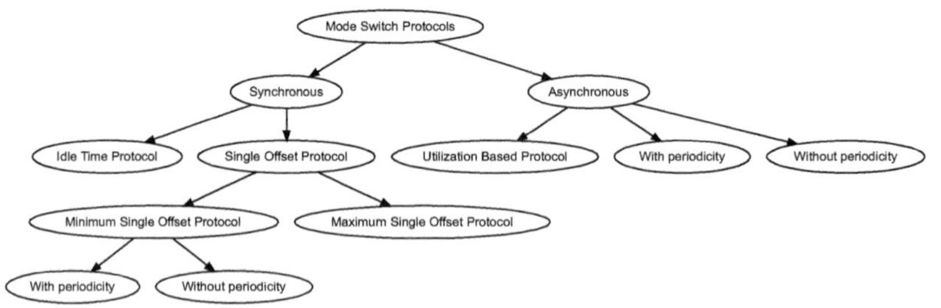

Fig. 3.12 Mode Switch Protocol Classification

Asynchronous protocols provide lower mode change latency but require a special schedulability analysis. These protocols allow the activation new-mode tasks while old-mode tasks are still running and are better suited for promptness. One example is the *Utilization Based Protocol*, which considers periodic tasks and shares processor capacity with a priority ceiling protocol [SRLR89, SG90]. *Tindell et al.* considered the protocol insufficient and proposed a protocol with a priority that lets tasks modify their parameters during mode changes [TBW92]. *Andersson* extended the Utilization Based Protocol for *Fixed-Task-Priority* schedulers to *Earliest-Deadline-First (EDF)* [And08]. Further protocols without periodicity were presented by *Pedro & Burns* [PB98, Ped99].

Scheduling multi-mode applications in heterogeneous multi-processor platforms is more complex than on single-processor platforms and cannot be solved by similar techniques. There are scheduling anomalies upon multiprocessors. Several protocols have been extended to multi-processor platforms. Based on the work of *Real & Crespo*, *Nelis et al.* propose two protocols without periodicity for identical multi-processor platforms and extended it for uniform platforms [RC04, MYNG10, NGA09]: The *Synchronous Multiprocessor Minimal Single Offset* protocol and the *Asynchronous Multiprocessor Minimal Single Offset* protocol. To provide periodicity, the *Synchronous Multiprocessor Maximum Deadline Offset* protocol was developed [NAMP11]. It is similar to the work of *Block et al.*, where tasks are reweighted at runtime, and uses the EDF algorithm for scheduling the modes on identical multiprocessors [BAD08]. For supporting mode-independent tasks, it is necessary to perform a schedulability test on the whole system.

3.8 Failure, Safety, and Security Modes

A common process for safety critical systems like nuclear power plants is the *Failure Mode and Effects Analysis (FMEA)*. According to the *ISO 24765:2017 - Systems and Software Engineering Vocabulary*, *failure modes* are "physical or functional manifestations of a failure", e.g., "slow operation, incorrect outputs, or complete termination of execution" [ISO17a, p. 179]. Reviewers should analyze every system component and identify potential failure modes and their causes [ISO17a]. Also, they check the effect on the overall system during this analytical procedure. After that, actions are planned to avoid or mitigate the impact of failures. In IEC 60812:2018 FMEA was extended to the *Failure Mode Effects and Criticality Analysis (FMECA)*, which includes the probability and severity of the consequences of failure modes.

Avizienis et al. provide a detailed taxonomy of failure modes [ALRL04]. They differentiate between the domain, detectability, consistency, and consequences. While the failure domain includes content and timing failures, the detectability addresses signaled and unsignaled failures. In that context, the authors mention degraded modes, that can still provide some services for the users. Failures can be consistent for all users or inconsistent for only some users. The consequences can range from minor to catastrophic failures. *McNelles et al.* extended that work with a focus on control systems and analyzed the failure modes on different levels of abstraction from basic components, modules, instrumentation and control, divisions, to systems [MZR+17].

In case of errors, faults, or failures of hardware or software, systems should be able to detect that abnormal behavior and switch to a *safe or safety mode*. This so-called fail-safe behavior keeps systems running despite faults. For example, partial or degraded modes or modes for emergency cooling or shutdown to prevent hazards. Safety modes are designed to mitigate failures and prevent harm to people or the environment. While mode-switching has been known in various domains, to our knowledge, it was rarely used in the security domain. This fact has motivated us to conduct a SLR [RS20]. The first step in doing just that is making systems more resilient and engineering systems to continue functioning correctly under malicious attacks. The idea was to utilize multiple modes to increase security and make systems more resilient.

In 1994, *Browne* used modes to organize system behavior and proposed a framework called "Mode Security" for covert channel suppression in computer systems [Bro94]. He considered each mode entirely secure when isolated from all other modes. Similar to the ideas of MTD, every mode switch interrupts covert channels and attackers.

While *Browne* focused on covert channel attacks, we are considering *security modes* to mitigate other attacks too and as a way to reduce the risk in the *Window of Vulnerability (WoV)* (see Sect. exrefsec:WindowOfVulnerability). The WoV opens when a vulnerability becomes known and can only be closed with a patch. Manufacturers control this process, and shortening the WoV is the goal. Currently, users of medical and industrial IoT devices have little influence. For example, medical devices can either be secure or insecure, depending on the existence and severity of vulnerabilities, regardless of their sensitivity, impact, or exposure. To classify devices from a privacy and safety perspective, security scores were introduced [SS17]. A defense-in-depth strategy with security modes is suggested for these devices, enabling a reduction of exposure when vulnerabilities are detected or exploited. Switching modes and, thus, the device's security scores by the manufacturer, HDO or even patients can reduce the attack surface. Thereby, modes can provide resilience during the WoV. If devices have sufficient processing power and power supply, they can also react to their environment, consider the context, and switch to a more secure mode upon detecting potential intrusion. The same is valid for industrial IoT devices.

Context awareness comes in various forms [CK00]. Active context awareness involves adapting the behavior of a device based on the discovered context, such as the device behaving differently in case of an emergency. Passive context awareness involves informing device users about alerts for a specific brand or model. IoT devices can directly contact a server to obtain security context, but resource constraints might make this challenging. Edge devices and home monitoring systems can provide this service and propagate the information to a machine or an implanted device. Customers, like patients, can confirm access to a medical device with a button on the home station. The device or home station can notify the healthcare organization and/or device manufacturer about mode switches. Anomaly detection is the most effective protection through active context awareness, where a device recognizes abnormal behavior and initiates countermeasures, like switching to a more secure mode [CRR22, LSL15, LL19].

The *European Telecommunications Standards Institute (ETSI)* consider in their technical specification on baseline cyber security requirements for consumer IoT that a device will switch between several states during its lifetime [ETS22]: from the factory default to initialized with and without personal data to decommissioned. That supports the idea to switch system behavior with predefined modes (see Chap. exrefchap:SmartModeSwitching). According to *NIST SP 800-160*, systems "must be able to continue to operate in a secure manner, as necessary, in accordance with defined system capabilities and limitations across all identified contingency situations" [RMCO16, p. 142]. Therefore, *Ross et al.* consider systems with "the ability to operate in a by-design degraded or limited capacity secure mode" [RMCO16, p. 15]. Similarly, ISO/IEC/IEEE 15288:2023(E) recommends modes for normal and contingency operations to enhance resilience against cybersecurity threats and attacks [ISO23].

Depending on the balance between security and performance, some security functionality may be reduced or sacrificed for higher critical goals. "Methods such as restart, recovery, reconstitution, reconfiguration, adaptation, or failover" help to switch back to the normal mode [RMCO16, p. 15]. This self-healing process must be implemented carefully to prevent abuse. For example, TPMs increase healing time after each failed attempt to make brute-force attempts more difficult [TCG22].

3.9 Mode Confusion & Errors

The usage of modes supports and simplifies the *Human–Computer Interaction (HCI)*. Modes can increase user's awareness of the current state of the system and its components. Additionally, short mnemonic mode names can make communication about systems easier. For example, *Heninger* considered aircraft pilots more likely to remember the mode in stressful circumstances than various settings of switches and digits on the aircraft's panel [Hen80].

However, misinterpretation or wrong awareness by the user in which mode the system is, can lead to frustration and serious problems, as the following airplane crashes show. These so-called mode errors and mode confusion are important topics in the context of modes [Nor83, DoD07]:

- According to the *Aviation Safety Network*, Indian Airlines Flight 605 crashed into a golf course near the Bangalore airport in 1990, eventually because of the wrong selected idle/open descent mode [Ran90].
- In 1991, an uncommanded thrust reverser deployment possibly caused the crash of Lauda Air Flight 004 in Thailand. After that, the U.S. Federal Aviation Administration issued directives to prevent the activation of that functionality in the flight mode [FAA91].
- Air Inter Flight 148 crashed into the French mountains in 1992, eventually because of confusion between the heading/vertical speed mode and the track/flight path angle mode [Ran92]: After input of 33 for a 3.3° descent angle, which corresponds to 800 feet/min descent rate, the input resulted in a high descent rate of 3,300 ft/min.
- In 2004, Flash Airlines Flight 604 crashed at Sharm el Sheikh, Egypt, eventually based on mode confusion by the captain and first officer [Eva05]: Both believed the autopilot mode was on, but due to unfulfilled pre-conditions, it was not activated.

As explained by *Monk*, there are three critical components for mode errors [Mon86]:

- the action by the user,
- the user's expectation that the action should have on the system,
- and the actual system effect.

Leveson et al. identified six potential design flaws that can lead to mode errors and mode confusion [LPS+97, pp. 140]:

- "Interface Interpretation Errors": If users think the system is in a specific mode, although the system is in another mode.
- "Inconsistent Behavior": If users anticipate a certain behavior, but the system behaves differently.
- "Indirect Mode Changes": The system automatically switches modes without human commands, e.g., triggered by controllers, sensors, or changed values.
- "Operator Authority Limits": On the one hand, lockouts and interlocks prevent hazardous states. On the other hand, these mechanisms may limit the operator's ability to act in emergencies.
- "Unintended Side Effects": Changed values may lead to a mode switch or mode switches change information, which at first glance are not connected.
- "Lack of Appropriate Feedback": If users possess insufficient knowledge to forecast or anticipate mode switches.

Vogelsang et al. investigated the source of defects in natural language requirements specifications of automotive software [VFW16]. In their findings, "46 % of all considered defects address modes" leading to misunderstandings and ambiguities. Therefore, they argue that it is necessary to specify modes explicitly.

To reduce mode errors and confusion, *Monk*, *Leveson et al.*, and others suggest the following design strategies [Mon86, LPS+97]:

- To reduce mode ambiguity, use clearly understandable terms and limit the number of modes to only absolutely necessary ones.
- Increase the user's awareness that a system has modes, signal what mode is active and if there is a mode switch, and train them [RMCO16, p. 140]. During graceful degradation, appropriate feedback is essential. Visual displays, sounds, or other measures that affect sensory perception must signal if the operator's mode switch is successful, ignored, or limited.
- Restrict available functions to control the consequences and provide a fault-tolerant environment with fail-safe or fail-secure mechanisms.

3.10 Summary

Modes have been used in many application areas and systems like airplanes, cars, and power plants. This chapter defined key terms, presented characteristics and principles, showed examples, and provided a comprehensive insight into the literature. Modes describe the specific behavior of systems or components and provide a way to partition required functionality. Multiple modes allow systems to be more flexible and adaptive to changing environmental conditions and different situations. However, there is a long and ongoing controversial discussion about the definitions of modes, states, phases, and other related terms. In some cases, the terms are used interchangeably, while in others, they are strictly separated. There has been a presentation of foundational research and a discussion of the differences and relationships between the terms. Several programming and specification languages and models have been presented on how to describe and visualize modes. Multi-Mode Systems use protocols to manage the different modes and possible switches. Events and timers trigger mode switches that result in a different system behavior. Differentiation can be made between failure, safety, and security modes. Failure modes refer to system malfunctions engineers use to prevent and mitigate them. In contrast, safety and security modes limit functionality with different goals. Safety modes prevent harm caused by systems to the people and environment by limiting critical situations. Security modes focus on system protection by limiting access and enforcing policies to increase resilience. Confusion about modes and unintentionally activated modes have led to serious consequences. In the context of Human–Computer Interaction, clear user interfaces with visual and, eventually, acoustic indicators and user-friendly documentation and training are needed to prevent errors and confusion.

4 Security Modes & Mode-Switching for Resilience

Vulnerabilities can affect critical systems, impacting processes, workflows, privacy, and safety. When a vulnerability becomes known, affected systems are at risk until appropriate updates become available and eventually deployed. This period can last from a few days to several months, during which attackers can develop exploits and take advantage of the vulnerability. It is tedious and time-consuming to keep track of vulnerabilities manually and react appropriately to each of them.

In this chapter, a novel approach is proposed for mode-switching of software systems to support security professionals, system administrators, operators, customers, and manufacturers in effectively dealing with vulnerabilities and reducing the risk of exposure. Model-driven techniques are employed, utilizing a multi-modal architecture to respond to discovered vulnerabilities and provide automated contingency support. A dedicated *Domain-Specific Language (DSL)* has been developed to describe potential mitigation through mode switches.

Parts of this chapter are published in several publication venues, including the *International Workshop on Cyber-Security and Functional Safety in Cyber-Physical Systems (IWCFS)* [RSR23], the *International Conference on Industry 4.0 and Smart Manufacturing (ISM)* [RS21b], the *Journal of Systems and Software (JSS)* [RSVW23a], the final research report to the *Austrian Marshall Plan Foundation* [Rie22], and the *IEEE/ACM International Symposium on Software Engineering for Adaptive and Self-Managing Systems (SEAMS)* [RSV23].

M. Riegler, *Mode-Switching for Resilient Security*,
https://doi.org/10.1007/978-3-658-49306-6_4

After introducing the top-level design problem in Sect. 4.1, the definitions of modes, *Multi- Mode Systems (MMSs)*, and other important terms are provided in Sect. 4.2. Then, Sect. 4.3 presents a motivating example of the mode-switching approach. The multi-modal architecture and the mode-switching framework were developed, extended, and improved in three design iterations. Section 4.4 describes the architecture of the first iteration focusing on single systems. To further explore and to simplify the usage, a framework was developed as detailed in Sect. 4.5. This framework integrates a *Mode Domain-Specific Language (MDSL)* to facilitate the definition of modes. Sections 4.6 and 4.7 adopt a broader perspective and extend the architecture and the framework in the second iteration for multiple systems. In the third iteration, self-protective features and the *MAPE-K* loop were added to the architecture and the framework as presented in Sects. 4.8 and 4.9. Finally, Sect. 4.10 summarizes this chapter.

4.1 Top-Level Design Problem & General Approach

In the context of cybersecurity for medical and industrial IoT devices, the *Window of Vulnerability (WoV)* is critical (see Sect. 2.2). As a consequence of increased connectivity, it is only a matter of time before systems get attacked. Even air-gapped systems are susceptible to attacks. As a prominent example, the computer worm Stuxnet spread via USB flash drives and caused widespread damage [FOMC10, KRT10]. A common method is the use of classic risk management, as mentioned in Chap. 2:

- Avoid mistakes by secure development
- Reduce or mitigate the risk by a defense-in-depth approach with multiple layers of protection
- Share/transfer the risk with a third party or insurance
- Accept the residual risk

However, reality shows that vulnerabilities are commonplace. To mitigate the potential risk of damage, optimizing and speeding up patch development (t2–t4) is important, see the orange arrow in Fig. 4.1. According to statistics, closing a vulnerability takes an average of 77.5 days [Edg19]. Developing and deploying patches within a very short period of time is quite a challenge, even for companies with large development teams. Google's Project Zero reports that 95.8 % of found security bugs are resolved before the 90-day deadline [Cim19]. After this deadline, Google discloses the vulnerability to the public.

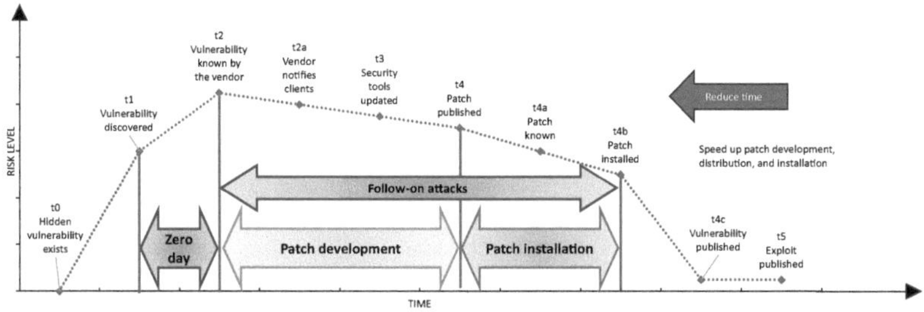

Fig. 4.1 Time Reduction of the Window of Vulnerability

Updating medical and industrial IoT devices can be challenging due to their long lifespan, the need for legacy systems, and the lack of automatic update mechanisms. Patches must be tested extensively due to safety, certification, and warranty requirements. Some industrial and medical devices need to be certified through quality assurance processes, e.g., *European Machinery Directive* and *European Medical Device Regulation (MDR)* [Eur06, Eur17]. However, these certification processes take time. System integrators and asset owners have to wait until product or component suppliers are able to release updates. If integrators and asset owners change something in the hardware or software on their own, the warranty may expire. Some product manufacturers, component suppliers, and software vendors hope that by not disclosing known vulnerabilities, they develop and provide updates timely before exploits become active. It is also essential to consider that vendors may discontinue product support (reaching the end of life) or go bankrupt. In the meantime and after the disclosure, attackers may independently discover vulnerabilities, write exploits, and take advantage.

After providing a patch, attackers can analyze the changes to previous versions, discover the vulnerabilities, and attack unpatched systems. Consequently, patch distribution and installation (t4–t4b) must be accelerated to reduce the time period for possible attacks, as illustrated by the green arrow in Fig. 4.1. The WoV is only closed when the patch is installed (t4b). For example, an unaddressed vulnerability led to a significant data breach affecting 143 million customers of the credit reporting agency Equifax [Whi17]. The time problem is even exacerbated for industrial and medical IoT devices since availability is one of the top priorities. Stopping a machine or a pacemaker is the worst-case scenario, as it typically means losing money or leading to a life-threatening situation for patients. In such systems, patches typically cannot be applied automatically. Before installing patches, they must be tested extensively, and it is essential to create a system backup [ENI13]. For that, development environments or redundant systems may be necessary. Due to the failure rate of patches, some customers may wait to apply them because they fear interruptions and other problems, e.g., former update crashes affected millions of users [ENI13, Kel20]. Patches may unintentionally change the system and affect process stability. There is always a potential for a complete loss of a device's functionality after an update. After applying the patch, a re-certification of the system might be required [ENI13].

We propose another method of reducing the WoV highlighted in Fig. 4.2. It is still necessary to speed up the whole process of developing and distributing patches (x-axis). In addition, it is important to invest more time in the initial system design and development process to reduce the risk level (y-axis) if vulnerabilities are discovered, published, and exploited.

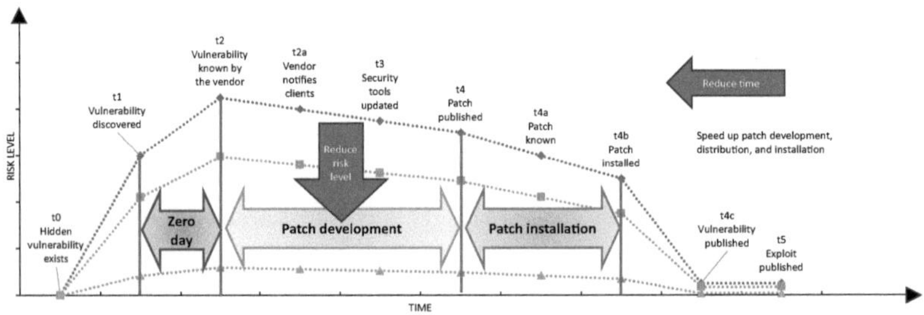

Fig. 4.2 Time and Risk Level Reduction of the Window of Vulnerability

Threats and vulnerabilities must be addressed over the entire lifespan of medical and industrial IoT devices. A lifespan of 10–20 years makes security by design and defense in depth challenging. However, the objective is to improve the security and resilience of stakeholders' systems, ensuring critical business functionality and safeguarding human lives.

Mode-switching is a method to reduce the risk level until a patch is provided and installed. If the vendor quits service or no longer exists, the customer can still reduce risk, for example, by switching from an online to an offline mode. Mode-switching can reduce attackers' range of activity and the attack surface during the WoV until a patch is provided and installed.

4.2 Definitions

As shown in Chap. 3, the notion of modes has been defined distinctively in the literature as states, behaviors, functions, or schedules. Systems are divided into logical and tangible modes of operation to control them. Modes have different purposes, as well as unique behavior. For example, airplanes have modes for parking, taxiing, take-off, manual and automatic cruising flight, and landing. A set of functionalities and configurations characterizes such modes. They can contain a plethora of different actions to be executed at various levels of abstraction, as modes are highly dependent on the domain and system they represent. In addition, some actions can be prohibited, like thrust reversal in the take-off mode of the airplane to prevent a crash.

Besides functionality and safety, we look at modes from a security perspective. In this work, a *mode* is defined as a system state where a specific configuration is active and specific functionality is provided that leads to a desired system behavior for a period of time. Depending on the implementation, the interfaces, the components, the libraries, and other factors, each mode faces a different security risk and offers more or less attack areas. Triggered by a timer or a specific event, a *mode switch* is used to change the behavior. In the literature, a change between modes is often referred to as "mode transition", "mode change", or "mode switch". In this work the latter term is mainly used. While a mode switch refers to the action of changing from one mode to another, the term *mode switching* refers to the broader system concept of switching between different modes. This concept is used in *Multi-Mode Systems (MMSs)* to operate in different modes and to reduce the attack surface and attackers' range of activity.

The *multi-modal architectures*, presented in Sects. 4.4, 4.6 and 4.8, are a design approach to extending typical software architectures with modes and integrating the functionality to switch between various modes. The design iterations of the proposed *mode-switching framework* facilitate the architectures and provide a general software framework for use in a wide variety of application areas. For further details, see Sects. 4.5, 4.7 and 4.9.

While *internal modes* like debugging, development, or production are used within the same system, *external modes* like online or offline relate to system interfaces or subsystem parts. Modes can be organized in a hierarchical structure. Using inheritance, a *sub mode* can extend a *super mode* with additional functionality and override configurations. *Alternative modes* provide similar functionality up to functional equivalent functionality in order to support the idea of MTD and mode rotation.

4.3 Motivating Example

In the following, a motivating example from the domain of *Database Management Systems (DBMSs)* illustrates how mode-switching can significantly reduce a software system's exposure to vulnerabilities. Additionally, automation support is demonstrated to ease the task of switching from one mode to another. Typically, applications use only one DBMS type, such as *MySQL* or an *Oracle Database*. If vulnerabilities are detected and subsequently reported, e.g., to CVE databases, system administrators who monitor these databases become aware of them. Additionally, some software vendors actively inform their customers about vulnerabilities. Based on this information, they can determine whether their systems are affected. Specific vulnerabilities may only affect certain versions of systems or specific components thereof. Administrators have to wait for the vendors and package maintainers to provide fixes or advisories to mitigate the vulnerabilities. It can usually take days and even months before fixes become available. For example, Oracle typically provides critical patch updates only four times a year, making their software more manageable for administrators, but eventually a longer WoV [Ora23]. The same scenario holds for operating systems, web servers, and other software components. Attackers may randomly choose targets and do not care about update frequencies. Therefore, some administrators may perform manual changes, update configurations, or roll back to previous versions at their own risk. Manual adaptations are tedious and time-consuming, especially if several systems are affected. Some patches even have to be applied manually since changes in configuration files are necessary or workarounds are needed to resolve update failures. Hastily applied changes to mitigate potential threats are prone to errors.

Things get worse when a vulnerability like *Log4Shell* affects multiple versions of software components, e.g., web servers in combination with databases [HNSW22]. When attackers exploit vulnerabilities, it is crucial to have short-term protection ready at hand. Automation support is highly desirable. Therefore, we envision support for quickly switching to an alternative, less susceptible mode without manually changing configurations, component versions, or services. An alternative can be a previous component version, another configuration, container, or even a completely different application that provides similar functionality. Since multiple vulnerabilities may exist simultaneously, the ultimate goal is automatically selecting the alternative mode exhibiting the lowest risk.

Figure 4.3 provides a high-level schematic overview of the envisioned multi-modal architecture with four example modes (M0 to M3) in the green box. In the form of a *System Configuration* (orange), operators (blue) should be empowered to describe several system modes, the components that run in specific modes, the configurations, and actions to start or stop a specific mode.

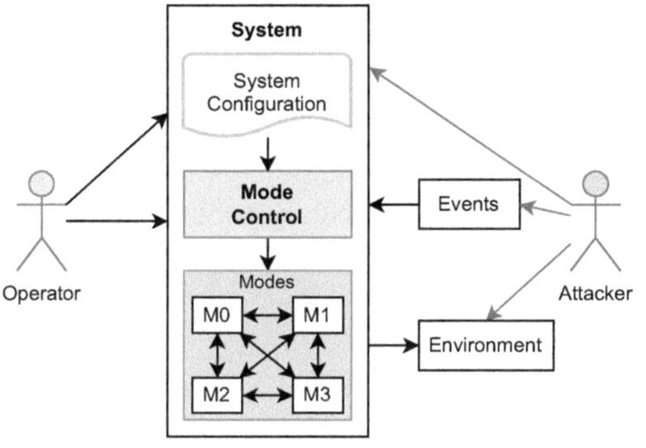

Fig. 4.3 Schematic Overview of the Multi-Modal Architecture

A central *Mode Control* component (blue box) is necessary to:

- utilize the system configuration with predefined modes,
- collect and analyze events
- calculate a risk level for each mode,
- switch among modes according to the risk level (green rectangle),
- and allow the operators to manually turn modes on or off if needed, e.g., for maintenance or in case of an emergency.

Instead of analyzing abstract weaknesses and attack patterns to calculate the risk, concrete vulnerabilities and patches from public databases are considered in this work [PM20]. If risks change, the mode will be switched to change the system's attack surface and affect the environment. Once a patch is provided and installed, the system can return to the previous mode. Considerations include that attackers (red) may try to exploit the system's vulnerabilities, manipulate events, and influence the environment. All the architectures of this dissertation follow the same structure and color scheme.

4.4 Multi-Modal Architecture for Single Systems

This section provides further details about the first design iteration of the multi-modal architecture, its components, and the workflow. Common software architectures typically focus on one mode of operation, e.g., the normal mode to provide all functionality. Multi-modal architectures extend typical software architectures and allow them to automatically and manually switch between several modes, e.g., configuration, normal, and debug. In general, modes are highly domain-specific and can be implemented at various levels of abstraction. For example, a complete software system uses modes describing its different phases of operation: pre-mission, mission, and post-mission. Modes can also describe a set of components operating within this system. In the latter case, switching to a different mode is possible by exchanging or modifying software components, modules, and configurations that are part of the system. For example, a concrete web server implementation providing certain functionality is exchangeable with a different implementation in many circumstances. While the implementation details and software components differ, the basic functionality (result for the user) is equal or similar enough. Therefore, the application remains operational regardless of the used implementation, i.e., the mode. The concept of multiple modes is flexible and can be adapted in many different ways.

Concerning security, different modes serve as a means to reduce potential risks and maintain functionality during deployment when parts of the system become compromised by a detected vulnerability. In addition, modes further allow isolating critical functionalities into "fail-safe" modes, i.e., a minimal set of required operational components to provide a specific functionality. This principle has been successfully applied to medical devices, e.g., to apply different modes when malware was suspected [SRLO15]. Similarly, the trustworthy multi-modal framework from *Rao et al.* incorporates a composite risk model into the system architecture at every hierarchical level [RRLS18]. Compared to traditional single-mode software architectures, the attack surfaces, vulnerabilities, and potential exploits of the modes within a multi-modal architecture are different. This difference can be utilized to choose the less risky option at runtime. Thus, the architecture mitigates vulnerabilities, wards off attacks, and ensures business continuity.

Based on these concepts, a multi-modal architecture has been created, that allows defining modes at different abstraction levels. This functionality allows us to switch modes by deactivating software modules and starting other ones as a replacement during runtime. When performing such mode switches, the objective is to reduce the overall risk of the system. For example, consideration is given to documented risks found in public vulnerability databases. These databases indicate the vulnerabilities and their severity scores known at a given time for the running system configuration (see Sect. 2.1).

Figure 4.4 presents a conceptual overview, offering insights into the proposed multi-modal architecture and the workflow. The architecture comprises three main components: the *Configuration Part* (enclosed in an orange box) the *Mode Control Part* (enclosed in a blue box), and the *Modes Part* (enclosed in a green box).

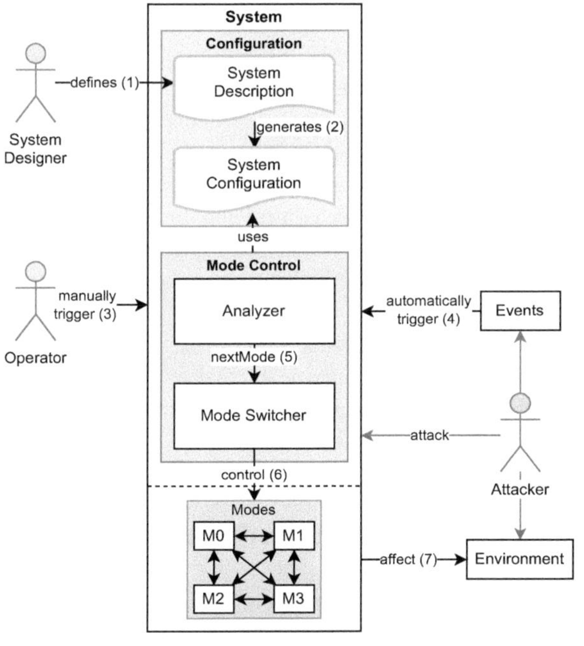

Fig. 4.4 Multi-Modal Architecture and Workflow for Single Systems (Iteration 1)

In the initial phase, the system designer defines (1) the desired modes in the form of a *System Description*. The description contains simplified information about the different system parts, including actions to be performed to switch from one mode to another. Based on the system description, the system automatically generates (2) a *System Configuration*. The configuration contains system-specific information about all the software components that run in various modes.

The mode control uses the system configuration to initialize the modes and is responsible for analyzing events and executing mode switches as defined in the system description.

While the system is operational, the operator can manually trigger (3) the mode control when specific modes are at risk. Additionally, the operator can proactively enable or disable specific modes to reduce the attack surface. Similarly, events automatically trigger (4) the mode control. These events can come from the system itself or external sources. Based on specific scores, the *Analyzer* decides whether measures need to be taken, such as suggesting the next mode (5), or whether the current mode remains the most suitable one.

If a mode switch needs to be performed, the new mode is selected, and the *Mode Switcher* stops and starts services and executes the actions (6) defined in the system description associated with the specific mode. Switching modes may affect the environment, e.g., increasing or decreasing the attack surface.

The dashed line between the mode control and the modes (green box) indicates that modes can either be directly integrated into a system (internal modes) or be part of a subsystem (external modes). In the latter case, the mode control sits on top of a system and act like a supervisor.

The architecture part with the events is adaptable and expandable, e.g., by collecting vulnerabilities and patches, runtime threat detection, and integrating IDS and IDPS. This adaptability should make systems robust and resilient against evolving cybersecurity challenges.

4.5 Mode-Switching Framework for Single Systems

Based on the related work on mode modeling and specification, see Sect. 3.6, this section presents a framework on how to model and run systems with the proposed multi-modal architecture in Sect. 4.4. For the configuration part (cf. Fig. 4.4), a MDSL was developed to structure and simplify the usage of the framework. The modes defined by the MDSL comprise specific services and start and stop actions that must be executed when a mode is activated. For instance, these actions can use shell commands, e.g., to start or stop different services, execute programs, adapt configuration files to reflect the respective mode, or change specific properties and settings within an application. The mode-switching framework builds on a set of components and modes, which facilitates the use of common OTS software and defines security modes around it. The following subsections give an overview of DSLs, explain mode modeling, the components of the framework, and show the prototypical implementation.

Overview of Domain-Specific Languages

In contrast to *General-Purpose Languages (GPLs)*, used for multiple domains, DSL are designed for particular application domains. GPLs like XML, UML, C, and Java are highly flexible and applied to a wide range of problems. Both are used for programming and specification. A DSL like SQL provides custom semantics and syntax and focuses on a particular target, the management of relational databases. Compared to a GPL, a DSL is typically more compact, easier to understand, and reduced to the specific problem, making it accessible to non-programmers. DSLs often replace configuration code and are combined or used with GPL and other DSLs to build a system. Over time, and depending on the usage of the languages, some DSLs became GPLs and vice versa. [MHS05, Fow10]

DSLs have been used to describe system components or adaptation strategies on various occasions. In the domain of self-adaptive systems, *Cheng & Garlan* present *Stitch*, a language for architecture-based self-adaptation that focuses on repair strategies and related business objectives [CG12]. Similarly, *Arcaini et al.* propose the *MAPE Specification Language*, a DSL for adaptation patterns as well as cost-benefit considerations [AMRS18, MRS19]. While these approaches also allow changing the system at runtime, security aspects are the primary concern of this work. In security-related work, DSLs have also been successfully used to define constraints for IDSs or in the context of SPLs [SA10, BLLV20]. Additionally, DSLs were able to handle multiple security requirements in one system [NKKT15, NAY17]. *Liu & Bruel* propose MoSt, a DSL for modeling and verification of natural language requirements based on states and modes [LB24]. This work aims to simplify the mode description when using the mode-switching framework.

Mode Modeling with a Mode Domain-Specific Language

The MDSL makes it possible to describe different modes, the components used, and specific actions to be performed for mode-switching. The mode descriptions are easy to modify and reuse, e.g., by system designers, administrators, and software engineers. Inheritance allows to extend classes of the MDSL and supports code reusability. The MDSL provides a way to specify the system definition and alternative modes in case of attacks, vulnerabilities, or other events. The MDSL is embedded in the mode-switching framework that generates code fragments from the system definition.

Manually activating and deactivating software components or creating multiple scripts executed for a mode switch is typically cumbersome and prone to errors, especially when multiple versions or configurations are involved. A declarative specification of modes, components, and actions facilitates this task. Such a specification further eases subsequent maintenance of the modes. The framework can validate the system description and automatically generate scripts and executable code (model-to-code transformation). This supports the adoption of new software components into the system and the definition of additional modes or updated versions of already existing components without manually updating any code parts or scripts.

Figure 4.5 depicts an overview of the meta-model of the MDSL. The five main elements, the *System* that is modeled, constituent *Software* components, the *Modes*, *Services* as well as *Actions*, are explained in more detail below. Every element has a unique name and a description (see Declaration).

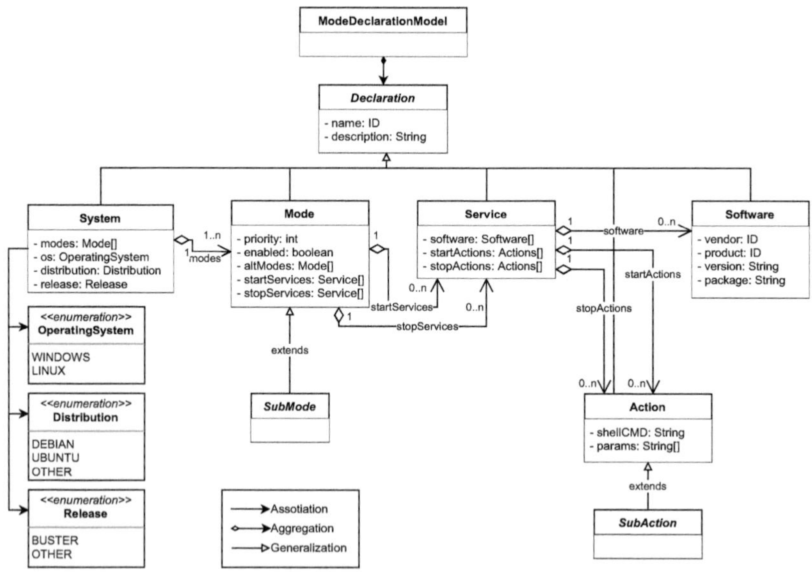

Fig. 4.5 Mode Domain-Specific Language Meta-Model

- **System:** This element represents the system for which modes are defined. For example, an operating system or a web server consisting of multiple different parts, i.e., software components. Every system has at least one mode. Some software and actions highly depend on the operating system, the distribution, and the release. These attributes are also relevant for the search for vulnerabilities and patches.
- **Software:** Each software is described by its name, vendor, product, and version. This information later serves as input for the automated CPE generation (cf. Sect. 4.5). Furthermore, the package name is needed to identify software patches of the distribution's package maintainers. The package name highly depends on the naming convention of the distribution. For example, the package name for the Apache HTTP Server in Red Hat is `httpd`, and in Debian it is `apache2`. Software components serve as the building blocks that enable switching between different modes. Examples are DBMS, libraries, or specific applications. They can be activated/deactivated, i.e., started or stopped, by executing specific actions like appropriate scripts.
- **Mode:** After the software components have been defined, the various modes, their combinations, and component configurations can be specified. A mode represents a specific system configuration, running specific software components. Each mode is defined by its unique name, textual description, priority, and whether it is enabled. For example, disabled modes can be used for maintenance or recovery reasons. Priorities are used to select a mode when two (or more) potential candidate modes are available. A mode uses a specific software set and relates to specific start and stop services executed at the mode switch. In order to support the ideas of MTD (cf. Sect. 2.6), specified alternative modes (altModes) with similar functionality can be switched regularly (mode rotation). Inheritance allows to extend the superclass Mode. The subclass SubMode inherits all attributes and methods from the superclass and can override and extend them. Thereby, a mode hierarchy can be defined that allows us to start and stop the services of the superclass before the services of its subclass. For example, the SubMode *ExctendedFunctionality* can extend the Mode *DefaultFunctionality* with additional functionalities.
- **Service:** To combine several actions in the correct sequence, needed services are used to start or stop specific software, e.g., a database server. Services are executed when the respective mode they are associated with is activated. Additionally, detailed data about the software used are linked.

- **Action:** In the simplest form, an action is a shell command executed to activate or deactivate a software component, e.g., the command `& "C:\script.ps1"` to run a PowerShell-script on Windows, or `./script.sh` to run a shell script on Linux. For defining more complex actions, additional parameters can be used. Actions also support inheritance, such that the subclass SubAction extends the superclass Action. Thereby, the action inherits basic commands of the superclass and adds specific parameters.

Modes and Mode Control

As highlighted in Fig. 4.4, mode control can either be directly integrated into a system to control internal modes or sit on top of a system to act as a supervisor and control modes from external. The external control and central analysis for multiple systems and sub-components is a precursor for the extended mode-switching framework described in Sect. 4.9.

The mode control part of the system uses the *System Configuration* generated from the MDSL to identify relevant software components (as specified and used in the different modes). The *Analyzer* component processes events of the operator, the system, or the environment. Based on this information, the *Analyzer* calculates a risk score for each mode and suggests the least risky mode. Priorities are used to choose among modes with identical scores. For example, system designers and operators can use priorities to specify that they prefer to run a certain software component and only run the alternative if its risk is lower than the preferred one. If mode rotation is enabled, and one of the alternative modes has the same risk as the current mode, the alternative is suggested [TMMI16].

The *Mode Switcher* executes the mode switch suggested by the *Analyzer*. It stops the services and actions of the current mode and starts the predefined services and actions of the new mode. If the *Analyzer* suggests an already running mode, then the *Mode Switcher* has nothing to do, and the current mode remains active. If certain services are used in the old and the new mode, the *Mode Switcher* keeps them running when switching from one mode to another. This partial stop and start reduces the possible downtime between a mode switch. Furthermore, the operator can change the mode priorities and manually enable or disable modes as a precaution, e.g., when vulnerabilities or exploits become known but are not yet documented or something unforeseen happens.

Prototypical Implementation

Based on the presented multi-modal architecture, see Fig. 4.4, a PoC of the mode-switching framework was developed to facilitate experimentation. The comprehensive Java implementation includes all elements mentioned in the architecture and workflow. Following the concepts and functionality outlined in Sect. 4.5, all elements of the MDSL meta-model have been implemented as Java classes in the framework: the DSL for defining modes, services, actions, a validation component, the code generator, the mode control components, and runtime mode-switching support. A brief report is provided on the implementation details, with a focus on the model-based generation of the system configuration and mode control.

Listing 1 shows an excerpt of the MDSL grammar based on the meta-model (cf. Fig. 4.5). The MDSL is implemented based on the Eclipse Modeling Framework (EMF) as a meta-model, with a textual syntax provided through the *Xtext* framework [Ecl23b]. The grammar employs a set of formal rules and constructs to define various system components, such as software, modes, services, and actions. The characteristics of a `System` include its name, supported modes, the operating system, distribution, and release. Mandatorily, each `System` must have at least one associated `Mode`, although the specification of multiple modes is possible. Certain attributes, such as the unique `name` of the system and the mode, are obligatory, while others, like the `operatingSystem`, are optional. Enumerations like `SUPPORTED_OPERATING_SYSTEMS` are utilized for the `operatingSystem` attribute to restrict user input. Similarly, the optional `enabled` attribute enables or disables modes. Also, attribute data types, such as `STRING` for the `description` and `INT` for the modes' `priority`, restrict user input. Then, attributes such as name, vendor, product, version, and package can be specified for each `Software`. The `Mode` rule includes name, description, priority, start and stop services, enabled status, and alternative modes. A `Service` is a combination of one or more `Software` components and actions to start and stop this service. In addition to the entities defined in the meta-model (cf. Fig. 4.5), the `ActionWithParams` rule has been incorporated in the grammar to support the reuse of actions with different parameters similar to the command pattern. As indicated with the `extends` keyword, modes and actions can inherit attributes from higher-level modes.

```
grammar org.xtext.mdsl.Mdsl with org.eclipse.xtext.common.Terminals
generate mdsl "http://www.xtext.org/mdsl/Mdsl"

Model: (declarations+=Declaration)*;
Declaration : System | Software | Mode | Service | Action;

System :
        'System' name=ID 'has' 'following' 'modes'
        (modes+=[Mode] (',' modes+=[Mode])*)?
        ('operatingSystem' operatingSystem=OperatingSystem)?
        ('distribution' distribution=Distribution)?
        ('release' release=Release)?;

enum OperatingSystem: WINDOWS|LINUX;
enum Distribution: DEBIAN|UBUNTU|OTHER;
enum Release: BUSTER|OTHER;

Software :
        'Software' name=ID
        ('vendor' vendor=STRING)? ('product' product=STRING)?
        ('version' version=STRING)? ('package' package=STRING)?;

Mode :
        'Mode' name=ID ('extends' superclass=[Mode])?
        'description' description=STRING
        'priority' priority=INT
        'startServices' (startServices+=[Service]
                (',' startServices+=[Service])*)?
        'stopServices' (stopServices+=[Service]
                (',' stopServices+=[Service])*)?
        (enabled=Enabled)?
        ('exchangeable' 'with' alternativeModes+=[Mode]
                (',' alternativeModes+=[Mode])*)?;

enum Enabled: TRUE | FALSE; //default: TRUE. FALSE for e.g. maintenance mode

Service :
        'Service' name=ID
        ('usesSoftware' usesSoftware+=[Software] (',' usesSoftware+=[Software])*)?
        'startActions' startActions+=ActionWithParams
                (',' startActions+=ActionWithParams)*
        'stopActions' stopActions+=ActionWithParams
                (',' stopActions+=ActionWithParams)*;

ActionWithParams : action=[Action]'('(params+=STRING (',' params+=STRING)*)?')';

Action :
        'Action' name=ID ('extends' superclass=[Action])?
        ('description' description=STRING)?
        (('shellCmd' shellCmd=STRING)|('params' params+=STRING (',' params+=STRING)*));
```

LISTING 1: Xtext Grammar for the Mode Domain-Specific Language

For example, Listing 2 shows a system description of a database server using the MDSL. The `DatabaseServer` utilizes the mode `MySQL` with the software `mysql` and related services to start and stop the mode. Using the *Eclipse Editor* for defining the system description provides syntax highlighting similar to the listing and other functionalities like error markers, code folding, code completion, outline view, and refactoring. To ensure valid system configurations, an implemented validation component checks whether the description conforms to the specified Xtext grammar (cf. Listing 1), prevents duplicates and logical errors like self-inheritance, and verifies package names according to the detected or specified operating system. Predefined default actions, such as starting and stopping specific services, are provided to simplify defining system descriptions without manually defining these commands for each operating system.

```
System DatabaseServer has following modes MySQL operatingSystem Linux distribution Debian
...
Software mysql package "mysql-8.0"
...
Mode MySQL
startServices mysql2Service stopServices mysql2Service
...
Service mysql2Service usesSoftware mysql
startActions startService("mysql")
stopActions stopService("mysql")
...
Action startService shellCmd "/usr/bin/systemctl start"
Action stopService shellCmd "/usr/bin/systemctl stop"
```

LISTING 2: Excerpts of a MDSL Database Server System Description

Once a system description with the extension `.mdsl` has been modified and validated, the respective system configuration can be automatically created within *Eclipse* or by running the mode-switching framework on the command line. *Xtend* is leveraged to generate Java code from the system description file, which is subsequently compiled into bytecode class files [Ecl20]. Listing 3 provides an excerpt of the Xtend implementation for generating the system configuration. The generator filters the declarations of the system description to get modes, services, software, and actions. Then, the items are added to the respective list in the system configuration using for-loops. During this process, specific attributes of the declarations are considered and translated into valid Java code, such as package names, versions, and the parameters of actions. If..else statements and conditions are used to set attributes, like `enabled` for the mode.

```
public static SystemConfiguration create() {
  List<Mode> modes  = new ArrayList<Mode>();
  List<Software> software = new ArrayList<Software>();
  ...

  <<FOR s : model.declarations.filter(Software)>>
  <<IF s.package !== null>>Software <<s.name.toFirstLower>> = new Software("<<s.package>>");
  <<ELSE>>Software <<s.name.toFirstLower>> = new Software("<<s.name>>");
  <<ENDIF>>
  software.add(<<s.name.toFirstLower>>);
  <<ENDFOR>>

  <<FOR a : model.declarations.filter(Action)>>
  <<IF a.superclass !== null>><<a.shellCmd = a.superclass.shellCmd>><<ENDIF>>
  <<a.params.forEach[param, index| a.shellCmd = a.shellCmd.replace("$" + index, param)]>>
  Action <<a.name.toFirstLower>> = new Action("", "<<a.shellCmd>>");
  <<ENDFOR>>

  <<FOR s : model.declarations.filter(Service)>>
  Service <<s.name.toFirstLower>> = new Service("<<s.name>>",
    Arrays.asList(<<s.usesSoftware.map[name.toFirstLower].join(', ')>>),
    Arrays.asList(<<s.startActions.map[action.name.concat(".withParams(\"")
      .concat(params.join('", "')).concat("\")")].join(', ')>>),
    Arrays.asList(<<s.stopActions.map[action.name.concat(".withParams(\"")
      .concat(params.join('", "')).concat("\")")].join(', ')>>)
  );
  <<ENDFOR>>

  <<FOR m : model.declarations.filter(Mode)>>
  Mode <<m.name.toFirstLower>> = new Mode("<<m.name.toFirstLower>>",
    "<<m.description>>", <<m.priority>>, Arrays.asList(<<IF m.superclass !== null>>
    <<m.superclass.startServices.map[name.toFirstLower].join(', ')>>
    <<IF m.startServices !== null>>,<<ENDIF>><<ENDIF>>
    <<m.startServices.map[name.toFirstLower].join(', ')>>),
    Arrays.asList(<<IF m.superclass !== null>><<m.superclass.stopServices
      .map[name.toFirstLower].join(', ')>><<IF m.stopServices !== null>>,<<ENDIF>><<ENDIF>>
      <<m.stopServices.map[name.toFirstLower].join(', ')>>)
  );
  <<IF m.enabled !==null && m.enabled == false>>
  <<m.name.toFirstLower>>.setEnabled(false);
  <<ENDIF>>
  modes.add(<<m.name.toFirstLower>>);
  <<ENDFOR>>

  <<FOR m : model.declarations.filter(Mode)>>
  <<IF m.alternativeModes !==null>>
    <<m.name.toFirstLower>>.setAlternativeModes(Arrays.asList(<<m.alternativeModes
      .map[name.toFirstLower].join(', ')>>));
  <<ENDIF>>
  <<ENDFOR>>

  return new SystemConfiguration(modes, software);
}
```

LISTING 3: Excerpts of the Xtend System Configuration Generator

In Sect. 5.1, parts of the framework were investigated with a web shop sample scenario. For the evaluation with a web server use case study described in Sect. 5.2, a main program has been implemented to execute all necessary steps for generating the system configuration, monitoring vulnerabilities, and providing mode-switching support during runtime. For example, a vulnerability manager has been implemented to collect CVEs and patches for the corresponding software, which can then, in turn, trigger mode switches once a vulnerability is reported.

4.6 Multi-Modal Architecture for Multiple Systems

Systems like IoT devices are exposed to various attacks, such as bots and automated attack scripts trying to break into these devices and guess credentials with dictionary and brute-force attacks [MCZ+19]. Securing these devices is especially challenging due to resource constraints, such as limited storage, memory, CPU, and power [MCZ+19, XWP14b]. Common countermeasures in such cases are, for example, delaying login attempts or blocking accounts and IP addresses after several failed attempts [SKY19, GRGB19]. However, the focus on single systems in Sect. 4.5 needed an extension for multiple devices in the second iteration of the design cycle. It is not effective to monitor all events, e.g., new vulnerabilities and patches on all devices separately. Considering the tight resource constraints of the IoT devices leads to central monitoring and control. Nevertheless, the devices should be robust against attacks even if the connection is lost.

Motivating Example

Figure 4.6 depicts a simple example of a scenario where an attacker attempts to attack multiple IoT devices. On the left side, the IoT devices are managed independently (locally on the device only), and do not share any information with a central server. An implementation of a basic security concept could, for instance, block attackers on that device after two failed login attempts. However, with no central managing authority being aware of attacks on individual devices, each IoT device stands alone, and attackers can perform the same number of attempts on each device (1–3).

On the right side of Fig. 4.6, IoT devices share the attackers' information with a higher-level authority, i.e., a server. The server can consolidate information from multiple devices, and thus recognize attacks from one attacker on multiple IoT devices (4–5). Once an attack has been detected, the server can inform other connected devices to automatically block the attacker and prevent further attacks (6). In the scenario with three IoT devices ($n = 3$), sharing attackers' information halves the possible brute-force attempts from six ($2 * n$) to a maximum of three. The results

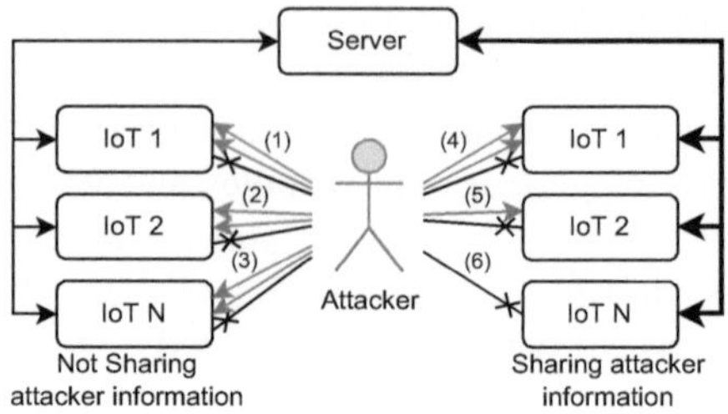

Fig. 4.6 Sharing Attacker Information among IoT Devices vs. Not Sharing [RSV23]

are even more severe in the real world, where dozens or even hundreds of devices are typically deployed. For example, with 100 devices ($n = 100$), an attacker would have 200 ($2 * n$) attempts on stand-alone devices. Sharing information limits attackers to only three attempts, which means a reduction of 98.5 % for $n = 100$.

The scenario is exacerbated when the attack is distributed and carried out by multiple attackers (DDoS) [VJ20]. Blocking the attackers and their IP addresses individually will no longer suffice, even if IoT devices share attacker information. Each IP address is limited to three attempts prior to being blocked. In an example with 100 devices ($n = 100$) and 100 coordinated attackers ($a = 100$), the attackers would have 300 ($3 * a$) attempts before they are blocked. If they focus on a single IoT device, e.g., brute-forcing a password, their chances of success are even higher. A promising approach is to arm such devices with multiple modes that can be activated and switched on demand [RSVW23a, RSS22]. Each mode provides specific functionality and has a specific attack surface. The server, as well as the IoT device itself, can trigger a mode switch. For example, in a *Mode Blocklist*, the device accepts connection attempts from any IP address, but after two failed attempts, the IP address is added to the blocklist and banned from further attempts. Sharing the attacker's information with a server and other IoT devices can enrich this mode. However, the blocklist may overflow if there are hundreds or even thousands of blocked IP addresses. Consequently, switching to a *Mode Allowlist* allows only predefined IP addresses to connect, changing the behavior and the attack surface. This mode provides less flexibility but more security.

Another challenge is represented by vulnerabilities [RSVW23a]. Once detected, attackers can exploit them until a patch is provided and installed. Analyzing the vulnerabilities' status on individual IoT devices is impractical. On the one hand, there is a need to detect local incidents and share them with a central server. On the other hand, it is possible to reduce the overall attack surface by centrally collecting and analyzing information about local incidents, vulnerabilities, and patches and by sharing this information with connected devices. This strengthens the need for distributed security. But for IoT devices, simply monitoring and sending events to a server is insufficient. If an IoT device has no connection, e.g., due to a deliberate attack, the server cannot relay information and trigger a mode switch or adaptation, yet the device should still be able to protect itself.

Attackers can attack medical and industrial IoT devices on the hardware, software, and network layer. Therefore, such devices should be designed to detect and respond to anomalies and potential cyberattacks and recover from them. Manufacturers need more comprehensive measures beyond the conventional focus on individual devices. Anomaly and attack detection must be implemented on multiple levels. Considering events of several IoT devices provides a better overview and reduces false positives. By aggregating and correlating events from multiple sources it is possible to get a more comprehensive understanding of the monitored systems. This helps to filter out noise, distinguish between normal and abnormal behavior, and reduce the likelihood of false alarms. Additionally, cross-validation of events increases the accuracy of alerts. The reaction to attacks can be implemented on multiple layers of the systems.

Extended Multi-Modal Architecture

To address the complex security challenges of medical and industrial IoT devices, the multi-modal architecture presented in Sect. 4.4 was revised and expanded. The ultimate goal is to provide availability, e.g., the clinical function, while reducing the attack surface. Figure 4.7 depicts the schematic overview of the proposed extended multi-modal architecture in a generalized form. The architecture includes a *Configuration Part* and a *Mode Control Part* on both the *Managing System* and the *Managed Systems*. The principle of divide and conquer was employed. Decisions are made as decentralized as possible and centralized as necessary.

The *Managing System* uses its *Inventory* to continuously monitor connected systems to provide enhanced visibility and situational awareness for the *Operator*. Then, the treating doctor, personnel from the manufacturer, and IT security professionals, can analyze the situation, plan adaptations, and execute them to reach the system goals. For efficiency, repetitive tasks and decisions can be partially or fully automated. Additionally, the *Managing System* can centrally monitor external events like CVEs, advisories, safety communications, product alerts, warnings, recalls, and other events of public databases, as presented in previous work [RSR22, RSVW23a]. Based on the monitored events and the analysis, the *Managing System* can automatically send adaptation requests to the *Managed Systems* to change their

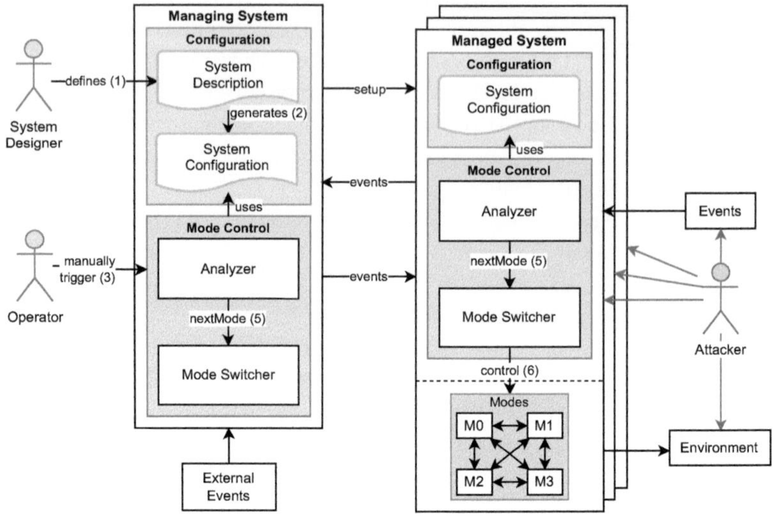

Fig. 4.7 Multi-Modal Architecture for Multiple Systems (Iteration 2)

behavior, e.g., block IP addresses or switch their mode. Likewise, the *Operator* and the *Customer/Patient* can trigger adaptations manually.

Managed Systems monitor and affect their environment like the customer/patient, attached sensors, and actuators. If *Managed Systems* are online, they can send data to the connected *Managing System* and forward information and decisions about further actions. For example, deep autoencoder models trained on typical non-malicious network packets/flows can help to detect anomalous traffic [RPLZ19]. Depending on the configuration, the *Managing System* automatically decides, or an *Operator* (human-in-the-loop) decides on further steps, like switching modes or turning on and off specific functionality. The reaction must also be implemented on multiple levels. When offline, the *Managed System* must determine its course of action autonomously and adjust itself accordingly. Predefined rules and actions on the *Managed System* help to achieve this goal. Additionally, software and hardware window watchdog timers may help to reset the system or components if the software crashes or is blocked by a DoS attack [SA00].

Attackers who try to crack passwords, login credentials, and encryption keys can be blocked after multiple wrong attempts with local firewall rules. However, if these attacks reach a specified amount, affect availability, or lead to battery depletion, the *Managed System* may adapt itself and switch to a more restrictive mode.

Battery-powered systems should have a low-power mode with limited functions to extend battery life and reduce the attack surface. An activity sensor or timer can trigger switching to a mode with more functionality. During the connection of the *Managed System* with the *Managing System*, switching to a high-security mode may provide an encrypted channel and make the device more resistant. In case of an attack, devices switch to degraded or failure mode, and self-healing capabilities can enable systems to recover from attacks [CRR22].

Sects. 5.3, 5.4 and 5.5, present sample scenarios with PACSs, insulin pumps and edge devices to evaluate the suggested extended multi-modal architecture.

4.7 Mode-Switching Framework for Multiple Systems

For implementing the extended multi-modal architecture presented in Sect. 4.6 for the widely used Linux operating systems, the system and service manager *systemd* was utilized [Sys23]. *Systemd targets* allow to configure system behavior and dependencies [Fre23b]. Targets are similar to *SystemVinit* runlevels, which were used in previous OS, and allow to specify and combine several services such as web servers, docker containers, and database servers [vS04]. Hard and soft dependencies to other targets can be specified to be started before. While a hard dependency is required to start a target, a soft dependency is wanted but not absolutely necessary. We extended *systemd targets* to define and manage modes.

Systemd only supports basic condition and assertion checks before a target or service becomes active. It can only check whether a file exists but cannot check whether a certificate is valid, a system is online, or up to date. In order to only start modes if certain conditions are met, the setting `ConditionPathExists` is used. The mode selection depends on whether a file is present or not. Therefore, files must be created or deleted to fulfill a condition.

Usually, the `multi-user.target` (on non-graphical systems) or the `graphical.target` are the default targets on Linux systems, starting at system boot. The `multi-user.target` (see Listing 4) builds up on the `basic.target` as specified with the `Requires` setting. The target conflicts with `rescue.service` and rescue.target, and should be started only after `basic.target`, `rescue.service`, and rescue.target have started. The setting `AllowIsolate=yes` allows to enter this target directly without rebooting. Targets are utilized as modes, and the following command is employed for switching between them: `systemctl isolate MODENAME.target`. When the command is executed, the system immediately switches to the desired mode, starts the specified services, and stops all other services.

```
#/lib/systemd/system/multi-user.target
[Unit]
Description=Multi-User System
Documentation=man:systemd.special(7)
Requires=basic.target
Conflicts=rescue.service rescue.target
After=basic.target rescue.service rescue.target
AllowIsolate=yes
```

LISTING 4: Systemd multi-user.target

For supporting inheritance similar to *Object-oriented Programming (OOP)*, modes should be able to extend other modes. `multi-user.target` serves as a parent or super mode for all other implemented sub modes, specified with the settings `Requires` and `After`.

Prototypical Implementation

Figure 4.8 shows the PoC of the mode-switching framework in the second iteration and consists of one *Managing System* and multiple *Managed Systems*, which are explained in more detail below.

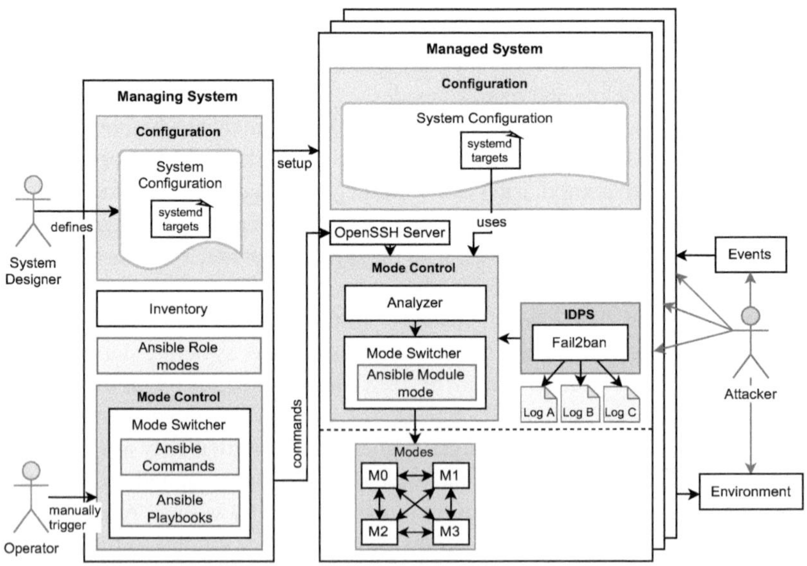

Fig. 4.8 Prototypical Implementation of the Mode-Switching Framework for Multiple Systems (Iteration 2)

• **Managing System:** In the deployment strategy, a dedicated package repository has been set up to deliver modes for multiple *Managed Systems* and facilitate the distribution of changes. This enables the installation of modes as if they belong to systemd. The packages can be checked, adapted, and activated individually.

In exploring system configuration management tools used for provisioning, application deployment, and infrastructure as code, *Ansible*, *Progress Chef*, and *Puppet* were analyzed [Red23]. Among these options, the focus centered on the open-source automation and orchestration tool *Ansible*due to its widespread adoption, versatility, and the presence of a built-in module to manage systemd units [Ans23]. The *Managed Systems* are organized in an inventory, and their software and configurations are centrally managed with Ansible [Red23]. Secure communication is established via the SSH protocol. Typically, several tasks to bring *Managed Systems* into the desired state are combined in so-called "playbooks", e.g., to install and configure software. Playbooks are written in the human-readable *YAML* format. Ansible modules simplify tasks, e.g., installing software with the package manager, starting or stopping services, copying files, replacing values, and executing shell commands. Unfortunately, the module does not support reading the currently running systemd target, setting the default systemd target, and isolating targets. To address these gaps, a custom Ansible module named `mode` was developed, specifically tailored to provide these functionalities and utilize systemd targets for mode-switching. The usage of the custom module in Ansible tasks and playbooks is illustrated in Listing 5, with task names designed to be self-explanatory.

```
---
- hosts: all
  tasks:
  - name: Get current mode information
    mode:

  - name: Apply default mode-0.target, will be applied immideately
    mode:
      name: mode-0.target
      state: default
      apply: True

  - name: Switch to mode-1.target
    mode:
      name: mode-1.target
      state: isolated
```

LISTING 5: Usage of the Custom Ansible Module Mode in Playbooks

The custom module `mode` helps to switch the mode of one or more systems depending on requirements. For example, Listing 6 shows how to switch all systems from group `devgroup1` to the *Mode 1* with a shell command.

```
ansible devgroup1 -m mode -a "name=mode-1.target state=isolated"
```

LISTING 6: Usage of the Custom Mode Module for Mode-Switching in Shell Commands

A custom Ansible role named `modes` was created to copy the mode configurations to managed systems. Listing 7 shows the usage.

```
---
- hosts: all
  tasks:
   - name: Install new modes and reload systemd
     import_role: name=modes
```

LISTING 7: Usage of the Custom Ansible Role Modes in Playbooks

• **Managed System:** A *OpenSSH Server* was installed and set up on the managed systems, to receive commands of the managing system. A shell script implements the *Mode Control* component. Modes can be automatically switched based on events or manually with the operator's commands. The script analyzes events triggered by the IDPS *Fail2ban (F2b)*, considers the current mode situation, checks the existence of mode-relevant files, and decides to execute a mode switch.

The python-based open-source component *F2b* is installed on the managed systems to monitor events, detect attacks, and execute actions [Fai23a]. *F2b* is mainly used to defend systems from brute-force attacks by blocking IP addresses of attackers after several failed login attempts within a specific timeframe [Fai23a]. Besides blocking, various less intrusive actions can be specified, such as sending an email to the operator.

Following the rules of *Event-Condition-Action (ECA)*, *F2b* checks new log entries with a *filter* of regular expressions to extract relevant information. *Jail* configurations become relevant, if there is a match (*event*). These configurations specify services or attack scenarios to be monitored. Jails consist of the log file, port or service, filter, time span and maximum number of failed attempts (*condition*), *actions* to be executed, and the ban time. For example, a jail configuration can be defined to block the IP address of a potential attacker after three failed SSH login attempts. Subsequently, an email notification is sent to the administrator. Then, after a ban time of 10 minutes, a new login attempt from the previously blocked IP address will be allowed again. *F2b* manuals provide more details about these mechanisms [Fai23b, Hes20].

The standard configuration of *F2b* includes filters for services like *Apache*, *sshd*, *vsftpd*, and *Postfix*. *F2b* serves not only to send notifications to the operator in case of abnormal behavior but also as a trigger for mode switches.

Section 5.5 shows an evaluation of the third iteration of the mode-switching framework with an industrial edge device sample scenario.

4.8 Multi-Modal Architecture for Multiple Systems using the MAPE-K Loop

In the third iteration, the multi-modal architecture should mitigate the following challenges:

- Limited system resources
- Vulnerabilities
- Coordinated attackers
- Offline attacks

Several requirements have been derived from the identified challenges for adapting to security threats and dynamic mode-switching for IoT devices. Subsequently, the architecture was extended.

- IoT devices require *lightweight data exchange protocols* to forward information about potential attacks and attackers (*transfer of knowledge*). The managing system should analyze this information from multiple devices, plan countermeasures, and send execution actions to all connected devices (*defense against coordinated attacks*). Then operators or IT security professionals are supposed to make decisions on the managing system side (*human-in-the-loop*).
- Likewise, the managing system should centrally *collect and analyze vulnerabilities*. Based on that analysis, adaptations can be planned, distributed, and executed.
- Even offline or without connection to the managing system, managed systems like IoT devices must be able to detect and analyze events and suspicious behavior independently (*offline monitoring*). For this purpose, ECA rules are employed, where an event triggers a rule, then the condition is checked, and if met, predefined actions are executed (*offline adaptation*). When the device regains connection, it can forward any event that occurred in the meantime.

Architecture Overview

The multi-modal architecture for multiple systems builds up on the *MAPE-K* control loop, as detailed in Sect. 2.7. It uses decentralized control patterns proposed by

Weyns et al., both at the managing and managed systems levels, with different levels of abstraction [WSG+13]. The architecture focuses on self-protecting activities and adopts a combination of the *Hierarchical Control Pattern (HCP)* and IBM's architectural blueprint, which separates the complexity of self-adaption into several layers [IBM05]. Each managed system hereby has its local environment, such as a company or manufacturing network. In contrast, the managing system has a more global and strategic perspective. Coordination between these levels is facilitated by the interaction of the monitoring and execution components at the managing system and managed system level. Figure 4.9 provides an overview of the *Multi-Modal Architecture for Multiple Systems using the* MAPE-K *Control Loop*. The individual parts are explained below.

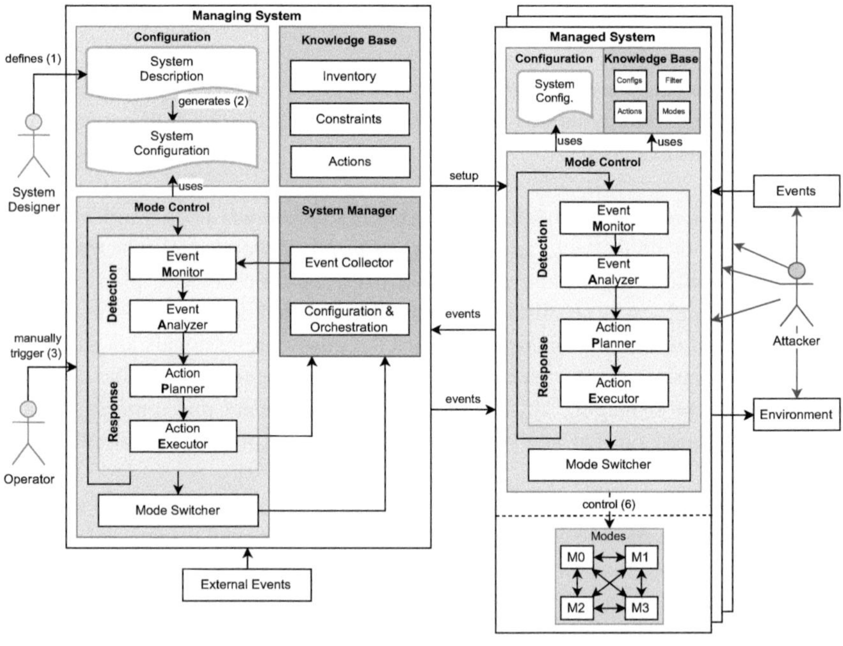

Fig. 4.9 Multi-Modal Architecture for Multiple Systems using the *MAPE-K* Control Loop (Iteration 3)

• **Managing System:** This building block implements the four activities of the *MAPE-K* control loop and stores constraints and actions in an ECA rules-based

Knowledge Base. The *Event Collector* Within *System Manager* collects events from the *Managed Systems* and forwards them to the *Event Monitor*. Then, the *Event Analyzer* checks these events against specified rules and provides information to the *Action Planner* if a condition is fulfilled. As part of these global checks, it is necessary to analyze not only individual events and the data attached to them but, more importantly, streams of events and temporal constraints, which are used to recognize patterns and detect security issues across various resources or time. The *Action Planner* considers the context, like the offline state, and selects appropriate actions from the *Knowledge Base* and notifies the *Action Executor* to execute those actions on all, on some, or on only affected *Managed Systems*. Thereby, the *Configuration & Orchestration* component of the *System Manager* helps to execute the actions to adapt the *Managed Systems* to meet security goals.

Examples of actions include sending notifications to the administrators, executing shell scripts, sending emails, sending HTTP requests, blocking IP addresses, or starting/stopping services. Furthermore, the *Managing System* can establish two-way communication with the *Managed Systems* to request their current state. Besides that, the *Managing System* leverages additional sources of information and external events such as CVEs, and updates/patches of the used operating systems, applications, and libraries from public databases like the NVD [NVD23a].

• **Managed Systems:** Each managed system implements the activities of the entire *MAPE-K* control loop to prevent delays and protect offline devices. Attacks, such as DDoS, may lead to performance issues and disrupt the connection to the managing system. Therefore, the necessary parts of the knowledge base are deployed to the managed systems.

Managed systems such as machines, smart factories, and smart homes are typically equipped with various soft- and hardware sensors for motion and temperature and actuators for electric switches and motors. Combined with these sensors and actuators, they affect the environment and monitor and evaluate the effects. For example, smart homes can monitor temperature, humidity, and occupancy and adjust heating, ventilation, and air conditioning. Additionally, managed systems have interfaces to interact with other devices and the managing system.

The framework focuses on managed system level *Event Monitoring* emphasizing security and self-protection. Dedicated *Filters*, with specific characters or patterns from the *Knowledge Base*, continuously observe the system logs, network interfaces, and critical applications for suspicious entries. These filters can log events at the hardware level, such as newly attached USB devices, connected network cables, or frequent restarts. At the system and software level, information related to authentication errors (i.e., failed authentication attempts that may occur locally or via the network), port scans, and HTTP errors can be monitored and collected. Further examples of filters and their application are provided in Sect. 5.6.

When a filter matches, the information is transformed into an event and a dedicated message notifying the *Event Analyzer*. This component observes various sources of device information to monitor the frequency and source of suspicious log entries and inform the *Action Planner* if limits w.r.t. maximal login attempts are exceeded. The *Action Planner*, in response, selects predefined countermeasures from the *Knowledge Base* and sends them to the *Action Executor*. Countermeasures include blocking an IP address, adapting configurations, switching the operating mode, or rebooting the system. While enhancing security, these measures may affect the managed system and its environment.

• **Transfer of Knowledge & Human-in-the-loop:** When managing multiple distributed systems, one important aspect is the administration of local rules, conditions, and actions. While this information has to be available locally (to enable the device to perform tasks independently), the knowledge bases for both the *Managing System* and the *Managed Systems* are centrally managed on the *Managing System*. This centralized approach facilitates creating and maintaining filters, constraints, actions, configurations, modes, and other settings. After testing new settings with single devices, the settings are deployed to the individual *MAPE-K* loops of the managed systems using infrastructure automation tools.

Another aspect is the integration of the human, e.g., a server administrator or engineer. While preventing potential attacks by blocking an attacker's IP address has to be performed automatically to avoid delays and efficiently react to attacks, other actions may require human confirmation or input. Before stopping a remotely *Managed System* and disrupting customers, the operator (human-in-the-loop) should validate and confirm this action. With frequent events and precise knowledge of the necessary actions, this process can be partially or fully automated to increase efficiency and responsiveness.

The following section provides more details about the framework and examples of its use. Afterward, details on the prototypical implementation are presented.

4.9 Mode-Switching Framework for Multiple Systems using the MAPE-K Loop

Figure 4.10 highlights the usage of the multi-modal architecture for multiple systems for blocking distributed brute-force login attempts. Each managed system monitors events of the authentication log. If the event analyzer detects two failed login attempts originating from the same IP address within five minutes, the planner component suggests an action to block the IP address for five minutes. Then, the executor automatically performs that action and sends an event about the failed login attempt to the managing system.

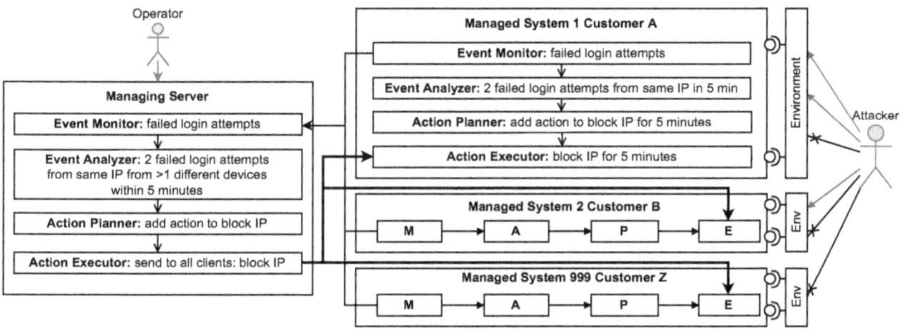

Fig. 4.10 Usage of the Multi-Modal Architecture (Iteration 3) for Blocking Brute-Force Login Attempts

On the managing system side, the events of multiple managed systems are continually monitored. If the event analyzer detects two failed login attempts from the same IP sent by two different managed systems within five minutes, the IP address is considered as an attacking bot. In response, an action is planned to block that IP on all managed systems. The managing systems' action executor distributes the action to the managed systems, and they run the action and block the identified IP address, preventing the attacker from executing further login attempts.

Prototypical Implementation

A PoC implementation of the third design iteration has been created for evaluation and to support further experimentation. Figure 4.11 shows the implemented framework consisting of a *Managing System* responsible for checking global constraints and relaying actions to managed systems, and an *Managed System* part that checks local data and performs local adaptations when needed.

• **Managed Systems:** *F2b* has been chosen as the main component part of the *MAPE-K* loop, serving as local IDPS and enacting adaptations of its policies and protection rules on the managed systems [Fai23a]. *F2b* has features of protecting systems and services against brute-force, dictionary, and DoS attacks, and a variety of other automated attacks. As specified in its configuration, *F2b* monitors log files with filters and subsequently analyzes the found entries. If there is a match, it plans

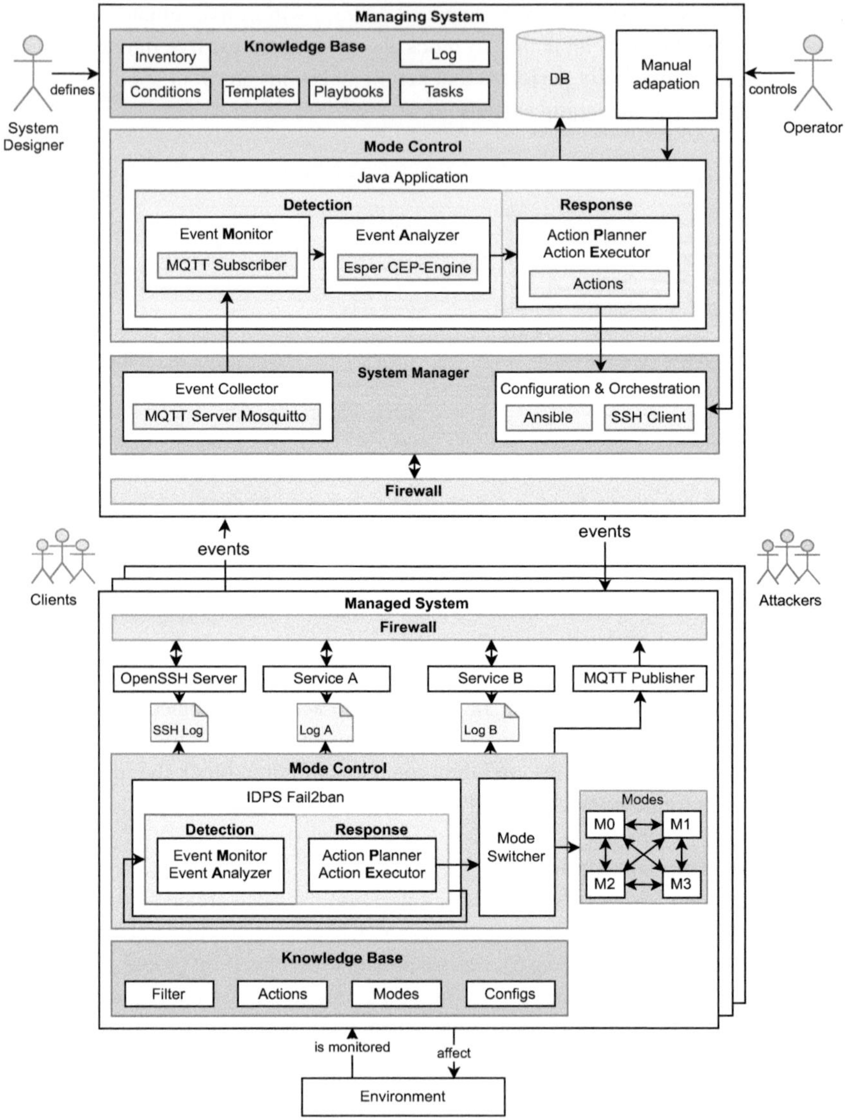

Fig. 4.11 Prototypical Implementation of the Mode-Switching Framework for Multiple Systems using the *MAPE-K* Control Loop (Iteration 3)

and executes specified actions. For each service to be monitored, filters and actions are combined in the so-called "jails" part of the *F2b* configuration. In the context of jails, service ports, filters, log paths, monitoring windows, ban actions, and times can be specified. If filters match, scripts are executed (actions), typically to temporarily ban the attacker's IP address for a specific time and send a notification to the system administrator.

There exists a set of predefined filters, e.g., for the web servers *Apache* and *nginx*, and the SSH Server *sshd*. As part of this work, additional filters have been created to detect non-IP-based attacks like failed login attempts on the terminal (physical device), power on/off cycles, changing network connections, and attached USB devices. *Regular Expression (Regex)* are used for defining filters to match specific patterns in log entries. For example, the pattern `Âuthentication failure*$` matches lines that start with "Authentication failure" and is commonly used to identify lines indicating failed login attempts. In addition to filters, custom actions have been developed, like sending notifications to the managing system and measures for offline adaptation. For configuring *F2b*, a custom *Ansible*module was extended to enhance the capabilities and simplify the framework's usage.

For the *Mode Control*, parts of the work on a model-based mode-switching based on security vulnerability scores have been integrated into the framework [RSVW23a]. This feature allows dynamic mode switches based on varying security conditions, e.g., DDoS attacks, new vulnerabilities, new patches, or changed severity scores. A mode switch can be triggered by local *F2b* actions, managing system actions, and manual actions of the operator. After the attack is over, the patch is available and installed, or on request of the operator, the managed systems switch back to the default mode.

• **Managing System:** On the managing system side, the *MAPE-K* loop is implemented as part of a custom *Java*-based server application. The communication infrastructure leverages the open-source MQTT message broker *Mosquitto* for lightweight and secure communication [Ecl23a]. Within the framework, the managed systems can publish messages for specific topics on the managing systems' message broker. In the *Java Server Application*, these topics are monitored, and events for the incoming messages are generated. For long-term and post-mortem analysis, the events are stored in the database. Additionally, the *Esper Esper Complex Event Processing (CEP) engine* is employed to analyze these events and check constraints [Esp23]. Esper offers a domain-specific *Event Processing Language (EPL)*, which extends the *SQL-92* standard and allows the analysis of high-frequency time-based event data.

Once a constraint is violated, the framework plans and executes predefined actions. For example, if a new vulnerability is found, the managing system notifies the *Managed Systems* to adapt their configuration and to protect them from further attacks. The managing system application also provides manual adaption capabilities for operators by using custom shell scripts, system commands, and other tools.

Section 5.6 presents a case study with IoT devices to evaluate the third design iteration of the mode-switching framework.

4.10 Summary

In this chapter, a novel approach for mode-switching is presented. After introducing the top-level design problem, key terms of the dissertation related to modes and *Multi-Mode System (MMS)* were defined. Critical vulnerabilities in software systems and their exploitation can disrupt the operations of organizations. Vulnerabilities, once identified, introduce a *Window of Vulnerability (WoV)* until appropriate updates are deployed and installed, during which attackers can develop and use exploits. Therefore, it is necessary to speed up patch development, distribution, and installation to reduce the WoV. Of course, acceleration is important, but with the mode-switching approach, we aim to reduce the risk level during the WoV and beyond.

In the *first iteration* of the design cycle, the multi-modal architecture was derived from a motivating example. That architecture supports in monitoring and managing vulnerabilities and reducing potential exposure during the WoV. Utilizing the architecture as foundation, a mode-switching framework was implemented, incorporating a dedicated *Mode Domain-Specific Language (MDSL)*. The MDSL describes MMS and potential mitigations through mode switches. Model-driven techniques were utilized to generate a system configuration from the MDSL descriptions. The system configurations, together with the mode-switching framework, are used to identify vulnerabilities and provide automated contingency support. For example, the framework switches the system to a less vulnerable mode if a new vulnerability is detected. All the material related to the multi-modal architecture, the mode-switching framework, and the MDSL are provided as open science material on Zenodo: https://doi.org/10.5281/zenodo.7603904 [RSVW23b]

In the *second design iteration*, the focus shifted from single to multiple decentralized systems, including IDPS. The multi-modal architecture and the mode-switching framework were improved, extended, and uses the connectivity for security. Centrally monitoring events like vulnerabilities and patches is more effective than monitoring them separately on every device. Collecting information about attack attempts and sharing it with connected devices can prevent further brute-force attacks. The PoC of the framework uses systemd targets, Ansible, and custom modules and roles for implementing and switching between modes.

In the *third design iteration*, the architecture and the framework were extended with the *MAPE-K* loop to provide self-adaptive behavior. The framework implements capabilities to monitor the device's environment and analyze the collected data. Based on that analysis, an action plan is created and executed. Finally, the knowledge base of the framework is updated. The framework allows to continuously adapt and improve the security of managed and managing systems, making them more resilient to potential threats and improving their self-protective ability. The framework provides features for automation, orchestration, information sharing, and offline adaption.

5 Evaluation

Four sample scenarios and two case studies were conducted to evaluate the three design iterations of the multi-modal architecture and mode-switching framework detailed in Chap. 4. Table 5.1 summarizes the evaluation with focus on single systems (Iteration 1), multiple systems (I2), and multiple systems with the *MAPE-K* loop (I3). The table connects the use cases and sample scenarios with the RQs and the publications. Excerpts from previously published work are included in this chapter.

Table 5.1 Evaluation Overview I: Iteration, A: Multi-Modal Architecture, F: Mode-Switching Framework

Use Cases	I1		I2		I3						
Sample Scenarios	A	F	A	F	A	F	RQ1	RQ2	RQ3	RQ4	Publication
5.1 Juice Shop	X	X					X			X	[RS21a]
5.2 Web Server	X	X					X			X	[RSVW23a]
5.3 PACS			X				X	X		X	[RSR22]
5.4 Insulin Pumps			X				X	X		X	[RSV23]
5.5 Edge Devices			X	X			X		X	X	[RSS22]
5.6 IoT Devices					X	X	X		X	X	[RSV23]

While Chaps. 3 and 4 answer the first part of RQ1 about mode switching, all evaluated use cases and sample scenarios contribute to the latter part of the RQ: how to apply mode switching to mitigate risks from a security perspective. In Sect. 5.1, the evaluation of the first iteration of the multi-modal architecture is discussed in

M. Riegler, *Mode-Switching for Resilient Security*,
https://doi.org/10.1007/978-3-658-49306-6_5

a sample scenario for web applications, exploring how modes can make the consciously insecure OWASP Juice Shop more secure [RS21a]. Section 5.2 presents a case study with different web servers and the application of the mode-switching framework and the MDSL [RSVW23a]. Historical vulnerabilities were analyzed to investigate the effectiveness of switching to less vulnerable modes in reducing the attack surface. Then a detailed examination of medical devices was conducted to evaluate the architecture of the second design iteration and answer RQ2. In Sects. 5.3 and 5.4, an evaluation was conducted with sample scenarios for *Picture Archiving and Communication Systems (PACSs)* and insulin pumps [RSR22, RSR23]. To answer RQ3 and how mode-switching works on industrial IoT devices, Section 5.5 investigates the application of the architecture and framework of the second design iteration to an industrial edge device sample scenario [RSS22]. In Sect. 5.6, the third design iteration of the architecture and framework were evaluated with an IoT case study [RSV23]. The evaluation forms the basis to discuss general risk mitigation strategies compared to strategies tailored to specific domains to answer RQ4. Section 5.7 reflects on threats to validity before Sect. 5.8 summarizes the evaluation results.

5.1 OWASP Juice Shop Sample Scenario

The objective of this sample scenario is to present how to use mode-switching to make web applications more secure and resilient. In order to demonstrate the usefulness, sample applications were sought. The *Open Worldwide Application Security Project (OWASP)* provides a directory of vulnerable web applications for testing security skills and tools for attack and defense. *Juice Shop*, one of the OWASP flagship projects, was chosen for the sample scenario [OWA23a]. Although it mimics a typical e-commerce application with features like product search, user registration, and checkout, it deliberately incorporates vulnerabilities outlined in the *OWASP Top Ten Security Risks* [OWA23c]. The vulnerabilities and common weaknesses of Juice Shop were analyzed, attacks were simulated and investigated how to detect them, and identify methods of how to react and mitigate them. This process involved utilizing the first iteration of the multi-modal architecture and parts of the mode-switching framework, see Sects. 4.4 and 4.5, to resiliently defend against attacks and reduce the risk until a vulnerability is fixed, see Sect. 4.1.

Juice Shop has been used in hackathons and capture-the-flag training events [LCC20]. The front end consists of a single page application with *Angular* and *Material Design*, while the back end consists of an *Express* application on a *Node.js* server [Kim21]. *Sequelize* and *finale-rest* are used for the connection to an *SQLite* database [Kim21]. Web developers can study the Juice Shop application for vulnerabilities, perform attacks, and sharpen security awareness. The project leader *Kimminich* has presented 94 vulnerabilities of the Juice Shop as challenges with difficulty levels between 1 and 6 [Kim21]. Detailed hints and step-by-step instructions are provided to guide users through the process of executing attacks. Kimminich recommends employing various hacking tools, such as browser development tools, *PostMan*, and *ZAP*, to identify and exploit vulnerabilities [Kim21]. Other common tools are *nmap* for scanning open ports and *sqlmap* to find and take advantage of SQL injections [Nma23].

Mode-Switching in OWASP Juice Shop

As mentioned in Sect. 4.1, vulnerabilities will emerge sooner or later, even if developers try to build software with security in mind. Security patches are not immediately available or operators delay installation for whatever reason. The implementation of mode-switching aims to enhance system resilience against security issues. This sample scenario aims to observe how the Juice Shop platform behaves differently when utilizing the multi-modal architecture and operating in different modes while exposing the same vulnerabilities.

Attacks can occur at different levels and in different system components, such as the front end, back end, and database. In the sample scenario, log file analysis was used to detect anomalies, attacks, and manipulations on the server side. For example, brute-force attacks on authentication and SQL injections can be detected, and after a specific number of failed attempts within a certain timeframe IP addresses are banned. However, for distributed attacks like a DDoS, using a *Mode Blocklist* where individual attackers are blocked is ineffective. A mode switch is anticipated in this scenario, for instance, after a predetermined number of blocked IP addresses, in order to restrict the number of connections and mitigate further attacks. After detecting anomalies, a mode with an extended log level is activated for further investigation. Additionally, unnecessary legacy encryption protocols are disabled, and third-party access is restricted in certain circumstances. In case of a detected injection, e.g., reading confidential data from a database, an immediate switch to a mode that prevents further data loss, GDPR violations, and reputational damage. Further extensions to improve security during authentication would be *captchas*, to determine whether the user is human, and automatically enforce the *Mode with 2FA Authentication*, when a user wants to log in from an unknown IP address. Also, modes can be switched when there is an unusually large number of failed access attempts.

Modes can be implemented at different layers, like operating systems, web servers, databases, web applications, and others. Modes can block or allow IP addresses at the network and firewall layer. From a financial perspective, modes can restrict access based on geographical location, e.g., countries with the most sales or countries where the most frequent attacks originate. On the application layer, modes may deny registration, restrict login, or set specific cash options. A *Mode Allowlist* may be implemented for scenarios with a limited number of known customers. This mode grants access to only specific IP addresses and users while locking out all others. Alternatively, a mode requiring an additional access code for the web shop could be utilized.

Concerning the MTD strategy, see Sect. 2.6, it is an option to switch among modes with different software implementations, versions, and configurations. Specific software components may have fewer functionalities but fewer bugs and vulnerabilities. Additionally, manipulated components like *Docker* containers can be exchanged with a fresh installation from a repository.

Evaluation of the Multi-Modal Architecture and the Mode-Switching Framework

Following the multi-modal architecture in Sect. 4.4 and the mode-switching framework Sect. 4.5, the Juice Shop platform was extended with modes as illustrated in Fig. 5.1. In order to include mode-switching behavior, it was necessary to establish mode-switch triggers and a mode control component to start and stop modes (enclosed in a blue box). Besides manually initiating mode switches by the operator, the IDPS *F2b* was installed to trigger automatic mode switches [Fai23b]. This tool analyzes log files with predefined filters for *sshd*, *apache*, *qmail*, *proftpd*, and other software to detect abnormal behavior and attacks [Fai23b]. Further detection filters can be defined with Regex. If filters match, the IDPS can execute shell scripts. Likewise, *OWASP Dependency Check* was employed to analyze project dependencies of Juice Shop for publicly disclosed vulnerabilities [OWA23b]. The features of both tools were used to trigger the mode control component and switch modes automatically. Thereby, configuration files, services, and permissions were changed. The mode control component was implemented as a shell script. The script analyzes the triggering requests considering the current mode information, executes the mode switches, and saves the mode information in an environment file (enclosed in an orange box of the system configuration).

Several useful modes have been identified at different levels for Juice Shop and are presented following the formalized description of MMS proposed by Burns as explained in Sect. 3.7 [Bur14].

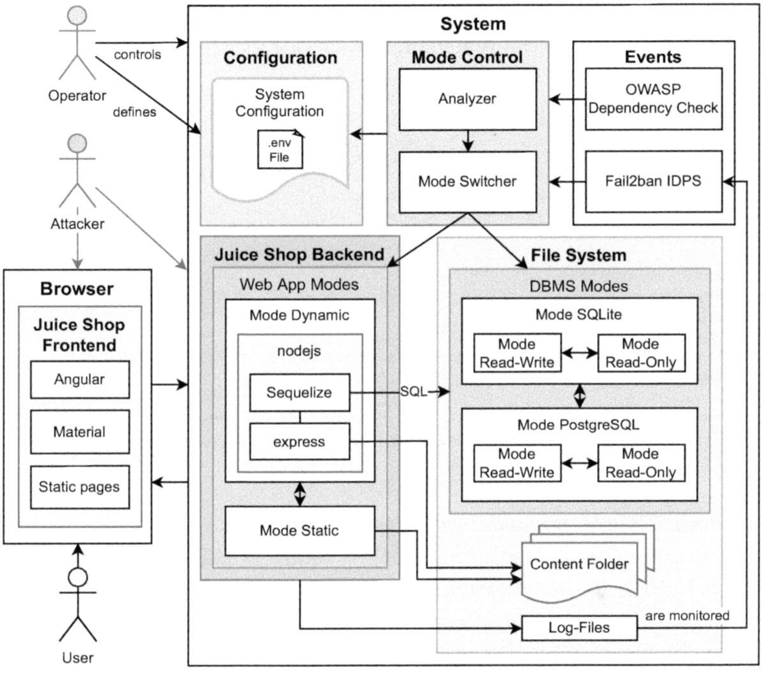

Fig. 5.1 OWASP Juice Shop using the Mode-Switching Framework (Iteration 1)

Web Application Modes

- **Type:** Two modes have been implemented at the web application layer: *Mode Dynamic* and *Mode Static*. The *Mode Dynamic* provides the full functionality of Juice Shop with dynamically generated pages based on user input and database entries. Conversely, the *Mode Static* serves as a degraded mode, presenting users only static content like HTML pages with contact information.
- **Trigger:** A mode switch from *Mode Dynamic* to *Mode Static* is automatically triggered by detecting vulnerabilities and brute force attacks on authentication. Additionally, the operator can manually trigger a mode switch.
- **Protocol:** An environment file is used to save the target web application mode. Listing 8 shows the shell script to start the specified target mode. Mode-switching involves modifying the target mode in the environment file and starting the shell script. The script stops the currently running container and starts the specified

container. In the event of a vulnerability or detected attack, the environment file is updated, the shell script is started, and a mode switch is executed to mitigate the WoV.

```
echo 'Stop the Currently Running Web Application Mode'
sudo docker container stop juice-shop-container
sudo docker container rm juice-shop-container
if [ $web_application_mode == 'STATIC' ]
then
  echo 'Start Mode Static'
  sudo docker run --name juice-shop-container --rm -d -it -p 3000:80 img-static-juice-shop
else
  echo 'Start Mode Dynamic'
  sudo docker run --name juice-shop-container --rm -d -it -p 127.0.0.1:3000:3000 \
    bkimminich/juice-shop
fi
```

LISTING 8: Switching Juice Shop Web Application Modes

- **Attributes:** *Mode Dynamic* serves as the default mode upon system startup because it provides the most functionality. Bidirectional mode switches are permitted, but only the mode switch from *Mode Dynamic* to *Mode Static* is currently automated. The switch from *Mode Static* to *Mode Dynamic* requires manual intervention by the operator. The criteria for triggering the switch to *Mode Static* is the detection of high-risk vulnerabilities, see Sect. 2.1.
- **Definition:** For supporting *Web Application Modes*, two *Docker* containers were implemented: one for *Mode Dynamic*, including Juice Shop and its components, and another for *Mode Static*, providing only static HTML pages. To provide auto-restart functionality on the Node.js server in the *Mode Dynamic*, Juice Shop is operated as a service of the production process manager PM2 [PM223]. Vulnerabilities are identified with OWASP Dependency Check [OWA23b].
- **Analysis:** During a mode switch, containers are stopped and started immediately in order to reduce the attack surface and increase security.

Database Management System Modes

- **Type:** Two functional equivalent modes have been implemented at the database layer: *Mode SQLite* and *Mode PostgreSQL*. Both modes encompass the corresponding DBMS and enable to create, read, update, and delete data. Security issues of SQLite may not impact PostgreSQL and vice versa. This consideration guides the switch to the mode with the fewest issues.
- **Trigger:** Modes can be switched as needed or on a regular basis. A mode switch is initiated in response to the publication of a new vulnerability or during the

patching process for the respective DBMS If a new vulnerability is published or patched for the respective DBMS a mode switch is triggered. To support the ideas of MTD, the DBMS modes can be switched timed-triggered.

- **Protocol:** An environment file saves the previous and the target DBMS mode. The previous mode is necessary to decide whether to synchronize the data between the modes. Listing 9 shows the shell script for switching to the specified target mode. If the previous mode was *SQLite* and the target mode is *PostgreSQL*, the script starts the PostgreSQL server and imports the data from the SQLite database into PostgreSQL using the command line tool `pgloader` [Fon23]. If the previous mode was *PostgreSQL* and the target mode is *SQLite*, the script exports the data from the PostgreSQL database using the included command line tool `pg_dump` [Pos23]. Only the data with all column names is exported. Lines with comments, set commands, and select commands are deleted. Then, the PostgreSQL server is stopped, and the SQLite database is truncated before importing the data. As the mode switch affects the *Mode Dynamic* of the web application, the PM2 service for Juice Shop is restarted.

```
if [ $dbms_mode_previous = 'SQLITE' ] && [ $dbms_mode = 'POSTGRESQL' ]
then
   echo "Switch DBMS Mode to PostgreSQL"
   echo "Start PostgreSQL Server"
   pg_ctl start
   echo "Import Data from SQLite"
   export PGPASSWORD=$PG_PWD
   echo $(pgloader -L $DATABASE_DIR/log.txt $SQLITE_DB \
      postgresql://$PG_USER:$PG_PWD@localhost/$PG_DB_NAME)
   Restart_PM2_Service_for_JuiceShop
elif if [ $dbms_mode_previous = 'POSTGRESQL' ] && [ $dbms_mode = 'SQLITE' ]
then
   echo "Switch DBMS Mode to SQLite"
   echo "Export Data from PostgreSQL and Replace Invalid Commands"
   export PGPASSWORD=$PG_PWD
   echo $(pg_dump --user=$PG_USER --column-inserts --data-only $PG_DB_NAME \
      | grep -v "^--" | grep -v "^SET" | grep -v "^SELECT" | sed 's/public.//g' \
      > $HOME_DIR"/posgresql_export.sql")
   echo "Stop PostgreSQL Server"
   pg_ctl stop
   echo "Truncate SQLite database"
   echo $(sqlite3 $SQLITE_DB_FILE < $DATABASE_DIR"/truncate_data.sql")
   echo "Import Data from PostgreSQL"
   echo $(sqlite3 $SQLITE_DB_FILE < $DATABASE_DIR"/posgresql_export.sql")
   Restart_PM2_Service_for_JuiceShop
fi
```

LISTING 9: Switching Juice Shop Database Management System Modes

- **Attributes:** *Mode SQLite* is the initial mode after the system starts. Switches are allowed in both directions, reflecting functional equivalence.
- **Definition:** By default, Juice Shop includes only the *SQLite* DBMS. To enable switching to an additional database implementation mode, *PostgreSQL* DBMS was installed. The schema (data definitions) of the *SQLite* database was exported, manually adapted, and then imported into *PostgreSQL*. As *Sequelize* is used by Juice Shop for the database connection, it was necessary to adapt the configuration to consider the currently active DBMS mode.
- **Analysis:** During a mode switch, the DBMS of the new mode is started immediately, but the DBMS of the old mode is not stopped immediately. In the protocol, it is necessary to transfer the data from the old mode database to the database of the new mode.

Database Access Modes

- **Type:** Database access modes consist of two functional states: *Mode Read-Write* and *Mode Read-Only*. These modes operate as sub-modes under the DBMS modes and manage access rights to the underlying database system. While in the *Mode Read-Write*, all major database operations are supported (create, read, update, and delete), only read is possible in the *Mode Read-Only*.
- **Trigger:** The switch to *Mode Read-Only* is automatically triggered after a specified number of SQL injection attacks. A *F2b* filter has been implemented to detect possible SQL injections in log files and execute countermeasures on matches. These measures include blocking the attacker's IP address, informing the operator, and switching the mode. To switch back to *Mode Read-Write*, manual intervention is required.
- **Protocol:** An environment file saves the target database access mode. Listing 10 shows the shell script for switching to the specified target mode. The commands initiate the necessary configurations and permissions adjustments. For SQLite, the access rights of the database file are changed. For PostgreSQL, the script uses the included command line tool `plsql` to set all transactions to read-only or read-write.
- **Attributes:** *Mode Read-Write* serves as the default mode upon system startup because it provides the most functionality. Mode switches are allowed in both directions. Currently, only the mode switch from *Mode Read-Write* to *Mode Read-Only* is automated. The switch from *Mode Read-Only* to *Mode Read-Write* requires manual intervention by the operator. The criteria for triggering the switch to *Mode Read-Only* is the detection of SQL injection attacks and the specified threshold.
- **Definition:** Implementing database access modes required no additional tools, as the required shell commands to change the access are already included in the

operating system and the DBMS. However, to prevent errors caused by writing attempts to the database, Juice Shop was adapted. For monitoring log files and automatic response, *F2b* was utilized.

```
if [ $db_access_mode = 'READONLY' ]
then
    # Set SQLite Database Read-Only
    chmod ugo-xw $SQLITE_DB_FILE
    # Set PostgreSQL Database Read-Only
    export PGPASSWORD=$PG_PWD
    psql -d $PG_DB_NAME -U $PG_USER -c "BEGIN; SET TRANSACTION READ WRITE;
    ALTER DATABASE $PG_DB_NAME SET default_transaction_read_only=on; COMMIT;"
else
    # Set SQLite Database to Read-Write
    chmod ugo+rw $SQLITE_DB_FILE
    # Set PostgreSQL Database to Read-Write
    export PGPASSWORD=$PG_PWD
    psql -d $PG_DB_NAME -U $PG_USER -c "BEGIN; SET TRANSACTION READ WRITE;
    ALTER DATABASE $PG_DB_NAME SET default_transaction_read_only=off; COMMIT;"
fi
```

LISTING 10: Switching Juice Shop Database Access Modes

- **Analysis:** The necessary configurations and permissions changes for a mode switch are executed immediately to reduce the attack surface and increase security in case of attacks.

Results

The introduction of different modes, such as *Mode SQLite* and *Mode PostgreSQL* at the database layer and *Mode Dynamic* and *Mode Static* at the web application layer, contributes to resilience. Switching between these modes in response to specific events like vulnerabilities, SQL injections, and brute force attacks on authentication, plays a crucial role in strengthening application security.

Simulated attacks revealed that mode-switching provides an immediate and effective response. After detecting an attack, the control component switches modes, leading to a reduction in functionality and an increase in security. This immediate adaptation mitigates specific vulnerabilities, limits the attack surface, and disrupts the attacker's strategies. For example, in case of a vulnerability in *Mode SQLite*, switching to *Mode PostgreSQL* immediately stops ongoing attacks and prevents further exploitation. Switching from database access *Mode Read-Write* to *Mode Read-Only* restricts data manipulation, providing an additional defense against potential data theft and further intrusion. Switching from the web application *Mode Dynamic* to *Mode Static* protects from data theft but brings a drastic reduction in functionality. Nevertheless, it mitigates ongoing attacks at the same time.

In conclusion, the systematic integration of a multi-modal architecture and the mode-switching framework in Juice Shop enhances its resilience by promptly responding to detected attacks, adapting to changing security requirements, and minimizing the impact of potential vulnerabilities.

5.2 Web Server Case Study

A case study was conducted using multiple web servers to demonstrate the expressiveness, succinctness, performance, and potential benefits of the mode-switching framework presented in Sect. 4.5. The objective was to assess whether mode switching can effectively mitigate vulnerabilities and increase resilience. Following the guidelines described by *Runeson & Hoest* [RH08], historical data spanning two years from February 2019 to February 2021 was analyzed and replayed.

This case study aimed to address RQ1, assessing how mode-switching can reduce the risk of a system being vulnerable and increase its resilience. The investigation focused on collecting and analyzing reported vulnerabilities for this system type recorded in the past to estimate the potential benefits of applying mode-switching.

Regarding RQ4, the study measured the framework efficiency in various tasks such as processing configurations, collecting vulnerabilities and patches, comparing mode risks, and executing mode-switching. Additionally, the study examined the capability of MDSL to accurately describe a real-world *System Configuration*, including its components, modes, and the actions required to switch from one mode to another. A comparison to manually defining the modes on the code level is provided.

Requirements: In order to accurately assess the expressiveness, succinctness, and performance of both the framework and MDSL, it is necessary to employ a system that can utilize several different modes and actions. The investigation explores how well MDSL can capture the different modes and how fast the framework processes the *System Configuration*. Additionally, the system should have well-documented vulnerabilities and patches, specifically the used software. This allows us to calculate the CVSS score sum for each mode, enabling informed decisions on whether to execute a mode switch.

Selected Use Case: Based on these requirements, the two most commonly used web servers *Apache* and *Nginx* were selected, combined with the *PHP: Hypertext Preprocessor (PHP)*. This combination was selected for the case study because it is frequently used by web applications and by common *Content Management Sys-*

tems (CMSs) such as *WordPress*, *Drupal*, or *Joomla* in the real world. Typically, an instance of a web application uses only a single web server. When a vulnerability becomes known, software vendors need to react quickly to provide a patch. Then, the system owner can test and install the patch, before attackers take advantage of the vulnerability. As the web servers and PHP were regularly affected by vulnerabilities in the past, there is a need for improvement. The use case fits the previously mentioned requirements as modes can be derived, and there is a long history of vulnerabilities and patches. Figure 5.2 depicts the use case with four modes using the multi-modal architecture, representing specific web server configurations and the functionality they can provide. The modes *Apache* and *Nginx* offer only static web content such as HTML pages. In addition to that, the modes *ApacheWithPhp*

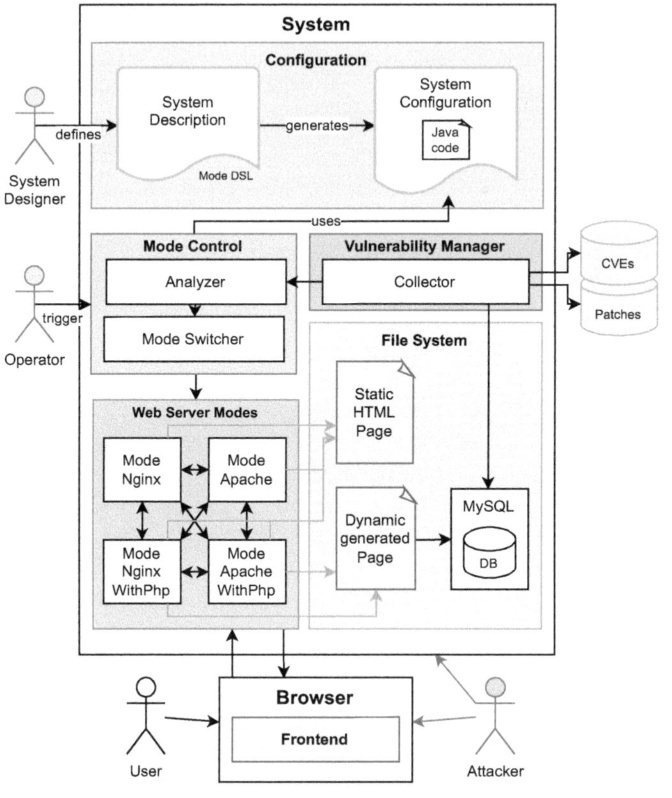

Fig. 5.2 Web Server Scenario using the Mode-Switching Framework (Iteration 1)

and *NginxWithPhp* extend the other modes with PHP functionality and can provide dynamic content from the MySQL database. The black arrows in the green box indicate possible mode switches. In this scenario, switching seamlessly from one mode to another is possible.

Evaluation Settings: For the evaluation on a virtualized Linux distribution Debian 10 (Buster) with 2 GB RAM on an Intel Core i7-8565U machine with 16 GB RAM, a *System Configuration* with two different web server implementations was created in their most recent versions: (*i*) Apache2 version 2.4.38 and (*ii*) nginx v1.14.2. Additionally, running on the web server, PHP v7.3 and FastCGI Process Manager are used to serve dynamic web content. Both web servers were selected because they provide similar functionality and are compatible with PHP. MariaDB v10.3.29 was selected as a MySQL DBMS and WordPress 5.8 as a CMS. The web content was saved to the default `/var/www` directory, so both web servers had access to it.

Using historical data, the aim was to demonstrate the effectiveness of mode switching in mitigating vulnerabilities and increasing resilience. From Feb. 2019 to Feb. 2021, the web server scenario with the software components and versions described above was simulated and replayed. This period was selected because there were no major software changes. Changes might have affected the measurements.

Software modules are identified by their CPE, e.g., `apache:http_server:2.4.38` [MIT23c]. Based on this information, NVD was used to retrieve relevant vulnerabilities (CVEs) [NVD23a]. For each CVE, NVD provides a summary, the severity with CVSS version 3.1, some additional links to the vendor and third-party advisories, as well as an assignment to the CWE for each vulnerability. However, NVD only provides the published date and the last modified date of CVEs. Therefore, manual inspection of vendor websites was necessary to determine the initial report dates to security teams and the release dates of updates and patches. This allowed the calculation of the duration until the update and the patch was provided. In addition, the time when CVEs were fixed in the corresponding packages of the stable release Buster was examined through the *Debian Security Tracker* [Deb23a].

While the unstable and testing releases of Debian may contain the latest software versions, they may lack extensive testing, potentially leading to bugs and compatibility issues with other packages [Deb23b]. Debian prioritizes production stability in stable releases and does not offer all the latest software versions [Deb23c]. Eventually, this approach reduces administrative effort for operators as the packages are tested extensively, and typically, no configuration adjustments are necessary after installing updates [Deb23b]. Nevertheless, outdated packages can contain vulnerabilities. To address this, the Debian security team backports actual fixes to older software versions and provides updates and advisories for about three years [Deb23d].

However, this may result in some delay compared to patches released directly by software vendors.

Evaluation Metrics: The evaluation metrics are further referenced with unique identifiers ranging from M1.1 to M3.3.

The evaluation focused on the expressiveness and succinctness of the mode-switching framework outlined in Sect. 4.5 for the selected case. Without using MDSL, a user must manually develop the *System Configuration*. Therefore, a comparison was made between the manually defined *MDSL System Description* (M1.1) and the automatically generated *System Configuration* (M1.2). The evaluation involved measuring the *Lines of Code (LoC)* and the *Halstead Software Complexity Measures* to objectively assess the complexity and quality of the MDSL [Hal77]. In the following, the metrics are briefly explained:

- The *loc* measure the size of the code by counting the lines.
- The *Vocabulary* (n) helps to understand the complexity and represents the total number of unique operators ($n1$) and unique operands ($n2$) used in the code: $n = n1 + n2$.
- The *Size* (N) or program length is the total number of elements (operators [$N1$] and operands [$N2$]): $N = N1 + N2$.
- The *Volume* (V) estimates the needed program space and is proportional to the size. It is calculated by the size multiplied with the 2-base logarithm of the vocabulary: $V = N * ld(n)$.
- The *Difficulty Level* (D) shows how difficult it is to handle the program. It is proportional to the number of unique operators and is calculated as $D = (n1/2) * (N2/n2)$.
- The *Effort* (E) is related to the mental activity needed to implement or understand the program and is calculated as the volume multiplied by the size: $E = D * V$.

Additionally, the execution time at design-time and runtime was measured. After defining the *System Description* with the MDSL, the time to validate it (M2.1) was measured, and how long it takes to generate (M2.2) and compile (M2.3) the *System Configuration*. At runtime, the time to fetch vulnerabilities and patches (M2.4), calculate and compare the CVSS score sum of the modes (M2.5), and switch the mode (M2.6) was measured. The mode-switching duration should be below the common HTTP timeouts from 30 to 120 seconds [TLG10]. If the server does not respond after the waiting time has expired, the user's browser will display an error message.

Based on real-world software vulnerabilities and patches, the CVSS score sum (M3.1) and the risky days (M3.2) were calculated for each mode and day manually. Finally, the results were compared with the automatically calculated results of the mode-switching framework (M3.3).

Web Server Modes

The following modes are presented based on the formalized description of MMS, further explained in Sect. 3.7.

- **Type:** *Web Server Modes* present specific web server configurations and the functionality they can provide. While the modes *Apache* and *Nginx* provide only static web content such as HTML pages, the sub modes *ApacheWithPhp* and *NginxWithPhp* extend the other modes with PHP functionality and can provide dynamic content from a database. Thus, the super modes and the sub modes are mostly functional equivalent. The super modes *Apache* and *Nginx* represent degraded modes as they provide less functionality compared to the sub modes *ApacheWithPhp* and *NginxWithPhp*.
- **Trigger:** The *Vulnerability Manager* regularly checks vulnerability and patch databases for changes, and automatically triggers the *Mode Control* component if the used software is affected. Time-triggered mode switching via cron jobs is supported using the alternative modes. Additionally, the operator can manually trigger a mode switch.
- **Protocol:** The system can only be in one *Web Server Mode* at a time. The *Mode Control* will suggest the enabled mode with the lowest total CVSS score and the highest priority. If there are no vulnerabilities at system startup, the initial mode is *ApacheWithPhp*. A mode switch involves stopping services of the old mode and starting services of the new mode. Still, the required services of the old mode in the new mode are left running. Only the services that are no longer required are stopped.
- **Attributes:** In the *MDSL Web Server System Description*, see Listing 11, the operating system, the distribution, and the supported modes are specified by the system designer. Also, the software used and the corresponding package can be specified. This data is needed for collecting vulnerabilities and patches. Mode attributes include a description and priorities to order the modes, e.g., from highest to lowest, based on functionality. Modes can consist of several services. These services must be started in the correct order when starting a mode. If the mode is exited again, the order to end the services must also be observed. Additionally, modes can be enabled or disabled, e.g., for maintenance, because of vulnerabilities or patches, or based on the operators' decision. Only enabled modes can be started. Modes can also have alternative modes to support MTD

and switching the modes regularly (`exchangeable with`). The modes are strongly connected, and the system permits mode switches from any enabled mode to another. Services use specific software and have actions to start and stop the services. These actions are typically shell commands. The criteria for triggering a mode switch are the total CVSS sum and the priority.

- **Definition:** The software components include the web servers *Apache* and *Nginx*, and the *PHP FastCGI Process Manager*. Vulnerabilities (CVEs) are collected from NVD. Based on the CVEs patches were fetched from OS or distribution vendor, e.g., the *Debian Security Tracker* [Deb23a].
- **Analysis:** A mode switch is executed immediately in order to reduce the attack surface and increase security.

MDSL Expressiveness & Succinctness

Research Design: For assessing the expressiveness and succinctness of the framework when defining components of a MMS, a *MDSL System Description* was manually created. This description outlined the different software components, modes, services, and actions, see Listing 11. Four modes were defined for the web server system and prioritized as follows: *ApacheWithPhp*, *NginxWithPhp*, *Apache* and *Nginx*. The operator can set the priority as desired. Typically, the mode with the most functionality has the highest priority. In the first step, one researcher created the different software components, each identified by a unique name, e.g., the corresponding Debian package name or the CPE. Afterwards, the respective modes were defined. For instance, mode *ApacheWithPhp* extends the general mode *Apache* with additional actions to enable required Apache modules and configurations and to start the Linux service for the PHP FastCGI Process Manager. The final step involves defining the specific actions required to switch to a new mode. This may include executing Linux-specific shell commands like (`systemctl start`) to start a service.

After manually defining the *System Description* with the MDSL, the Java-based *System Configuration* was automatically generated and compiled. Then, a comparison was made between their *LoC* and the *Halstead Software Complexity Measures*, i.e., vocabulary, size, volume, difficulty, and effort, see Table 5.2.

Table 5.2 Comparison of the LoC and the Halstead's Software Complexity Measures of the MDSL System Description and the Generated Java-Based System Configuration

Metric	*LoC*	*Vocabulary*	*Size*	*Volume*	*Difficulty*	*Effort*
Java-based System Config.	95	75	566	3525.51	38.75	136613
MDSL System Description	59	69	206	1258.36	29.25	36806
Reduction (rel)	38 %	8 %	64 %	64 %	25 %	73 %

```
System Webserver has following modes Apache,ApacheWithPhp,Nginx,NginxWithPhp
operatingSystem Linux distribution Debian

Software apache2
Software nginx
Software php7 package "php7.3"

Mode Apache description "Only static HTML pages" priority 2
startServices apache2Service stopServices apache2Service
exchangeable with NginxOnly

Mode ApacheWithPhp extends Apache
description "Static HTML and dynamic PHP pages" priority 4
startServices apacheModProxyFcgiService,apacheModSetenvifService,phpFpmService,
apacheConfPhpFpmService,apache2restart
stopServices apacheConfPhpFpmService,phpFpmService,apacheModSetenvifService,
apacheModProxyFcgiService,apache2restart
exchangeable with NginxWithPhp

Mode NginxOnly description "Only static HTML pages" priority 1
startServices nginxService stopServices nginxService
exchangeable with Apache

Mode NginxWithPhp extends NginxOnly
description "Static HTML and dynamic PHP pages" priority 3
startServices phpFpmService stopServices phpFpmService
exchangeable with ApacheWithPhp

Service apache2Service usesSoftware apache2
startActions startService("apache2") stopActions stopService("apache2")
Service  apacheModProxyFcgiService
startActions enableApacheMod("proxy_fcgi") stopActions disableApacheMod("proxy_fcgi")
Service  apacheModSetenvifService
startActions enableApacheMod("setenvif") stopActions disableApacheMod("setenvif")
Service  apacheConfPhpFpmService
startActions enableApacheConf("php7.3-fpm") stopActions disableApacheConf("php7.3-fpm")
Service  nginxService usesSoftware nginx
startActions startService("nginx") stopActions stopService("nginx")
Service  phpFpmService usesSoftware php7
startActions startService("php7.3-fpm") stopActions stopService("php7.3-fpm")
Service  apache2restart
startActions restartService("apache2") //to activate the new configuration
stopActions reloadService("apache2") //to refresh the changed configuration

Action startService shellCmd "/usr/bin/systemctl start"
Action stopService shellCmd "/usr/bin/systemctl stop"
Action restartService shellCmd "/usr/bin/systemctl restart"
Action reloadService shellCmd "/usr/bin/systemctl reload"
Action enableApacheMod shellCmd "/usr/sbin/a2enmod --quiet"
Action disableApacheMod shellCmd "/usr/sbin/a2dismod -f --quiet"
Action enableApacheConf shellCmd "/usr/sbin/a2enconf --quiet"
Action disableApacheConf shellCmd "/usr/sbin/a2disconf --quiet"
```

LISTING 11: MDSL Web Server System Description

Results: Describing a system using MDSL showed that only 59 LoC with 69 unique operators and operands vocabulary (M1.1) were needed to define software components, modes, and actions compared to 95 LoC and a vocabulary of 75 for the generated Java-based *System Configuration* (M1.2). MDSL provided a reduction of 36 LoC, 6 tokens in the vocabulary, and 360 tokens in the program size. Modes like *ApacheWithPhp* build a logical frame around the web server and additional described services. No further additions or modifications in the used software code were necessary. Inheritance allowed us to extend the mode *Apache* with PHP services as mode *ApacheWithPhp*. This mechanism reduces redundancy, lowers size and volume, and increases maintainability because changes like adapting actions or software versions only have to be made in one place. For the given use case, the MDSL allows focusing on the arrangement and declarative definition of modes and shows a 25 % reduction for the difficulty level. In addition to improving the evaluated metrics, it is worth noting that MDSL provides more functionality than a simple text file. Tool support provided by the Eclipse framework, such as code completion, syntax highlighting, and the configuration validation component, eases the task of creating modes and actions, thus increasing usability. The *System Description* is statically analyzed, and error messages occur if there are duplicates of mode names, priorities, or action names. If the Debian OS and the distribution are specified, the framework checks if the software and the packages are within the official package list. If not, warning messages are displayed.

The demonstrated capability of MDSL lies in its expressiveness to describe the given case and its modes at an abstract level, completed by features such as syntax highlighting, code completion, and validation through dedicated tool support. In addition, using MDSL leads to more succinct specifications than manual coding. Looking at the *Halstead Software Complexity Measures*, the *System Description* in MDSL shows a reduction of 38 % in LoC, 64 % in size and volume, and 73 % in the expected testing efforts when compared to the generated Java-based *System Configuration*.

Performance of the Mode-Switching Framework

The mode-switching framework was employed in the second part of the case study, and the performance was measured both at design-time and runtime.

Research Design: At design-time, the average runtime of 500 executions to validate the user-defined *System Description* and to generate and compile the *System Configuration* automatically was analyzed. As mentioned in Sect. 4.5, the Xtext parser is used, and the validation component provides functionality to check if the syntax conforms to the defined MDSL and if there are no logical errors. If successful, the framework automatically generates and compiles the Java-based *System Configuration* with Xtend. At runtime, the *Vulnerability Manager* fetched the CVEs from NVD and patches from the *Debian Security Tracker* based on the specified software (vendor, product, and version), and the time was measured [NVD23c, Deb23a]. Then, the appearance of CVEs and patches over the last two years was replayed. Therefore, the vulnerability scores for each software, mode, and day were calculated. For a plausibility check, a comparison was made with manual calculations to ensure the correct next mode was suggested based on the lowest risk and highest priority. Additionally, nearly 2,500 mode switches were simulated and analyzed. The analysis included examining the correct execution and overlaps between old and new modes. For example, old-mode services that are needed in the new mode continue to run, and only services that are no longer needed are stopped. Finally, the effects of mode-switching on currently running requests and logged-in users were analyzed. This involved studying the upload of a 50 MB file and the session handling at the WordPress CMS.

Results: The *MDSL System Description* and the generated *System Configuration* were successfully utilized in the framework. The validation of the *System Description* (M2.1) took about 3.1 milliseconds (ms), on average. While the generation of the *System Configuration* (M2.2) was finished in 1.3 ms, the compilation (M2.3) took longer with 204.2 ms. After starting the framework, the mode with the lowest CVSS score sum and the highest priority was selected. On average, the CVEs were fetched in 416.5 ms and the corresponding patches in 251.8 ms (M2.4). However, this runtime highly depends on the Internet connection, server availability, and rate limits of the online repositories. NVD provided some outdated and already fixed vulnerabilities for the CPEs or did not consider the used operating system or distribution. To compensate these inaccuracies, the framework ignores CVEs reported before the distribution's release year. In some rare cases, vulnerabilities were assigned to different packages or specific to Windows. For example, NVD assigned CVE-2019-9517 to CPE `apache:traffic_server` while the Debian Security Tracker used the package *apache2*. The other way round, NVD considered CVE-2019-19246 as PHP vulnerability, but Debian package maintainers assigned it to the PHP sub-package *libonig*. This complicates the process of connecting vulnerabilities to patches. To take this issue into account, the *Vulnerability Manager* considers CVEs from Debian Security Tracker with status resolved or urgency unimportant as fixed. In total, 29 of 74 CVEs from NVD were excluded because 23 were already resolved in the Debian distribution, three were vulnerabilities of other software parts, and three were specific to Windows. 45 vulnerabilities remained in the evaluation.

The calculation of the CVSS score sum for each mode and the comparison took 0.14 ms on average (M2.5). The next modes were automatically suggested, matching the manual evaluation. The mode switches were correctly executed, and partial start and stop functionality was used. Table 5.3 shows the average duration of 2500 mode switches (M2.6). It takes about 10.3 ms to start the mode *Apache*, whereas it takes 180.5 ms to start *ApacheWithPhp* and even longer (191.5 ms) to switch from *Nginx* to *ApacheWithPhp* and back (185.6 ms). Except for longer running requests like uploads, it became obvious that switching between the two modes with PHP had no negative side effects on users currently logged in to WordPress because the session information was stored in the database and not changed on the server and client.

At design-time, it takes 4.1 ms to validate the *System Description*, 1.3 ms to generate the *System Configuration*, and 204.2 ms to compile it. At runtime, it takes less than a second in total (345.41 ms) to collect vulnerabilities (416.5 ms) and patches (251.8 ms), to calculate and analyze the modes CVSS score sum (0.14 ms), and to perform the necessary mode switches (93.47 ms). The collection and analysis are performed in the background and do not affect current users. The mode switch takes from 10.3 ms (inactive to Apache) to 191.5 ms (Nginx to ApacheWithPhp), which is far below the required lower HTTP timeout of 30 seconds (30,000 ms) [TLG10].

Table 5.3 Average Web Server Mode Switch Duration in Milliseconds

To / From	*Apache*	*Apache WithPhp*	*Nginx*	*Nginx WithPhp*	*in- active*
Apache	–	176.9	64.7	116.3	13.8
ApacheWithPhp	121.6	–	185.6	163.7	141.4
Nginx	32.5	191.5	–	49.3	21.5
NginxWithPhp	48.1	148.8	16.7	–	36.2
inactive	10.3	180.5	51.7	98.2	–

Mitigating Vulnerabilities by Switching Modes

In the third part of the case study, the previously defined modes and the mode-switching framework were used for simulating a time span of two years with real vulnerabilities and patches in a web server scenario. The simulation shows the effects of previously reported vulnerabilities and to which extent mode-switching can lower the risk of being susceptible to potential attacks.

Research Design: The web server scenario consists of four different modes. In order to reduce the risk of the web server, the framework switches modes whenever there is another mode that provides less risk. The web server scenario was simulated for a time span of two years, from February 2019 to February 2021, and the risk was calculated for each mode and each day by retrieving information about unpatched vulnerabilities and their corresponding scores. Whenever another mode had less risk (a lower score), the framework switched to this mode. The score was re-calculated if a patch was provided, possibly leading to another mode switch.

Table 5.4 shows the number of CVEs for the used software, the CVSS scores, and how long it took the vendor respectively package maintainers to provide updates (Days4Update) and patches (Days4Patch). As mentioned in the evaluation settings, there is eventually a delay between the software vendor's update and the distribution maintainer's patch. The time span between vulnerability publication and the patch date provided by the Debian Package Tracker was considered. Software vendors provided updates for the 45 relevant vulnerabilities within an average of 77.2 days. This number is similar to the 77.5 days of vulnerability statistics by [Edg19]. The Google Project Zero [Cim19] reports that they resolve 95.8 % of found security bugs before the 90-day deadline of common public disclosure. In the given scenario, public disclosures of vulnerabilities did not occur prior to the release of updates by the vendors. Software vendors rarely publicly share reported new vulnerabilities or share information with customers before providing an update, so only the provided data by CVE sources was used. During the investigation period, the software nginx 1.14.2 had fewer vulnerabilities than Apache 2.4.38 and lower CVSS scores. PHP 7.3.5 had more CVEs than Apache and nginx combined, and the deployment of patches (Days4Patch) took 71.3 days on average. Overall, it took the Debian package maintainers 49.4 days on average, to analyze vendor updates and to provide patches for the used software. Once the patches are available, system owners can install them and close the WoV. Using this software resulted in a WoV of 126.6 days on average (Days4Update: 77.2 + Days4Patch: 49.4), from the report to the vendor's security team until the package maintainers provided the patch.

Table 5.4 Overview of the Different Web Server Components, the Number and Severity of the Vulnerabilities, and Time until Updates and Patches were available

Software	CVEs	CVSS 3.1		Days4Update		Days4Patch	
	#	*sum*	*avg*	*sum*	*avg*	*sum*	*avg*
apache 2.4.38	17	117.3	6.9	1318	77.5	467	27.5
nginx 1.14.2	4	26.8	6.7	284	71.0	46	11.5
php 7.3.5	24	169.3	7.1	1871	78.0	1712	71.3
total	45	313.4	7.0	3473	77.2	2225	49.4

Out of 45 vulnerabilities, five publicly available exploits were found on *Packet Storm* and the *Exploit Database* [Pac20, Off23]: four in Apache and one in PHP. On average, they were published 50 days after the vendor update and 32 days after the distribution patch. Three exploits were released very close to the patch date. However, there can be both unpublished vulnerabilities and unpublished exploits.

Results: Figure 5.3A shows the results of using only the modes *Apache* (red) or *Nginx* (green) compared with the mode-switching results (blue) for the analyzed time span of two years. The total CVSS 3.1 score for each mode and each day was calculated. New vulnerabilities increased the total score, and corresponding published patches reduced it, assuming that patches were installed as soon as they were deployed. The mode *Apache* (red) had several periods with a higher CVSS score in Apr. 2019, from Aug.-Oct. 2019, and from Apr.-Aug. 2020. The maximum CVSS score of 43.5 was on Aug. 14, 2019. The mode *Nginx* (green) had only two short periods in Aug. 2019 and Jan. 2020 with a higher CVSS score and a maximum of 21.5. The framework switched automatically to the mode with the lowest CVSS score and highest priority. This resulted in a risk reduction on 751 of 759 days (98.9 % of the time) to a zero CVSS score. Only eight days from Aug. 14-21, 2019 mode-switching had a higher CVSS score (max 21.5), because both modes *Apache* and *Nginx* had unpatched vulnerabilities. On Aug. 14, 2019, there was a mode switch from *Apache* (43.5) to *Nginx* (21.5). Because some *Apache* vulnerabilities were fixed and the CVSS score was reduced to 6.1 on Aug. 18, 2019, the framework switched back to *Apache*.

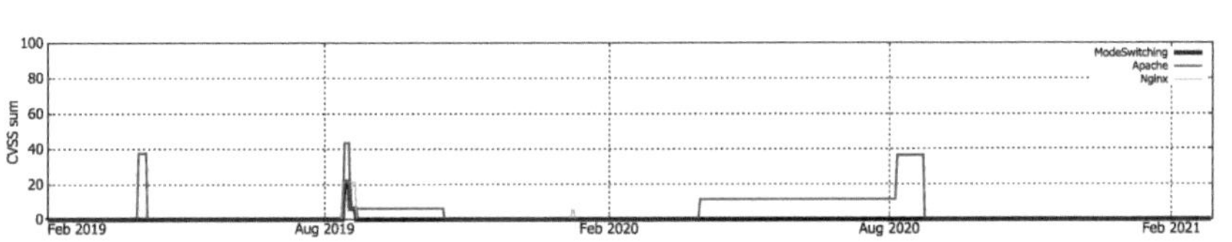

(A) CVSS Sum Mode-Switching (blue), Apache (red) and Nginx (green)

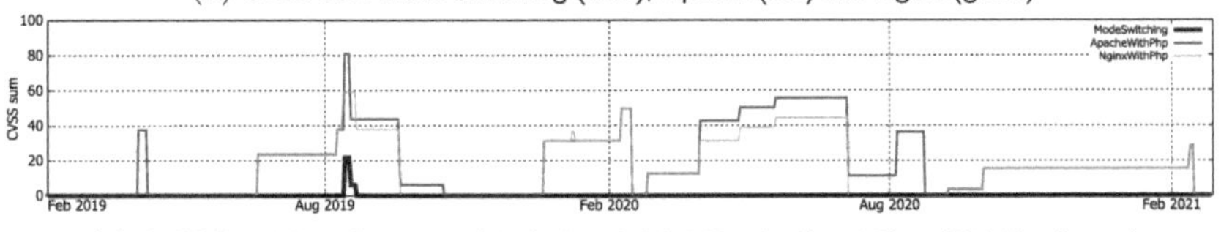

(B) CVSS Sum Mode-Switching (blue), ApacheWithPhp (red) and NginxWithPhp (green)

Fig. 5.3 Web Server Modes Total Vulnerability Scores

Without PHP the web servers *Nginx* and *Apache* can only provide static pages. Using PHP allows generating dynamic content with template systems, CMSs, or frameworks. Nevertheless, that also involves greater risks, as the results of the PHP modes *ApacheWithPhp* (red) and *NginxWithPhp* (green) in Fig. 5.4 indicates. The analysis was extended, and a comparison was made with mode-switching (blue). PHP and its vulnerabilities (cf. Table 5.4) increased the number of risky days of the PHP modes on 309 of 759 days (40.7 % of the time).

Table 5.5 compares the risky days and the CVSS score of the modes. Without mode-switching and using only one mode resulted in 11 (*Nginx*) to 536 (*Apache-WithPhp*) risky days (M3.2) and a maximum total CVSS score of from 21.5 (*Nginx*) to 81.2 (*ApacheWithPhp*). Using *ApacheWithPhp* 70.6 % of the analyzed time was risky, with an average total CVSS score of 18.5 (M3.1). Compared to that, mode-switching between all modes lead to only 1.1 % risky days and an average total CVSS score of 0.1. Switching only between the PHP modes *ApacheWithPhp* and *NginxWithPhp* did not reduce the risky days in the analyzed time span, because most of the vulnerabilities were caused by PHP. Using *NginxWithPhp* provides the same amount of 442 risky days (58.2 % of the time). However, mode-switching between the PHP modes leads to a lower CVSS score sum (11255.4) and lower average CVSS (14.8) instead of using only the mode *ApacheWithPhp* (14027.5, 18.5) or *NginxWithPhp* (11349.1, 15.0). At the critical PHP vulnerability CVE-2019-11043 with a CVSS of 9.8, mode-switching was tight, because the vendor update, the CVE, and the fix in the distribution were published on the same day as the exploit [Ler19]. Meanwhile, there is already a *Metasploit* module available, which makes it even easier to take advantage of the vulnerability. Mode switching helped in 4

Table 5.5 Comparison of the Web Server Modes

Mode	Risky days		CVSS 3.1		
	#	%	*sum*	*avg*	*max*
ApacheWithPhp	536	70.6	14027.5	18.5	81.2
NginxWithPhp	442	58.2	11349.1	15.0	59.2
Apache	217	28.6	2882.5	3.8	43.5
Nginx	11	1.4	204.1	0.3	21.5
Mode-Switching (All Modes)	8	1.1	110.4	0.1	21.5
Mode-Switching (PHP Modes)	442	58.2	11255.4	14.8	59.2

out of 5 (80 %) of the mentioned exploits to reduce the attack surface and helped to increase resilience in such cases where it was close to fixing a problem before it became publicly known. Compared to the modes with the maximum and minimum risky days, mode-switching between all modes has 3 (27 %) to 528 (98 %) fewer risky days than using only one mode (M3.3).

Table 5.6 shows mode-switching simulations with different priorities and different available modes, which resulted in different active days for each mode and a different number of mode switches (#). If a new vulnerability or patch is published, modes are switched depending on the total CVSS scores and the priorities. If the mode *Apache* had a higher priority as shown in Table 5.6A, in total, 17 mode switches were needed, and the modes *ApacheWithPhp* and *Apache* were running 70 % (29.4 + 41.2) of the time. Figure 5.4 illustrates the active *Web Server Modes* employing mode-switching with identical priority settings. Throughout the observed time span, the mode with the highest priority, *ApacheWithPhp* (depicted in red), was active 223 days. The mode *NginxWithPhp* (orange) was active only 94 days. The mode *Apache* (violet) was running 313 days, while the mode *Nginx* (green) was active for 129 days.

Table 5.6 Simulations with Different Mode Priorities and Available Modes Modes ordered by Priority Ascending, Number of Active Days, Percentage of Active Days (%), and Number of Switches to the Mode (#)

(A) All Modes - Priority Apache

Mode	***days***	***%***	***#***
ApacheWithPhp	223	29.4	5
NginxWithPhp	94	12.4	4
Apache	313	41.2	5
Nginx	129	17.0	3

(B) All Modes - Priority Nginx

Mode	***days***	***%***	***#***
NginxWithPhp	317	41.8	5
ApacheWithPhp	0	0.0	0
Nginx	435	57.3	7
Apache	7	0.9	2

(C) PHP Modes - Priority Apache

Mode	***days***	***%***	***#***
ApacheWithPhp	447	58.9	6
NginxWithPhp	312	41.1	5

(D) PHP Modes - Priority Nginx

Mode	***days***	***%***	***#***
NginxWithPhp	752	99.1	4
ApacheWithPhp	7	0.9	3

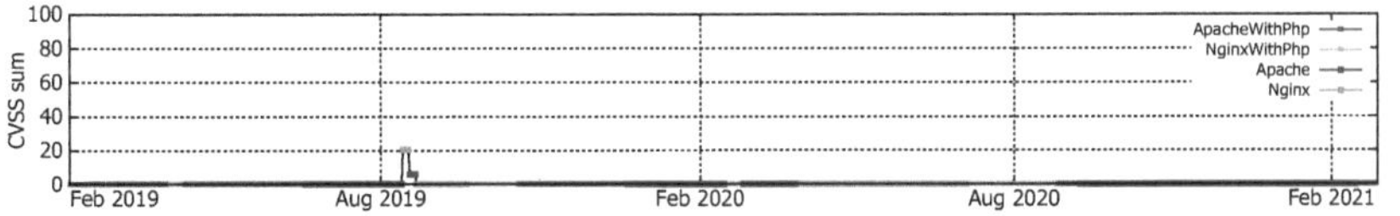

Fig. 5.4 Active Web Server Modes during Mode-Switching

If prioritizing the mode *Nginx* as shown in Table 5.6B, a total of 14 mode switches were required, and the modes *NginxWithPhp* and *Nginx* were active for 99.1 % of the time (41.8 + 57.3). Because CMSs like WordPress need PHP, using only the corresponding modes with PHP are shown in Table 5.6C and Table 5.6D. Preferring the mode *ApacheWithPhp* led to 11 mode switches in total and a nearly equal amount of active days (*ApacheWithPhp* 58.9 %, *NginxWithPhp* 41.1 %). If the mode *NginxWithPhp* had a higher priority, it was running 99.1 % of the time. However, seven mode switches were still executed, and *ApacheWithPhp* was active for seven days, because this mode had a lower CVSS score.

Mode-switching reduced the risk to a zero CVSS score in 98.9 % of the time. Only eight days (1.1 % of the time) had the lowest CVSS score sum from 6.1 to 21.5, because all the software used in the modes had unpatched vulnerabilities. If PHP is needed for specific CMSs, switching modes only between the PHP modes showed no improvement in the number of risky days. However, the results provided a lower total CVSS score sum (-93.7) and a lower average total CVSS score (-0.2) compared to using the mode *NginxWithPhp* only.

5.3 Picture Archiving and Communication System Sample Scenario

Not many CVE entries are available for medical IoT devices. However, the small publicly known number can be deceptive, as *Medical Device Manufacturers (MDMs)* may refrain from making CVEs public. Some MDMs keep that confidential or share detailed information only with registered customers. Therefore, it is hard to reconstruct when a CVE was discovered, shared with the vendor, publicly disclosed, and the patch provided. According to *Medcrypt*, *ICS Medical Advisories (ICSMAs)* have increased by 490 % since the FDA released their *Postmarket Cybersecurity Guidance* in 2016 [Med22a]. These advisories combine multiple CVEs, providing a summary, technical details, a risk evaluation, and possible mitigations [CIS23a]. The most common root causes of vulnerability disclosure are user authentication mismanagement and code defects, with researchers contributing to the identification of 71 % (87) of the 122 ICSMAs, disclosing 355 vulnerabilities [Med22a]. Among 15 device types, infusion pumps, imaging software, patient monitors, and cardiac rhythm management were the most affected and had nearly 50 % of the total found vulnerabilities [Med22a].

For the evaluation of the second iteration of the multi-modal architecture in Sect. 4.6, the focus was on analyzing self-adaptive restrictions provided by different modes and the self-healing process for medical IoT devices. The goal of this sample scenario was to monitor and control several medical IoT devices and investigate how to increase the security and resilience. Recognizing the limited computing power of devices such as pacemakers and insulin pumps for threat detection, the envisioned approach included triggering mode switches externally. Routine check-ups at the HDO, where the doctor checks the log for abnormalities and the battery state, optimizes the programming, and could initiate mode switches. Alternatively, home stations can trigger such switches more timely. As the hardware and software of medical IoT devices may not be readily accessible to researchers, the evaluation strategy involved modeling at least a subset of modes for a specific device, followed by a simulated run with real-world vulnerabilities.

Multi-Modal Architecture for Medical IoT Devices

In Fig. 5.5, the multi-modal architecture for multiple systems is applied to medical IoT devices. The architecture includes two main components: the *Managing System* and the PACS. The managing system includes *Configuration*, *Inventory*, *Mode Control*, and the *Vulnerability Manager*.

As a first step, the system designer defines desired modes (1) using a *System Description* in the MDSL. This description contains information about mode behavior and the actions necessary for switching modes. Based on this description, the *System Configuration* is automatically generated (2) and saved in the inventory for the specific setup of the medical devices. The mode control component manages the configuration, analyzes events, and executes mode switches. Events are triggered by the vulnerability manager and the PACSs. An IDS could also trigger mode switches in a further expansion stage.

The vulnerability manager automatically collects CVEs, patches, and exploits (3) from public databases and vendors, focusing on specified system parts in the system description. Information about new events or changed CVSS scores, is forwarded (4) to the mode control component. Depending on rules and policies, some events (5) will be automatically sent to the PACSs, while others require human-in-the-loop investigation (6). CVSS scores of all software components and modes are aggregated for determining if a mode switch is necessary. While new CVEs and exploits increase total CVSS scores, patches decrease them, assuming they have been installed. Thresholds are implemented to prevent frequent mode switches. If the current mode is no longer suitable, the *Mode Switcher* is called (7), and events are sent to the PACSs to switch their mode. For safety reasons and to reduce the attack surface, manual mode switches, triggered by the operator (3), can be necessary for a specific PACSs or a group of devices. The operator oversees (8) the inventory of multiple PACS and manages (9) the settings of the vulnerability manager.

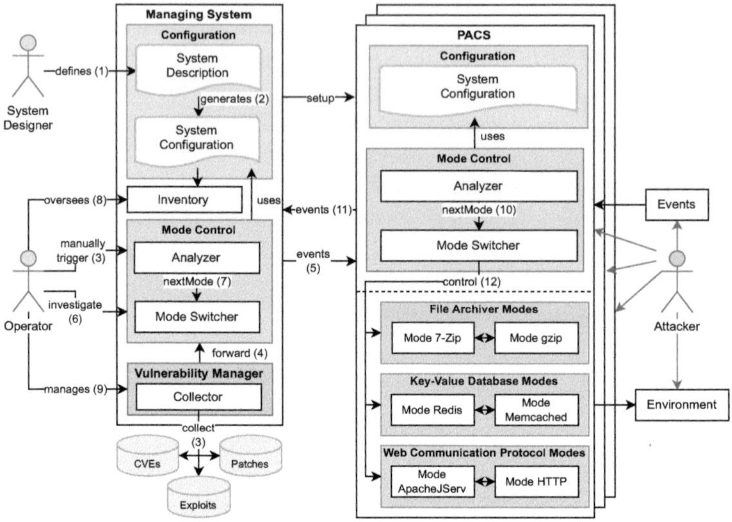

Fig. 5.5 PACS using the Multi-Modal Architecture for Multiple Systems (Iteration 2)

PACSs include components for the *Configuration*, the *Mode Control*, and the *Modes*. Similar to the managing system, mode control uses the configuration, analyzes events of the device itself and from the managing system, suggests (10), and executes mode switches. Events can occur from the PACS itself, the environment, and the managing system. Specific events about attacks and mode switches are sent to the managing system (11). More details about the PACS modes and how to control (12) them are provided in the following subsection.

Sample Mode Simulation

A thorough investigation of medical device vulnerabilities led us to focus on the medical advisory ICSMA-21-187-01 concerning the *Philips Vue Picture Archiving and Communication System (PACS)* [CIS22]. This advisory, dated 2022, highlighted a total of 17 vulnerabilities. Notably, some of these vulnerabilities have low attack complexity and can be exploited remotely. In contrast, many other ICSMAs included only one vulnerability [CIS23a].

Philips Vue PACS uses third-party software components such as *7-Zip*, *Redis*, the *Oracle Database Server* and protocols such as the *Apache JServ Protocol* [Pav24, Red24, ASF23]. Insecure third-party libraries and code defects introduced most vulnerabilities. For example, it took the MDM over nine years to fix the publicly known

CVE-2012-1708 of the *Oracle Database Server*. Even four third-party vulnerabilities with publicly known exploits were patched only up to four years later: CVE-2018-10115, CVE-2018-11218, CVE-2018-12326, CVE-2020-1938. On average, it took Philips more than two years to provide a fix.

The proposed multi-modal architecture and framework can detect vulnerabilities and offer a method to mitigate these vulnerabilities until a patch is provided. Information regarding CVEs and their severity scores, along with available patches, is gathered based on a BOM containing details of all software versions utilized. This information forms the basis for hypothetical mode switch decisions. In fact, the MDSL was used to define a multi-modal architecture.

Various programs and software components are available in typical scenarios to solve a specific task. Instead of relying solely on one type of software for file compression and decompression like 7-Zip or key-value databases like Redis, exploring alternative implementations is advisable. Similarly, considering alternative protocols for data transfer, such as *HTTP*, rather than exclusively depending on the *Apache JServ Protocol*, is recommended [IET24].

Modes for File Archiver, Key-Value Database, and Web Communication Protocols

The following structure of the sample mode simulation considers the formalized description of MMS elaborated in Sect. 3.7.

- **Types:** Functional equivalent modes are established for switching between file archivers, key-value databases, and web communication protocols. Each mode runs different software versions from different manufacturers, exhibiting different vulnerabilities. Thus, CVEs of 7-Zip may not impact the alternative file archiver *gzip* and vice versa [Fre20]. This information guides the selection on the mode with the fewest CVEs and the lowest total CVSS sum. When vulnerabilities arise in the currently running mode, switching to another mode with different software components or versions effectively mitigate these issues.

  - **File Archiver Modes**: *7-Zip* and *gzip*
  - **Key-Value Database Modes**: *Redis* and *Memcached* [Dor24]
  - **Web Communication Protocol Modes**: *ApacheJServ* and *HTTP*

- **Trigger:** Modes can be switched on demand or at regular intervals. Mode switching occurs in response to the disclosure of a new vulnerability or during the patching process for the respective software. To support the ideas of MTD, the modes can be switched also timed-triggered.

- **Protocol:** For implementing the protocol, it will be necessary to transfer the data of the current mode's key-value database to the alternate one, comparable to the sample scenario in Sect. 5.1.
- **Attributes:** Switches are bidirectional, reflecting the functional equivalence of file archivers, key-value databases, and web communication protocols.
- **Definition:** By default, PACS includes only the *7-Zip*, *Redis*, and *ApacheJServ*. The corresponding software must be installed to switch to equivalent implementation modes such as gzip, memcached, or HTTP.
- **Analysis:** Mode switches are executed immediately to increase security.

Results

For the comparison in Fig. 5.6, the sum of all CVSS scores with and without mode-switching for the ICSMA-21-187-01 have been calculated. The timeline shows, that mode-switching between file archiver modes, the key-value database modes, and the web communication protocol modes can reduce the CVSS sum on 1404 of 3760 days (37 % of the time). With mode-switching, the average CVSS sum decreases by 41 % from 28.2 to 16.6. Additional implementations and protocols make systems more complex but allow flexibility during operation. Mode-switching is not limited to only switching between implementation modes. Each implementation can have multiple configuration modes, making it possible to switch between them.

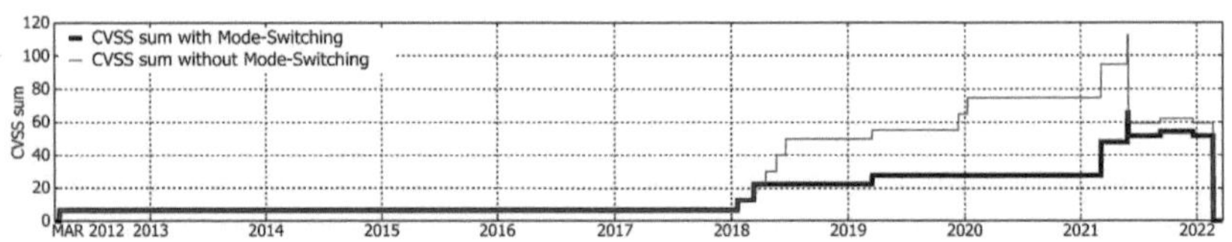

Fig. 5.6 CVSS Sum for Philips Vue PACS with Mode-Switching (blue) and without (red)

5.4 Insulin Pump Sample Scenario

In Sect. 5.3 and the previous work [RSR22], context-aware security modes were implemented and evaluated for medical IoT devices. These modes were switched based on vulnerability scores. For example, *Mode 0* provided core functionality, and *Mode 1-3* extended functionality like remote monitoring and control. Additionally, modes were switched between software components with equal or similar functionality to mitigate vulnerabilities. The emphasis was on securing single devices such as pacemakers and insulin pumps. Mode switching provides a method to reduce the attack surface. However, in light of the increasing number of medical IoT devices, protecting and securing them requires a broader focus than that of a single device. These devices are networked and can also be utilized for security purposes. Some anomalies and attacks can be detected only or easier with multiple devices and a central control component.

Multi-Modal Architecture for Medical IoT Devices

In this sample scenario, the evaluation focuses on the context of medical IoT devices using the multi-modal architecture for multiple systems presented in Sect. 4.6. Building upon previous work, a client-server perspective, multiple medical IoT devices, and an IDPS are introduced. The objective is to resiliently protect patients, medical IoT devices, and the environment by monitoring and automatically adapting when anomalies occur or vulnerabilities become known. Security and reactions to attacks are necessary on multiple layers. Depending on the context of the device, e.g., the connection state (online/offline), the reaction can be less or more restrictive.

Medical IoT devices can be used in hospitals and at home for patients with chronic diseases like diabetes. Manually measuring the blood glucose level, calculating the necessary medication dose, and self-administering insulin injections, especially between meals and overnight, is burdensome. Typically, *Automated Insulin Delivery (AID)* systems simplify this process and operate either partially or fully automated [Med23a]. They consist of wearable devices to monitor vital signs, continuous glucose monitors, wireless connected medicine pumps, and handheld devices or smartphones for local control and connection to the HDO. Based on device settings, patient history, and the current condition, the handheld analyzes the data and may adapt the settings. For example, if the blood glucose level changes, the system decides to increase or decrease the dose of medication. Within a certain threshold, this process is automated as a closed loop. A closed-loop system automatically considers feedback and sensor values to adapt to the system goal without human intervention [BH21]. According to *Boughton & Hovorka*, this process "reduced anxiety, improved sleep and confidence from improved overnight glucose control, less restrictive eating habits and 'time off' from the demands of diabetes management" [BH21, p. 1009]. There exist several commercial and non-commercial AID systems. In 2016, Medtronic MiniMed 670G was the first FDA-approved commercial AID system [FDA16a]. Before 2016, some affected patients with diabetes did not want to wait any longer and developed *Do-It-Yourself* closed-loop implementations and provided them open source, but, needless to say, without warranty.

According to the OpenAPS Community and its founder, Dana Lewis, over 2700 people still use this solution [Dan22].

In the context of medical IoT devices, the closed-loop scenario is extended with information transfer to the HDO for remote monitoring and reconfiguration by a physician, as described in *Rao et al.* [RCLR22]. During the information transfer, using asymmetric cryptography protocols for command and control messages, e.g., signed with a unique public key of each medical IoT device, can secure communication [ZDS21]. If measured values are outside predefined thresholds or if certain sequences of commands appear unusual and could potentially cause physical damage, as seen in the case of Stuxnet [CAN11], or pose a risk to harm patients, the *Operator* (human-in-the-loop) will get a notification. The data can then be checked manually, and the affected settings can be adjusted.

Sample Mode Simulation

The investigation focused on the vulnerabilities of the Medtronic MiniMed 600 series pump system and analyzed the potential benefits of implementing a multi-modal architecture. In 2022, Medtronic alerted patients about a vulnerability in the protection mechanism in their MiniMed 600 series [Med22b]: Exploitation can compromise communication, allow unauthorized users to change the insulin delivery, and "potentially lead to seizure, coma or death". According to the FDA recalls, over 600 000 products in commerce were affected [FDA22a, FDA22b]. The company recommended that patients should manually turn off the "Remote Bolus" feature on the pump, which was on by default. Simplifying this step for patients is beneficial. Using the multi-modal architecture for insulin pumps can make a substantial difference.

Figure 5.7 shows the proposed architecture with four *Connection Modes*. The architecture follows the principle of divide and conquer. Decisions are made as decentralized as possible and centralized as necessary. A *Managing System* will use its *Inventory* to continuously monitor connected *Insulin Pumps* to provide enhanced visibility and situational awareness for the *Operator*. Like an IT security professional, the *Operator*, can analyze the situation and plan and execute adaptations to reach the system goals. For efficiency, repetitive tasks and decisions can be partially or fully automated. Additionally, the *Managing System* centrally monitors CVEs, medical advisories, safety communications, product alerts, warnings, recalls, and other events of public databases, as presented in the previous work [RSR22, RSVW23a]. Based on that analysis, the *Managing System* can automatically send adaptation requests to the *Insulin Pumps* to change their behavior, e.g., block IP addresses or switch their mode. Likewise, the *Operator* and the *Patient* can do that manually at the *Managing System* and the *Insulin Pump*.

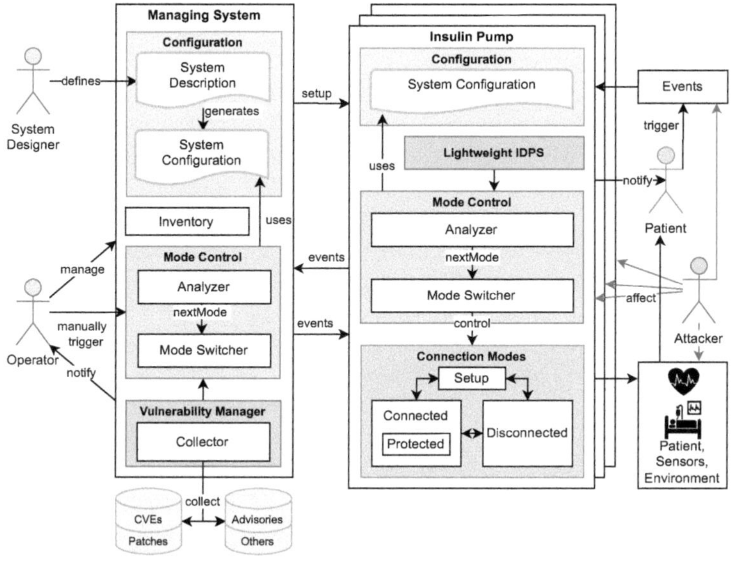

Fig. 5.7 Insulin Pumps using the Multi-Modal Architecture for Multiple Systems (Iter. 2)

Connection Modes

The following four identified valuable connection modes for insulin pumps are presented based on the formalized description of MMS explained in Sect. 3.7.

- **Type:** The *Mode Setup* is initially active after delivery or handover to the patient. Sensors, monitors, and control devices can be connected in this mode. This process should be done only in protected or private places and not in public. After the setup, the insulin pump can be switched to the *Mode Connected.* In this mode, the pump considers the information of connected sensors and automatically adapts the medication dose. While the *Super Mode Connected* allows to connect additional devices, the *Sub Mode Protected* allows only predefined connections to trusted devices to reduce the attack surface, especially in public. The pump works offline in the degraded *Mode Disconnected* and considers only predefined presets.
- **Trigger:** The pump supports automatic and manual triggers for switching modes. After the initial setup, the insulin pump can be manually switched from the *Mode Setup* to the *Mode Connected* by the patient, the doctor, or hospital staff. If the connection to sensors, monitors, or other devices gets lost, the pump switches

automatically from the *Mode Connected* to the *Mode Disconnected.* Likewise, when the battery has reached a specific limit, a switch to the *Mode Disconnected* is executed to save the battery. If the pump has enough battery, it can automatically check the accessibility of the sensors at specific intervals to reestablish the connection. Otherwise, the patient must manually press a connection button on the physical hardware to switch back to the *Mode Connected.* Mode switches can also be triggered by the *Managing System*, based on automatic procedures or manually by the *Operator*. The reasons for this are new vulnerabilities, security advisories, or other events. Additionally, manual changes by the patient using the switches on the physical hardware are possible.

- **Protocol:** The device configuration is used to save the current mode. For the patient, the current mode is shown on the pump display. The mode information is also sent to the *Managing System.*
- **Attributes:** The *Mode Setup* is initially active after delivery or handover to the patient. Bidirectional mode switches are permitted, but only the mode switch from *Mode Connected* to *Mode Disconnected* is currently automated. The other mode switches require manual intervention by the patient. Especially for security reasons, switching to the *Mode Connected* can be triggered only from the insulin pump. If the configuration setting "allow only predefined devices" is active, the *Sub Mode Protected* is used instead of the *Super Mode Connected.*
- **Definition:** For supporting *Connection Modes*, the pump needs mechanisms to manage them. The pump must be able to detect connection interruptions.
- **Analysis:** Modes are switched as immediately as possible. However, to bridge short-term connection interruptions, there is a defined threshold before switching from the *Mode Connected* to *Mode Disconnected.*

Results

The device operates offline in the *Mode Disconnected*, following predefined presets and allowing manual changes via physical switches. It serves as a fallback mechanism when connection to other devices is lost. Conversely, the device integrates data from connected sensors to adjust medication doses automatically in the *Mode Connected.*

In case of known vulnerabilities, the *Managing System* would have recognized that, notified the *Operator*, and may suggest actions to adapt the insulin pump, like installing patches or updates or switching from the *Mode Connected* to the *Mode Disconnected.* Considering events of multiple devices provides a better overview and reduces false positives. Security-concerned patients can manually switch to the *Mode Disconnected* in general or as needed, e.g., when updates are unavailable, security risks are high, or when patients are away from home or in foreign countries. The successor product, MiniMed 770G, offers similar functionality with auto and manual modes [Med23b].

The MDM recommends connecting or linking devices only in private places [Med22b]. Activating the *Sub Mode Protected* limits connections to trusted devices and reduces the attack surface. The lightweight IDPS can analyze the traffic and data from connected devices and detect abnormal behavior, automatically delete the suspicious device from the trusted list, and notify *Patients* and *Operators*. Sharing information about potential attacks like wrong connection attempts or abnormal behavior enriches the security visibility for the *Operator*. The *Managing System* can respond to attacks across multiple devices by automatically alerting *Operators*, warning other devices to increase the monitoring, and adjusting security settings, e.g., by switching modes.

5.5 Edge Device Sample Scenario

In this sample scenario, the security of edge devices is examined, focusing on how automatic and manual mode switches can reduce attack surfaces. For this purpose, a collaboration was established with the global manufacturer of injection molding machines and associated automation systems *ENGEL Austria GmbH*, to develop software prototypes. Edge devices are employed to connect industrial machines at the customers' site to the Internet and allow communication with the manufacturer, as depicted in Fig. 5.8.

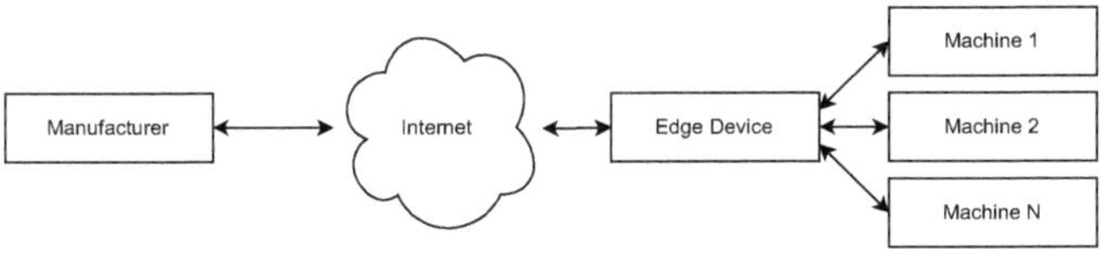

Fig. 5.8 Edge Device Sample Scenario

The increasing trend of industrial machine manufacturers evolving into service providers is evident, with the adoption of condition monitoring and big data analytics for predictive maintenance [Sta21]. In this context, edge devices play a crucial role by connecting physical devices such as machines, robots, and sensors to the virtual world, employing various protocols like MQTT, OPC UA, and others. Edge devices are widely used in CPS and IoT systems, and it is expected that their use will increase in the next decade [Tra21]. They provide on-site computing power for lightweight devices and build a bridge between OT and IT, typically in the cloud. Edge devices enable machine-to-machine communication, remote monitoring, and control across several locations, and increase availability and efficiency. Additionally, they contribute to load balancing, low latency, and service continuity in case of connection failures, optimize the network bandwidth to the cloud, and process the workload close to its occurrence [NKBC18].

As devices become increasingly interconnected, the complexity of systems grows, unleashing potential threats and vulnerabilities. Timely response to vulnerabilities is critical, but the gap between vulnerability identification and the availability of patches provides attackers with a window of opportunity to develop exploits and compromise systems. While workarounds may be offered to mitigate known vulnerabilities, manual interventions are time-consuming and, if performed hastily, may disrupt operations. In the worst case, services must be completely stopped or shut down.

Multi-Modal Architecture for Edge Devices

Based on the multi-modal architecture presented in Sect. 4.6, modes were designed to improve the security and resilience of edge devices. Figure 5.9 provides an overview of the different modes, associated services, and configurations using the second iteration of the mode-switching framework presented in Sect. 4.7. An edge device and its components can be in multiple modes. The following structure considers the formalized description of MMS explained in Sect. 3.7.

- **Type**: Four distinct *System Modes* have been implemented for edge devices, each serving different functionalities: *Mode Factory*, *Super Mode Onboarded*, *Sub Mode Updated*, and *Sub Mode Outdated. Mode Factory* provides only limited functionality, primarily configuring network settings to onboard the device, recognizing the potential risks of theft or tampering during delivery. *Super Mode Onboarded* is an abstract mode with two sub modes: *Sub Mode Updated* and *Sub Mode Outdated. Sub Mode Updated* offers all services, such as condition monitoring, big data analytics for predictive maintenance, and remote support. Conversely, *Sub Mode Outdated* provides degraded functionality for security reasons, e.g., certain software containers are stopped to reduce the attack surface. The implementation reflects normal functional modes (Factory, Onboarded, Updated) as well as degraded modes (Factory, Outdated).
- **Trigger**: Automatic mode switches are triggered by specific events, such as successful onboarding, detected potential attacks, and time-triggered update checks.

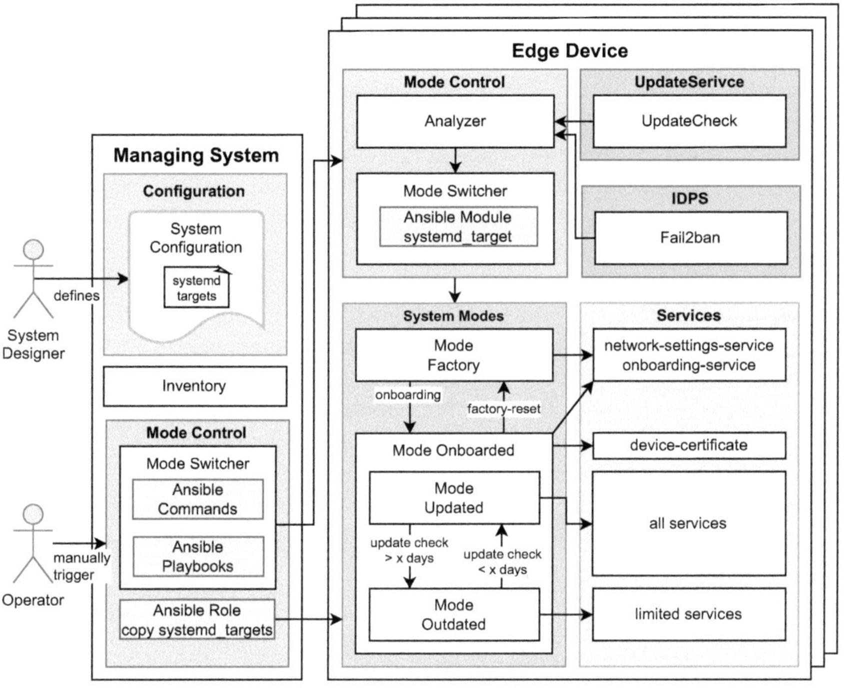

Fig. 5.9 Edge Device using the Mode-Switching Framework for Multiple Systems (Iter. 2)

Mode switches are triggered automatically by specific events, such as successful onboarding and detected potential attacks. Events can originate from the manufacturer or the device itself. Additionally, the operator on the managing system can trigger manual mode switches.

- **Protocol**: A cron job on the edge device periodically executes a shell script that checks if there are any changes. This script handles the onboarding process, updates, software repositories, and device certificate management. Once network settings are configured and the device successfully connects with the manufacturer's cloud service, an onboarding request is sent. Then, the customer can authenticate at the manufacturer's cloud service to view the request and get a security token to assign the edge devices to the customer, similar to the description of *Zoualfaghari & Reeves* [ZR19]. After successful onboarding, a device certificate is obtained, and the file `/var/lib/edgedevice/onboarded` is created. The file `/var/lib/edgedevice/up-to-date` is created if the last successful update check is within the specified time. Otherwise, the mentioned file is deleted if the check is not possible within limit, indicating potential outdated software. If the device certificate becomes invalid due to prolonged periods without updates, the file `/var/lib/edgedevice/onboarded` is deleted. The script triggers a mode switch at all events: successful onboarding, update check, and invalid device certificate. Additionally, log files are analyzed for specific events like DoS attacks. If there is a match, a mode switch is triggered. The mode control component analyzes the current mode situation, checks the existence of mode-relevant files, and decides to execute a mode switch.

- **Attributes**: Edge devices can be in multiple modes simultaneously, as the architecture supports modes and sub modes. Mode switches are permitted between *Mode Factory* and *Super Mode Onboarded* and between the sub modes *Sub Mode Updated* and *Sub Mode Outdated*. The criteria for triggering the mode-switches is the existence of files. Table 5.7 shows conditions, rules, and actions relevant to mode-switch decisions. Initially, *Mode Factory* is active when edge devices are delivered to customer factories or machines, since neither file exists (Rule 1). After successful onboarding, the file onboarding is created, leading to a switch to *Super Mode Onboarded* (R2/R4).

Table 5.7 Edge Device Mode Switch Decision Table

Conditions / Rules	R1	R2	R3	R4
File onboarded exists?	no	yes	no	yes
File up-to-date exists?	no	no	yes	yes
Actions				
Switch to Mode Factory	x		x	
Switch to Super Mode Onboarded		x		x
Switch to Sub Mode Updated				x
Switch to Sub Mode Outdated		x		

If the last successful update check is beyond the time limit, a switch to *Sub Mode Outdated* is indicated (R2). In this mode the update check frequency is increased to become updated and fully functional edge devices again. The system designer can specify the time limit for the validity of the update check in the system configuration. If the update check is within the limit, *Sub Mode Updated* becomes active (R4). The last resort is deleting the file onboarded, which results in a switch to *Mode Factory* and drastically reduces functionality and requires a new start of the onboarding process (R3).

- **Definition**: As part of the mode-switching framework presented in Sect. 4.7 the system and service manager *systemd* was used to implement modes for common edge devices. *Systemd targets* have been extended to define and manage modes [Fre23b].

As shown in Listing 12, the *Mode Factory* is defined as an extension of the `multi-user.target`. Before starting the *Mode Factory*, the system checks the setting `ConditionPathExists`. If the file `/var/lib/edgedevice/onboarded` does not exist (indicated with a call sign), the *Mode Factory* is started. Otherwise, if the onboarding process is successful, the file is created and builds the condition for switching to the *Super Mode Onboarded*.

```
#/etc/systemd/system/mode-factory.target
[Unit]
Description=Mode Factory
Requires=multi-user.target
Conflicts=
After=multi-user.target
AllowIsolate=yes
#Start target only when file does not exists
ConditionPathExists=!/var/lib/edgedevice/onboarded
```

LISTING 12: Mode Factory as Systemd Target

The *Mode Factory* includes services to define the network settings and for onboarding. Each of these services has its unit file and can be attached to one or multiple modes in their `[Install]` section with the command `WantedBy=MODENAME.target`. Additionally, services can be organized in folders for each target, e.g., `mode-factory.target.wants`. Notably, with the release of *systemd* version 250, special targets were introduced to trigger a factory reset, aligning with the conceptualization of modes [Fre23a].

Listing 13 shows *Sub Mode Updated* extends *Super Mode Onboarded* specified with the settings `Requires` and `After`. Additionally, the *Super Mode Onboarded* serves as parent mode for *Sub Mode Outdated*.

```
#/etc/systemd/system/mode-updated.target
[Unit]
Description=Mode Updated
Requires=mode-onboarded.target
Conflicts=
After=mode-onboarded.target
AllowIsolate=yes
#Start target only when file does exists
ConditionPathExists=/var/lib/edgedevice/up-to-date
```

LISTING 13: Mode Updated as Systemd Target

- **Analysis**: Mode switches are executed immediately.

Monitoring Events, Detecting Attacks, and Response

Additional filters have been implemented to detect a broader range of activities, such as failed and successful terminal logins, failed and successful SSH logins, port scans, HTTP errors, plugged-in LAN cables, attached USB devices, and lost connections. Fail2ban serves not only to send notifications to manufacturers' cloud services in case of abnormal behavior but also as a trigger for mode switches. While fail2ban typically blocks single IP addresses, it may not be suitable for blocking thousands of addresses in a DoS attack scenario. In such cases, switching to a more secure mode after a specific number of failed login attempts, where the attack is generally blocked, is more practical. The system modes, especially the *Mode Updated*, could be extended with a *Sub Mode DoS*. In this mode, all connection attempts could be blocked for some time. The option to block entire IP address ranges or even countries is also being considered.

A proactive response is initiated in the event of multiple failed login attempts on the console of the edge device. Especially when an edge device is offline and cannot be monitored remotely, switching to the degraded *Mode Outdated* can reduce the attack surface and protect the systems from further attacks. This approach is also employed when anomalies related to hardware attacks, such as sniffing or port scans, are detected.

Mode Management & Configuration

The mode control shell script on the edge devices responds to events triggered by the scheduled update check (outdated, updated) and the IDPS. Modes can be switched automatically based on these events or manually by the operator. With each executed mode switch, the script sends a notification to the manufacturer's cloud service, ensuring awareness of changes. Using the custom *Ansible* role `modes`, and the custom module `mode` simplified the setup and switching modes. The role was used to copy the modes to designated edge devices. The module allows to switch all edge devices, specific groups, or specified edge devices to a desired mode.

5.6 IoT Device Case Study

To demonstrate the applicability of the third iteration of the multi-modal architecture presented in Sect. 4.8 and assess the potential benefits of the mode-switching framework presented in Sect. 4.9, a case study has been conducted with multiple IoT devices. Monitoring, analyzing, and updating components have been applied in response to different security threats. The following use case requirements, the evaluation setup, and metrics are specified based on the guidelines described by *Runeson & Hoest* [RH08]. Based on this, RQ3 delves into mode-switching functionality within industrial IoT devices. Can the mode-switching framework be used to define and execute realistic test scenarios (applicability), and what effort is required to create and specify them? Additionally, the investigation included the framework's performance for detecting different security issues and the duration needed to apply a countermeasure.

Case Requirements

To assess the general applicability of the mode-switching framework and its performance, a suitable test scenario with an appropriate infrastructure is necessary. Both local and cloud analysis and execution of adaptations are required, as well as logging and analyzing events to evaluate the performed execution actions. This involves three main components:

1. A *Managing System*,
2. Multiple *Managed IoT Devices* connected to a network that can be subject to potential attacks and security threats, and
3. Simulation of attacks as described in previous research [DV17, CZL+18].

Selected Use Case

Considering the requirements, several test scenarios were discussed for the mode-switching framework at (i) the IoT device and (ii) managing system level. Then ECA rules were created, as shown in Table 5.8. Each rule outlined the threshold for the number of wrong attempts, a monitoring window in the condition, and actions to mitigate and prevent further attacks, e.g., blocking the attacker's IP, increasing the waiting time, rebooting the system, and adapting security policies.

Based on this initial set of scenarios, two common attack types were selected for the use case. Initially, *port scans* were chosen because attackers typically use tools like *Nmap* during the reconnaissance phase to scan for open ports on the network layer and running services [Nma23, SD13]. Then, they search for any potential service vulnerability to attack them. The second attack type for the use case is the simulation of *dictionary and brute-force attacks on authentication*. With this type of attack, attackers use lists of commonly used usernames and passwords and tools like *Hydra* to guess credentials and gain system access [Heu23]. Without proper countermeasures and suitable protection against these kinds of attacks, it is only a matter of time until attackers figure out credentials and ultimately take over a system. The duration is impacted by both the complexity of the password and the number of

Table 5.8 Test Scenarios for the Mode-Switching Framework

ID	Event	Condition	Action
		Managed IoT Device Scenarios	
D01	*Port Scan*	≥ 2 attempts in 1 min	block IP for 10 min
D02	*SSH Login Failure*	≥ 2 attempts in 5 min	block IP for 5 min
D03		≥ 4 attempts in 1 hour	block IP for 1 hour
D04		≥ 8 attempts in 1 day	block IP for 1 day
D05	*Terminal Login Failure*	≥ 2 attempts in 5 min	waiting time
D06		≥ 4 attempts in 1 hour	reboot system
D07		≥ 8 attempts in 1 day	halt system
D08	*HTTP 401 Unauthorized*	> 2 attempts in 5 min	block IP for 5 min
D09	*HTTP 404 Not Found*	> 10 attempts in 5 min	block IP for 5 min
D10	*HTTP 403 Forbidden*	> 5 attempts in 5 min	block IP for 5 min
D11	*HTTP 500 Internal Server Error*	> 5 attempts in 5 min	block IP for 5 min
D12	*Get/Lost Connection*	every attempt	adapt security
		Managing System Scenarios	
S01	*Blocked IP*	> 2 attempts in 5 min from ≥ 2 devices	block IP on all devices for 1 hour
S02	*IoT changed public IP*	every attempt	inform admin

parallel attempts per minute. Timely and fast detection and execution of appropriate countermeasures and necessary system adaptations are of key importance for both attack types. The *MAPE-K* loop is leveraged to monitor events related to security threats, analyze them through specific conditions, and plan and execute actions to mitigate and prevent further attacks. Several adjustments were necessary for the mode-switching framework to detect and appropriately adapt to the two attack types, which will be explained in the following sections. Figure 5.10 provides an overview of the IoT scenario with different modes, associated services, and configurations using the third iteration of the mode-switching framework presented in Sect. 4.9.

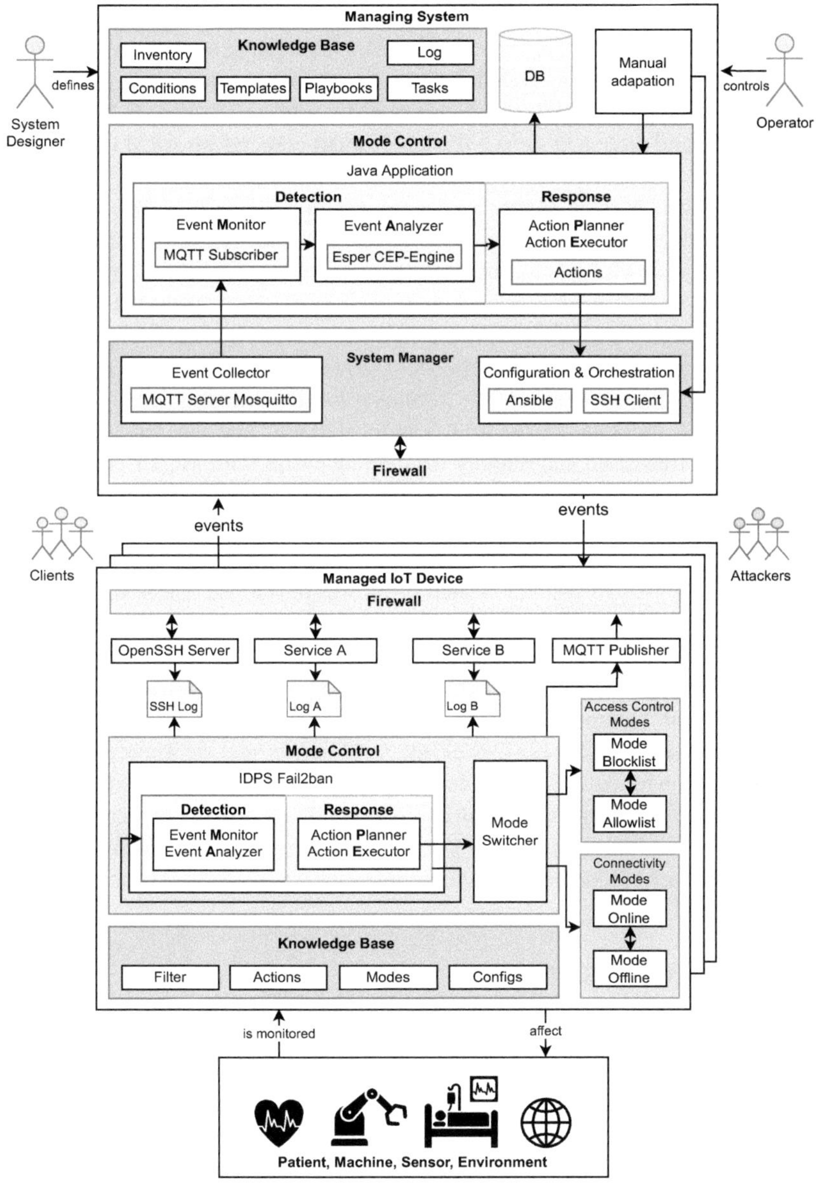

Fig. 5.10 IoT Scenario using the Mode-Switching Framework (Iteration 3)

Evaluation Settings

For the evaluation, a testbed was constructed to simulate a network with multiple connected IoT devices. This included a test environment containing five *Raspberry Pi (RPi) 4 Model B* (1.5 GHz quad-core ARM Cortex-A72, 4 GB RAM, Ubuntu 22.04.1 LTS) and three *RPI Zero 2 W* (1 GHz quad-core, 64-bit ARM Cortex-A53, 512 MB RAM, Raspbian 11).

An RPI 4 was used as *Managing System* running the *Java Server Application* with the *OpenJDK Runtime Environment* build 18.0.2, including the MQTT-Broker (Eclipse Mosquitto v2.0.11) for monitoring events of the IoT devices and the *Esper CEP Engine* 4.7.0 to analyze events and check constraints using the *Event Processing Language (EPL)* [Ecl23a, Esp23]. *Ansible* 2.10.8 was used for automation, orchestration and action execution [Red23]. Another RPi 4 was used as *Attacker.* The remaining six RPi represent the *Managed IoT Devices* and had an *OpenSSH-Server* 8.4-8.9, *Fail2ban (F2b)* 0.11.2 as local IDPS, and the *Mosquitto Client Package* 2.0.11 installed and running for sending events with MQTT [Fai23a].

Evaluation Metrics and Data Collection

The human effort to use the mode-switching framework for multiple systems and its applicability were measured. A shell script was developed to assess the deployment and installation, with ten executions conducted to calculate average values. Common security and penetration test tools were employed on the *Attacker Device.* A shell script was created to randomly select IoT devices, simulate attacks, measure timestamps, and document results in a log file. Subsequently, analysis of the log files encompassed the *Attacker Device*, the *IoT Devices*, and the *Managing System*, with the duration for detection and response measured. In the default configuration, system logs provide accuracy only in *Seconds (s).* To obtain more precise results in *milliseconds (ms)*, the *Rsyslog* configuration was altered for transferring log messages. Before each run, device time synchronization was performed to ensure precise measurements.

Defending Port Scans (D01)

During the reconnaissance phase, potential attackers will perform port scans. To prevent attacks on the IoT device, attackers should be discouraged by temporarily blocking their IP address for a specific duration (e.g., 10 minutes). Typically, requests during a port scan to unused ports are answered by the system with closed or denied and are not logged. To monitor these scans, the firewall (`iptables`) on the IoT devices was extended with additional IP packet filter rules. Listing 14 shows a playbook excerpt of the tasks to achieve this. It includes creating a user-defined chain named `PORTSCAN`, appending it to the bottom of the `INPUT` chain, and filtering and logging every request to unused ports before dropping it. Possible attackers get no reply, but this mechanism allows us to monitor the system logs with a regular expression filter, analyze port scans, and identify the attacker's IP address, see Listing 15.

```
- name: Create user-defined iptables chain PORTSCAN
    command: iptables -N PORTSCAN
- name: Log arriving packets into chain PORTSCAN
  iptables:
    chain: PORTSCAN
    action: append
    log_prefix: "PORT DENIED: "
    log_level: '4'
    comment: Log arriving packets chain PORTSCAN
- name: Log unreachable Ports (Portscan)
  iptables:
    chain: INPUT
    protocol: tcp
    match: tcp
    jump: PORTSCAN
    comment: Log unreachable Ports (Portscan)
- name: Drop portscan packets after logging
  iptables:
    chain: PORTSCAN
    action: append
    jump: DROP
    comment: Drop portscan packets after logging
```

LISTING 14: IoT Device: Playbook-Excerpt of the Firewall Adapation to Detect Port Scans

```
failregex = PORT DENIED: .* SRC=<HOST> .* DPT=<F-PORT/>
ignoreregex =
journalmatch = _TRANSPORT=kernel
```

LISTING 15: IoT Device: Fail2ban Filter for Monitoring Port Scans

For planning and executing the necessary actions to block and relay the information to the managing system, a *F2b* jail was defined, see Listing 16. The configuration includes the filter (`portscan`), the log path, and the actions to run if the conditions are met. These are the log monitoring window (`findtime`) and the number of found entries (`maxretry`).

```
- name: portscan
  logpath: /var/log/kern.log
  action: |
    iptables-allports[blocktype=DROP]
    publish_mqtt_message[topic="portscan",message="{\"ip\":\"<ip>\"}"]
    offline-adaptation
  maxretry: 2
  findtime: 1m
  bantime: 10m
```

LISTING 16: IoT Device: ECA Rule Excerpt for Port Scan Detection and Response

If there is a match, *F2b* runs the predefined actions: `iptables-multiport` to ban the IP temporarily for 10 minutes (`bantime`), `publish_mqtt_message` to notify the managing system via MQTT messages, and `offline-adaptation` to assess the connection state to the managing system and implement adaptions accordingly. The parameter values in the square brackets allow the customization of the actions' behavior. The `blocktype=DROP` parameter on `iptables-allports` drops all packets from the IP address without sending a message to the sender. In contrast, with the `blocktype=REJECT` parameter, an ICMP message (typically "Destination Unreachable") will be returned to the sender. For the action `publish_mqtt_message`, the parameter topic and message specify the corresponding settings, where `<ip>` is dynamically replaced by the IP address.

Defending Dictionary and Brute-Force Attacks on Authentication (D02-D04)
Failed authentication attempts are strong indicators of potential attacks, These attempts can occur via various different interfaces of the IoT device, such as SSH, Terminal, and HTTP. This evaluation focuses on SSH, which is commonly used for remote system administration and, thus, often publicly accessible via the internet. However, the same measures and threat mitigations could be applied to other protocols similarly. Dictionary and brute-force attacks on the SSH authentication, results in multiple login failures. Listing 17 shows an example rule for detecting these failures deployed on the IoT device. Similar to the port scan rule, the configuration specifies the log path, filter (`sshd`), and actions but with different conditions and ban times, as depicted in Table 5.8, D02–D04.

```
- name: ssh-login-failure
  logpath: /var/log/auth.log
  filter: sshd
  action: |
    iptables-allports[blocktype=DROP]
    publish_mqtt_message[topic="ssh-login-failure",message="{\"ip\":\"<ip>\"}"]
    offline-adapation
  maxretry: 2
  findtime: 1m
  bantime:  5m
```

Listing 17: IoT Device: ECA Rule Excerpt for Failed SSH Login Detection and Response

IoT Device Connectivity Modes
If the IoT device cannot connect to the managing system and attacks are ongoing, neither the automatic monitoring and control solutions on the managing system nor an operator can intervene. For this situation, connectivity modes provide self-protection methods for the devices.

- **Type:** The connectivity modes on the IoT devices are based on the network connectivity status with the managing system. The *Mode Offline* gets active when a device loses connection to the managing system. In this mode, the local *F2b* configuration is set to a formula that increments the ban time exponentially. With every wrong attempt, e.g., a SSH login failure, the ban time increases from 2 to 4, 8, 16, and so forth. This restricts ongoing attacks and hackers so they cannot take advantage of the offline situation. The *Mode Online* gets active when the IoT device is connected again, using a fixed ban time, e.g., 5 min.
- **Trigger:** Switching connectivity modes is determined by the connection status between the IoT device and the managing system. The loss of connection triggers the switch to *Mode Offline*. When the system regains connection, it triggers the switch back to *Mode Online*.
- **Protocol:** A periodic job executes the connectivity check to test the current network connection status to the managing system. According to the result, the mode control will suggest the corresponding mode. If the mode is already active, then there is no switch. Otherwise, the mode switcher changes configuration settings, enabling or disabling the ban time increment.
- **Attributes:** An IoT device can only be in one *Connectivity Mode* at a time. Mode switches are allowed in both directions. The shell script encompasses the IP address of the managing system, the file path of the *F2b* configuration, the network connection status, and the boolean variable for activating or deactivating the ban time increment.
- **Definition:** The connectivity check is implemented as shell script as depicted in Listing 18. The script uses the command *ping* for sending ICMP requests to the managing system. If there is no response within 5 seconds, the connection status is considered as offline. The script uses *grep* and regular expressions to get values from the config file and *sed* to replace values.
- **Analysis:** The periodic job for the connectivity check is planned every minute. With a 5-second delay for the check's timeout, a mode switch is executed immediately to reduce the attack surface and increase security.

```
#!/bin/bash
#NAME=IoT Device Connectivity Modes
#DESC=Determines if the managing system is online or offline and adapt measures based on that
SERVER_IP=10.10.10.211
FAIL2BAN_CONFIG_FILE=/etc/fail2ban/jail.local
server_available=0
bantime_increment=false
isServerIpAvailable(){
 if [ ping -c 1 -W 5 ${SERVER_IP} 1>/dev/null 2>&1 ]; then return 1; else return 0; fi
}
replaceConfigFileValue(){}
        CONFIG_FILE=$1 #jailtest.local
        TARGET_KEY=$2 #bantime.increment
        REPLACEMENT_VALUE=$3 #true
        sed -i "s/\(${TARGET_KEY} *= *\).*/\1${REPLACEMENT_VALUE}/" ${CONFIG_FILE}
}
getConfigFileValue(){
        CONFIG_FILE=$1 #jailtest.local
        TARGET_KEY=$2 #bantime.increment
        /usr/bin/grep -Po "(?<=${TARGET_KEY} = ).*" ${CONFIG_FILE}
        result=$?
        return $result
}
setFail2banBantimeIncrement(){
        TARGET_KEY=bantime.increment
        REPLACEMENT_VALUE=$1
        CONFIG_FILE=${FAIL2BAN_CONFIG_FILE}
        replaceConfigFileValue ${CONFIG_FILE} ${TARGET_KEY} ${REPLACEMENT_VALUE}
}
enableFail2banBantimeIncrement(){ setFail2banBantimeIncrement true }
disableFail2banBantimeIncrement(){ setFail2banBantimeIncrement false }
echoAndLog(){ logger $1; echo $1 }
echoAndLog "Adapt fail2ban config based on the server context (online/offline)"
isServerIpAvailable
server_available=$?
bantime_increment=$(getConfigFileValue "${FAIL2BAN_CONFIG_FILE}" "bantime.increment")
if [ $server_available -eq 1 ]; then
        if [ $bantime_increment = "true" ]; then
                echoAndLog "Disable fail2ban bantime.increment"
                disableFail2banBantimeIncrement
        else
                echoAndLog "fail2ban bantime.increment is already disabled"
        fi
else
        if [ $bantime_increment = "false" ]; then
                echoAndLog "Enable fail2ban bantime.increment"
                enableFail2banBantimeIncrement
        else
                echoAndLog "fail2ban bantime.increment is already enabled"
        fi
fi
```

LISTING 18: IoT Device: Connectivity Modes Shell Script

IoT Device Firewall Access Control Modes

For defending against attack waves and ensuring maintainability, firewall access control modes allow a shift in the security strategy.

- **Type:** The firewall access control modes on the IoT devices differentiate between the *Mode Blocklist* and the *Mode Allowlist*. In the *Mode Blocklist* all IP addresses can connect to the IoT device, except those on the blocklist. In the *Mode Allowlist* only IP addresses listed in the allowlist are granted access, while all others are blocked.
- **Trigger:** If the number of currently blocked addresses exceeds a predefined value, e.g., 1000, a mode switch to the *Mode Allowlist* is triggered. After the number falls below a certain threshold, e.g., 500, the system triggers a switch to *Mode Blocklist*.
- **Protocol:** A periodic job with a shell script checks the number of of currently banned IP addresses, informs the operator, and automatically triggers mode switches. Based on the notification, the operator can further analyze the situation and may decide to return to the previous mode, keep the current mode, or execute other manual or automated actions.
- **Attributes:** The number of currently blocked addresses changes since the IP addresses are not permanently blocked but only temporarily, e.g., for 10 minutes.
- **Definition:** For supporting the access control modes, the firewall of the IoT devices was extended by creating a user-defined chain called `BLOCKLIST`, as shown in Listing 19. IP addresses are added to this list after starting a port scan or several failed authentication attempts, as mentioned before. Requests from these addresses are blocked, and any further access attempts are rejected. The user-defined chain was appended at the bottom of the `INPUT` chain, ensuring that IP addresses not on the list are allowed. For the *Mode Allowlist*, the user-defined chain `ALLOWLIST` was added with trusted IP addresses, as shown in Listing 20. The placeholder `{{ item }}` is dynamically replaced with the IP addresses saved in the file allowlist.
- **Analysis:** A periodic job is planned every minute, and a mode switch is executed immediately to reduce the attack surface and increase security.

```
- name: Create user-defined iptables chain BLOCKLIST
  command: iptables -N BLOCKLIST
  when: "'Chain BLOCKLIST' not in iptables_rules.stdout"
- name: Add to INPUT chain check IPs on BLOCKLIST
  iptables:
    chain: INPUT
    protocol: tcp
    match: tcp
    jump: BLOCKLIST
    comment: Check IPs on BLOCKLIST
- name: Return from BLOCKLIST to INPUT
  iptables:
    chain: BLOCKLIST
    action: append
    jump: RETURN
```

LISTING 19: IoT Device: Playbook-Excerpt of the Firewall Adaptation for the Mode Blocklist

Setup of the Managing System and the IoT Devices

The applicability of the framework was manually tested on a single device before deploying the configurations and adaptions for the attack scenarios to all IoT devices. Listing 21 shows the *Ansible* playbook for the setup and configuration of *F2b* with the two attack types as services. The testing involved monitoring the topics `portscan` and `ssh-login-failure` with an MQTT subscriber and creating a `BlockedIpEvent` based on incoming messages. EPL constraints (see Listing 22) were specified to analyze these events and select IP addresses that were blocked on more than two different devices within five minutes. If the constraint is violated, the *Ansible* playbook `blockip`, see Listing 23 is executed to add the IP address to the user-defined chain `BLOCKLIST` on the IoT device's firewall, see Listing 24.

```
- name: Get existing iptables rules
  command: iptables -n -L
  changed_when: false
  register: iptables_rules
- name: Create user-defined iptables chain ALLOWLIST
  command: iptables -N ALLOWLIST
  when: "'Chain ALLOWLIST' not in iptables_rules.stdout"
- name: Add to INPUT chain check IPs on ALLOWLIST
  iptables:
    chain: INPUT
    protocol: tcp
    match: tcp
    destination_port: '9000'
    ctstate: NEW
    syn: match
    jump: ALLOWLIST
    comment: Check IPs on ALLOWLIST
- name: Allow packets from IPs on ALLOWLIST
  iptables:
    chain: ALLOWLIST
    action: append
    protocol: tcp
    source: "{{ item }}"
    jump: ACCEPT
  with_lines: cat allowlist
- name: Return from ALLOWLIST to INPUT
  iptables:
    chain: ALLOWLIST
    action: append
    jump: RETURN
```

LISTING 20: IoT Device: Playbook-Excerpt of the Firewall Adaptation for the Mode Allowlist

Attack Simulation

In the second step, the assessment of the mode-switching framework's response to attacks was conducted by measuring its performance. This involved (1) running port scans and (2) simulating wrong SSH authentication attempts using a shell script. Logs on the *Attacker Device*, the kernel logs and the *F2b* log entries on the IoT devices, and the logs on the *Managing System* were examined. Figure 5.11 shows the average detection and reaction time and how long it takes to switch modes.

```
---
- hosts: all
  roles:
    - ansible-fail2ban
  become: yes
  become_method: sudo
  vars:
    fail2ban_ignoreips:
      - 10.10.10.211
    fail2ban_bantime_increment: false
    fail2ban_banaction:
    fail2ban_banaction_allports:
    fail2ban_filterd_path: templates/etc/fail2ban/filter.d/
    fail2ban_actiond_path: templates/etc/fail2ban/action.d/
    fail2ban_root_path: templates/root/
    fail2ban_services:
      - name: ssh-login-failure
        logpath: /var/log/auth.log
        filter: sshd
        action: |
          iptables-multiport
                   publish_mqtt_message[topic="blockedip",message="{\"ip\":\"<ip>\"}"]
                   offline-adapation
        maxretry: 2
        findtime: 1m
        bantime: 1m
      - name: portscan
        logpath: /var/log/kern.log
        action: |
          iptables-multiport
                   publish_mqtt_message[topic="portscan",message="{\"ip\":\"<ip>\"}"]
                   offline-adaptation
        maxretry: 2
        findtime: 1m
        bantime: 1m
```

LISTING 21: Managing System: Ansible Playbook for Fail2ban Configuration

```
NAME = SameBlockedIpFromMoreThanOneHost
CONSTRAINT =
   select * from BlockedIpEvent.win:time(5 min) group by ip having count(*) >= 2
```

LISTING 22: Managing System: EPL Constraint for Detecting IP Addresses Blocked on Multiple Hosts Within a Certain Timeframe

```
ansible-playbook -i hosts blockIP.yml --extra-vars "ip2block=7.7.7.7"
```

LISTING 23: Managing System: Shell-Command to Run the blockIP Playbook

```yaml
---
- name: Block IP address
  hosts: all
  gather_facts: no
  become: yes
  become_method: sudo
  tasks:
  - name: Fail if "ip2block" is undefined
    fail: msg="This play requires variable ip2block"
    when: ip2block is undefined
  - name: Get existing iptables rules
    command: iptables -n -L
    changed_when: false
    register: iptables_rules
    when: ip2block is defined
  - name: Add IP to BLOCKLIST at the top
    iptables:
      chain: BLOCKLIST
      action: insert
      source: "{{ ip2block }}"
      jump: DROP
    when:
     - iptables_rules is defined
     - "'Chain BLOCKLIST' in iptables_rules.stdout"
```

LISTING 24: Playbook to Block an IP Address on All Managed IoT Devices

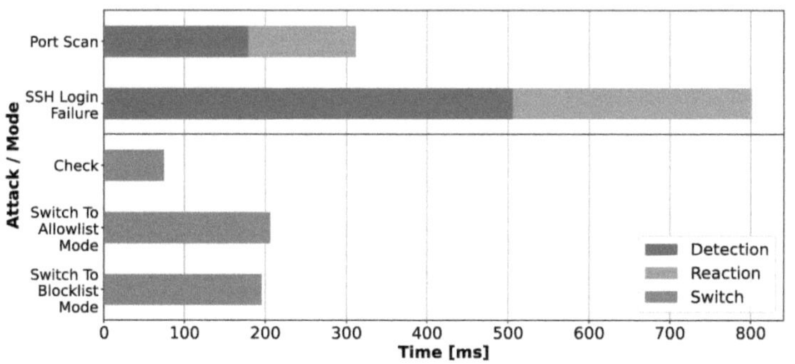

Fig. 5.11 Average Detection, Reaction, and Mode Switching Time on the IoT Devices

Port Scans (D01)

For the first attack scenario, *Nmap 7.80* was utilized on the *Attacker Device* to perform port scans. A total of 100 runs were conducted with a 60-second pause between each run. The execution time of the port scan was measured on the *Attacker Device*, while on the IoT device, a comparison was made between the kernel logs and the entries in the *F2b* log. Upon comparison, it can be concluded that the *F2b* filter pattern successfully detected all port scans. However, there was a slight delay in the monitoring process, with the *F2b* log timestamps lagging behind the kernel log entries being 179 ms.

Initially, it took the attacker script 200 ms per IoT device to scan ten ports before getting banned. The IoT devices took, on average, 86 ms to monitor and analyze the port scan and 247 ms to plan and execute predefined actions (333 ms in total). The attacker's IP address was blocked via iptable rules, and their information was forwarded to the managing system. Further attempts by the attacker were unsuccessful, as the connection was blocked. In 99 subsequent runs, the attacker script took 2s per IoT device and could only scan two ports before the recurring attacker IP address was banned 221 ms after the first monitored port scan.

SSH Login Failures (D02)

In the second scenario, a script was employed on the *Attacker Device* to attempt the connection to the SSH service of each IoT device using wrong credentials. Same as for the first scenario, 100 runs were executed, with three attempts per IoT device and a 60-second pause between each run. On the IoT device, authentication log events (auth.log) were compared with the entries in the *F2b* log file (fail2ban.log). The attack was logged in the file auth.log, and *F2b* monitored the pattern of the invalid user and the wrong password with its filter within 506 ms. After two wrong login attempts, it took 295 ms to analyze them and plan and execute the predefined actions. Finally, the attacker's IP address was blocked, and the attacker's information was forwarded to the managing system. Tests on the attacker side confirmed that further connection attempts were rejected.

IoT Device Connectivity Modes

The offline adaptation and the IoT device connectivity modes were tested by changing the managing system IP address so that IoT devices were unable to connect. Then, the switch to the *Mode Offline* on all IoT devices and the adaption of the *F2b* configuration were observed within 2 s. After repeated attacks, the ban time increased exponentially. However, this adaptation increased also the average processing time by 8 ms. After switching back to the original managing system IP, IoT devices switched to the *Mode Online* in 2 s.

IoT Device Firewall Access Control Modes

To test the switch from the *Mode Blocklist* to the *Mode Allowlist* and back, 500 runs were executed on the IoT devices with a shell script and randomly simulated the limit of 1000 blocked IP addresses. On average, it took 206 ms to switch to *Mode*

Allowlist, 195 ms to switch back, and 74 ms to check if the desired mode was already active, see Fig. 5.11. The RPI Zero took twice as long as the RPI 4, presumably due to limited hardware capabilities (RAM and CPU power).

Results

Using the custom *AnsibleF2b* module, the basic configuration was reduced to 15 LoC. Each ECA rule required 8-10 LoC, depending on the number of specified actions. The EPL constraint on the managing system needed only two LoC. The initial automated deployment of the software, the configuration files, and scripts to the seven managed IoT devices took less than 5 min (approx. 30 sec per device), including updating time, installing the MQTT-client mosquitto, configuring the firewall, and installing/configuring *F2b*. Checking the deployment on all devices was much faster and took only around 3 minutes. The human effort required to implement, test, and create filters, actions, configurations, modes, and deploy them to the IoT devices was approximately 30 hours.

Regarding the applicability of the mode-switching framework, it can be concluded that all necessary artifacts for the two selected scenarios were successfully created, and the overall effort was reduced compared to configuring all nine IoT devices manually. The deployment process took less than 5 minutes, while checks were completed in only 3 minutes.

In terms of performance, the mode-switching framework effectively detected attacks and responded promptly in the selected scenarios. By sharing attacker information, all remaining IoT devices were safeguarded within 14 seconds after the first attack. The attacker's information was automatically forwarded from the IoT devices to the managing system, with an average processing time of 102 ms. The *Java Server Application* monitored the blocked IP event in 3 ms, and the *CEP-Engine* analyzed the EPL constraints in 4 ms. After two blocked IP events from different IoT devices, the constraint was violated, and actions were planned and executed within one millisecond. This involved running an *Ansible* playbook to add the attacker's IP address to the blocklist on all IoT devices, which took another 13 s. Following this, further attacks on other IoT devices were denied. Additionally, adaptations allowed the protection of offline devices and the mitigation of ongoing attacks by exponentially increasing the ban time.

5.7 Threats to Validity

Like any other study or experiment, this work is subject to threats to validity [WRH+12]. These threats are addressed across different dimensions.

In terms of *conclusion validity*, the evaluation involved the calculation and comparison of total vulnerability scores with and without mode-switching. The results illustrated risk reduction for web applications and industrial and medical IoT devices. To further validate the approach, a usability study with industry engineers is planned to analyze the effectiveness of the MDSL and the concept of the multi-modal architecture and the mode-switching framework in an industrial context. The sample size could be further extended to other application areas, hardware, and software systems, as we could not test the frameworks on the hardware of real medical IoT devices.

Construct validity is concerned with the relationship between theory and observation. The prioritization of remediation efforts using CVSS aligns with the established industry standard. While the current metric focuses on the severity of vulnerabilities, enhancing the analysis is envisioned to consider factors such as the duration of vulnerability exposure, the frequency of vulnerabilities, and the presence of exploits. Regarding the MDSL evaluation, the LoC and the *Halstead Software Metrics* objectively measure software complexity and quality. However, further analysis is planned to included other factors, such as maintainability and portability.

In parts of the case studies and sample scenarios, it is assumed that the modes are functional and non-functional mostly equivalent. For example, in the Juice shop sample scenario presented in Sect. 5.1, the database modes *SQLite* and *PostgreSQL* provide nearly the same data access interface. Likewise, the modes *Apache* and *Nginx*, as well as *ApacheWithPhp* and *NginxWithPhp* in the web server case study showcased in Sect. 5.2, have almost similar functionality. However, the modes without PHP provide more security but less functionality. As part of our ongoing work, pre- and post-conditions are currently being added to the system description to verify whether the new suggested mode is adequate for the given scenario. This includes checking whether the mode provides specific features and whether the mode switch was carried out successfully.

For *internal validity*, the case studies and sample scenarios were conducted in a controlled virtual environment with historical CVE and patch data, ensuring that external factors did not influence the mode-switching evaluation. Other influences on the measurements were mitigated by conducting multiple measurements of the evaluation and calculating the average. For example, for the web server case study presented in Sect. 5.2, 2500 mode switches were analyzed.

Concerning *external validity*, commonly available hardware and open-source software were used to demonstrate the multi-modal approach. A realistic scenario reflecting real-world application settings was created by using several different devices as part of a network. The used tools like OWASP dependency check, fail2ban, Ansible, and others have a large user base. Millions of internet domains widely used the analyzed web servers, which had several CVEs and patches over time. Therefore, confidence exists that developers can use and adapt the MDSL and framework to systems with similar characteristics. However, the framework does not apply to any given environment or technology. A user study will be necessary to validate the feasibility of the framework. Nevertheless, the multi-modal architecture (c.f. Fig. 4.4) can be applied to different domains as long as events and modes are available. These events may encompass vulnerabilities, patches, exploits, or manual decisions by the operator and trigger mode switches. Modes represent either functional equivalent (alternative) or restricted (degraded) implementations and configurations. To demonstrate generalizability, future plans include expanding the scope by incorporating additional constraints and vulnerability scenarios.

For *reproducibility*, parts of work are publicly available. For example, the reproduction package for the web server case study is provided on Zenodo [RSVW23b]. It is possible to reuse the system description and configuration with the executable Java `jar-file` and to simulate and replay the two years of the case study to see the mode-switching and the reaction to CVEs as well as patches. There might be slight changes in the results due to possibly changed severity scores of CVEs in the meantime. For example, this can happen when new exploits become available for existing vulnerabilities.

5.8 Summary

Four sample scenarios and two case studies showed the practical usage of mode-switching to improve security and enhance resilience. For the consciously vulnerable OWASP Juice Shop, security enhancements were showcased using modes on the database and application layer. Similar vulnerabilities exist in real-world applications. Modes provide an additional layer of defense, according to the defense-in-depth concept. For web servers, the *Mode Domain-Specific Language (MDSL)* and the mode-switching framework were employed. A case study illustrated that switching to less vulnerable modes reduced the attack surface in 98.9 % of the analyzed time.

Regarding medical devices, scenarios involving *Picture Archiving and Communication Systems (PACSs)* and insulin pumps were examined. By implementing the mode-switching architecture, the study for Philips Vue PACS demonstrates a reduction in the CVSS sum by 41 %, resulting in improved security. Also, the insulin pump sample scenario with the Medtronic MiniMed 600 series highlights how the application of the multi-modal architecture can enhance device security, patient safety, and system adaptability.

A sample scenario with industrial edge devices of our collaboration partner ENGEL Austria GmbH showed how modes can be implemented in the Linux operating system to improve security and how configuration management tools can help to manage modes of multiple systems. The implementation involves custom Ansible modules and roles, systemd management, and fail2ban for monitoring events.

The applicability and performance of the mode-switching framework were analyzed through an IoT case study, which involved a managing system, multiple managed IoT devices, and simulations of attacks such as port scans and brute-force SSH login attempts. Various security measures were implemented to protect IoT devices, including sharing attacker information, offline adaptation, and mode-switching. The mode-switching framework was able to effectively detect and respond promptly to security threats with reduced human effort and rapid deployment (< 5 min). Within 14 s after the first attack, the framework warns and protects other IoT devices. Offline adaptation ensures self-protection during server unavailability by dynamically switching to a more restrictive mode with exponentially increasing ban times.

6 Discussion

The evaluation in Chapter 5 has confirmed that the three iterations of the multi-modal architecture and the mode-switching framework with the accompanying *Mode Domain-Specific Language (MDSL)* are capable of describing different system components, modes, and actions to be executed. Mode switches provide a resilient countermeasure during attacks. In case of vulnerabilities, mode switches reduce the attack surface.

This chapter is dedicated to the discussion of the findings, practical implications, and limitations of this dissertation. Parts of this chapter are published in several publication venues, including the *Journal of Systems and Software (JSS)* [RSVW23a], the *IEEE/ACM International Symposium on Software Engineering for Adaptive and Self-Managing Systems (SEAMS)* [RSV23], the *International Conference on Industry 4.0 and Smart Manufacturing (ISM)* [RS21b], the *International Workshop on Cyber-Security and Functional Safety in Cyber-Physical Systems (IWCFS)* [RS20, RS21a], and the final research report to the *Austrian Marshall Plan Foundation* [Rie22].

Section 6.1 discusses how switching modes can mitigate vulnerabilities and attacks. The multi-modal architecture and security modes are not a comprehensive solution to all security issues. Adaptions of the configurations are necessary to leverage mode switching for different application areas. However, the development and distribution of patches must be accelerated. Section 6.2 deals with implications arising from the multi-modal architecture and mode-switching framework. The focus is on possible side effects, false alarms, overreactions, and dealing with malicious insiders. Section 6.3 discusses the effects of using a MDSL. Then, Sect. 6.4 analyzes the effect of mode-switching for self-protection and the conflict between centralization and decentralization. Before Sect. 6.6 summarizes the discussion, Sect. 6.5 reflects the advantages and disadvantages of the presented framework.

M. Riegler, *Mode-Switching for Resilient Security*,
https://doi.org/10.1007/978-3-658-49306-6_6

6.1 Mitigating Vulnerabilities and Attacks by Switching Modes

The Systems Engineering Handbook from the *U.S. Air Force Space and Missile Systems Center* highlights a significant point: applying states and modes to an entire system can be challenging and may not always be practical or effective [USA13, p. 126]. Similarly, security modes are not a one-size-fits-all solution for all security issues. With regards to RQ1, this dissertation focused on applying mode-switching to mitigate risks from a security perspective. Vulnerabilities are common for medical and industrial IoT devices, and manufacturers often take a long time to develop and distribute patches, as mentioned in Chap. 2. If vulnerabilities are known and patches are unavailable, the mode-switching framework presented in Chap. 4 provides an important security advantage, as it helps to reduce the risk in the WoV.

In Sect. 5.1, an analysis was conducted on how mode-switching can improve the security of the intentionally insecure *OWASP Juice Shop*. In case of detected SQL injections, blocking individual IP addresses can be effective. However, this is not a comprehensive solution to all security issues. Blocking specific countries is another way to mitigate attacks, especially for businesses operating only in a specific country. The country IP zone files can be used to allow or deny specific IP addresses [IPd23]. Nevertheless, attackers may switch to another IP addresses or manipulate log files, e.g., adding IP addresses from legitimate users to wrong authentication attempts. Switching to other database modes proved advantageous to mitigate further SQL injection attacks, as some attacks target only specific implementations. As a further step, switching to *Mode Read-Only*, caused a functional restriction but provided security benefits.

Another approach by *Thompson et al.* rotates web servers to increase uncertainty and resilience, preventing fingerprinting, mitigating persistent attacks, and increasing the effort of attackers [TMMI16]. In addition to their approach, the mode-switching framework can respond to specific events, e.g., detected attacks, disclosed vulnerabilities, or known exploits.

The evaluation in Sect. 5.2 demonstrated that switching between multiple web server modes improved security compared to using a single web server. Mode-switching resulted in lower CVSS scores and thus in fewer days at risk. In contrast to simply turning a system on and off, mode-switching provides more flexibility to react to vulnerabilities. Although historical vulnerability data might recommend using only the *Mode Nginx*, caution is advised. Fewer vulnerabilities may indicate less frequent or comprehensive testing. Unknown vulnerabilities and exploits will still threaten systems. Nevertheless, mode-switching resulted in three fewer days of the system being at risk, a 93.7 lower CVSS score sum, provided more functionality, and added flexibility to react to future vulnerabilities that may not be known today. Even one day more risk may be enough for state-sponsored attackers to take over a web application. In the case study, the *Mode Apache* was suggested seven times in August 2019 and in January 2020 because of a lower CVSS sum of the *Mode Nginx*. The number of vulnerabilities in the software's past not only highlights the associated challenges but also underscores the substantial effort invested in addressing and resolving these issues. It does not necessarily imply that the software is of poor quality.

Use-Case Related Configurations

The test scenarios in Chap. 5 are just examples and should be adjusted based on practical experience and the specific use case. For example, in the web server case study CMSs like *WordPress* and *Joomla* need PHP. Switching only between the corresponding PHP modes is more practical. However, it is beneficial to have a fallback mode providing a static HTML page rather than a vulnerable dynamically generated PHP page or an HTTP error page like *404 Not Found*.

Handling risks may differ among various stakeholders, businesses, and contexts [RMCO16]. For example, security measures in online banking applications differ from measures in industrial systems and medical devices. The weighting of security goals influences the treatment of risks. If confidentiality and integrity are more important than availability, switching to a secure mode reduces the risk. In the mode-switching framework, the mode can be prioritized in the system description to choose the optimal mode with the lowest total severity score.

Accelerating Patch Development, Distribution, and Vulnerability Information Sharing

Using the mode-switching framework gives manufacturers more time to provide well-tested patches. Nevertheless, the performance of patch development and their distribution will remain critical. One extension of this work is to consider the number of days a vulnerability exists and is not fixed, similar to the temporal risk strategy by Rapid7 [Rap21]. In addition, new work on machine-readable CVE advisories can be added to decide whether a system is affected and the vulnerability exploitable [OAS20]. For example, by August 2023, Google started to provide security fixes for the Chrome browser every week [Res23]. The sooner the vulnerability is patched, the sooner the WoV closes. However, some operators do not install available patches promptly or at all. They may not have heard about them yet or may be worried about patch errors and business interruptions.

In scenarios where devices are not always connected, patch installation may require manual procedures, e.g., hospital appointments or service visits.

Over-the-air updates through edge devices or home stations can speed up distribution, but this also poses additional risks, as demonstrated by update errors from other areas. Mode-switching can offer flexibility before hospital appointments or the visits of a service employee. The mode-switching framework can have an even greater impact if vulnerability information is provided immediately after reporting to the manufacturer. The analyzing component does not need all vulnerability details to switch from one mode to another. It is sufficient to know that one mode is riskier than another.

6.2 Multi-Modal Architecture, Framework, and Security Modes

Mode-switching provides a method to secure systems between vulnerability disclosure and patch installation. It reduces the attack surface, enhances systems resilience, and enables manual risk reduction even if the manufacturer has stopped service. In fact, mode-switching is a security-by-design approach. However, this approach is not a silver bullet to solve all security problems for all possible systems and can introduce side effects. It may increase complexity, maintenance requirements, and the potential for unintended behavior. Putting together different mode components and eventually implementing functional equivalent software with different programming languages will make the design and development phase more extensive and expensive. For example, having two web servers installed and running, the *Mode Control* needs additional disk space, memory, and processing time. Moreover, the second and third design iterations of the framework require network communication.

Negative Side Effects

Preventing misuse of the mode-switching approach is crucial to avoid security failures like those experienced by *SolarWinds* and others, where the security solution became a problem [CIS23b]. For example, the *REvil* ransomware misused *Windows Safe Mode* in order to encrypt files on the hard drive [Abr21]. Therefore, it is essential to follow the *Secure Software Development Life Cycle (SSDLC)* and the guidelines for secure coding and development [ENI19, IEC18].

False Alarms and Overreactions

Once a vulnerability becomes known, potential intrusions are detected, or warnings are issued, the mode-switching framework provides the basis for automatically transitioning to a different mode, thus, for example, deactivating the endangered component or system, or activating more restrictive security policies as part of a configuration change. However, fake anomalies, false-positive detected attacks, and misunderstood vulnerabilities may trigger unnecessary mode switches, reduce functionality, and negatively impact business functions. To prevent business interruptions and false positives, e.g., IDPSs like fail2ban allow configuring ignore lists for IP/DNS, such as private networks, company and partner DNS hosts.

Another concern is related to the autonomy of the framework. Critical events can go undetected in highly automated settings amid the vast amount of data and an overreliance on the system. As indicated by insights from our research partners, full automation of mode-switching is not desirable under all circumstances. Therefore, automation should be introduced gradually and require human interaction (human-in-the-loop). With continuous *MAPE-K* loops on the server and the IoT devices, the system can adapt and improve over time. Manufacturers can use the central *Managing System* to communicate with their IoT devices, analyze their security status, alert customers, provide patches, and modify device settings as needed. Initially, IT security professionals and operators may need to be involved in monitoring, analyzing, planning, and executing to avoid false alarms or overreactions. As confidence in the system grows, manually suggested actions can be delegated and executed automatically by the framework in the future. Knowledge from past security events can improve staff productivity, efficiency, and effectiveness.

Malicous Insiders

Manual mode-switching by operators can be risky due to malicious insiders who might control the *Managing System* and harm people or the environment from this end. They could switch multiple systems to degraded or more vulnerable modes. Thus, both technical and organizational security measures are essential. Except for user authentication, the framework currently has no special authentication and authorization mechanisms on the server side. Indeed, these mechanisms may prevent insider attacks. In order to prevent potential abusive use of insiders, administrative capabilities must be limited, and all actions must be logged and monitored for real-time or post-mortem analysis. Procedural or administrative controls are needed to handle incidents and increase security awareness [NIS18]. The two-person rule may be necessary for significant changes. This organizational measure would not only make misuse more difficult but would also limit hasty and excessive actions.

Secure Design and Development

Secure system design and development remain crucial for producing high-quality products and avoiding device recalls. The framework aims not to make developers less concerned about security in system development. Security aspects must be considered from the beginning of the development process [ENI19, ISO09].

Medical and industrial IoT applications need an overall concept with several security layers (defense in depth), which considers the entire life cycle from architecture, design, and implementation up to end of life. It is important to identify and avoid security flaws at an early stage [TSSY20, CZ08]. According to the principles for resilient systems, the focus should be on mission-critical capabilities [Fir20b]. All types of threats should be considered. Besides security, factors such as robustness, safety, and interoperability are essential as well. Early detection of adversities, identification of protection needs, and prompt implementation of countermeasures are necessary. Systems should resist attacks and recover quickly from harm. The architecture and framework presented in this dissertation enable timely and tailored security responses, adapting system behavior, configurations, and functionalities in response to detected threats.

Testing

Before systems are put into production, they should be thoroughly tested. To find typical problems, static and dynamic application security testing can be implemented within the continuous integration and deployment (CI/CD) pipeline. Additionally, fuzzing and other black-box testing tools can reveal previously unknown vulnerabilities. Fixing vulnerabilities at their core is necessary. This process needs time, money, and attention. There should be fewer bugs and vulnerabilities after an update than before. Quality assurance with comprehensive testing is essential. As demonstrated in the Juice Shop sample scenario in Sect. 5.1, leveraging vulnerability scanners like OWASP Dependency Check detects publicly disclosed vulnerabilities of libraries and third-party software [OWA23b]. The results help not only in the development phase but also in operation. Therefore, scans should be carried out periodically to check whether the systems are affected by CVEs. The results can, in turn, lead to mode switches and other security measures. Future plans include analyzing and using other vulnerability scanners such as *OpenVAS* for automatically building up a software inventory and checking whether the software is affected by CVEs [Ope23].

6.3 A Domain-Specific Language for Mode-Switching

Regarding RQ1 on applying mode-switching, a MDSL was developed to simplify the steps for operators to define modes and enhance the capability of systems to manage security dynamically. The benefits of a multi-modal architecture and the resulting mode switches were studied using the use cases and sample scenarios presented in Chap. 5. The MDSL provides a flexible and systematic way to define modes and use different combinations of software products and versions, e.g., different web servers or system configurations. Within the *Eclipse Integrated Development Environment (IDE)*, the user-defined system description is analyzed by an Xtext validator to ensure that names are unique and inheritance is properly used. Code completion and syntax highlighting increase usability. Looking at the *Halstead Software Complexity Measures* of the web server case study, see Sect. 5.2, using the MDSL shows a reduction of 64% in size and volume and 73% in the expected testing efforts when compared to the generated Java-based system configuration. However, larger and more complex systems with multiple web servers, databases, and applications may require additional constraints and rules to be considered. Therefore, the support for rule sets is planned in the MDSL.

Besides switching modes in parts of a system when affected by a vulnerability, the framework can also adapt components on a more fine-grained level. For example, to change single properties of configurations to force secure login or adapt access control [MMF+10]. To further ease the creation of a MDSL system description, automatic detection of the currently installed software versions is desired. This involves scanning the target system to generate a software inventory.

As mentioned in Chap. 1, software product line concepts can complement the MDSL. For example, mode switches can be constructed upon variability models and dynamic product configuration at runtime and leverage capabilities to define features, dependencies between different components, or cross-tree constraints [ACL+11, MFB+08]. Furthermore, providing a *Graphical User Interface (GUI)* for defining the modes can complement the textual description of the MDSL. A GUI would increase usability and allow the definition of modes and actions for users less familiar with programming language concepts.

6.4 Decentralization & Self-Protection

A resilient security strategy should include decentralization and self-protection mechanisms. The mode-switching framework provides a one-way step-by-step degradation functionality to increase security. It also includes a self-adaptive defense based on the CVSS vulnerability scores. The ability of systems to self-protect by switching modes autonomously ensures that even in the absence of immediate external intervention, the system can defend itself against ongoing attacks. This decentralized approach not only reduces the dependency on continuous human oversight but also enhances the overall robustness of the system.

A self-healing functionality is employed as the system switches back to a mode with more functionality after installing an update or patch. This mechanism can also be helpful in case of failures or problems after updates, e.g., the applications are not working as expected. The system can then switch to another mode to provide business continuity while repairing the other mode. As an extension, switching between functional equivalent modes periodically or randomly makes attacks more complicated and relates to the idea of *Moving Target Defense (MTD)* presented in Sect. 2.6.

Prevent Misuse of Self-Protection Mechanisms

Instead of just having the option to turn devices on and off, the availability of multiple modes provides more flexibility. However, self-protection mechanisms may switch systems automatically to degraded modes. This step involves turning off critical system features. Especially with medical devices, reducing functionality is always challenging and must only be considered as a last resort. As shown in the case study with IoT devices in Sect. 5.6, a central *Managing System* allows manufacturers to communicate with connected IoT devices, analyze the security situation, notify customers (in specific cases), provide patches, and adapt device settings. The central control can pose new risks. Precautions must be taken to prevent the misuse of self-protection and self-healing mode mechanisms. For example, attackers may manipulate sensor values or the communication between IoT devices and the server. SSH is used for communication from the server to the IoT devices, e.g., for switching modes. However, the message queue communication of the IoT devices with the server was unencrypted for testing purposes. Attackers may abuse the self-protection mechanism, such that legitimate users are blocked, and unnecessary mode switches are triggered. If attackers or malicious insiders get control of the central *Managing System* component, they may harm people or the environment from this end. An encrypted channel in both directions is essential, along with authentication measures and monitoring to prevent unauthorized access. Both technical and organizational security measures are essential.

Centralization vs. Decentralization

In the second and third iteration of the mode-switching framework, vulnerabilities and patches are collected and analyzed centrally. This approach takes the load off the IoT devices and performs better overall. Likewise, the framework collects potential security issues from the connected IoT devices and responds based on the analysis. In Sect. 5.6, an analysis was conducted on applying the framework, which establishes a model for managing and controlling self-protective IoT devices. The framework was evaluated through simulations of potential security issues, such as port scans and unauthorized login attempts. Results confirmed that the framework can easily detect and mitigate diverse security threats. However, the server and the central components may become a single point of failure and, thereby, a malicious insider may manipulate all connected IoT devices, as already mentioned in Sect. 6.2.

If IoT devices are not connected to a central authority like the proposed manager, the options for monitoring and control are limited. A lightweight IDPS was installed on the IoT devices to address this. IDPS can decentrally detect and prevent suspicious activities and make adaptation decisions when the server is offline or compromised. Decentralization increases scalability, reliability, and robustness. Thus, incorrect or faulty detection can result in reduced functionality and usability. According to *Quin et al.* , decentralization can increase potential security risks [QWG21]. Similar to the discussion on the level of automation, a comprehensive solution will need both central and decentral components. Further research is necessary to consider this aspect.

6.5 Advantages & Disadvantages

It is crucial how the presented framework can be combined and merged with medical and industrial IoT devices and eventually already existing operation modes. Do the new modes increase the system's complexity and the risk of unintended interactions or negative side effects? How does the transition from one mode to another work? The multi-modal architecture should not become the new Achilles' heel or a new attack surface. Table 6.1 summarizes the advantages and disadvantages of mode-switching from a security point of view.

Table 6.1 Advantages and Disadvantages of Mode-Switching

Advantages	Disadvantages
Secure the time between vulnerability disclosure and the availability of security patches	Potential misuse of mode-switching framework to reduce functionality and availability
Context-specific reduction of attack surface	Increased system complexity
More resilience to existing vulnerabilities	More extensive and more expensive design and development phase
Means of risk reduction for system operator (also for unpatched systems and even if the manufacturer has stopped service)	Potential unintended negative side effects due to multi-modal architecture
Reduced machine downtimes due to cyber-attacks	
Use of the multi-modal architecture for other purposes, not only for security	

On the one hand, mode-switching will increase system complexity, make the design and development phase more extensive and expensive, and may lead to unintended negative side effects. Mode-switching may increase the software evolution overhead with regression testing and impact analysis. Additional maintenance may be necessary to control the modes. False-positive vulnerabilities and automatic mode-switching can also lead to unintended system behavior changes. From a customer perspective, it is highly unsatisfactory if a device switches the mode automatically and seemingly unexpectedly denies services and interrupts operational processes. Attackers could misuse the mode-switching framework in order to reduce functionality and availability. Insiders can also misuse the framework and the central control of multiple devices. For example, MITRE mentions in their ATT&CK framework the change of operating mode as a technique for ICS to gain additional access [MIT20].

On the other hand, mode-switching will secure the time between vulnerability disclosure and security patch installation, reduce the attack surface, and make the system more resilient. Manual risk reduction is possible with mode-switching even if a system's or module's manufacturer has stopped maintenance. For example, Cisco has not yet released and will not release software updates to address recently discovered vulnerabilities for end-of-life firewalls and routers [Cis20b]. From a security perspective, switching to a restricted mode as soon as possible is advantageous if attacks or abnormal behavior happen. The use of the framework can be simplified by providing several sample modes for common systems, like the case study for web servers. Furthermore, modes can be used to reduce system downtime by switching to another mode during the update or maintenance of software. However, complete automation of mode-switching is not preferable in every scenario. Handling risks may differ among various stakeholders, businesses, and contexts [RMCO16]. To reduce false-positive vulnerabilities and overreactions, it is suggested to relay information to a central service like a manufacturer's cloud and put a human in the loop. Thus, security professionals and operators can control online devices semi-automatically or manually. They can manually inspect the vulnerabilities and take action. Eventually, a CVE ignore list will be needed in the future to avoid false positive entries. An internal vulnerability database can reduce dependency on public databases and allow a differentiated view. In addition, the devices can learn over time about false positives of possible attacks, e.g., with machine learning techniques. If devices are offline for a long time, they must act autonomously and be stricter about switching modes. However, a specific threshold needs to be considered. Shutting down the device must be the last resort. Finally, an authorization and monitoring system is essential to prevent and detect insider attacks, e.g., switching multiple devices to a degraded or more vulnerable mode.

Modes can also be viewed more holistically. An individual IoT device, a cluster of IoT devices, the production line, or a factory can each have a separate mode, not only from a technical but also from an organizational perspective. Modes of several IoT devices affect the factory or global mode and vice versa. If several IoT devices are attacked, they switch to a different mode, which can lead to further actions by the operator or manufacturer after exceeding a certain threshold value.

6.6 Summary

The comprehensive discussion in this chapter explored various topics related to mode-switching, security, the framework, and the evaluation. Utilizing the MDSL provides a flexible and systematic way to specify modes and their different software components and versions. Code completion and syntax highlighting increase usability, particularly for people with less programming language experience. However, a GUI would make the framework accessible to a wider range of users. By leveraging the multi-modal architecture and security modes, developers can ensure that the system remains protected between the Window of Vulnerability and the implementation of patches, reducing the overall attack surface and enhancing system resilience. If security is considered from the very beginning of system design, it is easier to engineer secure and resilient systems that work correctly as expected, even during attacks. While the proposed system shows notable effectiveness in addressing specific security problems, it is important to recognize its limitations and the fact that it does not represent a solution for all security problems. Secure system design and development remain essential. The presented framework may increase system complexity and the risk of unintended negative side effects. It needs a multifaceted approach. The framework with the mapek feedback loop is designed for continuous improvement, initially with manual involvement from IT security professionals and operators. As confidence in the system grows, the framework can automate suggested actions, leveraging knowledge from past security events for increased productivity and efficiency. By adjusting security modes dynamically, systems can efficiently address emerging threats, mitigate vulnerabilities, and minimize potential risks. As demonstrated in this dissertation, mode-switching provides a robust tool for both reactive and proactive security measures.

Conclusion and Future Work 7

This work aims to contribute to the state of the art in the security of industrial and medical *Internet of Things (IoT)* devices. Smart mode-switching helps manufacturers and operators to be prepared for the unexpected. Modes provide a better overview and more transparency about several systems' current status and security situation for security experts, system administrators, operators, customers, and manufacturers. The case studies and sample scenarios in Chap. 5 showed threat detection with a lightweight host-based *Intrusion Detection and Prevention System (IDPS)*, which can analyze log files, do integrity checks, compare execution times, and execute actions. If an anomaly is detected, a mode switch changes the system behavior, adapts functionality, and modifies the configuration. The system can send notifications to inform the owner and manufacturer during the transition from one mode to another. Single IP addresses of brute-force attacks can be blocked permanently or temporarily. Communication can be restricted through blocklists, allowlists, or one-way communication. Furthermore, authentication methods can be increased, for instance, with Two-Factor Authentication. A mode-switch can lead to the stopping and starting of different software containers like docker instances and allow or prohibit specific functionality. Switching implementation modes during runtime refers to *Moving Target Defense (MTD)* and provides a way to reduce the attack surface. Switching modes is also a method to reduce the risk in the *Window of Vulnerability (WoV)*, while vulnerabilities exist in certain software versions until a patch is provided. As an extension, a combined mode based on several IoT devices, for example, a type mode, a factory mode, or a global mode, can lead to a quick overview of what is going on out there. The IoT devices send mode information to the higher-level mode or the server. If a certain threshold is reached, the higher-level mode is changed, which can affect the mode of all other devices, such as preventing malicious activities from spreading further, similar to the COVID-19 measures.

M. Riegler, *Mode-Switching for Resilient Security*,
https://doi.org/10.1007/978-3-658-49306-6_7

This chapter presents the contribution of this dissertation in Sect. 7.1 and provides an outlook for future research in Sect. 7.2.

7.1 Contributions

This dissertation presents a step towards increasing the security of industrial and medical IoT devices. Following the ideas of resilient security and the *NIST Cybersecurity Framework*, systems should identify, protect, detect, respond, and recover from disruptions [NIS18, NIS24]. The value of the three presented design iterations of the multi-modal architecture and the mode-switching framework in Chap. 4 becomes evident when considering the following core functions:

1. **Identify:** Dividing systems into controllable modes facilitates the identification of system risks.
2. **Protect:** The architecture enhances system protection by default, utilizing secure modes. Periodic mode switches of the framework interrupt and mitigate ongoing attacks, strategically reducing the attack surface. Switching to riskier modes is only done when necessary.
3. **Detect:** The framework's key feature is continuous monitoring of vulnerabilities, patches, and connected devices. Modes with increased log levels improve the detection of deviant behavior, intrusions, and attacks.
4. **Respond:** After detecting anomalies, the framework automatically responds by switching to a more secure mode, incorporating self-adaptation and self-defense mechanisms. Manual mode switches allow preventive measures, particularly when the manufacturer quits service.
5. **Recover:** The system can automatically or manually recover from attacks or degraded modes, switching back to full functionality (self-healing).

In concordance with the *U.S. National Science Foundation (NSF)* definition of a "smart service system", characterized by "learning, dynamic adaptation, and decision-making" capabilities, the frameworks address *specific* measures for enhancing the security of medical and industrial IoT devices [NSF14, p. 2]. This is particularly pertinent in overcoming the WoV and mitigating both online and offline attacks. Leveraging the *MAPE-K* control loop, the framework learns normal system behavior, extends the knowledge base, and identifies patterns that may indicate security risks. With publicly available databases of vulnerabilities, exploits, and patches, the collection is *assigned* to the framework. Following predefined rules, the framework can automatically decide countermeasures, such as switching to a more secure mode. Dynamic adaptation through mode switches leads to lower CVSS scores and *measurable* risk reduction. The case studies and sample scenarios presented in Chap. 5 underscore the framework's effectiveness, demonstrating *realistic* countermeasures in a *timely* manner, as mode switches are between timeouts.

The contributions of this dissertation can be summarized as follows:

1. **Multi-modal Architecture & Mode-Switching Framework:** The introduction of robust architectures and frameworks for single and multiple devices (three design iterations) reduces the attack surface by automatic or manual mode switches based on events like detected vulnerabilities, anomalies, or intrusions.
2. **Mode Domain-Specific Language (MDSL):** The development of a language to define modes and specify actions for starting and stopping them.
3. **Self-Protection Mechanisms:** The implementation of mechanisms ensuring system defense while sustaining operations. Additionally, the framework has functionality for sharing attack information and enabling offline adaptation.
4. **Evaluation:** Rigorous evaluation through two comprehensive case studies and four sample scenarios, demonstrating the effectiveness of the mode-switching approach and the practical usage of the architecture and framework iterations. Eleven different mode types with 28 modes have been identified during the evaluation, see Table 7.1.
5. **Discussion:** A comprehensive discussion on how mode-switching mitigates the WoV and reduces various attack threats.

Table 7.1 Evaluated Mode Types

Use Cases Sample Scenarios	Mode Types
5.1 Juice Shop	Database Management System Modes: SQLite, PostgreSQL
	Database Access Modes: Read-Write, Read-Only
	Web Application Modes: Static, Dynamic
5.2 Web Server	Web Server Modes: Apache, Nginx, ApacheWithPhp, NginxWithPhp
5.3 PACS	File Archiver Modes: 7-Zip, gzip
	Key-Value Database: Redis, Memcached
	Web Communication Protocols: ApacheJServ, HTTP
5.4 Insulin Pumps	Connection Modes: Setup, Connected, Protected, Disconnected
5.5 Edge Devices	System Modes: Factory, Onboarded, Updated, Outdated
5.6 IoT Devices	Access Control Modes: Blocklist, Allowlist
	Connectivity Modes: Online, Offline

Answering Research Questions

The answers to the RQs of this dissertation are provided in the subsequent paragraphs.

- **RQ1: What is mode-switching, and how can it be applied to mitigate risks from a security perspective?**

Based on the *Systematic Literature Review (SLR)* [RS20], Chap. 3 presents the key terms, characteristics, and principles of modes, and gives comprehensive insights into the history, examples, and application areas. The chapter explains mode-switching in-depth, presents existing work on mode-switching protocols, and how to adapt modes in the context of security.

By dividing systems into modes, they can perform multiple behaviors and functions as required. At design time, modes and their transitions were specified, considering specific requirements. This phase is important to prevent any mode confusion or mode errors during runtime. Based on a timer or an event, modes are switched automatically or manually based on an operator's decision.

Mode-switching mechanisms are used in *Multi-Mode System (MMS)* to facilitate the transition between different modes. Depending on the operation purpose, these protocols are often employed for safety, resource efficiency, and adaptability in various applications, including airplanes, pacemakers, and self-driving cars. Typically, MMS are used to prevent resource bottlenecks and provide flexibility for single- and multi-core platforms, especially for resource-constrained systems. Over time, the use of MMSs has expanded to manage complexity and provide flexible and quick adjustments of behavior. Modes are part of several programming, modeling, and specification languages like AADL, SysML, and UML and have been used to provide real-time adaptivity.

In this dissertation, a mode is defined as a system state in which a specific configuration is active and specific functionality is provided, leading to a desired system behavior for a period of time. Each mode faces different security risks and poses more or less attack surface. Modes are switched to adapt system behavior, reduce the attack surface, and minimize attackers' range of activity.

With the theoretical findings as a foundation, mode-switching was applied to mitigate risks from a security perspective. The multi-modal architecture and the mode-switching framework were developed and extended in three design iterations, as detailed in Chap. 4. A MDSL was created to simplify the definition of modes and specific actions to start and stop them.

Manufacturers often take a long time to develop and distribute patches for their software. In the meantime, attackers can discover vulnerabilities and take advantage. After providing patches, the situation is exacerbated, as attackers can analyze them and attack unpatched systems. Some systems are not always connected, and installing patches may require manual procedures. As shown in Chap. 5 and discussed in Sect. 6.1, using modes helps to reduce the risk in the WoV and provides resilience between vulnerability disclosure and patch installation. If a vulnerability is detected, modes are switched automatically to mitigate the risk until software manufacturers provide patches and system administrators install them. Switching modes enable manual risk reduction, even if the manufacturer has stopped service.

The web server case study in Sect. 5.2 showed that compared to a single server, using a multi-mode server system reduced the attack surface to zero known risk in 98.9% of the analyzed time. Mode-switching resulted in fewer days at risk, lower CVSS score, more functionality, and flexibility to react to vulnerabilities. The multi-modal architecture and mode-switching framework present a solution to mitigate risks by offering a range of modes instead of just on/off options. This gives manufacturers more time to develop and test patches and customers more time to test and install them.

Answering RQ1, mode-switching is a valuable strategy to mitigate security risks by offering a range of operational modes. These modes can be used manually by operators and automatically by systems to dynamically adjust behavior and effectively minimize the attack surface, especially in scenarios where timely patches are unavailable. The presented multi-modal architecture and framework offer a structured approach to risk mitigation, equipped with a comprehensive set of tools, including a MDSL for defining modes and specific actions.

- **RQ2: How does mode-switching work on medical IoT devices?**

Medical IoT devices have special threats and provide safety, security, and privacy challenges, see Chap. 2. Conventional IT security measures may not always be suitable for these devices. Medical devices have constrained resources and do not support resource-intensive security measures. The patient's well-being must be ensured. Medical devices must be operational and provide clinical function all the time. Updates and patches are rare and need long planning. Comprehensive testing and validation are necessary for every change. Stopping medical in case of potential or real threats would lead to a highly secure state (fail-secure) but could be life-threatening. Therefore, safety and availability are considered more important for these life-critical systems than security, confidentiality, and integrity. Especially in emergencies, safety-critical modes are prioritized over security modes. Not many CVE entries are available for medical devices. Some manufacturers provide detailed vulnerability information only for their registered customers or keep incidents confidential. However, the number of CVEs is increasing yearly.

Mode-switching was explored in two medical sample scenarios in Sects. 5.3 and 5.4. The second iteration of the multi-modal architecture presented in Sect. 4.6

was practically showcased with *Picture Archiving and Communication Systems (PACSs)* and insulin pumps. Code defects and insecure third-party libraries caused nine years of risks for PACSs before manufacturers provided a patch. The suggested MMS with *Modes for File Archiver, Key-Value Database, and Web Communication Protocols* has discovered the vulnerabilities and provided a way to mitigate them until a patch was provided and installed. Instead of using only one type of software, using modes with different libraries provides the same functionality and flexibility during operation. Mode-switching reduced the CVSS sum on 1404 of 3760 days. The average CVSS sum decreases by 41%.

In Sect. 5.4, the multi-modal architecture was evaluated for insulin pumps with a real-world example. Vulnerabilities make it possible to alter the insulin dose of a pump and may cause seizure, coma, or even death. The manufacturer suggested manually adapting settings to turn off remote functionality. The approach with multiple *Connection Modes* simplifies this step for *Healthcare Delivery Organizations (HDOs)* and affected patients. The system notifies the operator about the vulnerability and suggests interventions. If no patch is available or the risk for patient safety is too high, the device can switch manually or automatically to the *Mode Disconnected*. In this mode, the insulin pump works only with predefined presets, providing a fallback if sensors are broken or communication is lost. Additionally, lightweight IDPS on the medical IoT devices can analyze traffic, notify patients and HDOs, and switch modes in case of anomalies or attacks.

Answering RQ2, the exploration of mode-switching in sample scenarios with PACS and insulin pumps demonstrated the application of the multi-modal architecture. Considering patient safety and continued clinical functionality as the top priorities, using modes with predefined presets provides a fail-safe mechanism in case of sensor malfunctions, communication loss, vulnerabilities, or attacks.

- **RQ3: How does mode-switching work on industrial IoT devices?**

In Chap. 2, threats and security issues of industrial IoT devices were analyzed in the context of Industry 4.0. If left unaddressed, severe consequences like personal injury or environmental damage can result. In contrast to medical IoT devices, industrial IoT devices have a greater computational capacity but also expose a larger attack surface because of multiple interfaces, including USB ports, Ethernet connections, and debug ports. Industrial IoT devices have a life cycle from 5 to 20 years, while availability is a top priority.

The investigation included applying the presented multi-modal architecture and mode-switching framework of Chap. 4 and analyzing its usability for industrial IoT and edge devices. The second and third iterations of the architecture and framework were evaluated in a sample scenario and a case study to determine their potential to enhance the next generation of these devices.

In the context of edge devices, four *System Modes* were introduced to manage the device's behavior throughout the life cycle, see Sect. 5.5. Initially, devices are deployed in the *Mode Factory*, offering limited functionality, which safeguards against potential theft and manipulation. After successful onboarding, where the device connects to the manufacturer's cloud service, it switches into the *Mode Onboarded*, granting full operational capabilities. The framework continuously checks for software updates. If the update check is too long ago, it automatically switches from *Mode Updated* to *Mode Outdated* with restricted functionality, reducing the attack surface. This automation ensures that devices remain up-to-date and quickly respond to emerging threats or vulnerabilities by switching to more secure modes, reducing the attack surface, and ensuring minimal downtime during mode switches. The implementation involved leveraging system, service, configuration management, and application deployment tools. Potential challenges and limitations of the approach were discussed, emphasizing the trade-offs between security and availability.

In Sect. 5.6, a case study was conducted to demonstrate the applicability of the third iteration of the mode-switching framework for industrial IoT devices. The goal was to monitor, analyze, and update components in response to different security threats and assess the framework's potential benefits. The introduction of *Access Control Modes* and sharing information about attackers among IoT devices and a central managing system improve the ability to defend against attacks collectively. In the *Mode Blocklist*, devices automatically block IP addresses after a few failed login attempts and inform the managing system. This system detects attacks on multiple devices and shares the list of blocked IP addresses. This method reduces the attack surface of the entire system. Alternatively, the devices can switch to *Mode Allowlist*, allowing only predefined IP addresses to connect, increasing security at the expense of flexibility. Mode-switching can be triggered automatically or manually, adapting to security threats and the device's environment. If the device cannot connect to the managing system and attacks are ongoing, neither the managing system nor an operator can intervene. Therefore, *Connectivity Modes* provide self-protection methods to restrict attacks during the offline situation.

The framework also considers vulnerabilities and provides adaptation mechanisms. It employs lightweight data exchange protocols for information sharing, event-driven adaptations, and human-in-the-loop interactions. The framework encompasses components at both the managing server and IoT device levels, using a combination of decentralized control patterns and self-adaptive mechanisms based on the *MAPE-K* loop.

Throughout this research, feedback from our industrial partner *ENGEL Austria GmbH* was incorporated multiple times. This collaboration facilitated a better understanding of the perspectives and requirements of professionals and how the results contribute to addressing their everyday challenges to improve the security of industrial IoT devices and draw conclusions for science. It was discovered that automatic mode switches are not always desirable. When introducing such a framework, a human-in-the-loop may be necessary initially to balance security and user satisfaction. When enough confidence has been established, switching from manual operation to automatic processing can be considered. If devices are offline, they must act autonomously and be stricter about switching modes.

Answering RQ3, the sample scenarios and case studies with industrial IoT and edge devices showcased the effective utilization of mode-switching. Implementing self-protective countermeasures ensures prompt responses in online and offline scenarios, effectively minimizing attack surfaces. However, it is important to consider that mode-switching should not interrupt critical processes.

- **RQ4: What are general risk mitigation strategies as compared to strategies tailored to specific domains?**

As web servers and web applications are used in multiple domains, in Sects. 5.1 and 5.2 mode-switching was investigated in a sample scenario with the consciously insecure *OWASP Juice Shop* and a case study with the popular web servers *Apache* and *Nginx* in combination with *PHP*.

For the Juice Shop, several modes were identified at different levels to handle specific types of attacks. Examples include detection and response at the database level by blocking the IP address of SQL injection attacks, notifying the operator, and switching between *Database Management System Modes* like *Mode SQLite* and *Mode PostgreSQL*. Additionally, switching the *Database Access Modes* from *Mode Read-Write* to *Mode Read-only* mitigates ongoing attacks and reduces the risk of further breaches. Similarly, *Web Application Modes* have been implemented using software containers. Risk mitigation strategies also encompass combating brute force attacks by blocking single IP addresses, potentially using CAPTCHAs, and enforcing 2FA when login attempts exhibit suspicious behavior. Through simulated attacks, findings indicate that these measures enhance Juice Shop's security and resilience by promptly reducing functionality when attacks are detected.

The web server case study evaluated the expressiveness and succinctness of the MDSL for defining system configurations. The results were compared with manual code-level definitions. The MDSL reduced the lines of code by 38% and improved maintainability. Secondly, the performance of the mode-switching framework was assessed, including system description validation, system configuration generation, and mode-switching. The framework proved efficient, with quick validation and runtime operations far below the relevant timeouts. A time span of two years was replayed with real vulnerabilities and patches, and the reaction of the mode-switching framework was analyzed. The study showed that using the *Web*

Server Modes reduced the WoV from 536 to 8 days and improved system security, especially when prioritizing modes with lower risk.

Answering RQ4, mode-switching was examined with a web application sample scenario and a web server case study. The key components of the MDSL and the framework are broadly applicable, offering functionalities for mode management, vulnerability and attack detection and response. Modes can be deployed across various layers, enhancing the overall security of the system.

Contributing Papers

Figure 7.1 shows the relationship between the published articles contributing to this dissertation. The SLR [RS20] provides the foundation for the other papers. We investigated the security of web applications and the surrounding environment [RS21a]. Subsequent papers focused on industrial IoT and edge devices [RS21b, RSS22], and

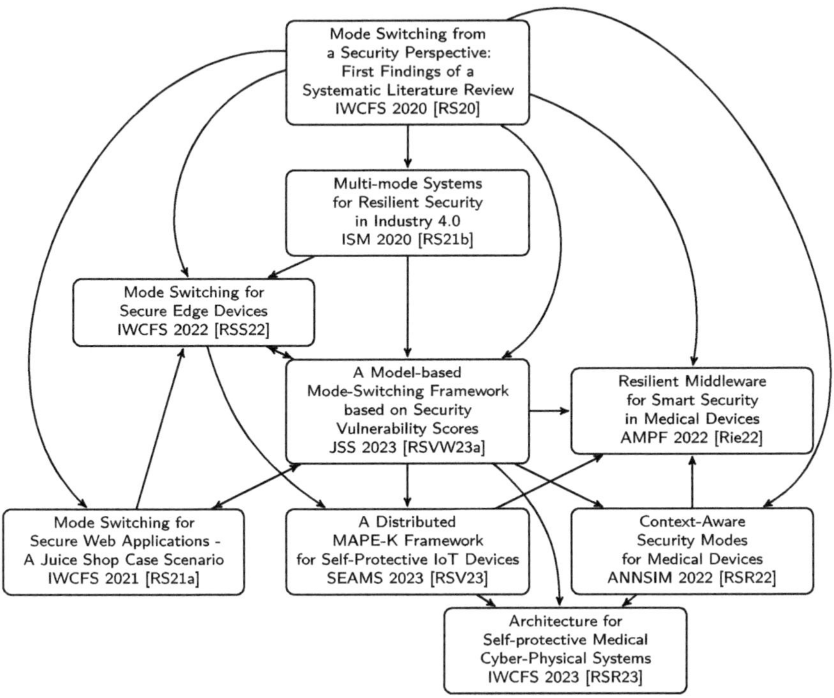

Fig. 7.1 Relations Between the Articles Contributing to this Dissertation

medical IoT devices followed [RSR22, Rie22, RSR23]. A multi-modal architecture and a mode-switching framework were developed [RSVW23a], with an extension for decentralized systems [RSV23].

7.2 Future Work

The security of industrial and medical IoT devices has received significant attention over the last few years. However, there are still unresolved research problems. Some points have already been raised in the discussion in Chap. 6. This section provides an overview of potential research directions for future work in the context of mode-switching.

Mode Description and Modeling—Further validation of the MDSL to textually describe modes will be necessary. Larger and more complex systems may require additional constraints and rules. User studies, interviews, and surveys with domain experts could contribute to this aspect. Code completion, syntax highlighting, and code validation have been implemented to increase usability. However, a graphical interface could complement the textual mode description with the MDSL, increase usability, and allow the definition of modes and actions for users less familiar with programming language concepts. Also the application of LLMs may be beneficial.

Deeper Mode Integration—Modes could be integrated deeper into software, e.g., frameworks, components, kernel, and be combined with the ideas of Software Product Line. Modes and mode-switching could be built around variability models and dynamic product configuration during runtime. Capabilities could be utilized to specify features and analyze dependencies across various components.

Architecture and Framework Extension—In the course of the research, the architecture and framework were revised several times. The first design iteration focused on security modes for single devices, like industrial IoT and edge devices, and medical devices like pacemakers and insulin pumps. Due to the increasing interconnectivity of these devices, a broader perspective was adopted. In the second iteration, the focus was expanded to multiple devices. As many devices are already connected, it makes sense to configure and control them centrally. In the third iteration, the *MAPE-K* loop was built in to enable self-adaptive behavior.

The architecture and framework could be expanded further. A combined mode based on several IoT devices, for example, a type mode, a factory mode, or a global mode, could lead to a quick overview for operators and enhance the presented frameworks. Leveraging the interconnected topology of IoT devices can be beneficial. Components and devices send their mode information to the higher-level mode or the managing system. Once a certain threshold is reached, the higher-level mode is changed, affecting the mode of all other components and devices. Mode-switching helps to prevent malicious activities from spreading further. When user confidence in the system increases, manually requested mode switches can be assigned to be automatically carried out by the framework.

Edge, Fog, and Cloud Computing—Edge and fog computing could further enrich the performance and increase the robustness of the overall system. Centrally analyzing all log files and incidents would require considerable disk space and processing power. Analyzing and detecting threats and automatic reactions at the edge will increase response time and enable mode-switching for real-time systems. The plan is to extend this work with existing configuration management systems like *Ansible* and *Puppet* to support several operating systems and simplify provisioning and application deployment on multiple systems. Thereby, entry barriers to mode-switching could be reduced, and the framework could become a broader scope of application. Cloud computing may support central analysis of vulnerabilities and patches. However, it is essential to investigate how much centralization and decentralization are beneficial.

Detection and Prediction of Vulnerabilities and Patches—Automated software and version detection simplifies the search process for vulnerabilities and patches. Also, additional databases for vulnerabilities and exploits and other information sources for operating systems, distributions, and software could be considered. If vulnerability information is provided immediately after reporting to the manufacturer, the reaction of customers will be faster. The patches in the framework were explicitly collected for Debian and must be adapted when Windows or another Linux distribution is used. Future research should investigate how mode-switching can mitigate threats and vulnerabilities in the context of software supply chain security. Another interesting aspect is whether predicting vulnerabilities based on knowledge from past security events is possible [LRW20]. If rainy weather is forecast, it is advisable to take an umbrella with you as a precaution. Likewise, modes can be switched precautionary. Perhaps the use of AI could be considered here.

Additional Mode-Switch Triggers—Based on the vulnerabilities and patches collected, the analysis component of the framework can automatically trigger mode switches. Usually, vulnerability scanners only inform system administrators, and they decide how to proceed. The framework uses the vulnerability scores to decide which mode is more or less risky. When exploits become known, the vulnerability scores are rising. Additional triggers are planned to be incorporated into the framework, which is not restricted to reported vulnerabilities. Considering exploit databases may help to reduce the attack surface. In addition, IDS [KHG+21, KAD21, TCD+21, ASVAG21] and log file analyzer can trigger mode-switching. Analyzing and correlating issues from multiple IoT devices could further reveal anomalies and attacks that a single device would not have detected. Early warning systems could trigger switches to modes with more restrictive security policies [APPN21, ABFM09]. Rapidly increasing or decreasing number of online users, network traffic, or other events should also be considered, eventually to prevent a DoS attack. Future work could use machine learning approaches to analyze the data, detect any abnormal behavior, and trigger mode switches.

Single Reactions vs. Mode-Switching—It is not always clear if single reactions, e.g., to drop packets or block IP addresses, are sufficient to ward off attacks. At what point are more extensive countermeasures like mode-switching necessary to adjust behavior, rules, policies, and configurations? Should both approaches be combined? In which circumstances do single reactions lead to more extensive countermeasures? In which cases is mode-switching better? Further case studies and sample scenarios will contribute to that aspect. Furthermore, surveys and interviews can provide further insights.

False Positives and Insider Threats—The framework provides support for secure software design. However, developers should consider security when developing and executing systems. Artificial anomalies or false-positive vulnerabilities may cause needless mode switches and decrease functionality. Therefore, care must be taken to ensure that self-protection, self-healing, and other automatic mechanisms are not abused. As highlighted in the limitations, malicious insiders are not currently taken into account. Manual mode-switching by operators can cause disruptions. Eventually, separation of duties with technical and organizational measures is needed, and more than one person is required for extensive mode switches. Critical events should not be undetected. Therefore, authorization and monitoring systems are essential.

Additional Use Cases and Sample Scenarios—There is optimism that the architecture and framework can be relevant beyond the specific domains examined in this dissertation. It would be interesting to see how mode-switching can be applied to other use cases and sample scenarios and how effective it is in diverse environments and for other attack types.

Bibliography

[AAA+22] Sunday Adeola Ajagbe, Joseph Bamidele Awotunde, Ademola Olusola Adesina, Philip Achimugu, and T. Ananth Kumar. *Internet of Medical Things (IoMT): Applications, Challenges, and Prospects in a Data-Driven Technology*, page 299–319. Springer Nature, Singapore, 2022.

[AAS17] F. Alsubaei, A. Abuhussein, and S. Shiva. Security and Privacy in the Internet of Medical Things: Taxonomy and Risk Assessment. In *2017 IEEE 42nd Conference on Local Computer Networks Workshops (LCN Workshops)*, page 112–120, Oct 2017.

[AB01] L. Abeni and G. Buttazzo. Hierarchical QoS management for time sensitive applications. In *Proceedings of the seventh IEEE Real-Time Technology and Applications Symposium*, page 63–72, May 2001.

[ABB+15] Simon Allier, Olivier Barais, Benoit Baudry, Johann Bourcier, Erwan Daubert, Franck Fleurey, Martin Monperrus, Hui Song, and Maxime Tricoire. Multitier Diversification in Web-Based Software Applications. *IEEE Software*, 32(1):83–90, 2015.

[ABFM09] Martin Apel, Joachim Biskup, Ulrich Flegel, and Michael Meier. Towards early warning systems–challenges, technologies and architecture. In *International Workshop on Critical Information Infrastructures Security*, pages 151–164. Springer, 2009.

[Abr21] Lawrence Abrams. REvil ransomware has a new 'Windows Safe Mode' encryption mode. Bleeping Computer. https://www.bleepingcomputer.com/news/security/revil-ransomware-has-a-new-windows-safe-mode-encryption-mode, 2021. Accessed: 15.03.2024.

[ACL+11] Mathieu Acher, Philippe Collet, Philippe Lahire, Sabine Moisan, and Jean-Paul Rigault. Modeling Variability from Requirements to Runtime. In *16th IEEE International Conference on Engineering of Complex Computer Systems*, page 77–86, Apr 2011.

[AFB+92] Thomas A. Alspaugh, Stuart R. Faulk, Kathryn H. Britton, R. A. Parker, and David L. Parnas. Software Requirements for the A-7E Aircraft. Technical Report NRL/FR/5530-92-9194, Naval Research Laboratory Washington, DC, Aug 1992.

[AG93] J. M. Atlee and J. Gannon. State-Based Model Checking of Event-Driven System Requirements. *IEEE Trans. Softw. Eng.*, 19(1):24–40, Jan 1993.

M. Riegler, *Mode-Switching for Resilient Security*,
https://doi.org/10.1007/978-3-658-49306-6

[AIKR13] Homa Alemzadeh, Ravishankar K. Iyer, Zbigniew Kalbarczyk, and JaiRaman. Analysis of safety-critical computer failures in medical devices. *IEEE Security Privacy*, 11(4):14–26, Jul 2013.

[All19] Allianz. Allianz Risk Barometer 2019 Appendix. Allianz Global Corporate & Specialty SE. https://www.agcs.allianz.com/content/dam/onemarketing/agcs/agcs/reports/Allianz-Risk-Barometer-2019-Appendix.pdf, 2019. Accessed: 15.03.2024.

[All20] Allianz. Allianz Risk Barometer 2020 Appendix. Allianz Global Corporate & Specialty SE. https://www.agcs.allianz.com/content/dam/onemarketing/agcs/agcs/reports/Allianz-Risk-Barometer-2020-Appendix.pdf, 2020. Accessed: 15.03.2024.

[All21] Allianz. Allianz Risk Barometer 2021 Appendix. Allianz Global Corporate & Specialty SE. https://www.agcs.allianz.com/content/dam/onemarketing/agcs/agcs/reports/Allianz-Risk-Barometer-2021-Appendix.pdf, 2021. Accessed: 15.03.2024.

[All22] Allianz. Allianz Risk Barometer 2022 Appendix. Allianz Global Corporate & Specialty SE. https://www.agcs.allianz.com/content/dam/onemarketing/agcs/agcs/reports/Allianz-Risk-Barometer-2022-Appendix.pdf, 2022. Accessed: 15.03.2024.

[All23] Allianz. Allianz Risk Barometer 2023 Appendix. Allianz Global Corporate & Specialty SE. https://www.agcs.allianz.com/content/dam/onemarketing/agcs/agcs/reports/Allianz-Risk-Barometer-2023-Appendix.pdf, 2023. Accessed: 15.03.2024.

[All24] Allianz. Allianz Risk Barometer 2024 Appendix. Allianz Global Corporate & Specialty SE. https://www.agcs.allianz.com/content/dam/onemarketing/agcs/agcs/reports/Allianz-Risk-Barometer-2024-Appendix.pdf, 2024. Accessed: 15.03.2024.

[ALRL04] A. Avizienis, J.-C. Laprie, B. Randell, and C. Landwehr. Basic concepts and taxonomy of dependable and secure computing. *IEEE Transactions on Dependable and Secure Computing*, 1(1):11–33, 2004.

[AMRS18] Paolo Arcaini, Raffaela Mirandola, Elvinia Riccobene, and Patrizia Scandurra. A DSL for MAPE patterns representation in self-adapting systems. In *European Conference on Software Architecture*, pages 3–19. Springer, 2018.

[And08] Björn Andersson. Uniprocessor EDF Scheduling with Mode Change. In Theodore P. Baker, Alain Bui, and Sébastien Tixeuil, editors, *Principles of Distributed Systems*, Lecture Notes in Computer Science, page 572–577. Springer, 2008.

[ANS99] ANSI/RIA. Industrial Robots and Robot Systems—Safety Requirements. Standard ANSI/RIA R15.06-1999, American National Standard Institute/Robotic Industries Association (ANSI/RIA), 1999.

[Ans23] Ansible Project Contributors. ansible.builtin.systemd_service module—Manage systemd units. https://docs.ansible.com/ansible/latest/collections/ansible/builtin/systemd_service_module.html, Nov 2023. Accessed: 15.03.2024.

[APPN21] Mazen Azzam, Liliana Pasquale, Gregory Provan, and Bashar Nuseibeh. Grounds for Suspicion: Physics-based Early Warnings for Stealthy Attacks on Industrial Control Systems. arXiv:2106.07980, 2021.

[ARR20] Ibrahim Almazyad, Aakarsh Rao, and Jerzy Rozenblit. A framework for secure data management for medical devices. In *SpringSim 2020*, page 1–12. 2020 Spring Simulation Conference, May 2020.

[ARR22] Nasser S. Albalawi, Michael Riegler, and Jerzy W. Rozenblit. Towards Strategies for Secure Data Transfer of IoT Devices with Limited Resources. In G. Kotsis et al., editor, *Database and Expert Systems Applications—DEXA 2022 Workshops*, Communications in Computer and Information Science, page 331–336, Cham, 2022. Springer International Publishing.

[ARS15] Paolo Arcaini, Elvinia Riccobene, and Patrizia Scandurra. Modeling and Analyzing MAPE-K Feedback Loops for Self-Adaptation. In *2015 IEEE/ACM 10th International Symposium on Software Engineering for Adaptive and Self-Managing Systems*, pages 13–23, 2015.

[ASB+10] Habtamu Abie, Reijo M Savola, John Bigham, Ilesh Dattani, Domenico Rotondi, and Giorgio Da Bormida. Self-healing and secure adaptive messaging middleware for business-critical systems. *International Journal on Advances in Security*, 3(1&2), 2010.

[ASF23] ASF. Tomcat Home The Apache Software Foundation The Apache Tomcat Connectors—AJP Protocol Reference. https://tomcat.apache.org/connectors-doc/ajp/ajpv13a.html, 2023. The Apache Software Foundation (ASF), Accessed: 11.02.2024.

[ASVAG21] Maryam Raiyat Aliabadi, Margo Seltzer, Mojtaba Vahidi Asl, and Ramak Ghavamizadeh. ARTINALI#: An Efficient Intrusion Detection Technique for Resource-Constrained Cyber-Physical Systems. *International Journal of Critical Infrastructure Protection*, 2021.

[AUT17] AUTOSAR. Guide to Mode Management. https://www.autosar.org/fileadmin/user_upload/standards/classic/4-3/AUTOSAR_EXP_ModeManagementGuide.pdf, Dec 2017. Accessed: 15.03.2024.

[BAD08] Aaron Block, James H. Anderson, and UmaMaheswari C. Devi. Task reweighting under global scheduling on multiprocessors. *Real-Time Systems*, 39(1):123–167, 2008.

[Bad19] Ronan Baduel. *An integrated model-based early validation approach for railway systems*. Theses, Université Toulouse le Mirail – Toulouse II, 2019.

[Bai93] CM Bailey. Hard real-time operating system kernel. Investigation of mode change. Task 14 Deliverable on ESTSEC Contract 9198/90/NL/SF. Technical report, British Aerospace Systems Ltd, 1993.

[BBOD18] Ronan Baduel, Jean-Michel Bruel, Iulian Ober, and Eddy Doba. Definition of states and modes as general concepts for system design and validation. In *12e Conference Internationale de Modelisation, Optimisation et Simulation (MOSIM 2018)*, Toulouse, France, Jun 2018. HAL.

[BCC+21] Elaine Bochniewicz, Melissa P. Chase, Steven M. Christey Coley, Kyle Wallace, Matt Weir, and Margie Zuk. Playbook for Threat Modeling Medical Devices. Technical report, The MITRE Corporation and the Medical Device Innovation Consortium (MDIC), Nov 2021.

[BCRF12] Wayne Burleson, Shane S. Clark, Benjamin Ransford, and Kevin Fu. Design Challenges for Secure Implantable Medical Devices. In *DAC 2012*, pages 12–17, San Francisco, California, USA, Jun 2012.

[BDBB18] A. Burns, R. I. Davis, S. Baruah, and I. Bate. Robust Mixed-Criticality Systems. *IEEE Transactions on Computers*, 67(10):1478–1491, 2018.

[BH21] Charlotte K. Boughton and Roman Hovorka. New closed-loop insulin systems. *Diabetologia*, 64(5):1007–1015, 2021.

[BHP09] Etienne Borde, Gregory Haik, and Laurent Pautet. Mode-based reconfiguration of critical software component architectures. In *Automation Test in Europe Conference Exhibition 2009 Design*, page 1160–1165, 2009.

[BJH16] A. J. Burns, M. Eric Johnson, and Peter Honeyman. A brief chronology of medical device security. *Communications of the ACM*, 59(10):66–72, Sep 2016.

[BKFL17] Cornel Barna, Hamzeh Khazaei, Marios Fokaefs, and Marin Litoiu. Delivering elastic containerized cloud applications to enable DevOps. In *2017 IEEE/ACM 12th International Symposium on Software Engineering for Adaptive and Self-Managing Systems (SEAMS)*, pages 65–75, 2017.

[BLLV20] Maurice H. ter Beek, Axel Legay, Alberto Lluch Lafuente, and Andrea Vandin. Variability meets security: qantitative security modeling and analysis of highly customizable attack scenarios. In *Proceedings of the 14th International Working Conference on Variability Modelling of Software-Intensive Systems*, VAMOS '20, page 1–9. ACM, Feb 2020.

[BM15] Benoit Baudry and Martin Monperrus. The multiple facets of software diversity: Recent developments in year 2000 and beyond. *ACM Comput. Surv.*, 48(1), September 2015.

[Bro94] Randy Browne. Mode security: An infrastructure for covert channel suppression. In *Proceedings of the 1994 IEEE Symposium on Security and Privacy*, SP '94, page 39, USA, 1994. IEEE Computer Society.

[Bro10] Manfred Broy. Multifunctional software systems: Structured modeling and specification of functional requirements. *Science of Computer Programming*, 75(12):1193–1214, 2010.

[BSI18] BSI. Cyber security requirements for network-connected medical devices. German Federal Office for Information Security (BSI). https://www.bsi.bund.de/SharedDocs/Downloads/EN/BSI/ICS/Medical_Devices_CS-E_132.html, 2018. Accessed: 15.03.2024.

[BSI22] BSI. Industrial Control System Security—Top 10 threats and countermeasures 2022. German Federal Office for Information Security (BSI). https://www.allianz-fuer-cybersicherheit.de/SharedDocs/Downloads/Webs/ACS/DE/BSI-CS/BSI-CS_005.html?nn=133710, 2022. Accessed: 15.03.2024.

[BT18] Edwin J. Beggs and John V. Tucker. Analogue-digital systems and the modular decomposition of physical behaviour. *CoRR*, 2018.

[BT21] Edwin J. Beggs and John V. Tucker. A model of systems with modes and mode transitions. *CoRR*, 127, 2021.

[BT22] Edwin J. Beggs and John V. Tucker. A model of systems with modes and mode transitions. *Journal of Logical and Algebraic Methods in Programming*, 127:100774, 2022.

[Bue09] Dennis M. Buede. *The Engineering Design of Systems*. John Wiley & Sons, Ltd, 1 edition, 2009.

[Bur14] A. Burns. System Mode Changes—General and Criticality-Based. In L. Cucu-Grosjean and R. Davis, editors, *Proceedings 2nd Workshop on Mixed Criticality Systems (WMC), RTSS*, page 3–8, 2014.

[BVEN17] Stéphane Bonnet, Jean-Luc Voirin, Daniel Exertier, and Véronique Normand. Modeling system modes, states, configurations with Arcadia and Capella: method and tool perspectives. *INCOSE International Symposium*, 27(1):548–562, 2017.

[CAN11] Thomas M. Chen and Saeed Abu-Nimeh. Lessons from Stuxnet. *Computer*, 44(4):91–93, 2011.

[CCF+20] Betty H. C. Cheng, Robert Jared Clark, Jonathon Emil Fleck, Michael Austin Langford, and Philip K. McKinley. AC-ROS: assurance case driven adaptation for the robot operating system. In *Proceedings of the 23rd ACM/IEEE International Conference on Model Driven Engineering Languages and Systems*, MODELS '20, page 102–113. ACM, Oct 2020.

[CEN14] CENELEC. Railway applications—Train Modes functional interface specification. Technical Report CLC/TR 50610:2014-09, European Committee for Electrotechnical Standardization (CENELEC), 2014.

[CG12] Shang-Wen Cheng and David Garlan. Stitch: A language for architecture-based self-adaptation. *Journal of Systems and Software*, 85(12):2860–2875, 2012.

[Che65] H. Chestnut. *Systems Engineering Tools*. Wiley, 1965.

[Cim19] Catalin Cimpanu. Google Project Zero: 95.8% of all bug reports are fixed before deadline expires. ZDNet. https://www.zdnet.com/article/google-project-zero-95-8-of-all-bug-reports-are-fixed-before-deadline-expires/, 2019. Accessed: 15.03.2024.

[CIS20a] CISA. Alert (AA20-099A)—COVID-19 Exploited by Malicious Cyber Actors. U.S. Cybersecurity and Infrastructure Security Agency (CISA). https://us-cert.cisa.gov/ncas/alerts/aa20-099a, 2020. Accessed: 15.03.2024.

[Cis20b] Cisco. Cisco Small Business RV110W, RV130, RV130W, and RV215W Routers Management Interface Remote Command Execution Vulnerability. https://tools.cisco.com/security/center/content/CiscoSecurityAdvisory/cisco-sa-rv-rce-q3rxHnvm, 2020. Accessed: 15.03.2024.

[CIS22] CISA. ICS Medical Advisory (ICSMA-21-187-01) Philips Vue PACS (Update A). U.S. Cybersecurity and Infrastructure Security Agency (CISA). https://www.cisa.gov/uscert/ics/advisories/icsma-21-187-01, 2022. Accessed: 15.03.2024.

[CIS23a] CISA. Cybersecurity Alerts & Advisories. U.S. Cybersecurity and Infrastructure Security Agency (CISA). https://www.cisa.gov/news-events/cybersecurity-advisories, 2023. Accessed: 15.03.2024.

[CIS23b] CISA. Emergency Directive 21-01. U.S. Cybersecurity and Infrastructure Security Agency (CISA). https://www.cisa.gov/emergency-directive-21-01, 2023. Accessed: 15.03.2024.

[CIS23c] CISA. Known Exploited Vulnerabilities Catalog. U.S. Cybersecurity and Infrastructure Security Agency (CISA). https://www.cisa.gov/known-exploited-vulnerabilities-catalog, 2023. Accessed: 15.03.2024.

[Cis23d] Cisco. Cisco ACI Multi-Site CloudSec Encryption Information Disclosure Vulnerability. https://sec.cloudapps.cisco.com/security/center/content/CiscoSecurityAdvisory/cisco-sa-aci-cloudsec-enc-Vs5Wn2sX, Jul 2023. Accessed: 15.03.2024.

[CK00] Guanling Chen and David Kotz. A Survey of Context-Aware Mobile Computing Research. Computer Science Technical Report TR2000-381, Dartmouth College, 2000.

[CKK19] Stephan Cejka, Felix Knorr, and Florian Kintzler. Edge Device Security for Critical Cyber-Physical Systems. In *2nd Workshop on Cyber-Physical Systems Security and Resilience (CPS-SR)*, 04 2019.

[Cla22] Claroty. State of XIoT Security Report. https://claroty.com/press-releases/iot-vulnerability-disclosures-grew-57-percent-from-2h21-to-1h22, Aug 2022. Accessed: 15.03.2024.

[Cla24] Claroty. State of CPS Security Report: Healthcare 2023. https://claroty.com/resources/reports/state-of-cps-security-report-healthcare-2023, 2024. Accessed: 15.03.2024.

[Con16] Wm. Arthur Conklin. IT vs. OT Security: A Time to Consider a Change in CIA to Include Resilience. In *2016 49th Hawaii International Conference on System Sciences (HICSS)*, page 2642–2647, Jan 2016.

[CP18] Tianyang Chen and Linh Thi Xuan Phan. SafeMC: A System for the Design and Evaluation of Mode-Change Protocols. In *2018 IEEE Real-Time and Embedded Technology and Applications Symposium (RTAS)*, page 105–116, Apr 2018.

[CRR22] Ana S. Carreon-Rascon and Jerzy W. Rozenblit. Towards Requirements for Self-Healing as a Means of Mitigating Cyber-Intrusions in Medical Devices. In *2022 IEEE International Conference on Systems, Man, and Cybernetics (SMC)*, page 1500–1505, Oct 2022.

[CSMC19] Eugenia Ana Capota, Cristina Sorina Stangaciu, Mihai Victor Micea, and Daniel-Ioan Curiac. Towards mixed criticality task scheduling in cyber physical systems: Challenges and perspectives. *Journal of Systems and Software*, 156:204–216, 2019.

[CVG03] S. Cherukuri, K.K. Venkatasubramanian, and S.K.S. Gupta. Biosec: a biometric based approach for securing communication in wireless networks of biosensors implanted in the human body. In *2003 International Conference on Parallel Processing Workshops, 2003. Proceedings.*, page 432–439, Oct 2003.

[CZ08] J. Cabot and N. Zannone. Towards an integrated framework for model-driven security engineering. In J. Whittle, J. Jürjens, B. Nuseibeh, and G. Dobson, editors, *Proceedings of the 1st Modeling Security Workshop*, CEUR Workshop Proceedings. CEUR-WS.org, 2008.

[CZCM11] Liang Cai, Kai Zeng, Hao Chen, and Prasant Mohapatra. Neighbor: Ad Hoc Pairing of Nearby Wireless Devices by Multiple Antennas. In *In Proceedings of the 18th Annual Network & Distributed System Security Conference (NDSS 2011*, page 16532, 2011.

[CZL+18] Kejun Chen, Shuai Zhang, Zhikun Li, Yi Zhang, Qingxu Deng, Sandip Ray, and Yier Jin. Internet-of-Things Security and Vulnerabilities: Taxonomy, Challenges, and Practice. *Journal of Hardware and Systems Security*, 2(2):97–110, 2018.

[Dan22] Dana Lewis and the OpenAPS Community. OpenAPS Outcomes. https://openaps.org/outcomes/, 2022. Accessed: 15.03.2024.

[Deb23a] Debian. All information about the Security Bug Tracker in JSON format. https://security-tracker.debian.org/tracker/data/json, 2023. Accessed: 15.03.2024.

[Deb23b] Debian. Chapter 3. Choosing a Debian distribution. https://www.debian.org/doc/manuals/debian-faq/choosing.en.html, 2023. Accessed: 15.03.2024.

[Deb23c] Debian. DebianReleases. https://wiki.debian.org/DebianReleases, 2023. Accessed: 15.03.2024.

[Deb23d] Debian. Security Information. https://www.debian.org/security/index.en.html, 2023. Accessed: 15.03.2024.

[DFK08] Tamara Denning, Kevin Fu, and Tadayoshi Kohno. Absence makes the heart grow fonder: New directions for implantable medical device security. In *Proceedings of the 3rd Conference on Hot Topics in Security*, HOTSEC'08, USA, 2008. USENIX Association.

[DHS11] DHS. Cyber security assessments of industrial control systems. Technical report, U.S. Department of Homeland Security (DHS), 2011.

[DHS16] DHS. Recommended Practice: Improving Industrial Control System Cybersecurity with Defense-in-Depth Strategies. Technical report, U.S. Department of Homeland Security (DHS)'s National Cybersecurity, Communications Integration Center (NCCIC) and Industrial Control Systems Cyber Emergency Response Team (ICS-CERT), September 2016.

[DJI20] DJI. How To Protect Your Drone Data When Using DJI Products. https://enterprise-insights.dji.com/blog/how-to-protect-drone-data, 2020. Accessed: 15.03.2024.

[DJI23a] DJI. Description of Remote Controller Modes. https://support.dji.com/help/content?customId=en-us03400006544&spaceId=34&re=US&lang=en#!, 2023. Accessed: 15.03.2024.

[DJI23b] DJI. The Era of Intelligent Flight is Now. https://www-v1.dji.com/intelligent-flight-modes.html, 2023. Accessed: 15.03.2024.

[DK95] A. Degani and A. Kirlik. Modes in human-automation interaction: initial observations about a modeling approach. In *1995 IEEE International Conference on Systems, Man and Cybernetics. Intelligent Systems for the 21st Century*, volume 4, page 3443–3450, Vancouver, BC, Canada, 1995. IEEE.

[DoD88] DoD. Security Requirements for Automated Information Systems (AISs). Directive AD-A227 815 5200.28, U.S. Department of Defense (DoD), 1988.

[DoD00] DoD. System/Subsystem Specification (SSS). Standard DI-IPSC-81431, U.S. Department of Defense (DoD), 2000.

[DoD07] DoD. Unmanned Systems Safety Guide for DoD Acquisition. https://apps.dtic.mil/sti/citations/ADA472102, 2007. U.S. Department of Defense (DoD), Office Undersecretary of Defense, Acquisition, Technology and Logistics, Washington, DC.

[Dof19] Zak Doffman. Critical "Update Now" Warning Issued For VxWorks OS Inside 2 Billion IoT Devices. Forbes. https://www.forbes.com/sites/zakdoffman/2019/07/29/warning-as-2-billion-medical-industrial-and-enterprise-iot-devices-at-risk-of-attack/, Jul 2019. Accessed: 15.03.2024.

[Dor24] Dormando. memcached—a distributed memory object caching system. https://memcached.org/, 2024. Accessed: 15.03.2024.

[DRANH13] Glynis Dsouza, Gabriel Rodriguez, Youssif Al-Nashif, and Salim Hariri. Building resilient cloud services using DDDAS and moving target defence. *International Journal on Cloud Computing*, 2(2/3), 2013.

[DSK99] Asaf Degani, Michael Shafto, and Alex Kirlik. Modes in Human-Machine Systems: Constructs, Representation, and Classification. *The International Journal of Aviation Psychology*, 9(2):125–138, 1999.

[Dut15] Kevin J. Dutcher. The Linkage Between States & Modes and Performance Requirements. In *NDIA Ground Vehicle Systems Engineering and Technology Symposium*, Novi, Michigan, 2015.

[DV17] Jyoti Deogirikar and Amarsinh Vidhate. Security attacks in IoT: A survey. In *2017 International Conference on I-SMAC (IoT in Social, Mobile, Analytics and Cloud) (I-SMAC)*, page 32–37, Feb 2017.

[Ecl20] Eclipse Foundation. Xtend—Modernized Java. https://www.eclipse.org/xtend, 2020. Accessed: 15.03.2024.

[Ecl23a] Eclipse Foundation. Mosquitto MQTT Broker. https://mosquitto.org, 2023. Accessed: 15.03.2024.

[Ecl23b] Eclipse Foundation. Xtext—Language Engineering Framework. https://www.eclipse.org/Xtext, 2023. Accessed: 15.03.2024.

[Edg19] Edgescan. Vulnerability Stats Report 2019. https://www.edgescan.com/wp-content/uploads/2019/02/edgescan-Vulnerability-Stats-Report-2019.pdf, 2019. Accessed: 15.03.2024.

[Edw03] Michael Thomas Edwards. A Practical Approach to State and Mode Definitions for the Specification and Design of Complex Systems. In *Practical Approaches for Complex Systems Conference*, Systems Engineering Test and Evaluation (SETE) Conference, Canberra, Australia, 2003. Systems Engineering Society of Australia (SESA).

[EM19] Chuek Easttom and Nagi Mei. Mitigating implanted medical device cybersecurity risks. In *2019 IEEE 10th Annual Ubiquitous Computing, Electronics Mobile Communication Conference (UEMCON)*, page 0145–0148, 2019.

[ENI13] ENISA. Window of exposure... a real problem for scada systems? European Union Agency for Cybersecurity (ENISA). https://www.enisa.europa.eu/publications/window-of-exposure-a-real-problem-for-scada-systems, Dec 2013.

[ENI19] ENISA. Good Practices for Security of IoT—Secure Software Development Lifecycle. European Union Agency for Cybersecurity (ENISA). https://www.enisa.europa.eu/publications/good-practices-for-security-of-iot-1, Nov 2019.

[ENI20] ENISA. *Guidelines for securing the Internet of Things: secure supply chain for IoT*. European Union Agency for Cybersecurity (ENISA). Publications Office, 2020.

[ENI23] ENISA. Enisa threat landscape: Health sector. European Union Agency for Cybersecurity (ENISA). https://www.enisa.europa.eu/publications/health-threat-landscape, Jul 2023.

[EPV17] Jessica Rodriguez Espinoza, Vladimir Sanchez Padilla, and W Velasquez. IoT Generic Architecture Proposal Applied to Emergency Cases for Implanted Wireless Medical Devices. In *Proceedings of the International Multi-Conference of Engineers and Computer Scientists*, Hong Kong, 2017.

[Esp23] EsperTech. Esper—Complex Event Processing & Stream Analytics. https://www.espertech.com, 2023. Accessed: 15.03.2024.

[ETA+21] Mohamed Elhoseny, Navod Neranjan Thilakarathne, Mohammed I. Alghamdi, Rakesh Kumar Mahendran, Akber Abid Gardezi, Hesiri Weerasinghe, and Anuradhi Welhenge. Security and Privacy Issues in Medical Internet of Things: Overview, Countermeasures, Challenges and Future Directions. *Sustainability*, 13(2121):11645, Jan 2021.

[ETS22] ETSI. Cyber security for consumer internet of things: Baseline requirements. Technical Specification TS 103 645 V2.1.2, European Telecommunications Standards Institute (ETSI), Jun 2022.

[Eur06] European Union. Directive 2006/42/EC of the European Parliament and of the Council of 17 May 2006 on machinery, and amending Directive 95/16/EC. http://data.europa.eu/eli/dir/2006/42/oj, 2006. Accessed: 15.03.2024.

[Eur16] European Union. Regulation (EU) 2016/679 of the European Parliament and of the Council of 27 April 2016 on the protection of natural persons with regard to the processing of personal data and on the free movement of such data, and repealing Directive 95/46/EC (General Data Protection Regulation). https://eur-lex.europa.eu/eli/reg/2016/679/oj, 2016. Accessed: 15.03.2024.

[Eur17] European Union. Regulation (EU) 2017/745 of the European Parliament and of the Council of 5 April 2017 on medical devices, amending Directive 2001/83/EC, Regulation (EC) No 178/2002 and Regulation (EC) No 1223/2009 and repealing Council Directives 90/385/EEC and 93/42/EEC. http://data.europa.eu/eli/reg/2017/745/oj, 2017. Accessed: 15.03.2024.

[Eur19] European Union. Regulation (EU) 2019/881 of the European Parliament and of the Council of 17 April 2019 on ENISA (the European Union Agency for Cybersecurity) and on information and communications technology cybersecurity certification and repealing Regulation (EU) No 526/2013 (Cybersecurity Act). http://data.europa.eu/eli/reg/2019/881/oj, 2019. Accessed: 15.03.2024.

[Eur22] European Union. Directive (EU) 2022/2555 of the European Parliament and of the Council of 14 December 2022 on measures for a high common level of cybersecurity across the Union, amending Regulation (EU) No 910/2014 and Directive (EU) 2018/1972, and repealing Directive (EU) 2016/1148 (NIS 2 Directive). http://data.europa.eu/eli/dir/2022/2555/oj, 2022. Accessed: 15.03.2024.

[Eur23] European Commission. EU Cyber Resilience Act. https://digital-strategy.ec.europa.eu/en/policies/cyber-resilience-act, 2023. Accessed: 15.03.2024.

[Eva05] David Evans. Avionics International. https://www.aviationtoday.com/2005/07/01/safety-mode-confusion-timidity-factors/, Jul 2005. Accessed: 15.03.2024.

[FAA91] FAA. Lessons Learned | Transport Airplane | Accidents | Boeing 767-300ER | Lauda Air Flight NG 004, OE-LAV. U.S. Federal Aviation Administration (FAA). https://www.faa.gov/lessons_learned/transport_airplane/accidents/OE-LAV, 1991. Accessed: 15.03.2024.

[Fai23a] Fail2Ban. fail2ban. https://www.fail2ban.org, 2023. Accessed: 15.03.2024.

[Fai23b] Fail2ban. Manual Fail2ban 0.8. https://www.fail2ban.org/wiki/index.php/MANUAL_0_8, 2023. Accessed: 15.03.2024.

[FDA16a] FDA. FDA approves first automated insulin delivery device for type 1 diabetes. U.S. Food and Drug Administration (FDA). https://www.fda.gov/news-events/press-announcements/fda-approves-first-automated-insulin-delivery-device-type-1-diabetes, 2016. Accessed: 15.03.2024.

[FDA16b] FDA. Postmarket Management of Cybersecurity in Medical Devices - Guidance for Industry and Food and Drug Administration Staff. Technical Report FDA-2015-D-5105, U.S. Food and Drug Administration (FDA), Dec 2016.

[FDA17] FDA. Firmware Update to Address Cybersecurity Vulnerabilities Identified in Abbott's (formerly St. Jude Medical's) Implantable Cardiac Pacemakers. http://www.fda.gov/medical-devices/safety-communications/firmware-update-address-cybersecurity-vulnerabilities-identified-abbotts-formerly-st-jude-medicals, Aug 2017.

[FDA22a] FDA. Class 2 Device Recall Medtronic MiniMed 600 Series Insulin Pump Systems. U.S. Food and Drug Administration (FDA). https://www.accessdata.fda.gov/scripts/cdrh/cfdocs/cfRES/res.cfm?id=196205, 2022. Accessed: 15.03.2024.

[FDA22b] FDA. Class 2 Device Recall Medtronic MiniMed 600 Series Insulin Pump Systems. U.S. Food and Drug Administration (FDA). https://www.accessdata.fda.gov/scripts/cdrh/cfdocs/cfRES/res.cfm?id=196183, 2022. Accessed: 15.03.2024.

[FDA23a] FDA. Content of Premarket Submissions for Device Software Functions—Guidance for Industry and Food and Drug Administration Staff. Technical Report FDA-2021-D-0775, U.S. Food and Drug Administration (FDA), Jun 2023.

[FDA23b] FDA. Recognized Consensus Standards: Medical Devices. U.S. Food and Drug Administration (FDA). https://www.accessdata.fda.gov/scripts/cdrh/cfdocs/cfStandards/results.cfm?start_search=1&sortcolumn=pdd&productcode=&category=Software&title=Security&supportingdocsyn=off&ascapilotyn=off&organization=&referencenumber=&regulationnumber=&recognitionnumber=&effectivedatefrom=&effectivedateto=&PAGENUM=100, 2023. Accessed: 15.03.2024.

[Fir19] Donald Firesmith. System Resilience: What Exactly is it? Software Engineering Institute, Carnegie Mellon University. https://insights.sei.cmu.edu/sei_blog/2019/11/system-resilience-what-exactly-is-it.html, 2019. Accessed: 15.03.2024.

[Fir20a] Donald Firesmith. System Resilience Part 5: Commonly-Used System Resilience Techniques. Software Engineering Institute, Carnegie Mellon University. https://insights.sei.cmu.edu/blog/system-resilience-part-5-commonly-used-system-resilience-techniques/, 2020. Accessed: 15.03.2024.

[Fir20b] Donald Firesmith. System Resilience Part 7: 16 Guiding Principles for System Resilience. Software Engineering Institute, Carnegie Mellon University. https://insights.sei.cmu.edu/sei_blog/2020/04/part-7-16-guiding-principles-for-system-resilience.html, 2020. Accessed: 15.03.2024.

[FIR23] FIRST. Common Vulnerability Scoring System SIG. Forum of Incident Response and Security Teams (FIRST). https://www.first.org/cvss, 2023. Accessed: 15.03.2024.

[FMB+21] Michael Fagan, Jeffrey Marron, Kevin Brady, Barbara Cuthill, Katerina Megas, Rebecca Herold, David Lemire, and Brad Hoehn. IoT Device Cybersecurity Guidance for the Federal Government: Establishing IoT Device Cybersecurity Requirements. Special Publication (SP) SP 800-213, U.S. National Institute of Standards and Technology (NIST), Nov 2021.

[FOMC10] Nicolas Falliere, Liam O Murchu, and Eric Chien. W32.Stuxnet Dossier. Symantec. https://www.wired.com/images_blogs/threatlevel/2010/11/w32_stuxnet_dossier.pdf, 2010. Accessed: 15.03.2024.

[Fon23] Dimitri Fontaine. PGLOADER. https://pgloader.io/, 2023. Accessed: 15.03.2024.

[Fow10] Martin Fowler. *Domain-Specific Languages*. Addison-Wesley Signature Series (Fowler). Pearson Education, 2010.

[FR07] Peter Feiler and Ana Rugina. Dependability Modeling with the Architecture Analysis & Design Language (AADL). Technical Report CMU/SEI-2007-TN-043, Software Engineering Institute, Carnegie Mellon University, Pittsburgh, PA, 2007.

[Fre20] Free Software Foundation. Gzip—GNU Project—Free Software Foundation. https://www.gnu.org/software/gzip/, 2020. Accessed: 15.03.2024.

[Fre23a] Freedesktop.org. systemd.special — Special systemd units. https://www.freedesktop.org/software/systemd/man/latest/systemd.special.html, 2023. Accessed: 15.03.2024.

[Fre23b] Freedesktop.org. systemd.target — Target unit configuration. https://www.freedesktop.org/software/systemd/man/systemd.target.html, 2023. Accessed: 15.03.2024.

[Gar21] Gartner. Gartner Predicts By 2025 Cyber Attackers Will Have Weaponized Operational Technology Environments to Successfully Harm or Kill Humans. https://www.gartner.com/en/newsroom/press-releases/2021-07-21-gartner-predicts-by-2025-cyber-attackers-will-have-we, 2021. Accessed: 15.03.2024.

[GBWG19] Christopher Greer, Martin Burns, David Wollman, and Edward Griffor. Cyber-Physical Systems and Internet of Things, Mar 2019.

[GCW20] Matheus E. Garbelini, Sudipta Chattopadhyay, and Chundong Wang. SweynTooth: Unleashing Mayhem over Bluetooth Low Energy. In *Proceedings of the 2020 USENIX Conference on Usenix Annual Technical Conference*, pages 911–925. USENIX Association, 2020.

[Gem08] Mike Gemünde. *Evaluation Environment for AUTOSAR-Autocode in Motor Control Units*. Diploma thesis, University of Kaiserslautern, Kaiserslautern, 2008.

[GGVCS24] Miguel García-Gordillo, Joan J. Valls, Javier Coronel, and Sergio Sáez. Mode change management for adaptive cyber-physical systems. *ACM SIGAda Ada Letters*, 43(2):84–88, 2024.

[GHR+11] Shyamnath Gollakota, Haitham Hassanieh, Benjamin Ransford, Dina Katabi, and Kevin Fu. They can hear your heartbeats: Non-invasive security for implantable medical devices. In *Proceedings of the ACM SIGCOMM*

2011 Conference, SIGCOMM '11, page 2–13, New York, NY, USA, 2011. Association for Computing Machinery.

[GMV06] S.K.S. Gupta, T. Mukherjee, and K. Venkatasubramanian. Criticality aware access control model for pervasive applications. In *Fourth Annual IEEE International Conference on Pervasive Computing and Communications (PERCOM'06)*, page 5 pp. – 257, Mar 2006.

[Goo20] Bill Goodwin. Travelex hackers shut down German car parts company Gedia in massive 'cyber attack'. Computer Weekly. https://www.computerweekly.com/news/252477247/Travelex-hackers-shut-down-German-car-parts-company-Gedia-in-massive-cyber-attack, 2020. Accessed: 15.03.2024.

[GRGB19] Akshay Kumar Goel, Ajay Rose, Jayesh Gaur, and Bharat Bhushan. Attacks, countermeasures and security paradigms in IoT. In *2019 2nd International Conference on Intelligent Computing, Instrumentation and Control Technologies*, volume 1, pages 875–880, 2019.

[GSP+19] Ilias Gerostathopoulos, Dominik Skoda, Frantisek Plasil, Tomas Bures, and Alessia Knauss. Tuning self-adaptation in cyber-physical systems through architectural homeostasis. *Journal of Systems and Software*, 148:37–55, 2019.

[Gup13] Sanjay Gupta. "60 Minutes"—Dick Cheney's heart. CBS News. https://www.cbsnews.com/news/dick-cheneys-heart/, October 2013. Accessed: 15.03.2024.

[HACT04] H. Hansson, M. AAkerholm, I. Crnkovic, and M. Torngren. SaveCCM—a component model for safety-critical real-time systems. In *Proceedings. 30th Euromicro Conference, 2004.*, page 627–635, 2004.

[Hal77] Maurice H. Halstead. *Elements of Software Science*. Computer Science Library. Elsevier, 1977.

[Har87] David Harel. Statecharts: a visual formalism for complex systems. *Science of Computer Programming*, 8(3):231–274, 1987.

[HBR+08] Daniel Halperin, Thomas S. Heydt Benjamin, Benjamin Ransford, Shane S. Clark, Benessa Defend, Will Morgan, Kevin Fu, Tadayoshi Kohno, and William H. Maisel. Pacemakers and Implantable Cardiac Defibrillators: Software Radio Attacks and Zero-Power Defenses. In *Proceedings of the 2008 IEEE Symposium on Security and Privacy*, pages 129–142, 2008.

[HD11] Xiali Hei and Xiaojiang Du. Biometric-based two-level secure access control for Implantable Medical Devices during emergencies. In *2011 Proceedings IEEE INFOCOM*, page 346–350, Apr 2011.

[HE07] R. Henia and R. Ernst. Scenario Aware Analysis for Complex Event Models and Distributed Systems. In *28th IEEE International Real-Time Systems Symposium (RTSS 2007)*, page 171–180, Dec 2007.

[Hen80] K.L. Heninger. Specifying Software Requirements for Complex Systems: New Techniques and Their Application. *IEEE Transactions on Software Engineering*, SE-6(1):2–13, Jan 1980.

[Hes20] Ken Hess. Linux security: Protect your systems with fail2ban. Red Hat. https://www.redhat.com/sysadmin/protect-systems-fail2ban, Jun 2020. Accessed: 15.03.2024.

[Heu23] Marc Heuse. Hydra. https://github.com/vanhauser-thc/thc-hydra, 2023. Accessed: 15.03.2024.

[HHBF+08] D. Halperin, T. S. Heydt-Benjamin, K. Fu, T. Kohno, and W. H. Maisel. Security and Privacy for Implantable Medical Devices. *IEEE Pervasive Computing*, 7(1):30–39, 2008.

[HHK01] Thomas A. Henzinger, Benjamin Horowitz, and Christoph Meyer Kirsch. Giotto: A Time-Triggered Language for Embedded Programming. In *Embedded Software*, page 166–184. Springer, Berlin, Heidelberg, Oct 2001.

[HKMU06] Dan Hirsch, Jeff Kramer, Jeff Magee, and Sebastian Uchitel. Modes for Software Architectures. In *Software Architecture*, page 113–126. Springer, Berlin, Heidelberg, Sep 2006.

[HLLL17] Abdulmalik Humayed, Jingqiang Lin, Fengjun Li, and Bo Luo. Cyber-Physical Systems Security—A Survey. *IEEE Internet of Things Journal*, 4(6):1802–1831, 2017.

[HMTN+08] Kaj Hanninen, Jukka Maki-Turja, Mikael Nolin, Mats Lindberg, John Lundback, and Kurt-Lennart Lundback. The Rubus component model for resource constrained real-time systems. In *2008 International Symposium on Industrial Embedded Systems*, page 177–183, Jun 2008.

[HNSW22] Raphael Hiesgen, Marcin Nawrocki, Thomas C Schmidt, and Matthias Wählisch. The Race to the Vulnerable: Measuring the Log4j Shell Incident. *arXiv preprint* arXiv:2205.02544, 2022.

[HP87] Derek Hatley and Imtiaz Pirbhai. *Strategies for Real-Time System Specification*. Dorset House Publishing Co., Inc., USA, 1987.

[HSC24a] HSCC. Health Industry Cybersecurity Strategic Plan 2024-2029. Healthcare and Public Health Sector Coordinating Council (HSCC). https://healthsectorcouncil.org/jsp2/, February 2024.

[HSC24b] HSCC. Medical Device and Health IT Joint Security Plan version 2. Healthcare and Public Health Sector Coordinating Council (HSCC). https://healthsectorcouncil.org/jsp2/, March 2024.

[HWMK20] Allen Householder, Garret Wassermann, Arthur Manion, and Christopher King. CERT®Guide to Coordinated Vulnerability Disclosure. Technical report, Carnegie Mellon University, 2020.

[IBM05] IBM. An architectural blueprint for autonomic computing. Florida International University. http://users.cis.fiu.edu/~sadjadi/Teaching/Autonomic%20Grid%20Computing/CIS-6612-Summer-2006/AC-Blueprint-WhitePaper-V7.pdf,Jun 2005. Accessed: 15.03.2024.

[IEC18] IEC. Security for industrial automation and control systems—Part 4-1: Secure product development lifecycle requirements. International Standard ISO/IEC 27000:2018, International Electrotechnical Commission (IEC), 2018.

[IET24] IETF HTTP Working Group. HTTP Documentation. Internet Engineering Task Force (IETF). https://httpwg.org/specs/, 2024. Accessed: 15.03.2024.

[IMD20] IMDRF. Principles and Practices for Medical Device Cybersecurity. International Medical Device Regulators Forum (IMDRF). https://www.imdrf.org/sites/default/files/docs/imdrf/final/technical/imdrf-tech-200318-pp-mdc-n60.pdf, 2020.

[IMD23a] IMDRF. Principles and Practices for Software Bill of Materials for Medical Device Cybersecurity. International Medical Device Regulators Forum (IMDRF). https://www.imdrf.org/documents/principles-

and-practices-software-bill-materials-medical-device-cybersecurity, April 2023.

[IMD23b] IMDRF. Principles and Practices for the Cybersecurity of Legacy Medical Devices. International Medical Device Regulators Forum (IMDRF). https://www.imdrf.org/documents/principles-and-practices-cybersecurity-legacy-medical-devices, April 2023.

[INC23] INCOSE. *Systems Engineering Handbook*. John Wiley & Sons Ltd, Hoboken, NJ, fifth edition edition, 2023. International Council On Systems Engineering (INCOSE).

[IPd23] IPdeny Project. IPdeny country block downloads. https://www.ipdeny.com/ipblocks/, 2023. Accessed: 15.03.2024.

[ISO09] ISO. Information technology — Security techniques — Information security management systems — Overview and vocabulary. International Standard ISO/IEC 27000:2018, International Organization for Standardization (ISO), 2009.

[ISO17a] ISO. Systems and software engineering—Vocabulary. International Standard ISO/IEC/IEEE 24765:2017(E), International Organization for Standardization (ISO), 2017.

[ISO17b] ISO. Systems and software engineering—Software life cycle processes. International Standard ISO/IEC/IEEE 12207:2017, International Organization for Standardization (ISO), 2017.

[ISO18] ISO. Systems and software engineering—Life cycle processes—Requirements engineering. International Standard ISO/IEC/IEEE 29148:2018(E), International Organization for Standardization (ISO), 2018.

[ISO23] ISO. Systems and software engineering—System life cycle processes. International Standard ISO/IEC/IEEE 15288:2023(E), International Organization for Standardization (ISO), 2023.

[Jen11a] Joe Jenney. 6.3.2.2 Define Life Cycle System Modes. The Manager's Guide. http://themanagersguide.blogspot.com/2011/01/6322-define-life-cycle-system-modes.html, Jan 2011. Accessed: 15.03.2024.

[Jen11b] Joe Jenney. Methods for Verifying Functional Architecture. The Manager's Guide. http://themanagersguide.blogspot.com/2011/03/methods-for-verifying-functional.html, Mar 2011. Accessed: 15.03.2024.

[JM94] F. Jahanian and A.K. Mok. Modechart: a specification language for real-time systems. *IEEE Transactions on Software Engineering*, 20(12):933–947, 1994.

[JRW+20] Sharmin Jahan, Ian Riley, Charles Walter, Rose F. Gamble, Matt Pasco, Philip K. McKinley, and Betty H. C. Cheng. MAPE-K/MAPE-SAC: An interaction framework for adaptive systems with security assurance cases. *Future Generation Computer Systems*, 109:197–209, 2020.

[KAD21] Ipsita Koley, Sunandan Adhikary, and Soumyajit Dey. An RL-Based Adaptive Detection Strategy to Secure Cyber-Physical Systems. arXiv:2103.02872 *[cs]*, 2021.

[Kap11] Dan Kaplan. Insulin pumps can be hacked. SC Magazine. http://www.scmagazine.com/black-hat-insulin-pumps-can-be-hacked/article/209106/, Aug 2011. Accessed: 15.03.2024.

[KBEI22] Rick Kazman, Philip Bianco, Sebastián Echeverría, and James Ivers. Robustness. Technical Report CMU/SEI-2022-TR-004, Software Engineering Institute, Carnegie Mellon University, Pittsburgh, PA, 2022.

[KC03] J.O. Kephart and D.M. Chess. The vision of autonomic computing. *Computer*, 36(1):41–50, 2003.

[Kel20] G. Kelly. Microsoft Confirms New Windows 10 Update Crashes, Boot Failures. Forbes. https://www.forbes.com/sites/gordonkelly/2020/03/26/microsoft-windows-10-warning-crashes-boot-problems-no-internet-wifi-upgrade-windows-10-free/, Mar. 28, 2020. Accessed: 15.03.2024.

[KHG+21] Amir Khazraei, Spencer Hallyburton, Qitong Gao, Yu Wang, and Miroslav Pajic. Learning-Based Vulnerability Analysis of Cyber-Physical Systems. arXiv:2103.06271 *[cs, eess]*, Mar 2021.

[Kim21] Björn Kimminich. *Pwning OWASP Juice Shop*. OWASP Foundation, Inc., 2021.

[KRT10] Paul K. Kerr, John Rollins, and Catherine A. Theohary. The Stuxnet Computer Worm: Harbinger of an Emerging Warfare Capability. U.S. Congressional Research Service. https://fas.org/sgp/crs/natsec/R41524.pdf, 2010. Accessed: 15.03.2024.

[KSA07] Xu Ke, Krzysztof Sierszecki, and Christo Angelov. COMDES-II: A Component-Based Framework for Generative Development of Distributed Real-Time Control Systems. In *13th IEEE International Conference on Embedded and Real-Time Computing Systems and Applications (RTCSA 2007)*, page 199–208, 2007.

[KSNW19] Narges Khakpour, Charilaos Skandylas, Goran Saman Nariman, and Danny Weyns. Towards Secure Architecture-Based Adaptations. In *Proceedings of the IEEE/ACM 14th International Symposium on Software Engineering for Adaptive and Self-Managing Systems*, page 114–125, May 2019.

[KTGM22] Mohan Krishna Kagita, Navod Thilakarathne, Thippa Reddy Gadekallu, and Praveen Kumar Reddy Maddikunta. *A Review on Security and Privacy of Internet of Medical Things*, page 171–187. Internet of Things. Springer International Publishing, Cham, 2022.

[LB22] Yinling Liu and Jean-Michel Bruel. Modeling of Natural Language Requirements based on States and Modes. In *2022 IEEE 30th International Requirements Engineering Conference Workshops (REW)*, page 190–194, Aug 2022.

[LB24] Yinling Liu and Jean-Michel Bruel. Modeling and verification of natural language requirements based on states and modes. *Formal Aspects of Computing*, February 2024.

[LCC20] Yunkai Liu, Jeremy C. Cannell, and John H. Coffman. Gannon University Hackathon: A Combination of Virtual and Onsite Education Event to Recruit High-School Students within Cybersecurity Major. In *2020 IEEE Frontiers in Education Conference (FIE)*, page 1–4, 2020.

[Lec18] Nadica Hrgarek Lechner. Developing a Compliant Cybersecurity Process for Medical Devices. In *Proceedings of the 29th Central European Conference on Information and Intelligent Systems*, Varaždin, Croatia, Sep 2018.

[Ler19] Emil Lerner. PHP-FPM + Nginx—Remote Code Execution. Exploit Database. https://www.exploit-db.com/exploits/47553, 2019. Accessed: 15.03.2024.

[LFK+14] Heiner Lasi, Peter Fettke, Hans-Georg Kemper, Thomas Feld, and Michael Hoffmann. Industry 4.0. *Business & Information Systems Engineering*, 6(4):239–242, Aug 2014.

[LGM+13] Rogério de Lemos, Holger Giese, Hausi A Müller, Mary Shaw, Jesper Andersson, Marin Litoiu, Bradley Schmerl, Gabriel Tamura, Norha M Villegas, Thomas Vogel, et al. Software engineering for self-adaptive systems: A second research roadmap. In *Software engineering for self-adaptive systems II*, pages 1–32. Springer, 2013.

[LL19] Sixing Lu and Roman Lysecky. Data-driven Anomaly Detection with Timing Features for Embedded Systems. *ACM Transactions on Design Automation of Electronic Systems*, 24(3):1–27, 2019.

[LPS+97] Nancy G. Leveson, L. Denise Pinnel, Sean D. Sandys, Shuichi Koga, and Jon Damon Reese. Analyzing Software Specifications for Mode Confusion Potential. In *Proceedings of a workshop on human error and system development*, page 132–146. Glasgow Accident Analysis Group, 1997.

[LRJ11] Chunxiao Li, Anand Raghunathan, and Niraj K. Jha. Hijacking an insulin pump: Security attacks and defenses for a diabetes therapy system. In *IEEE 13th International Conference on e-Health Networking, Applications and Services*, page 150–156. IEEE, Jun 2011.

[LRS+24] Roman Lysecky, Jerzy Rozenblit, Johannes Sametinger, Aakarsh Rao, and Nadir Carreón. US11868479B2 Runtime adaptive risk assessment and automated mitigation. European Patent Organisation. https://worldwide.espacenet.com/patent/search/family/070461918/publication/US11868479B2?q=pn%3DUS11868479B2, Jan 2024.

[LRW20] Eireann Leverett, Matilda Rhode, and Adam Wedgbury. Vulnerability Forecasting: In theory and practice. arXiv:2012.03814 *[cs]*, Dec 2020.

[LSL15] Sixing Lu, Minjun Seo, and Roman Lysecky. Timing-based anomaly detection in embedded systems. In *The 20th Asia and South Pacific Design Automation Conference*, page 809–814, Jan 2015.

[Mar97] James N. Martin. *Systems engineering guidebook: a process for developing systems and products*. Systems engineering series. CRC Press, Boca Raton, Fla, 1997.

[MB08] Paulo Martins and Alan Burns. On the meaning of modes in uniprocessor real-time systems. In *Proceedings of the 2008 ACM symposium on Applied computing*, SAC '08, page 324–325, Fortaleza, Ceara, Brazil, Mar 2008. Association for Computing Machinery.

[MBFD12] Sébastien Mosser, Mireille Blay-Fornarino, and Laurence Duchien. A Commutative Model Composition Operator to Support Software Adaptation. In *Modelling Foundations and Applications*, volume 7349 of *Lecture Notes in Computer Science*, page 4–19. Springer, 2012.

[McG04] Gary McGraw. Software security. *IEEE Security and Privacy*, 2(2):80–83, mar 2004.

[MCZ+19] Francesca Meneghello, Matteo Calore, Daniel Zucchetto, Michele Polese, and Andrea Zanella. IoT: Internet of threats? A survey of practical security vulnerabilities in real IoT devices. *IEEE Internet of Things Journal*, 6(5):8182–8201, 2019.

[MDC19] MDCG. Guidance on Cybersecurity for medical devices. Medical Device Coordination Group (MDCG), European Commission. https://ec.europa.eu/docsroom/documents/41863/attachments/1/translations/en/renditions/native, 2019.

[Med22a] Medcrypt. What the medical device industry can learn from past cybersecurity vulnerability disclosures. https://www.medcrypt.co/whitepaper_resources/MedCrypt_Vuln_Disclosures_2022.pdf, 2022. Accessed: 15.03.2024.

[Med22b] Medtronic. URGENT MEDICAL DEVICE CORRECTION: MiniMed™ 600 Series Pump System Communication Issue. https://www.medtronicdiabetes.com/customer-support/product-and-service-updates/notice19-letter, 2022. Accessed: 15.03.2024.

[Med23a] Medtronic. MiniMed 670G System Discontinuation of New Sales. https://www.medtronicdiabetes.com/products/minimed-670g-insulin-pump-system, 2023. Accessed: 15.03.2024.

[Med23b] Medtronic. The MiniMed 630G and 770G Insulin Pumps. https://www.medtronic.com/us-en/healthcare-professionals/therapies-procedures/diabetes/education/diabetes-digest/minimed-insulin-pumps.html, 2023. Accessed: 15.03.2024.

[MFB+08] Brice Morin, Franck Fleurey, Nelly Bencomo, Jean-Marc Jézéquel, Arnor Solberg, Vegard Dehlen, and Gordon Blair. An Aspect-Oriented and Model-Driven Approach for Managing Dynamic Variability. In *Model Driven Engineering Languages and Systems*, volume 5301 of *Lecture Notes in Computer Science*, page 782–796. Springer Berlin Heidelberg, 2008.

[MHS05] Marjan Mernik, Jan Heering, and Anthony M. Sloane. When and how to develop domain-specific languages. *ACM Comput. Surv.*, 37(4):316–344, 2005.

[Mic14] Patrice Micouin. *Model?Based Systems Engineering*. John Wiley & Sons, Ltd, 2014.

[MIT20] MITRE. Change Operating Mode, Technique T0858—ICS | MITRE ATT&CK. The MITRE Corporation. https://attack.mitre.org/techniques/T0858/, 2020. Accessed: 15.03.2024.

[MIT23a] MITRE. ATT&CK—Adversarial Tactics, Techniques, and Common Knowledge. The MITRE Corporation. https://attack.mitre.org/, 2023. Accessed: 15.03.2024.

[MIT23b] MITRE. CAPEC—Common Attack Pattern Enumerations and Classifications. The MITRE Corporation. https://capec.mitre.org/, 2023. Accessed: 15.03.2024.

[MIT23c] MITRE. CPE—Common Platform Enumeration. The MITRE Corporation. https://cpe.mitre.org/, 2023. Accessed: 15.03.2024.

[MIT23d] MITRE. CVE ? CWE Mapping Guidance. The MITRE Corporation. https://cwe.mitre.org/documents/cwe_usage/guidance.html, 2023. Accessed: 15.03.2024.

[MIT23e] MITRE. CVE® Program Mission. The MITRE Corporation. https://cve.org/, 2023. Accessed: 15.03.2024.

[MIT23f] MITRE. CWE Top 25 Most Dangerous Software Weaknesses. The MITRE Corporation. https://cwe.mitre.org/top25/, 2023. Accessed: 15.03.2024.

[MIT23g] MITRE. Frequently Asked Questions (FAQ). The MITRE Corporation. https://cwe.mitre.org/about/faq.html, 2023. Accessed: 15.03.2024.

[MMF+10] Brice Morin, Tejeddine Mouelhi, Franck Fleurey, Yves Le Traon, Olivier Barais, and Jean-Marc Jézéquel. Security-driven model-based dynamic adaptation. In *Proceedings of the IEEE/ACM International Conference on Automated Software Engineering—ASE'10*, pages 205–214. ACM Press, 2010.

[MMU+16] P. S. Martins, F. R. Massaro, E. L. Ursini, M. G. Carvalho, and J. Real. Chapter 11—Minimizing the mode-change latency in real-time image processing applications. In Xin-She Yang and João Paulo Papa, editors, *Bio-Inspired Computation and Applications in Image Processing*, page 239–268. Academic Press, Jan 2016.

[MMV+11] Suhas Mathur, Robert Miller, Alexander Varshavsky, Wade Trappe, and Narayan Mandayam. ProxiMate: Proximity-based Secure Pairing Using Ambient Wireless Signals. In *Proceedings of the 9th International Conference on Mobile Systems, Applications, and Services*, MobiSys '11, page 211–224, New York, NY, USA, 2011. ACM.

[Mon86] Andrew Monk. Mode errors: a user-centred analysis and some preventative measures using keying-contingent sound. *International Journal of Man-Machine Studies*, 24(4):313–327, 1986.

[MR98] Florence Maraninchi and Yann Rémond. Mode-automata: About modes and states for reactive systems. In Chris Hankin, editor, *Programming Languages and Systems*, volume 1381 of *Lecture Notes in Computer Science*, page 185–199. Springer Berlin Heidelberg, Berlin, Heidelberg, 1998.

[MR03] Florence Maraninchi and Yann Rémond. Mode-Automata: a new domain-specific construct for the development of safe critical systems. *Science of Computer Programming*, 46(3):219–254, 2003.

[MRS19] Raffaela Mirandola, Elvinia Riccobene, and Patrizia Scandurra. Self-accounting in architecture-based self-adaptation. In *Proceedings of the 13th European Conference on Software Architecture-Volume 2*, pages 14–17, 2019.

[MSB+18] Mahshid Helali Moghadam, Mehrdad Saadatmand, Markus Borg, Markus Bohlin, and Björn Lisper. Adaptive runtime response time control in PLC-based real-time systems using reinforcement learning. In *2018 IEEE/ACM 13th International Symposium on Software Engineering for Adaptive and Self-Managing Systems (SEAMS)*, pages 217–223. IEEE, 2018.

[MSG+16] Eduard Marin, Dave Singelée, Flavio D. Garcia, Tom Chothia, Rik Willems, and Bart Preneel. On the (in)Security of the Latest Generation Implantable Cardiac Defibrillators and How to Secure Them. In *Proceedings of the 32Nd Annual Conference on Computer Security Applications*, ACSAC '16, page 226–236, New York, NY, USA, 2016. ACM.

[MVL+20] Matt McCormack, Amit Vasudevan, Guyue Liu, Sebastián Echeverría, Kyle O'Meara, Grace Lewis, and Vyas Sekar. Towards an Architecture for Trusted Edge IoT Security Gateways. In *3rd USENIX Workshop on Hot Topics in Edge Computing (HotEdge 20)*. USENIX Association, Jun 2020.

[MYNG10] Patrick Meumeu Yomsi, Vincent Nelis, and Joël Goossens. Scheduling multi-mode real-time systems upon uniform multiprocessor platforms. In

15th IEEE International Conference on Emerging Technologies and Factory Automation (ETFA'10), page 1–8, 2010.

[MZR+17] Phillip McNelles, Zhao Chang Zeng, Guna Renganathan, Marius Chirila, and Lixuan Lu. Failure mode taxonomy for assessing the reliability of field programmable gate array based instrumentation and control systems. *Annals of Nuclear Energy*, 108:198–228, Oct 2017.

[NAMP11] V. Nelis, B. Andersson, J. Marinho, and S. M. Petters. Global-EDF Scheduling of Multimode Real-Time Systems Considering Mode Independent Tasks. In *2011 23rd Euromicro Conference on Real-Time Systems*, page 205–214, Jul 2011.

[NAS17] NASA. Systems Engineering Handbook. Special Publication (SP) 2016-6105 Rev2, U.S. National Aeronautics and Space Administration (NASA), 2017.

[NAY17] Phu H. Nguyen, Shaukat Ali, and Tao Yue. Model-based security engineering for cyber-physical systems: A systematic mapping study. *Information and Software Technology*, 83:116–135, 2017.

[Nay22] Saryu Nayyar. Is It Time For A U.S. Version Of GDPR? Forbes. https://www.forbes.com/sites/forbestechcouncil/2022/02/01/is-it-time-for-a-us-version-of-gdpr/, Feb. 1, 2022. Accessed: 15.03.2024.

[New11] Lily Hay Newman. The Log4J Vulnerability Will Haunt the Internet for Years. Wired. https://www.wired.com/story/log4j-log4shell/, 2011. Accessed: 15.03.2024.

[NGA09] Vincent Nelis, Joël Goossens, and Björn Andersson. Two Protocols for Scheduling Multi-mode Real-Time Systems upon Identical Multiprocessor Platforms. In *Proceedings—Euromicro Conference on Real-Time Systems*, page 151–160, 2009.

[NIS18] NIST. Framework for Improving Critical Infrastructure Cybersecurity, Version 1.1. Cybersecurity White Paper (CSWP) 04162018, U.S. National Institute of Standards and Technology (NIST), Gaithersburg, MD, Apr 2018.

[NIS22] NIST. Security and Privacy Controls for Information Systems and Organizations. Special Publication (SP) 800-53A, Rev. 5, U.S. National Institute of Standards and Technology (NIST), Joint Task Force Interagency Working Group, Gaithersburg, MD, Jan 2022.

[NIS24] NIST. The NIST Cybersecurity Framework (CSF) 2.0. Cybersecurity White Paper (CSWP) NIST CSWP 29, U.S. National Institute of Standards and Technology (NIST), Gaithersburg, MD, February 2024.

[NKBC18] Shadi Noghabi, John Kolb, Peter Bodik, and Eduardo Cuervo. Steel: Simplified Development and Deployment of Edge-Cloud Applications. In *10th USENIX Workshop on Hot Topics in Cloud Computing*, HotCloud'18, USA, Jul 2018. USENIX Association.

[NKKT15] Phu H. Nguyen, Max Kramer, Jacques Klein, and Yves Le Traon. An extensive systematic review on the Model-Driven Development of secure systems. *Information and Software Technology*, 68:62–81, Dec 2015.

[NLR21] Arne Nordmann, Ralph Lange, and Francisco Martín Rico. System Modes—Digestible System (Re-)Configuration for Robotics. In *2021 IEEE/ACM 3rd International Workshop on Robotics Software Engineering (RoSE)*, page 19–24, Jun 2021.

[Nma23] Nmap. Nmap: Network Security Scanner. https://nmap.org, 2023. Accessed: 15.03.2024.

[Nor83] Donald A. Norman. Design rules based on analyses of human error. *Communications of the ACM*, 26(4):254–258, Apr 1983.

[NRC95] NRC. Standard technical specifications: Babcock and Wilcox Plants. Revision 1. Technical Report NUREG–1430-Vol.1-Rev.1, 87064, U.S. Nuclear Regulatory Commission (NRC), Washington, DC., U.S. Department of Energy Office of Scientific and Technical Information (OSTI), Apr 1995.

[NSF14] NSF. Partnerships for innovation: Building innovation capacity (pfi:bic). Program Announcements & Information 14–610, U.S. National Science Foundation (NSF), Sep 2014.

[NVD23a] NVD. National Vulnerability Database. U.S. National Vulnerability Database (NVD). https://nvd.nist.gov, 2023. Accessed: 15.03.2024.

[NVD23b] NVD. Official Common Platform Enumeration (CPE) Dictionary. U.S. National Vulnerability Database (NVD). https://nvd.nist.gov/products/cpe, 2023. Accessed: 15.03.2024.

[NVD23c] NVD. Retrieve CPE information. U.S. National Vulnerability Database (NVD). https://nvd.nist.gov/developers/products, 2023. Accessed: 15.03.2024.

[OAS20] OASIS. Common Security Advisory Framework (CSAF). Organization for the Advancement of Structured Information Standards (OASIS). https://oasis-open.github.io/csaf-documentation/, 2020. Accessed: 15.03.2024.

[OED23] OED. mode, n. meanings, etymology, pronunciation and more. Oxford English Dictionary (OED), https://www.oed.com/dictionary/mode_n?tab=factsheet&tl=true, 2023. Accessed: 15.03.2024.

[Off23] Offensive Security. Exploit Database. https://www.exploit-db.com, 2023. Accessed: 15.03.2024.

[OMG06] OMG. SysML Glossary (Draft—April 3, 2006). Object Management Group (OMG). https://www.sysml.org/.res/docs/specs/SysML-v1-Glossary-06-03-04.pdf, Apr 2006.

[OMG17] OMG. Unified modeling language—version 2.5.1. Object Management Group (OMG). https://www.omg.org/spec/UML/, December 2017.

[OMG19] OMG. UML Profile for MARTE: Modeling and Analysis of Real-Time Embedded Systems—Version 1.2. Object Management Group (OMG). https://www.omg.org/spec/MARTE/, Apr 2019.

[Ope23] OpenVAS. Open Vulnerability Assessment Scanner. https://openvas.org, 2023. Accessed: 15.03.2024.

[Ora23] Oracle. Critical Patch Updates, Security Alerts and Bulletins. https://www.oracle.com/security-alerts/, 2023. Accessed: 15.03.2024.

[OWA20] OWASP. Web Security Testing Guide 4.2. Open Worldwide Application Security Project (OWASP). https://owasp.org/www-project-web-security-testing-guide/v42/, Dec 2020. Accessed: 15.03.2024.

[OWA23a] OWASP. Juice Shop. Open Worldwide Application Security Project (OWASP). https://owasp.org/www-project-juice-shop/, 2023. Accessed: 15.03.2024.

[OWA23b] OWASP. OWASP Dependency-Check. Open Worldwide Application Security Project (OWASP). https://owasp.org/www-project-dependency-check/, 2023. Accessed: 15.03.2024.

[OWA23c] OWASP. OWASP Top 10 Web Application Security Risks. Open Worldwide Application Security Project (OWASP). https://owasp.org/www-project-top-ten/, 2023. Accessed: 15.03.2024.

[Pac20] Packet Storm. Apache 2.4.43 mod_http2 Memory Corruption. https://packetstormsecurity.com/files/160392/Apache-2.4.43-mod_http2-Memory-Corruption.html, 2020. Accessed: 15.03.2024.

[Pac23] Packet Storm. Packet Storm. https://packetstormsecurity.com, 2023. Accessed: 15.03.2024.

[Pal21] Palo Alto Networks. Observing Attacks Against Hundreds of Exposed Services in Public Clouds. Unit42. https://unit42.paloaltonetworks.com/exposed-services-public-clouds/, Nov 2021. Accessed: 15.03.2024.

[Pal22] Palo Alto Networks. Infusion Pump Vulnerabilities: Common Security Gaps. Unit42. https://unit42.paloaltonetworks.com/infusion-pump-vulnerabilities/, Mar 2022. Accessed: 15.03.2024.

[Pav24] Pavlov, Igor. 7-Zip—File archiver. https://7-zip.org/, 2024. Accessed: 15.03.2024.

[Pay96] Stephen Paynter. Real-time mode-machines. In Bengt Jonsson and Joachim Parrow, editors, *Formal Techniques in Real-Time and Fault-Tolerant Systems*, volume 1135 of *Lecture Notes in Computer Science*, page 90–109. Springer Berlin Heidelberg, Berlin, Heidelberg, 1996.

[PB98] P. Pedro and A. Burns. Schedulability analysis for mode changes in flexible real-time systems. In *Proceeding. 10th EUROMICRO Workshop on Real-Time Systems (Cat. No.98EX168)*, page 172–179, Jun 1998.

[PDL+16] Subhav Pradhan, Abhishek Dubey, Tihamer Levendovszky, Pranav Srinivas Kumar, William A. Emfinger, Daniel Balasubramanian, William Otte, and Gabor Karsai. Achieving resilience in distributed software systems via self-reconfiguration. *Journal of Systems and Software*, 122:344–363, Dec 2016.

[PdOG+20] Flávia Pisani, Fabíola Martins Campos de Oliveira, Eduardo S. Gama, Roger Immich, Luiz F. Bittencourt, and Edson Borin. Fog Computing on Constrained Devices: Paving the Way for the Future IoT. *Advances in Parallel Computing*, Volume 35: Advances in Edge Computing: Massive Parallel Processing and Applications:20–60, 2020.

[Ped99] Paulo Sergio Martins Pedro. *Schedulability of mode changes in flexible real-time distributed systems.* PhD thesis, University of York, Department of Computer Science, 1999.

[PH16] Jesus Pacheco and Salim Hariri. IoT Security Framework for Smart Cyber Infrastructures. In *Proceedings of the 2016 IEEE 1st International Workshops on Foundations and Applications of Self* Systems*, pages 242–247, 2016.

[PHM21] Jeremy Page, Drew Hinshaw, and Betsy McKay. In Hunt for Covid-19 Origin, Patient Zero Points to Second Wuhan Market. *Wall Street Journal*, February 2021.

[PKK11] Nathanael Paul, Tadayoshi Kohno, and David C. Klonoff. A Review of the Security of Insulin Pump Infusion Systems. *Journal of Diabetes Science and Technology*, 5(6), Nov 2011.

[PL11] Linh T.X. Phan and Insup Lee. Towards a Compositional Multi-modal Framework for Adaptive Cyber-physical Systems. In *2011 IEEE 17th International Conference on Embedded and Real-Time Computing Systems and Applications*, page 67–73. IEEE, Aug 2011.

[PM20] Gabriel Pedroza and Guillaume Mockly. Method and framework for security risks analysis guided by safety criteria. In *Proceedings of the 23rd ACM/IEEE International Conference on Model Driven Engineering Languages and Systems: Companion Proceedings*, MODELS '20, page 1–8. ACM, Oct 2020.

[PM223] PM2. PM2. https://pm2.io, 2023. Accessed: 15.03.2024.

[Pon22] Ponemon Institute. Cyber Insecurity in Healthcare: Cost & Impact on Patient Care. https://www.proofpoint.com/sites/default/files/threat-reports/pfpt-us-tr-cyber-insecurity-healthcare-ponemon-report.pdf, Jul 2022.

[Pos23] PostgreSQL. Documentation: 16: pg_dump. https://www.postgresql.org/docs/current/app-pgdump.html, 2023. Accessed: 15.03.2024.

[PZBH17] Jesus Pacheco, Xiaoyang Zhu, Youakim Badr, and Salim Hariri. Enabling Risk Management for Smart Infrastructures with an Anomaly Behavior Analysis Intrusion Detection System. In *Proceedings of the 2017 IEEE 2nd International Workshops on Foundations and Applications of Self* Systems*, pages 324–328, 2017.

[QWG21] Federico Quin, Danny Weyns, and Omid Gheibi. Decentralized Self-Adaptive Systems: A Mapping Study. In *2021 International Symposium on Software Engineering for Adaptive and Self-Managing Systems*, page 18–29, 2021.

[Ran90] Harro Ranter. ASN Aircraft accident Airbus A320-231 VT-EPN Bangalore-Hindustan Airport (BLR). Aviation Safety Network. https://aviation-safety.net/database/record.php?id=19900214-2, 1990. Accessed: 15.03.2024.

[Ran92] Harro Ranter. ASN Aircraft accident Airbus A320-111 F-GGED Strasbourg-Entzheim Airport (SXB). Aviation Safety Network. https://aviation-safety.net/database/record.php?id=19920120-0, 1992. Accessed: 15.03.2024.

[Rap21] Rapid7. Risk Strategies—InsightVM Documentation. https://docs.rapid7.com/insightvm/working-with-risk-strategies-to-analyze-threats/, 2021. Accessed: 15.03.2024.

[RC04] Jorge Real and Alfons Crespo. Mode Change Protocols for Real-Time Systems: A Survey and a New Proposal. *Real-Time Systems*, 26(2):161–197, Mar 2004.

[RCHBC09] Kasper Bonne Rasmussen, Claude Castelluccia, Thomas S. Heydt-Benjamin, and Srdjan Capkun. Proximity-based Access Control for Implantable Medical Devices. In *Proceedings of the 16th ACM Conference on Computer and Communications Security*, CCS '09, page 410–419, New York, NY, USA, 2009. ACM.

[RCLR22] Aakarsh Rao, Nadir A. Carreón, Roman Lysecky, and Jerzy Rozenblit. FIRE: A Finely Integrated Risk Evaluation Methodology for Life-Critical Embedded Systems. *Information*, 13(1010):487, Oct 2022.

[RDP20] Stefan Resmerita, Patricia Derler, and Wolfgang Pree. Timing Definition Language (TDL) Modeling in Ptolemy II. Technical Report 21, Department of Computer Science, University of Salzburg, 2020.

[Rea00] Jorge Real. *Protocolos de cambio de modo para sistemas de tiempo real (mode change protocols for real time systems)*. PhD thesis, Universitat Politècnica de València, 2000.

[Red23] Red Hat. Ansible. https://www.ansible.com, 2023. Accessed: 15.03.2024.

[Red24] Redis. Remote Dictionary Server FAQ. https://redis.io/docs/get-started/faq/, 2024. Accessed: 15.03.2024.

[Rep21] Reports And Data. Market value of the internet of medical things worldwide in 2019 and 2027. Statista. https://www.statista.com/statistics/1264333/global-iot-in-healthcare-market-size/, 2021. Accessed: 15.03.2024.

[Res23] Amy Ressler. An update on Chrome Security updates—shipping security fixes to you faster. Google Security Blog. https://security.googleblog.com/2023/08/an-update-on-chrome-security-updates.html, August 2023. Accessed: 15.03.2024.

[RH08] Per Runeson and Martin Höst. Guidelines for conducting and reporting case study research in software engineering. *Empirical Software Engineering*, 14(2):131, 2008.

[Rie22] Riegler, Michael. Resilient middleware for smart security in medical devices. Austrian Marshall Plan Foundation. https://www.marshallplan.at/images/All-Papers/mp-2022/Michael_Riegler_1048.PDF, 2022. Accessed: 15.03.2024.

[RJK13] Masoud Rostami, Ari Juels, and Farinaz Koushanfar. Heart-to-heart (H2H): authentication for implanted medical devices. In *Proceedings for ACM SIGSAC Conference on Computer & Communications Security*, page 1099–1112. ACM Press, 2013.

[RMCO16] Ron Ross, Michael McEvilley, and Janet Carrier Oren. Systems Security Engineering: Considerations for a Multidisciplinary Approach in the Engineering of Trustworthy Secure Systems. Special Publication (SP) 800-160, U.S. National Institute of Standards and Technology (NIST), Gaithersburg, MD, Nov 2016.

[RO14] Mike Ryan and Anthony Olver. On a Useful Taxonomy of Phases, Modes, and States in Systems Engineering. In *Systems Engineering/Test and Evaluation (SETE) Conference*, Adelaide, Australia, 2014. Systems Engineering Society of Australia (SESA).

[RPLZ19] Shahadate Rezvy, Miltos Petridis, Aboubaker Lasebae, and Tahmina Zebin. Intrusion Detection and Classification with Autoencoded Deep Neural Network. In Jean-Louis Lanet and Cristian Toma, editors, *Innovative Security Solutions for Information Technology and Communications*, Lecture Notes in Computer Science, page 142–156, Cham, 2019. Springer International Publishing.

[RRLS18] Aakarsh Rao, Jerzy Rozenblit, Roman Lysecky, and Johannes Sametinger. Trustworthy multi-modal framework for life-critical systems security. In *Proceedings of the Annual Simulation Symposium*, ANSS '18, page 1–9, Baltimore, Maryland, Apr 2018. Society for Computer Simulation International.

[RS20] Michael Riegler and Johannes Sametinger. Mode Switching from a Security Perspective: First Findings of a Systematic Literature Review. In G. Kotsis et al., editor, *Database and Expert Systems Applications*, Communications in Computer and Information Science, page 63–73, Cham, 2020. Springer International Publishing.

[RS21a] Michael Riegler and Johannes Sametinger. Mode Switching for Secure Web Applications—A Juice Shop Case Scenario. In G. Kotsis et al., editor,

Database and Expert Systems Applications—DEXA 2021 Workshops, Communications in Computer and Information Science, page 3–8, Cham, 2021. Springer International Publishing.

[RS21b] Michael Riegler and Johannes Sametinger. Multi-mode Systems for Resilient Security in Industry 4.0. *Procedia Computer Science*, 180:301–307, 2021. Proceedings of the 2nd International Conference on Industry 4.0 and Smart Manufacturing (ISM 2020).

[RSR22] Michael Riegler, Johannes Sametinger, and Jerzy W. Rozenblit. Context-Aware Security Modes For Medical Devices. In *2022 Annual Modeling and Simulation Conference (ANNSIM)*, pages 372–382, 2022.

[RSR23] Michael Riegler, Johannes Sametinger, and Jerzy W. Rozenblit. Architecture for Self-protective Medical Cyber-Physical Systems. In G. Kotsis et al., editor, *Database and Expert Systems Applications—DEXA 2023 Workshops*, Communications in Computer and Information Science, pages 57–66, Cham, 2023. Springer Nature Switzerland.

[RSS22] Michael Riegler, Johannes Sametinger, and Christoph Schönegger. Mode Switching for Secure Edge Devices. In G. Kotsis et al., editor, *Database and Expert Systems Applications—DEXA 2022 Workshops*, page 347–356, Cham, 2022. Springer International Publishing.

[RSV23] Michael Riegler, Johannes Sametinger, and Michael Vierhauser. A Distributed MAPE-K Framework for Self-Protective IoT Devices. In *2023 IEEE/ACM 18th Symposium on Software Engineering for Adaptive and Self-Managing Systems (SEAMS)*. IEEE, May 2023.

[RSVW23a] Michael Riegler, Johannes Sametinger, Michael Vierhauser, and Manuel Wimmer. A Model-based Mode-Switching Framework based on Security Vulnerability Scores. *Journal of Systems and Software*, 200, 2023.

[RSVW23b] Michael Riegler, Johannes Sametinger, Michael Vierhauser, and Manuel Wimmer. Supplemental Material for a Mode Switching Framework with Web Server Case Study. Zenodo, https://doi.org/10.5281/zenodo.7603904, 2023.

[SA00] Frank Stajano and Ross Anderson. The Grenade Timer: Fortifying the Watchdog Timer Against Malicious Mobile Code. In *Proceedings of 7th International Workshop on Mobile Multimedia Communications (MoMuC 2000)*, Waseda, Tokyo, Japan, 2000.

[SA10] Pedro D. Salgueiro and Salvador P. Abreu. A DSL for intrusion detection based on constraint programming. In *Proceedings of the 3rd International Conference on Security of Information and Networks*, page 224–232. ACM, Sep 2010.

[Sch00] Bruce Schneier. Crypto-gram: September 15, 2000—Schneier on Security. https://www.schneier.com/crypto-gram/archives/2000/0915.html, Sep 2000. Accessed: 15.03.2024.

[Sch07a] Bruce Schneier. Full Disclosure of Security Vulnerabilities a "Damned Good Idea"—Schneier on Security. https://www.schneier.com/essays/archives/2007/01/schneier_full_disclo.html, Jan 2007. Accessed: 15.03.2024.

[Sch07b] Martin Schoeberl. Mission Modes for Safety Critical Java. In Roman Obermaisser, Yunmook Nah, Peter Puschner, and Franz J. Rammig, editors, *Software Technologies for Embedded and Ubiquitous Systems*, Lecture Notes in Computer Science, page 105–113. Springer, 2007.

[Sch10] Stuart Schechter. Security that is Meant to be Skin Deep: Using Ultraviolet Micropigmentation to Store Emergency-Access Keys for Implantable Medical Devices. Microsoft Research. https://www.microsoft.com/en-us/research/publication/security-that-is-meant-to-be-skin-deep-using-ultraviolet-micropigmentation-to-store-emergency-access-keys-for-implantable-medical-devices/, Apr 2010.

[SCL+18] Wencheng Sun, Zhiping Cai, Yangyang Li, Fang Liu, Shengqun Fang, and Guoyan Wang. Security and Privacy in the Medical Internet of Things: A Review. *Security and Communication Networks*, 2018:9, 2018.

[SD13] HP Sanghvi and MS Dahiya. Cyber reconnaissance: an alarm before cyber attack. *International Journal of Computer Applications*, 63(6), 2013.

[SE19] Vijaya Kumar Sundar and Arvind Easwaran. A Practical Degradation Model for Mixed-Criticality Systems. In *2019 IEEE 22nd International Symposium on Real-Time Distributed Computing (ISORC)*, page 171–180, 2019.

[Seg22] Edward Segal. 1 Year Later: Actions Taken, Lessons Learned Since The Colonial Pipeline Cyberattack. Forbes. https://www.forbes.com/sites/edwardsegal/2022/05/07/1-year-later-actions-taken-lessons-learned-since-the-colonial-pipeline-cyberattack/, 2022. Accessed: 15.03.2024.

[SFM16] J. Kenneth Shultis, Richard E. Faw, and Douglas S. McGregor. *Fundamentals of Nuclear Science and Engineering; 3rd Edition*. CRC Press, 2016.

[SG90] Lui Sha and John B. Goodenough. Real-Time Scheduling Theory and Ada. *Computer*, 23(4):53–62, 1990.

[Sha23] Parth Shah. How to Use Modes and Routines on Samsung Galaxy Phones. Guiding Tech Media. https://www.guidingtech.com/how-to-use-modes-and-routines-on-samsung-galaxy-phones/, Feb 2023. Accessed: 15.03.2024.

[Sho14] Adam Shostack. *Threat modeling: designing for security*. Wiley, Indianapolis, IN, 2014.

[SK20] Ali Shafaat and C. Robert Kenley. Model-based design of project systems, modes, and states. *Systems Engineering*, 23(2):165–176, Mar 2020.

[SKY19] Mohammad M. Shurman, Rami M. Khrais, and Abdulrahman A. Yateem. IoT denial-of-service attack detection and prevention using hybrid IDS. In *2019 International Arab Conference on Information Technology*, pages 252–254, 2019.

[SR16] Johannes Sametinger and Jerzy Rozenblit. Security Scores for Medical Devices. In *9th International Conference on Health Informatics*, page 533–541, Rome, Italy, Feb 2016. In Proceedings of the 9th International Joint Conference on Biomedical Engineering Systems and Technologies (BIOSTEC 2016)—Volume 5: HEALTHINF.

[SRJ20] Vishnu Suresh, Jino Ramson, and Dr Jegan. Internet of Medical Things (IoMT)—An overview. In *2020 5th International Conference on Devices, Circuits and Systems (ICDCS)*, pages 101–104, Mar 2020.

[SRLO15] Johannes Sametinger, Jerzy Rozenblit, Roman Lysecky, and Peter Ott. Security Challenges for Medical Devices. *Communications of the ACM*, 58(4):74–82, 2015.

[SRLR89] Lui Sha, Ragunathan Rajkumar, John Lehoczky, and Krithi Ramamritham. Mode change protocols for priority-driven preemptive scheduling. *Real-Time Systems*, 1(3):243–264, 1989.

[SRS23] Marco Stadler, Michael Riegler, and Johannes Sametinger. Towards Increasing Safety in Collaborative CPS Environments. In G. Kotsis et al., editor, *Database and Expert Systems Applications—DEXA 2023 Workshops*, Communications in Computer and Information Science, page 79–85, Cham, 2023. Springer Nature Switzerland.

[SRS24] Marco Stadler, Michael Riegler, and Johannes Sametinger. Cyber-Resilient Edge Computing: A Holistic Approach with Multi-Level MAPE-K Loops, 2024. Accepted for the 21st IEEE International Conference on Software Architecture (ICSA), 2024, Hyderabad, India.

[SRSS24] Marco Stadler, Michael Riegler, Johannes Sametinger, and Christoph Schönegger. Improve CPSoS Security: An enhanced Anomaly-based Intrusion Detection Architecture, 2024. Submitted for publication.

[SRTS08] Hans Søndergaard, Anders Peter Ravn, Bent Thomsen, and Martin Schoeberl. *A Practical Approach to Mode Change in Real-Time Systems*. Technical Report. Department of Computer Science, Aalborg University, 2008.

[SS17] Johannes Sametinger and Clemens Steinwender. Resilient Context-Aware Medical Device Security. In *International Conference on Computational Science and Computational Intelligence, Symposium on Health Informatics and Medical Systems (CSCI-ISHI)*, page 1775–1778, Las Vegas, NV, USA, Dec 2017. IEEE Computer Society.

[Sta21] Statista. In-depth: Industry 4.0 2021. https://www.statista.com/study/66974/in-depth-industry-40/, Jun 2021. Accessed: 15.03.2024.

[STB+16] Farzad Samie, Vasileios Tsoutsouras, Lars Bauer, Sotirios Xydis, Dimitrios Soudris, and Jörg Henkel. Computation offloading and resource allocation for low-power IoT edge devices. In *2016 IEEE 3rd World Forum on Internet of Things (WF-IoT)*, page 7–12, Dec 2016.

[STB+19] Farzad Samie, Vasileios Tsoutsouras, Lars Bauer, Sotirios Xydis, Dimitrios Soudris, and Jörg Henkel. Oops: Optimizing operation-mode selection for iot edge devices. *ACM Transactions on Internet Technology*, 19(2):22:1–22:21, Mar 2019.

[SWdH+18] Robert Mark Seepers, Wenjin Wang, Gerard de Haan, Ioannis Sourdis, and Christos Strydis. Attacks on Heartbeat-Based Security Using Remote Photoplethysmography. *IEEE Journal of Biomedical and Health Informatics*, 22(3):714–721, May 2018.

[Sys23] Systemd. System and Service Manager. https://systemd.io/, 2023. Accessed: 15.03.2024.

[SYST18] Chi-Sheng Shih, Chang-Min Yang, Wei-Lun Su, and Pei-Kuei Tsung. OSAMIC: Online Schedulability Analysis of Real-time Mode Change on Heterogeneous Multi-core Platforms. In *Proceedings of the 2018 Conference on Research in Adaptive and Convergent Systems*, RACS '18, page 205–212. ACM, 2018.

[TA96] Ken Tindell and Alejandro Alonso. A very simple protocol for mode changes in priority preemptive systems. Technical report, Universidad Politécnica de Madrid, 1996.

[TA21] Nicole M. Thomasian and Eli Y. Adashi. Cybersecurity in the Internet of Medical Things. *Health Policy and Technology*, 10(3):100549, Sep 2021.

[TBW92] K. W. Tindell, A. Burns, and A. J. Wellings. Mode Changes in Priority Pre-emptively Scheduled Systems. In *Proceedings of the Real Time Systems Symposium*, page 100–109, 1992.

[TCBM18] Ivana Tomić, Po-Yu Chen, Michael J Breza, and Julie A McCann. Antilizer: run time self-healing security for wireless sensor networks. In *Proceedings of the 15th EAI International Conference on Mobile and Ubiquitous Systems: Computing, Networking and Services*, pages 107–116, 2018.

[TCD+21] Soumyadeep Thakur, Anuran Chakraborty, Rajonya De, Neeraj Kumar, and Ram Sarkar. Intrusion detection in cyber-physical systems using a generic and domain specific deep autoencoder model. *Computers & Electrical Engineering*, 91:107044, May 2021.

[TCG22] TCG. Trusted Platform Module Library Specification, Family "2.0", Level 00, Revision 01.59. Trusted Computing Group (TCG). https://trustedcomputinggroup.org/wp-content/uploads/TCG_TPM2_r1p59_Part1_Architecture_pub.pdf, Mar 2022. Accessed: 15.03.2024.

[TDJ+14] Ashish Tiwari, Bruno Dutertre, Dejan Jovanovi?, Thomas de Candia, Patrick D. Lincoln, John Rushby, Dorsa Sadigh, and Sanjit Seshia. Safety envelope for security. In *in Proceedings of the 3rd international conference on High confidence networked systems*, HiCoNS '14, page 85–94. Association for Computing Machinery, 2014.

[The20] The Brussels Times. Cyber attack sees Picanol shares suspended. https://www.brusselstimes.com/news-contents/economic/89253/cyber-attack-sees-picanol-shares-suspended/, 2020. Accessed: 15.03.2024.

[TLG10] Martin Thomson, Salvatore Loreoto, and Wilkins Greg. Hypertext Transfer Protocol (HTTP) Timeouts. Internet Engineering Task Force (IETF). https://tools.ietf.org/id/draft-thomson-hybi-http-timeout-00.html, 2010. Accessed: 15.03.2024.

[TMMI16] M. Thompson, M. Mendolla, M. Muggler, and M. Ike. Dynamic Application Rotation Environment for Moving Target Defense. In *2016 Resilience Week (RWS)*, page 17–26, Aug 2016.

[Tou24] Toulas, Bill. German battery maker Varta halts production after cyberattack. Bleeping Computer. https://www.bleepingcomputer.com/news/security/german-battery-maker-varta-halts-production-after-cyberattack/, 2024. Accessed: 15.03.2024.

[Tra21] Transforma Insights. Number of edge enabled internet of things (IoT) devices worldwide from 2020 to 2030, by market. Statista. https://www.statista.com/statistics/1259878/edge-enabled-iot-device-market-worldwide/, Aug 2021. Accessed: 15.03.2024.

[TSSY20] Katja Tuma, Laurens Sion, Riccardo Scandariato, and Koen Yskout. Automating the early detection of security design flaws. In *Proceedings of the 23rd ACM/IEEE International Conference on Model Driven Engineering Languages and Systems*, page 332–342, Oct 2020.

[Tun21] Liam Tung. Kaseya ransomware attack: 1,500 companies affected, company confirms. ZDNet. https://www.zdnet.com/article/kaseya-ransomware-attack-1500-companies-affected-company-confirms/, 2021. Accessed: 15.03.2024.

[Tur20] William Turton. Hackers used a little-known IT vendor to attack U.S. agencies. Fortune. https://fortune.com/2020/12/15/solarwinds-hackers-u-s-agencies/, Dec 2020. Accessed: 15.03.2024.

[USA13] USAF SMC. *SMC Systems Engineering Primer & Handbook: : Concepts, Processes, and Techniques*, volume 4. U.S. Air Force (USAF) Space and Missile Systems Center (SMC), Los Angeles, Mar 2013.

[VFW16] Andreas Vogelsang, Henning Femmer, and Christian Winkler. Take care of your modes! an investigation of defects in automotive requirements. In Maya Daneva and Oscar Pastor, editors, *Requirements Engineering: Foundation for Software Quality*, Lecture Notes in Computer Science, page 161–167, Cham, 2016. Springer International Publishing.

[VHFST15] Birgit Vogel-Heuser, Alexander Fay, Ina Schaefer, and Matthias Tichy. Evolution of software in automated production systems: Challenges and research directions. *Journal of Systems and Software*, 110:54–84, Dec 2015.

[VJ20] Ruchi Vishwakarma and Ankit Kumar Jain. A survey of DDoS attacking techniques and defence mechanisms in the IoT network. *Telecommunication systems*, 73(1):3–25, 2020.

[VLPSJ21] Dimitri Van Landuyt, Liliana Pasquale, Laurens Sion, and Wouter Joosen. Threat modeling at run time: the case for reflective and adaptive threat management (NIER track). In *SEAMS'21: Proceedings of the 16th International Symposium on Software Engineering for Adaptive and Self-Managing Systems*, pages 203–209, 2021.

[Voi18a] Jean-Luc Voirin. 14—system supervision, states and modes. In Jean-Luc Voirin, editor, *Model-Based System and Architecture Engineering with the Arcadia Method*, page 211–232. Elsevier, 2018.

[Voi18b] Jean-Luc Voirin. 18—Concepts of States and Modes. In Jean-Luc Voirin, editor, *Model-Based System and Architecture Engineering with the Arcadia Method*, page 305–308. Elsevier, 2018.

[vOvdLKM00] R. van Ommering, F. van der Linden, J. Kramer, and J. Magee. The Koala component model for consumer electronics software. *Computer*, 33(3):78–85, 2000.

[vS04] Miquel van Smoorenburg. init, telinit—process control initialization. Debian. https://manpages.debian.org/testing/sysvinit-core/init.8.en.html, Jul 2004. Accessed: 15.03.2024.

[VS19] Maurizio Volterrani and Barbara Sposato. Remote monitoring and telemedicine. *European Heart Journal Supplements: Journal of the European Society of Cardiology*, 21(Suppl M):M54–M56, Dec 2019.

[Was05] Charles S. Wasson. *System analysis, design, and development: concepts, principles, and practices*. Wiley-Interscience, Hoboken, N.J, 2005.

[Was14] Charles S. Wasson. 3.3.1 System Phases, Modes, and States: Solutions to Controversial Issues. *INCOSE International Symposium*, 21(1):279–294, Jun 2014.

[Was15] Charles S. Wasson. *System Engineering Analysis, Design, and Development*. Wiley Series in Systems Engineering and Management. John Wiley & Sons, Nashville, TN, 2 edition, 2015.

[WGR+22] Thomas Witte, Raffaela Groner, Alexander Raschke, Matthias Tichy, Irdin Pekaric, and Michael Felderer. Towards Model Co-evolution Across Self-Adaptation Steps for Combined Safety and Security Analysis. In *Proceedings of the 2022 International Symposium on Software Engineering for Adaptive and Self-Managing Systems*, pages 106–112, 2022.

[Whi17] Zack Whittaker. Equifax confirms Apache Struts security flaw it failed to patch is to blame for hack. ZDNet. https://www.zdnet.com/article/equifax-confirms-apache-struts-flaw-it-failed-to-patch-was-to-blame-for-data-breach/, 2017. Accessed: 15.03.2024.

[Wie14] Roel J. Wieringa. *Design Science Methodology for Information Systems and Software Engineering*. Springer Berlin Heidelberg, Berlin, Heidelberg, 2014.

[Wra93] Richard B. Wray. An Analysis of the Modes and States for Generic Avionics. Technical Report NAS 1.26:188253, Lockheed Engineering and Sciences Co. Houston, TX, United States, Jul 1993.

[WRH+12] Claes Wohlin, Per Runeson, Martin Höst, Magnus C. Ohlsson, Björn Regnell, and Anders Wesslén. *Experimentation in Software Engineering*. Springer Berlin, Heidelberg, 2012.

[WSG+13] Danny Weyns, Bradley Schmerl, Vincenzo Grassi, Sam Malek, Raffaela Mirandola, Christian Prehofer, Jochen Wuttke, Jesper Andersson, Holger Giese, and Karl M. Göschka. *On Patterns for Decentralized Control in Self-Adaptive Systems*, pages 76–107. Springer Berlin Heidelberg, 2013.

[Wym93] A. Wayne Wymore. *Model-based systems engineering: an introduction to the mathematical theory of discrete systems and to the tricotyledon theory of system design*. CRC Press, 1993.

[XQT+11] Fengyuan Xu, Zhengrui Qin, Chiu C. Tan, Baosheng Wang, and Qun Li. IMDGuard: Securing Implantable Medical Devices with the External Wearable Guardian. In *2011 Proceedings IEEE INFOCOM*, page 1862–1870, 2011.

[XWP14a] Teng Xu, James B. Wendt, and Miodrag Potkonjak. Matched Digital PUFs for Low Power Security in Implantable Medical Devices. In *2014 IEEE International Conference on Healthcare Informatics*, page 33–38. IEEE, Sep 2014.

[XWP14b] Teng Xu, James B Wendt, and Miodrag Potkonjak. Security of IoT systems: Design challenges and opportunities. In *2014 IEEE/ACM International Conference on Computer-Aided Design*, pages 417–423, 2014.

[YAA19] T. Yaqoob, H. Abbas, and M. Atiquzzaman. Security Vulnerabilities, Attacks, Countermeasures, and Regulations of Networked Medical Devices—A Review. *IEEE Communications Surveys Tutorials*, 21(4):3723–3768, 2019.

[YEM14] Eric Yuan, Naeem Esfahani, and Sam Malek. A Systematic Survey of Self-Protecting Software Systems. *ACM Transactions on Autonomous and Adaptive Systems*, 8(4):17:1–17:41, Jan 2014.

[YH13] Hang Yin and Hans Hansson. Handling Multiple Mode Switch Scenarios in Component-Based Multi-mode Systems. In *2013 20th Asia-Pacific Software Engineering Conference (APSEC)*, volume 1, page 404–413, Dec 2013.

[YH14] Hang Yin and Hans Hansson. Handling Emergency Mode Switch for Component-Based Systems. In *2014 21st Asia-Pacific Software Engineering Conference*, volume 1, page 151–158, Dec 2014.

[YH18] Hang Yin and Hans Hansson. Fighting CPS Complexity by Component-Based Software Development of Multi-Mode Systems. *Designs*, 2(4):39, 2018.

[YHO+16] Hang Yin, Hans Hansson, Daniele Orlando, Francesco Miscia, and Simone Di Marco. Component-based software development of multi-mode systems—An extended report. Technical Report 312, Mälardalen University, Embedded Systems, 2016.

[Yin15] Hang Yin. *Introducing Mode Switch in Component-Based Software Development*. PhD thesis, Mälardalen University, 2015.

[YMS+13] Eric Yuan, Sam Malek, Bradley Schmerl, David Garlan, and Jeff Gennari. Architecture-based self-protecting software systems. In *Proceedings of the 9th International ACM Sigsoft Conference on Quality of Software Architectures*, page 33–42, New York, NY, USA, Jun 2013. Association for Computing Machinery.

[YNN+20] Jean-Paul A. Yaacoub, Mohamad Noura, Hassan N. Noura, Ola Salman, Elias Yaacoub, Raphaël Couturier, and Ali Chehab. Securing internet of medical things systems: Limitations, issues and recommendations. *Future Generation Computer Systems*, 105:581–606, Apr 2020.

[ZDS21] Sherali Zeadally, Ashok Kumar Das, and Nicolas Sklavos. Cryptographic technologies and protocol standards for Internet of Things. *Internet of Things*, 14:100075, Jun 2021.

[Zet23] Kim Zetter. The Untold Story of the Boldest Supply-Chain Hack Ever. Wired. https://www.wired.com/story/the-untold-story-of-solarwinds-the-boldest-supply-chain-hack-ever/, 2023. Accessed: 15.03.2024.

[ZR19] Mohammad Hossein Zoualfaghari and Andrew Reeves. Secure & zero touch device onboarding. In *Living in the Internet of Things (IoT 2019)*, page 1–3, May 2019.

MIX
Papier aus verantwortungsvollen Quellen
Paper from responsible sources
FSC® C105338
FSC
www.fsc.org

If you have any concerns about our products,
you can contact us on
ProductSafety@springernature.com

In case Publisher is established outside the EU,
the EU authorized representative is:
Springer Nature Customer Service Center GmbH
Europaplatz 3, 69115 Heidelberg, Germany

Printed by Libri Plureos GmbH
in Hamburg, Germany